www.brookscole.com/astronomy

Discover the Brooks/Cole Galaxy is the heart of our
Astronomy Resource Center. It includes a wealth of
information and links to all Brooks/Cole astronomy
titles. At brookscole.com/astronomy you can find
out about supplements, demonstration software, and
student resources. You can also send e-mail to authors
and preview new publications and exciting new
technologies.

Astronomy

About the Author

Mike Seeds is the John W. Wetzel Professor of Astronomy at Franklin and Marshall College as well as Director of the College's Joseph R. Grundy Observatory. His research interests focus on peculiar variable stars and the automation of astronomical photometry. He is the Principal Astronomer in charge of the Phoenix 10, the first fully robotic telescope located in southern Arizona. In 1989 he received the Christian R. and Mary F. Lindback Award for Distinguished Teaching. In addition to writing textbooks, Seeds pursues an active interest in astronomical research, creates education tools for use in computer-smart classrooms, and continues to develop his upper-level courses in Archaeoastronomy and in the History of Astronomy. He has also published educational software for preliterate toddlers! Seeds was Senior Consultant in the creation of the 26-episode telecourse, *Universe: The Infinite Frontier.* He is author of two other books published by Brooks/Cole, *Horizons: Exploring the Universe,* Sixth Edition, and *Foundations of Astronomy,* Sixth Edition.

Astronomy
THE SOLAR SYSTEM AND BEYOND
Second Edition

Michael A. Seeds
Joseph R. Grundy Observatory
Franklin and Marshall College

Brooks/Cole
Thomson Learning™

Australia • Canada • Mexico • Singapore • Spain • United Kingdom • United States

Sponsoring Editors: *Gary Carlson and Jennifer Huber*
Marketing Manager: *Steve Catalano*
Project Development Editor: *Marie Carigma-Sambilay*
Marketing Communications: *Laura Hubrich*
Marketing Assistant: *Christina DeVito*
Editorial Assistant: *Samuel Subity*
Production Editor: *Kirk Bomont*
Production Service: *Heckman & Pinette*

Manuscript Editor: *Margaret Pinette*
Interior Design: *Jeanne Calabrese*
Cover Design: *Vernon T. Boes and Bob Western*
Interior Illustration: *Precision Graphics*
Print Buyer: *Vena Dyer*
Typesetting: *Thompson Type*
Printing and Binding: *Transcontinental Printing Inc.*

For more information, contact:
BROOKS/COLE
511 Forest Lodge Road
Pacific Grove, CA 93950 USA
www.brookscole.com

For permission to use material from this work, contact us by
Web: www.thomsonrights.com
fax: 1-800-730-2215
phone: 1-800-730-2214

Printed in Canada

10 9 8 7 6 5 4 3 2 1

The Library of Congress has cataloged the student edition as follows:
Library of Congress Cataloging-in-Publication Data
Seeds, Michael A.
    Astronomy : the solar system and beyond / Michael A. Seeds.—2nd ed.
       p.   cm.
    Includes index.
    ISBN 0-534-37764-5 (pbk.: alk. paper)
    1. Astronomy.   2. Astrophysics.   I. Title.

QB43.2.S425 2001
520—dc21   00-022548

For Cleo F. "Shorty" Seeds

Brief Contents

Contents

Manuscript Reviewers

Parviz Ansari	*Seton Hall University*	Bernie McNamara	*New Mexico State University*
Erik Bodegom	*Portland State University*	J. Ward Moody	*BYU*
Marten denBoer	*Hunter College of The City University of New York*	Jack C. Straton	*Portland State University*
		Steven Velasquez	*Heidelberg College*
Bernd Enders	*College of Marin*	William R. Wharton	*Wheaton College*
David J. Griffiths	*Oregon State University*	Richard Williamson	*Emory University*
Steven L. Kipp	*Minnesota State University, Mankato*	Louis Winkler	*Penn State University*
Michael C. Lo Presto	*Henry Ford Community College*		

A Note to the Student

From Mike Seeds

Astronomy is about us. Although an astronomy course covers planets, and stars, and galaxies, it is really about us. Astronomy helps us answer the ultimate question of human existence. What are we? Great minds like Plato, Beethoven, and Hemingway have tried to give their personal answers, but we must each find our own answer. Astronomy helps us understand the meaning of our own existence.

Of course, part of our personal answer must include philosophical and cultural issues. We define ourselves by what we create, what we worship, what we admire, and what we expect of each other. But a major part of our answer is imbedded in the physical universe. We cannot hope to find a complete answer until we understand how the birth of the universe, the origin of the galaxies, the evolution of stars, and the formation of planets created and brought together the atoms of which we are made. A basic understanding of astronomy is fundamental to knowing what we are. Astronomy is exciting because it is about us.

Part of the excitement of astronomy is the discovery of new things. Astronomers expect to be astonished. That excitement is reflected in this book in the form of new images, new discoveries, and new understandings that take us, in an introductory course, to the frontier of human knowledge. Here we explore the new evidence of an ancient ocean on Mars and a modern ocean on Jupiter's moon Europa. We will study the newest images of dying stars, and we will struggle to understand new evidence that the expansion of the universe is speeding up. The Hubble Space Telescope, spacecraft orbiting Mars and Jupiter, the new Chandra X-ray observatory, and giant Earth-based telescopes on remote mountaintops provide us with a daily dose of excitement that goes far beyond sensationalism. These new discoveries in astronomy are exciting because they are about us. They tell us more and more about what we are.

As a teacher, my quest is simple. I want you to understand your place in the universe—not just your location in space, but your location in the unfolding history of the physical universe. By the end of this book, I want you to know that the universe is very big, but that it is described by a small set of rules and that we have found a way to figure out the rules—a method called science.

The most important concept in introductory astronomy is the process of science, the process by which scientists ask questions of nature and gradually puzzle out the beautiful secrets of the physical world. You should not memorize facts or believe principles because some astronomer says it is so. Of course not. To be a participant in the adventure and not a mere observer, you need to understand how evidence and hypotheses interact to create new understanding. Those who understand how science works are part of the adventure. Those who do not are baggage.

Another reason you need to understand how science works is that you live in a technological world that is changing rapidly through the practical application of new scientific discoveries. To control your life in such a world, to guide and care for others, you need to understand the power and the limitation of scientific study.

Science is based on the interplay of evidence and hypothesis, and that interplay is the principal organizing theme for this book at every level from the individual sentences to the order of the chapters. We survey basic observational facts, synthesize hypotheses, and discuss further observations as evidence supporting or contradicting those hypotheses. A good example is how a survey of the solar system in Chapter 5 leads to theories to explain the origin of the solar system, which is followed by further evidence in Chapters 6 through 9 that helps us test and generalize our theories. At a different level, the paragraphs that discuss Jupiter's rings show how basic observations allow us to propose theories for the origin of the ring particles.

One big advantage of emphasizing evidence and hypothesis is that it unifies the mass of facts that discourage so many nonscience readers. You must understand science not as facts to be memorized but as logical arguments to be understood, and the interplay of evidence and theory is the core of every logical argument. Of course, this makes the facts easier to remember because they make sense, but the goal of science isn't remembering. Rather it is understanding. Once you begin to understand nature through logical arguments, you will have gained an analytical way of thinking that will serve you in many other areas beyond science.

To help you learn to think about science in logical arguments, I have created a number of features for this book.

- **Windows on Science** provides a parallel commentary on how science works. For example, the Windows point out where we are using statistical evidence, where we are reasoning by analogy, and where we are building a scientific model in place of a scientific hypothesis.

- **Guideposts** at the opening of each chapter help you see the organization of the book by connecting the chapter with previous and succeeding chapters. Organization is the key to understanding and the Guideposts emphasize the overall organization of our study.

- **Review/Critical Inquiry** at the end of each text section is a carefully designed question to help you review and synthesize concepts and understand scientific arguments. A short answer follows to show how scientists construct logical arguments from observations, evidence, theories, and natural laws that lead to conclusions. A further question then gives you a chance to construct your own argument on a related issue.

- **End of Chapter Review Questions** are designed to help you review and test your understanding of the material.

- **End of Chapter Discussion Questions** go beyond the text and invite you to think critically and creatively about scientific arguments. These questions could be used in class for discussion or by individual students as part of a review process.

- **By the Numbers** are short math boxes that occasionally present an equation and an example or two. The mathematical aspects of science are integral to our understanding, so it is important for you to see these ideas. But because students in introductory science courses have different levels of math experience, arguments in the text do not depend entirely on mathematical reasoning.

- **Critical Inquiries for the Web** conclude each chapter by challenging you to use the World Wide Web to explore further and to think creatively and analytically about astronomy.

- **Exploring The Sky** are experiments you can perform with the software *The Sky*, experiments that will allow you to see the sky and celestial bodies in new ways on your own computer screen.

Do not be humble. Astronomy tells us that the universe is vast and powerful, but it also tells us that we are astonishing creatures. We humans are the parts of the universe that think. The human brain is the most complex piece of matter known, so as you explore the universe, remember that it is your human brain that is capable of understanding the depth and beauty of the cosmos.

To appreciate what you are in this vast universe, you must learn more than just the facts of astronomy. You must understand what we are, where we are, and how we know. Every page of this book reflects that ideal.

Mike Seeds
m_seeds@acad.fandm.edu

Acknowledgments

In 1973, I started writing astronomy textbooks, and over the years I have had the guidance of a great many people who care about astronomy and teaching. I would like to thank all of the students and teachers who have responded so enthusiastically to *Astronomy: The Solar System and Beyond.* Their comments and suggestions have been very helpful in shaping the book. I would especially like to thank the many reviewers whose careful analysis and thoughtful suggestions have been invaluable in completing this new edition.

The following institutions have provided the many photos and diagrams in this book: Anglo-Australian Telescope Board, Arecibo Observatory, Association of Universities for Research in Astronomy, The Astronomical Journal, The Astrophysical Journal, AT&T Archives, Arizona State University Computer Science and Geography Departments, Brookhaven National Laboratories, Calar Alto Observatory, California Association for Research in Astronomy, California Institute of Technology, Carnegie Institute of Washington, Celestron International, Chandra X-Ray Center, Cornell University, European Southern Observatory, Galileo Imaging Team, Gemini 8-m Telescope Project, Goddard Space Flight Center, Granger Collection, Grundy Observatory, Hobby-Eberly Telescope Project, Hubble Heritage Team, Infrared Processing and Analysis Center, Institute for Astronomy, Japan/US Yohkoh Team, Johns Hopkins Applied Physics Lab, Joint Astronomy Center, JPL, Lick Observatory, Lowell Observatory, Lunar and Planetary Institute, Max-Planck Institute for Extraterrestrial Physics, McDonald Observatory, NASA, National Astronomy and Ionospheric Center, National Geophysical Data Center, National Museum of Natural History, National Oceanic and Atmospheric Administration, National Optical Astronomy Observatory, National Radio Astronomy Observatory, New England Meteorical Services, North Museum Lancaster PA, Sac Peak Observatory, SETI Institute, Smithsonian Astrophysical Observatory, The Solar and Heliospheric Observatory, Space Telescope Science Institute, Steward Observatory, Tektronix Inc., University of Hawaii, U.S. Geological Survey, William Keck Observatory, Yerkes Observatory.

I would especially like to thank the following individuals who went out of their way to help me locate materials for *Astronomy:* H. A. Abt, Andy Aperala, Dana Backman, Sam Barden, Thomas G. Barns, Ginny Beal, Allen Beechel, Roger Bell, Dana Berry, David Bradstreet, Michael Briley, Jack Burns, Adam Burrows, Bruce Campbell, Chip Clark, Don Davis, Frank Drake, R. J. Dufour, Richard Fluck, Sidney W. Fox, James Gelb, Margaret J. Geller, Dan Gezari, Bronagh Glaser, Daniel Good, Bradford Greeley, Randall Grubbs, William Hartmann, John Hayes, Helen Horstman, John Hughes, Vincent Icke, George Jacoby, Rebecca Johnson, William Keel, Charles Keller, R. Kempton, Stephen Larson, Tod Lauer, Kathy Lestition, Deborah Levine, Kenneth Libbrecht, Kwok-Yung Lo, John W. Mackenty, David Malin, Jack Marling, Ursula B. Marvin, Maureen McLean, David Miller, Stanley Miller, Reinhard Mundt, Lynne Nakauchi, Harold Nations, Steven Ostro, David Parker, Richard Perley, Carolyn C. Porco, Kathleen Powell, Vera Rubin, Laurence Rudnick, Anneila Sargent, Rudolph Schild, Maartin Schmidt, J. William Schopf, Francois Schweizer, M. Seldner, Nigel Sharp, Seth Shostak, L. D. Simon, Steve Snowden, Hyron Spinrad, Alan Stockton, J. Trumper, Tony Tyson, Joseph Ververka, Steven Vogt, Mike Werner, Michael J. West. Gart Westerhout, Richard E. White, Simon D. M. White, Robin Witmore, Dennis Zaritsky.

Certain unique diagrams in Chapters 9 and 10 are based on figures I designed for my article "Stellar Evolution," which appeared in *Astronomy,* February 1979.

It has been a great pleasure to work with and learn from all of the Brooks/Cole team. Special thanks go to all of the people who have contributed to the project, including Marie Carigma-Sambilay, Vernon Boes, Kirk Bomont, Jeanne Calabrese, Steve Catalano, Christina De Veto, Heather Dutton, Laura Hubrich, Sam Subity, and Pat Waldo. I have enjoyed working with Margaret Pinette and Bill Heckman of Heckman & Pinette on production, and I appreciate their understanding of my multifaceted failings. I would especially like to thank my editor, Gary Carlson, for all his help and advice and guidance on all of my books and for his efforts to develop the entire astronomy list at Brooks/Cole.

Most of all, I would like to thank my wife, Janet, and my daughter, Kate, for putting up with "the books." They know all too well that textbooks are made of time.

Mike Seeds

Astronomy

Exploring *The Sky*

1. Install and open *The Sky*. Take this opportunity to famil-
    iarize yourself with this program that turns your com-
    puter screen into a window through which you are view-
    ing a virtual sky. *The Sky* can show your local sky at any
    time and date (well, within limits . . .) as well as many
    objects you can only see with binoculars and telescopes.
    You can face different directions by clicking on the **Look
    North, Look East, Look South, Look West, Look Up** but-
    tons on your **Orientation Toolbar.** Or you can simply
    drag the slider on the bottom of your screen. Also experi-
    ment with changing between the **Daytime Sky Mode** and
    Nighttime Sky Mode by clicking on the second button in
    the **View Toolbar.**

    Before you experiment much further, it is useful to de-
    fine some settings.

    a. First, set *The Sky* for your location, i.e.. your latitude
        and longitude. Click on **Data,** then on **Site Informa-
        tion.** The easiest way to input your location is simply
        to chose a large city that is closest to you—for our pur-
        pose, that will most likely be close enough to your ac-
        tual location. Also check to make sure that the chosen
        time at which you'll view *The Sky* is the computer
        clock (the default)—check to see that a check mark ap-
        pears next to **Use Computer's Clock** within the **Site In-
        formation** window.

    b. It will also be useful to change the **Status bar** at the
        bottom of your screen to show date and time. To do
        this, click on **View,** then **Status Bar...,** then check **date**
        and **time** in **Status Bar window.**

The Scale of the Cosmos

The longest journey begins

with a single step.

Confucius

Guidepost

How can we study something so big it includes everything, even us? How do we start thinking about the universe? Perhaps the best way is to grab a quick impression as we zoom from things our own size up to the largest things in the entire universe.

That cosmic zoom, the subject of this introduction, gives us our first glimpse of the subject of the rest of this book—the universe. In the next chap-ter we will return to Earth to think about the appearance of the sky, and subsequent chapters will discuss other aspects of the universe around us.

While we study the universe, we must also observe the process by which we learn. That process, sci-ence, gives us a powerful way to un-derstand not only the universe but also ourselves.

We are going on a voyage out to the end of the universe. Marco Polo journeyed east, Columbus west, but we will travel away from our home on Earth, out past the moon, sun, and other planets, past the stars we see in the sky, and past billions more that we cannot see without the aid of the largest telescopes. We will journey through great whirlpools of stars to the most distant galaxies visible from Earth—and then we will continue on, carried only by experience and imagination—looking for the structure of the universe itself.

Besides journeying through space, we will also travel in time. We will explore the past to see the sun and planets form and search for the formation of the first stars and the origin of the universe. We will also explore forward in time to watch the sun die and Earth wither. Our imagination will become a scientific time machine searching for the ultimate end of the universe.

Though we may find an end to the universe, a time when it will cease to exist, we will not discover an edge. It is possible that our universe is infinite and extends in all directions without limit. Such vastness dwarfs our human dimensions, but not our intelligence or imagination.

Astronomy—this imagined voyage—is more than the study of stars and planets. It is the study of the universe in which we exist. Our personal lives are confined to a small corner of a small planet circling a small sun drifting through the universe, but astronomy can take us out of ourselves and thus help us understand what we are.

Our study of astronomy introduces us to sizes, distances, and times far beyond our common experience. The comparisons in this chapter are designed to help us grasp their meaning.

How big is a star? The answer—roughly 1 million miles in diameter—is meaningless. Such a large number tells us nothing. How can we humans, only 5 or 6 feet tall, hope to understand the vastness of the universe? The secret lies in the single word *scale*.

To understand the universe, we must understand the relative scale of planets, stars, galaxies, and the universe as a whole. Only when we can relate our own body size to the astronomical universe around us can we begin to understand nature on the grandest scale.

To illustrate the scale of astronomical bodies—to fit ourselves into the universe—we will journey from a campus scene to the limits of the cosmos in 12 steps. Each successive picture in this chapter will show a region of the universe that is 100 times wider than the preceding picture. That is, each step will widen our view by a factor of 100.

Figure I-1 shows a region about 52 feet across. It is occupied by a human being, a sidewalk, and a few trees—all objects whose size we can understand. Only 12 steps separate this scene from the universe as a whole.

In Figure I-2, we increase our field of view by a factor of 100 and see an area 1 mile across. The area of the preceding photograph is shown by the arrow. Individual people, trees, and sidewalks vanish, but now we can see a college campus and the surrounding streets and houses.

These dimensions are familiar. We have been in houses, crossed streets, and walked or run a mile. We

Figure I-1
(Michael A. Seeds)

Figure I-2
(USGS)

Figure I-3
(NASA infrared photograph)

see that it, like Earth's surface, is evolving. Change is the norm in the universe.

Green foliage of different kinds appears as shades of red in this infrared photo, a reminder that our human eyes see only a narrow range of colors. As we explore the universe, we will learn to use a wide range of "colors," from X rays to radio waves, to reveal sights invisible to our unaided eyes.

The next step in our journey (Figure I-4) shows our entire planet. Earth is 12,756 km in diameter and rotates on its axis once a day.

Figure I-4
(NASA)

have personal experience with such dimensions, and we can relate them to the scale of our bodies.

Although we have begun our adventure using feet and miles, we should use the metric system of units because it makes arithmetic simpler (see Appendix A). To convert 1 mile to inches, we must multiply $5280 \times 12 = 63,360$ inches. To convert 1 kilometer to centimeters, we simply multiply 1000 meters by 100 to get 100,000 cm.

One mile equals 1.609 km, so our photograph is about 1.6 km across. (See Appendix A for other conversion factors.) Only 11 more steps of 100 separate us from the largest dimensions in the universe.

In Figure I-3, our view spans 160 km, about 100 miles. The college campus is now invisible, and we see few signs of human activity. Cities are small blotches and farmlands tiny rectangles. The suburbs of Philadelphia are visible at the lower right. At this scale, we see the natural features of Earth's surface. The Allegheny Mountains of southern Pennsylvania cross the image in the upper left, and the Susquehanna River flows southeast into Chesapeake Bay. A few puffs of clouds dot the area.

These features remind us that we live on the surface of a changing planet. Forces in Earth's crust pushed the mountain ranges up into parallel folds, like a rug wrinkled on a polished floor. The clouds remind us that Earth's atmosphere is rich in water, which falls as rain and erodes the mountains, washing material down the rivers and into the sea. The mountains and valleys that we know are only temporary features; they are constantly changing. As we explore the universe, we will

This image shows most of the daylight side of the planet, with the sunset line at the extreme right. Earth's rotation carries us eastward across the daylight side, and as we cross the sunset line into darkness, we say the sun has set. Thus, the rotation of our planet causes the cycle of day and night.

We know that Earth's interior is made mostly of iron and nickel and that its crust is mostly silicate rocks. Only a thin layer of water makes up the oceans, and the atmosphere is only a few hundred miles deep. On the scale of this photograph, the depth of the atmosphere on which our lives depend is less than the thickness of a piece of thread.

Our **solar system** consists of the sun, its family of nine planets, and some smaller bodies such as moons and comets. Within our solar system, our Earth is the only planet with the air and liquid water that make life possible. Only recently (1996) have astronomers found planets orbiting other stars, and those planets seem quite unlike Earth. Other Earthlike planets probably exist, but they are too small and too distant to detect. So far as we know, we are the only life in the universe.

Again we enlarge our field of view by a factor of 100 (Figure I-5), and we see a region 1,600,000 km wide. Earth is the small white dot in the center, and our moon, only one-fourth Earth's diameter, is an even smaller dot along its orbit 380,000 km from Earth.

These numbers are so large that it is inconvenient to write them out. Astronomy is the science of big numbers, and we will use numbers much larger than these to discuss the depths of the universe. Rather than write out these numbers, it is convenient to write them in **scientific notation.** This is nothing more than a simple way to write big numbers without writing a great many zeros.

In scientific notation, we would write 380,000 as 3.8×10^5. The 5 tells us to move the decimal point five places to the right. The 3.8 then becomes 380,000. In the same way, we would write 1,600,000 as 1.6×10^6. Notice that we could also write this as 16×10^5 or 0.16×10^7. All represent the same quantity.

We can also use scientific notation to write very small numbers. If you are not familiar with sci-

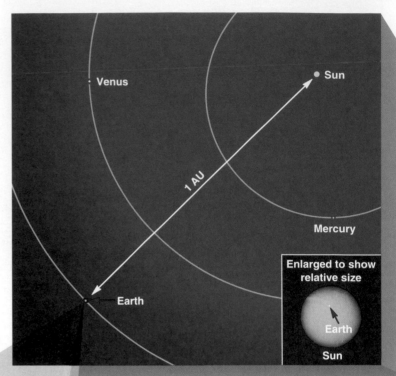

Figure I-6

entific notation, consult Appendix A. The universe is too big to discuss without using scientific notation.

When we once again enlarge our field of view by a factor of 100 (Figure I-6), Earth, our moon, and the moon's orbit all lie in the small red box at lower left. But now we see the sun and two other planets.

Venus is about the size of Earth, and Mercury is a bit larger than our moon. On this diagram, they are both too small to be seen as anything but tiny dots. The sun is 109 times larger in diameter than Earth (inset), but it too is nothing more than a dot on this diagram.

This figure spans 1.6×10^8 km. One way to deal with such large numbers is to define new units. Astronomers use the average distance from Earth to the sun as a unit of distance called the **astronomical unit (AU).** Using this unit, we can say that the average distance from Venus to the sun is about 0.7 AU. The average distance from Mercury to the sun is about 0.39 AU.

The orbits of the planets are not perfect circles, and this is particularly apparent for Mercury. Its orbit carries it as close to the sun as 0.307 AU and as far away as 0.467 AU. Earth's orbit is more circular, and its distance from the sun varies by only 1.7 percent.

Our first field of view was only 52 feet (about 16 m) in width. After only six steps of enlarging our field of view by a factor of 100 at each step, we now see the entire solar system (Figure I-7). Our

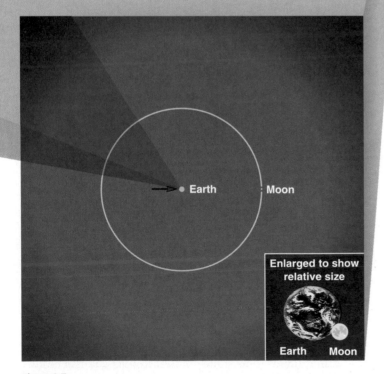

Figure I-5

field of view is 1 trillion (10^{12}) times wider than in our first view.

The details of the preceding figure are now lost in the tiny square at the center of this diagram. We see only the brighter, more widely separated objects as we back away. The sun, Mercury, Venus, and Earth lie so close together that we cannot separate them at this scale. Mars, the next outward planet, lies only 1.5 AU from the sun.

In contrast, the outer planets Jupiter, Saturn, Uranus, Neptune, and Pluto are so far from the sun that they are easy to place in this diagram. These planets are cold worlds far from the sun's warmth. Light from the sun takes over 4 hours to reach Neptune, which is slightly over 30 AU distant. In contrast, sunlight reaches Earth in only 8 minutes.

Notice that Pluto's orbit is so elliptical that Pluto can come closer to the sun than Neptune. In fact, Pluto moved closer to the sun than Neptune in 1979, not re-

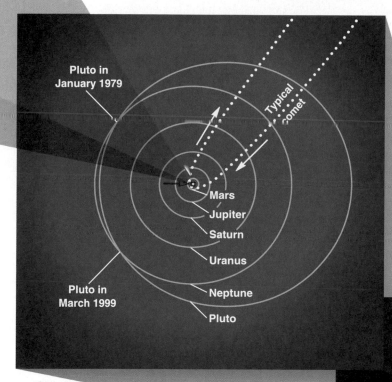

Figure I-7

claiming its crown as the most distant planet in our solar system until March 14, 1999.

When we again enlarge our field of view by a factor of 100 (Figure I-8), our solar system vanishes. The sun is only a point of light, but all the planets and their orbits are now crowded into the small square at the center. None of the sun's family of planets are visible. They are too small and reflect too little light to be visible so near the brilliance of the sun.

Nor are any stars visible except for the sun. The sun is a fairly typical star, and it seems to be located in a fairly average neighborhood in the universe. Although there are many billions of stars like the sun, none are close enough to be visible in this diagram of only 11,000 AU in diameter. The stars are typically separated by distances about ten times larger than this diagram. We will see stars in our next field of view, but except for the sun at the center this diagram is empty.

It is difficult to image the isolation of the stars. If the sun were represented by a golf ball in New York City, the nearest star would be another golf ball in Chicago. Except for the widely scattered stars and a few atoms of gas drifting between the stars, the universe is nearly empty.

In Figure I-9, our field of view has now expanded to a diameter of a bit over 1 million AU. The sun is located at the center, and we see a few of the nearest stars.

These stars are so distant that it is not reasonable to give their distances in astronomical units. We must define a new unit of distance, the light-year. One **light-year (ly)** is the distance that light travels in 1 year, roughly 10^{13} km or 63,000 AU. The nearest star to the sun is Proxima Centauri at a distance of 4.2 ly. Light from Proxima Centauri takes 4.2 years to reach Earth. The diameter of our field of view is now 17 ly.

Although these stars are roughly the same size as the sun, they are so far away that we cannot see them as anything but points of light. Even with the largest telescopes on Earth, we still see only points of light when we look at stars. In Figure I-9, the sizes of the dots represent not the size of the stars but their brightness. This is the custom

Figure I-8

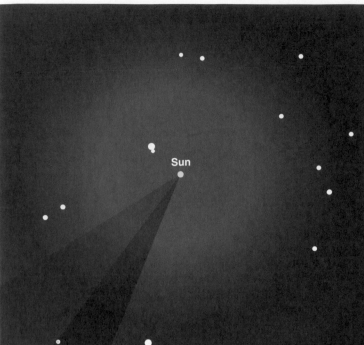

Figure I-9

space to photograph our galaxy, so this photo shows a similar galaxy with an arrow pointing to a representative location for the sun.

Our sun and the neighboring stars of the previous figure would be lost among the 100 billion stars of the galaxy. Most of the stars are smaller and fainter than our sun, but some are larger and more luminous. Most are cooler, but a few are much hotter. Why some stars are larger, more luminous, or hotter than others is one of the mysteries of the universe that we will explore.

Typical of our galaxy are the graceful **spiral arms** marked by clusters of bright stars and clouds of gas. We will discover that stars are born in great clouds of gas and dust passing through these spiral arms.

in astronomical diagrams, and it is also how star images are recorded on photographs. Bright stars make larger spots on a photograph than faint stars. Thus, the size of a star image in a photograph tells us not how big the star is but only how bright the star looks.

In Figure I-10, we expand our field of view by another factor of 100, and the sun and its neighboring stars vanish into the background of thousands of stars. The field of view is now 1700 ly in diameter.

Of course, no one has ever journeyed thousands of light-years from the sun to photograph the solar neighborhood, so we use a representative photo of the sky. The sun is a relatively faint star, so we could no longer locate it on such a photo.

We notice a tendency for stars to occur in clusters. A loose cluster of stars lies in the lower left quarter of the photograph. We will discover that many stars are born in clusters and that both old and young clusters exist in the sky. Star clusters are forming right now.

What we do *not* see is critically important. We do not see the thin gas that fills the spaces between the stars. Although those clouds of gas are thinner than the best vacuum on Earth, it is those clouds that give birth to new stars. Our sun formed from such a cloud about 5 billion years ago. We will see more star formation in our next view.

If we expanded our field of view by a factor of 100 (Figure I-11), we would see our own **Milky Way Galaxy.** Of course, no one can journey far enough into

Figure I-10
This box ■ represents the relative size of the previous frame. *(NOAO)*

Our galaxy is roughly 100,000 ly in diameter, and until the 1920s astronomers thought it was the entire universe—an island universe of stars in an otherwise empty vastness. Now we know that our galaxy is not unique. Indeed, ours is only one of many billions of galaxies scattered throughout the universe.

As we expand our field of view by another factor of 100 (Figure I-12), our galaxy appears as a tiny luminous speck surrounded by other specks. This diagram includes a region 17 million ly in diameter. Each of these dots represents a galaxy.

We see that our galaxy (arrow) is one of a small cluster of galaxies. This **Local Group** consists of roughly

Figure I-13

This box ■ represents the relative size of the previous frame. *(Detail from galaxy map from M. Seldner, B. L. Siebers, E. J. Groth, and P. J. E. Peebles,* Astronomical Journal *82 [1977])*

If we again expand our field of view (Figure I-13), we see that our Local Group of galaxies is part of a large supercluster—a cluster of clusters. Other galaxies are not scattered at random throughout the universe but lie in clusters within larger superclusters.

To represent the universe at this scale, we use a diagram in which each dot represents the location of a single galaxy. At this scale, we see superclusters linked to form long filaments outlining voids that seem nearly empty of galaxies. These appear to be the largest structures in the universe. Were we to expand our field of view yet another time, we would probably see a uniform sea of filaments and voids. When we puzzle over the origin of these structures, we are at the frontier of human knowledge.

Our problem in studying astronomy is to keep a proper sense of scale. Remember that each of the billions of galaxies contains billions of stars. Most of those stars probably have families of planets like our solar system, and on some of those billions of planets liquid-water oceans and a protective atmosphere may have spawned life. It is possible that some other planets in the universe are inhabited by intelligent creatures who share our curiosity and our wonder at the scale of the cosmos.

two dozen galaxies scattered throughout a region about 6 million ly in diameter.

Among the galaxies we see here, a few are as large as our own galaxy, but most are smaller. A few have the beautiful spiral features we see in our galaxy, but most do not. Among more distant galaxies, we see a few galaxies twisted into peculiar shapes or wracked by violent eruptions. Although astronomers understand why stars differ from one another, it is not entirely clear why one galaxy differs from another. We will find some clues to the mystery when we compare the clusters of galaxies.

Figure I-11

(© Anglo-Australian Telescope Board)

Figure I-12

Summary

Our goal has been to preview the scale of astronomical objects. To do so, we journeyed outward from a familiar campus scene by expanding our field of view by factors of 100. Only 12 such steps took us to the largest structures in the universe.

The numbers in astronomy are so large it is not convenient to express them in the usual way. Instead, we use the metric system to simplify our calculations and scientific notation to more easily write big numbers. The metric system and scientific notation are discussed in Appendix A.

We live on the rotating planet Earth, which orbits a rather typical star we call the sun. We defined a unit of distance, the astronomical unit, to be the average distance from Earth to the sun. Of the eight other planets in our solar system, Mercury is closest to the sun, and Pluto is the most distant.

The sun, like most stars, is very far from its neighboring stars, and this leads us to define another unit of distance, the light-year, the distance light travels in 1 year. The nearest star to the sun is Proxima Centauri at a distance of 4.2 ly.

As we enlarged our field of view, we discovered that the sun is only one of 100 billion stars in our galaxy and that our galaxy is only one of many billions of galaxies in the universe. Galaxies appear to be grouped together in clusters, superclusters, and filaments, the largest structures known.

As we explored, we noted that the universe is evolving. Earth's surface is evolving, and so are stars. Stars form from the gas in space, grow old, and eventually die. We do not yet understand how galaxies form or evolve.

Among the billions of stars in each of the billions of galaxies, many probably have planets, but detecting such planets is very difficult. Although astronomers can now detect planets orbiting other stars, we know very little about the nature of these planets. Yet we suppose that there must be many planets in the universe and that some are like Earth. We wonder if a few are inhabited by intelligent beings like ourselves.

New Terms

solar system

scientific notation

astronomical unit (AU)

light-year (ly)

Milky Way Galaxy

spiral arm

Local Group

Review Questions

1. What is the largest dimension you have personal knowledge of? Have you run a mile? Hiked 10 miles? Run a marathon?

2. Why are astronomical units more convenient than miles or kilometers for measuring some astronomical distances?

3. In what ways is our planet changing?

4. What is the difference between our solar system, our galaxy, and the universe?

5. Why do all stars, except for the sun, look like points of light as seen from Earth?

6. Why are light-years—rather than miles or kilometers—more convenient units for some astronomical distances?

7. Why is it difficult to see planets orbiting other stars?

8. Which is the outermost planet in our solar system? Why does that change?

9. Why can't we measure the diameters of stars from the size of the star images on photographs? What does the diameter of a star image really tell us about the star?

10. How long does it take light to cross the diameter of our solar system, of our galaxy, of the Local Group?

11. What are the largest known structures in the universe?

12. How many planets inhabited by intelligent life do you think the universe contains? Explain your answer.

Problems

1. If 1 mile equals 1.609 km and the moon is 2160 miles in diameter, what is its diameter in kilometers? in meters?

2. If sunlight takes 8 minutes to reach Earth, how long does moonlight take?

3. How many suns would it take, laid edge to edge, to reach the nearest star?

4. How many kilometers are there in a light-minute? (*Hint:* The speed of light is 3×10^5 km/s.)

5. How many galaxies like our own, laid edge to edge, would it take to reach the nearest large galaxy (which is 2×10^6 ly away)?

Exploring *The Sky*

1. Locate and center one example of each of three different types of objects:

   a. A planet, such as Saturn. Find its rising and setting time. Such objects have distances measured in **astronomical units (AU).**

      How to proceed: Decide on the object to locate. Then find and center it by pressing the **Find** button on the **Object Toolbar.** The second method is to press the **F** key. The third is to click **Edit,** then **Find.** Once you have the **Object Information** window, press the **center** button.

   b. A star. All stars in *The Sky* belong to our Milky Way Galaxy. Give the star's name, its magnitude, and its distance in light-years.

      How to proceed: Click on any star, which brings up an **Object Information** window.

   c. A galaxy; give its name and/or its designation.

      How to proceed: Click on the **Galaxies button** in the **Object Toolbar,** then click on any galaxy. Distances to galaxies are millions and billions of light years.

Go to the Brooks/Cole Astronomy Resource Center (www.brookscole.com/astronomy) for critical thinking exercises, articles, and additional readings from InfoTrac College Edition, Brooks/Cole's online student library.

CHAPTER ONE

The View from Earth

The Southern Cross I saw every night abeam. The sun every morning came up astern; every evening it went down ahead. I wished for no other compass to guide me, for these were true.

Captain Joshua Slocum
Sailing Alone Around the World

Guidepost

Astronomy is about us. As we learn about the universe, we learn about ourselves. We search for an answer to the question "What are we?" The quick answer is that we are thinking creatures living on a planet that circles a star we call the sun. In this chapter, we begin trying to understand that answer. What does it mean to live on a planet?

The introduction gave us a quick overview of the universe, and chapters later in this book will discuss the details. This chapter will help us understand what the universe looks like seen from our spinning planet.

What we see is caused by the cycles of Earth, its moon, and the sun. We will see in the next chapter how Renaissance astronomers analyzing these cycles first realized that we do indeed live on a planet.

9

Stonehenge was built by space aliens who landed in UFOs, according to some so-called experts. Others will tell you Stonehenge was built by King Arthur's magician, Merlin, who floated the multi-ton stones into place by magic. In fact, Stonehenge is more amazing than any of these tabloid explanations. Stonehenge was built roughly 4000 years ago by Stone Age people using their hands, their backs, and simple tools such as antler picks. The entire structure is oriented to face the rising sun on the longest day of the year—the first day of summer (Figure 1-1).

All around the world, ancient peoples oriented their sacred temples astronomically in recognition that they lived within a cosmos of order and beauty, a cosmos ruled by cycles of vast power. The daily cycle of the sun, the monthly cycle of our moon, and the yearly cycle of the seasons created and maintained their world.

What are these great cycles and how do they arise? We begin our voyage into the universe by looking at the sky from Earth and seeing with modern eyes what our ancestors saw with superstition and wonder.

of stars and individual stars and by specifying the brightness of individual stars.

Constellations

All around the world, ancient cultures gave names to groups of stars—**constellations**—to honor gods, heroes, and animals (Figure 1-2a). Of course, each culture created its own unique set of constellations. The constellations we are familiar with in Western culture originated in Mesopotamia over 5000 years ago, with other constellations added by Babylonian, Egyptian, and Greek astronomers during the classical age. Of these ancient constellations, 48 are still in use today. In recent centuries, astronomers added 40 modern constellations to fill gaps and to mark the southern sky not visible to the ancient astronomers. In 1928 the International Astronomical Union established 88 official constellations with clearly defined boundaries (Figure 1-2b). Thus a constellation is now represented by a precise region of the sky.

1-1 The Sky

On a dark night far from city lights, we can see a few thousand stars in the sky. Like ancient astronomers, we will try to organize what we see by naming groups

a

b

Figure 1-1
(a) Ancient Stonehenge on Salisbury Plain in England appears on this stamp issued to mark the return of Comet Halley. (b) On the morning of the longest day of the year, an observer standing at the center of Stonehenge sees the sun rise (dashed line) behind the Heelstone in the center of the avenue extending to the northeast (c).

c

In addition to the 88 official constellations, the sky contains a number of less formally defined groupings called **asterisms.** The Big Dipper, for example, is a well-known asterism, but it is only part of the larger constellation Ursa Major (the Great Bear). Another asterism is the Great Square of Pegasus (Figure 1-2b), which includes three stars from Pegasus and one from Andromeda. The star charts at the end of this book will introduce you to the brighter constellations.

Although we name constellations and asterisms, keep in mind that they are made up of stars that are usually not physically associated with one another. Some stars may be many times farther away than others and moving through space in different directions (Figure 1-3). The only thing they have in common is that they lie in approximately the same direction from Earth.

The names of constellations come from Latin, the language of science during the Renaissance, but most of the brighter stars have names derived from ancient Arabic. The name of Betelgeuse, the bright red star in Orion, comes from the Arabic *yad al-jawza,* meaning "hand of Jawza [Orion]."

Within a constellation, stars are assigned Greek letters in approximate order of brightness. Thus the brightest star is usually α (alpha), the second brightest β (beta) and so on (Figure 1-2b). Thus Betelgeuse is also known as Alpha Orionis. The custom is to use the Greek letter followed by the genitive (possessive) form of the constellation name. In some constellations, by tradition or mistake, the letters don't follow brightness exactly, so this is not a totally reliable key to brightness.

The Brightness of Stars

Hipparchus, a Greek astronomer who lived about 2100 years ago, divided the stars into six classes. The brightest were first-class stars, and those slightly fainter were second-class stars. Continuing down to the faintest stars he could see, the sixth-class stars, he recorded his classifications in a great star catalogue that became a basic reference in ancient astronomy. His method, slightly modified, is still in use today.

In spite of its value, Hipparchus's method may seem a bit confusing. First, when early astronomers translated the catalogue into Latin, they used the word *magnitudo,* meaning "size." In English, this became *magnitude,* even though it refers to the brightness of the stars and not to their size. Thus the **magnitude scale** is the astronomer's brightness scale.

The second source of confusion is that the fainter the star, the larger the magnitude number. For example, 6th-magnitude stars are fainter than 1st-magnitude stars. This may seem backwards at first, but think of it as Hipparchus did. The brightest stars are

a

b

Figure 1-2
(a) The ancients placed their great gods, goddesses, and symbolic animals in the sky as constellations with only rough boundaries (dotted lines). *(From Duncan Bradford,* Wonders of the Heavens, *Boston: John B. Russell, 1837)* (b) Modern constellations have boundaries precisely defined by international agreement. Thus the star Alpheratz, the eye of Andromeda, is not part of the constellation Pegasus although it is included in the asterism called the Great Square of Pegasus.

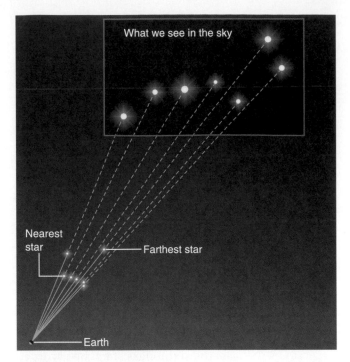

Figure 1-3
The stars we see in the Big Dipper—the brighter stars in the constellation Ursa Major, the Great Bear—are not the same distance from Earth. In fact, the most distant star is not the faintest in the Big Dipper and the nearest star Is not the brightest. We see stars in a group in the sky because they lie in the same general direction as seen from Earth. The size of the star dots in the star chart represents the apparent brightness of the stars. (Based on distances from the Hipparcos satellite.)

first-class stars, and the fainter stars are second- and third-class, and so on.

Modern astronomers can measure the brightness of stars to high precision, so instead of saying that the star known by the charming name Chort (Theta Leonis) is 3rd magnitude, they can say its magnitude is 3.34. If we measure magnitudes, we discover that some stars are brighter than 1.0. For example, Vega (α Lyrae) is so bright that its magnitude, 0.04, is almost zero. A few are so bright the magnitude scale must extend into negative numbers (Figure 1-4). On this scale, Sirius, the bright-

est star in the sky, has a magnitude of −1.47. If we use a telescope to search for very faint stars, we can find stars much fainter than the limit for the unaided eye. Thus the magnitude system has been extended to numbers larger than 6th magnitude to include fainter stars.

These measures are called **apparent visual magnitudes** because they refer to how bright a star looks. Of course, distance is also important. A powerful star at a great distance will look faint. Apparent visual magnitudes ignore the distance to the stars and refer only to how the stars look to our eyes.

Brightness is quite subjective depending on both the physiology of our eyes and the psychology of perception. To be accurate we should refer to intensity—a measure of the light energy from a star that hits one square meter in one second. A simple relationship connects apparent visual magnitudes and the intensity of star light (By the Numbers 1-1). Thus modern astronomers can measure the brightness of stars to high precision while still making comparisons to observations of apparent visual magnitude that go back to the time of Hipparchus.

REVIEW **Critical Inquiry**

Compare Chort and Vega. What can we conclude about these stars?

The given names of these stars tell us nothing except that the stars are not extremely faint. We know that Vega is Alpha Lyrae, so we can guess that it is probably the brightest star in that constellation. On the other hand, Chort is Theta Leonis, so it is probably not among the brightest stars. In fact, the apparent visual magnitude of Vega is 0.04, very bright, and the apparent visual magnitude of Chort is 3.34, much fainter than Vega.

Sometimes the names of stars can give us clues to their brightness, but the magnitude system is more dependable. How many times brighter is Vega compared to Chort?

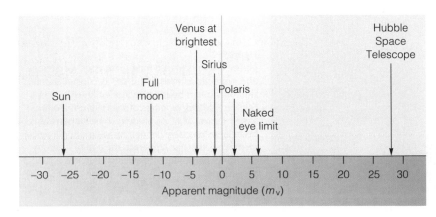

Figure 1-4
The scale of apparent visual magnitudes extends into negative numbers to represent the brighter objects, and to positive numbers larger than 6 to represent objects fainter than the human eye can see.

Astronomers use a simple formula to convert magnitudes to intensities. If two stars have intensities I_A and I_B, then the ratio of their intensities is I_A/I_B. Modern astronomers have defined the magnitude scale so that two stars that have an intensity ratio of exactly 100 differ by 5 magnitudes. Then two stars that differ by 1 magnitude must have an intensity ratio of the 5th root of 100, $\sqrt[5]{100}$, which equals 2.512—that is, the light of one star must be 2.512 times more intense. Two stars that differ by 2 magnitudes will have an intensity ratio of 2.512 × 2.512, or about 6.3, and so on (Table 1-1).

Example A: Suppose star C is 3rd magnitude and star D is 9th magnitude. What is the intensity ratio? *Solution:* The magnitude difference is 6 magnitudes, and the table shows the intensity ratio is 250. Therefore light from star C is 250 times more intense than light from star D.

A table is convenient, but for more precision we can express the relationship as a simple formula. The intensity ratio I_A/I_B is equal to 2.512 raised to the power of the magnitude difference $m_A - m_B$:

$$\frac{I_A}{I_B} = (2.512)^{(m_B - m_A)}$$

Example B: If the magnitude difference is 6.32 magnitudes, what is the intensity ratio? *Solution:* The intensity ratio must be $2.512^{6.32}$. A pocket calculator tells us the answer: 337.

When we know the intensity ratio and want to find the magnitude difference, it is convenient to solve the formula for the magnitude difference:

$$m_A - m_B = 2.5 \log(I_B/I_A)$$

Example C: Suppose that light from Sirius is 24.2 times more intense than light from Polaris. What is the magnitude difference? *Solution:* The magnitude difference is 2.5 log(24.2). Our pocket calculator tells us the logarithm of 24.2 is 1.38, so the magnitude difference is 2.5 × 1.38 = 3.4 magnitudes.

Table 1-1 Magnitude and Intensity

Magnitude Difference	Intensity Ratio
0	1
1	2.5
2	6.3
3	16
4	40
5	100
6	250
7	630
8	1600
9	4000
10	10,000
⋮	⋮
15	1,000,000
20	100,000,000
25	10,000,000,000
⋮	⋮

1-2 The Daily Cycle of the Sky

Ancient astronomers thought the sky was a great sphere with the stars stuck on the inside like thumbtacks in the ceiling. They imagined that Earth was located at the center of this **celestial sphere.** Of course, modern astronomers know that the stars are scattered at different distances through space. Nevertheless, the celestial sphere is still a useful model to help us think about the motions of the sky (Window on Science 1-1).

A Model of the Sky

The sky we see at any moment is the half of the celestial sphere above the **horizon,** the dividing line between earth and sky. The top of the sky, the **zenith,** is directly overhead. Because Earth rotates eastward once a day, the celestial sphere appears to rotate westward (Figure 1-5a).

"To rotate" means to turn on an axis, while "to revolve" means to follow a path around an external point. Thus we say that Earth rotates on its axis but revolves around the sun.

Figure 1-5b shows how stars in the northern sky appear to revolve around a point called the **north celestial pole,** the point on the sky directly above Earth's North Pole. The star Polaris (the North Star) happens to lie very near the north celestial pole and thus hardly moves as Earth rotates. (Locate Polaris on the star charts at the end of the book.)

Constellations, asterisms, star names, and apparent visual magnitudes give us convenient ways to describe the sky, but we need now to look at one of the great sky cycles—the daily rotation of the sky.

In everyday language, we use the word *model* in various ways—fashion model, model airplane, model student—but scientists use the word in a very specific way. A **scientific model** is a carefully devised mental conception of how something works, a framework that helps scientists think about some aspect of nature. For example, astronomers use the celestial sphere as a way to think about the motions of the sky, sun, moon, and stars.

Although we will think of a scientific model as a mental conception, it can take many forms. Some models are quite abstract—the psychologist's model of how the human mind processes visual information into images, for instance. But other models are so specific that they can be expressed as a set of mathematical equations. For example, an astronomer might use a set of equations to describe in detail how gas falls into a black hole. We could refer to such a calculation as a model. Of course, we could use metal and plastic to build a celestial globe, but the thing we build wouldn't really be the model any more than the equations are a model. A scientific model is a mental conception, an idea in our heads, that helps us think about nature.

On the other hand, a model is not meant to be a statement of truth. The celestial sphere is not real; we know the stars are scattered through space at various distances, but we can imagine a celestial sphere and use it to help us think about the sky. A scientific model does not have to be true to be useful. Chemists, for example, think about the atoms in molecules by visualizing them as balls joined together by bonds. This model of a molecule is not fully correct, but it is a helpful way to think about molecules; it gives chemists a framework within which to organize their ideas.

Because scientific models are not meant to be totally correct, we must always remember the assumptions on which they are based. If we begin to think a model is true, it can mislead us instead of helping us. The celestial sphere, for instance, can help us think about the sky, but we must remember that it is only a mental crutch. The universe is much larger and much more interesting than this ancient scientific model of the heavens.

Figure 1-5
Motion in the sky. (a) Stars rising in the northeast make streaks in this time exposure photograph taken from a road near an observatory dome. *(Photo by Dean Ketelsen, NOAO)* (b) Looking north, we would see stars circling the north celestial pole. (c) In the eastern sky, stars rise along straight lines, and in the southern sky, the stars follow arcs as they circle the south celestial pole hidden below the southern horizon.

The **south celestial pole** is the point directly above Earth's South Pole. No easily visible star marks the south celestial pole. The south celestial pole lies below the southern horizon as seen from the United States, so we see stars in the southern sky following curved paths westward as they circle the south celestial pole (Figure 1-5d).

Earth's rotation defines the directions. The **north point** is the point on the horizon directly below the north celestial pole, and the **south point** is directly opposite the north point. The **east** and **west points** are halfway between the north and south points. The directions we use every day are defined astronomically.

The **celestial equator** is an imaginary line directly above Earth's equator. It touches the horizon at the east point and the west point and divides the celestial sphere into a two equal hemispheres (Figure 1-6).

Angles on the Sky

We might use the celestial sphere to locate objects. Just as we might tell a friend that we live 10 kilometers north of Flagstaff, Arizona, we might want to tell our friend to look for an interesting star a certain distance north of the bright star Sirius. We can't measure such distances in kilometers, so we need to measure angles on the celestial sphere.

Astronomers express angles in degrees, minutes, and seconds, where there are 60 **minutes of arc** in a degree and 60 **seconds of arc** in a minute of arc.

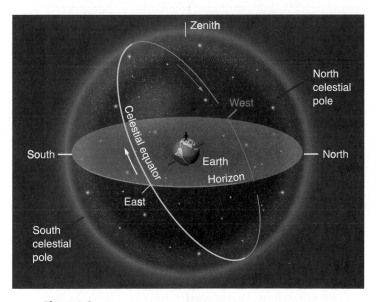

Figure 1-6
From our location on Earth, the horizon divides the celestial sphere into the half we can see and the half we cannot. The celestial equator circles the celestial sphere directly above Earth's equator and touches the horizon at the east point and the west point (yellow). Earth's rotation carries us eastward and makes stars rise in the east and set in the west. Earth's axis of rotation (red) extended out to the celestial sphere defines the celestial poles (red), the pivots around which the sky appears to rotate.

Astronomers might say, for instance, that the angular distance between two stars is 8°, meaning that if we point at the two stars with our arms, the angle between our arms is 8°. We may know that one star is much further away than the other, but we can imagine them painted on the celestial sphere. Then we can think of their separation as an "angle *on* the sky." We can give the angular separation of two stars even when their true distances are not known.

The angular diameter of an object is the angular distance across the object. For example, if we point at the left edge of the full moon with one arm and the right edge with the other arm, the angle between our arms is the angular diameter of the moon, about 0.5°. The sun also has an angular diameter of about 0.5°.

Being able to measure angles on the sky is a great help as we discuss the appearance of the sky as seen from different locations on Earth.

Latitude and the Sky

Our discussion so far has assumed we live at a latitude representative of North America, that is, about 40° north of Earth's equator. If we live at a different latitude, then our zenith and horizon will be quite different. Notice in Figure 1-6 how someone in Australia would have a different zenith and horizon.

What we see in the sky depends on our latitude. If we live in Australia, the south celestial pole would be above our horizon, and the north celestial pole would be hidden below the northern horizon. To be precise, the angular distance from the horizon to the north celestial pole equals the latitude of the observer.

Figure 1-7 shows the effect of latitude on what we see in the sky by taking us on an imaginary journey southward from Earth's North Pole. We begin in the ice and snow at Earth's North Pole where our latitude is 90° N, and the north celestial pole is directly overhead, 90° from the horizon. As we walk southward, our latitude decreases and the north celestial pole sinks closer to the northern horizon. As we cross Earth's equator, as shown in Figure 1-7d, our latitude is 0°, and the north celestial pole lies on our northern horizon. If we continue our journey south of Earth's equator, our latitude becomes negative, and the north celestial pole sinks below the northern horizon. That is, its angular distance from the horizon becomes negative.

This relationship between latitude and the distance from the horizon to the north celestial pole is a great aid to sailors. We can find our latitude by measuring the north celestial pole's distance above the northern horizon. (For precision, we would want to use navigation tables to correct for the fact that Polaris is almost a degree from the true pole.) Of course, if we were voyaging through southern oceans, we could never see Polaris, and then we would have to observe the

Figure 1-7
A journey beginning at Earth's north pole (a) shows how our latitude L determines what we see in the sky. As we travel south, our latitude decreases, and the north celestial pole moves down toward the northern horizon. The angular distance from the northern horizon up to the north celestial pole always equals our latitude. As we cross Earth's equator, our latitude becomes negative, the north celestial pole sinks below the horizon, and the south celestial pole rises above our southern horizon.

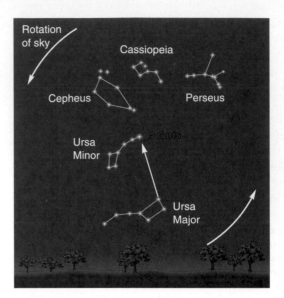

Figure 1-8
The brighter north circumpolar constellations visible from the latitude of the United States never set. The pointer stars at the front of the cup of the Big Dipper always point toward Polaris.

south celestial pole, which is more difficult because it lacks a bright star as a marker.

Our latitude determines what constellations we can see. From northern latitudes, constellations near the north celestial pole never set. These **north circumpolar constellations,** such as Ursa Minor and Cassiopeia, are always above the horizon (Figure 1-8). Constellations near the south celestial pole never rise for an observer in the Earth's northern hemisphere. These **south circumpolar constellations** are never visible from the United States and similar latitudes. How many constellations are circumpolar depends on your latitude; if you live at Earth's equator, you have no circumpolar constellations at all.

Many of the technical terms that describe the celestial sphere are not common in our daily lives, but we do not learn them here for their own sake. Merely naming things is not the same as understanding (Window on Science 1-2). Our goal in science is to understand nature, so we can now use this powerful vocabulary of the sky to analyze the motions of Earth and heavenly bodies.

Precession

Over 2000 years ago, Hipparchus compared his star positions with those made nearly two centuries before and realized that the celestial poles and equator were slowly moving across the sky. Later astronomers understood that this motion is caused by Earth's toplike motion.

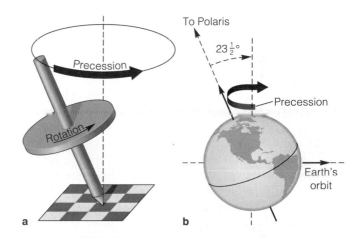

Window on Science 1-2

Understanding Versus Naming: The True Goal of Science

One of the fascinations of science is that it can reveal things we normally do not sense, such as the precessional motion of Earth's axis. Science can even take our imagination into realms beyond our experience, from the inside of an atom to the inside of a star. Because these experiences are so unfamiliar, we need a vocabulary of technical terms just to talk about them, and that leads to a common confusion between naming things and understanding things.

The first step in understanding something is naming the parts, but it is only the beginning. For example, we need to know terms such as *celestial equator* and *celestial pole* in order to discuss the apparent motion of the sky. That can lead us to understand precession, something we do not sense in our daily lives.

We must know the words, but real understanding comes from being able to use the vocabulary to discuss the way nature works. We must be able to tell the stories that scientists call theories and hypotheses and use the technical words as needed to cite the evidence that makes us think those stories are true. That goes far beyond merely memorizing the terminology; that is real understanding.

In folklore, naming a thing gives us power over it—recall the story of Rumpelstiltskin. In science, naming things is only the beginning. Our goal lies in understanding, not just naming.

If you have ever played with a gyroscope or top, you have seen how the spinning mass resists any change in the direction of its axis of rotation. The more massive the top and the more rapidly it spins, the more difficult it is to change the direction of its axis of rotation. But you probably recall that the axis of even the most rapidly spinning top sweeps around in a conical motion. The weight of the top tends to make it tip, and this combines with its rapid rotation to make its axis sweep around in a conical motion called **precession** (Figure 1-9).

Earth spins like a giant top, but it does not spin upright in its orbit; it is tipped 23.5° from vertical. Earth's large mass and rapid rotation keep its axis of rotation pointed toward a spot near the star Polaris. The axis would not wander if Earth were a perfect sphere. However, Earth, because of its rotation, has a slight bulge around its middle, and the gravity of the sun and moon pull on this bulge tending to twist Earth upright in its orbit. The combination of these forces and Earth's rotation causes Earth's axis to precess in a conical motion, taking about 26,000 years for one cycle (Figure 1-9b).

Because the celestial poles and equator are defined by Earth's rotational axis, precession moves these reference marks. We notice no change at all from night to night or year to year, but precise measurements reveal the precessional motion of the celestial poles and equator.

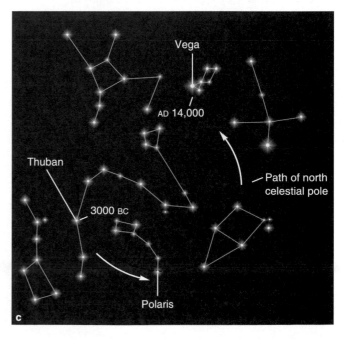

Figure 1-9

Precession. (a) A spinning top precesses in a conical motion around the perpendicular to the floor because its weight tends to make it fall over. (b) Earth precesses around the perpendicular to its orbit because the gravity of the sun and moon tend to twist it upright. (c) Precession causes the north celestial pole to drift among the stars, completing a circle in 26,000 years.

Over centuries, precession has dramatic effects. Egyptian records show that 4800 years ago the north celestial pole was near the star Thuban (α Draconis). The pole is now approaching Polaris and will be closest to it about 2100. In about 12,000 years, the pole will have moved to within 5° of Vega (α Lyrae). Figure 1-9c shows the path followed by the north celestial pole.

R E V I E W **Critical Inquiry**

Does everyone see the same circumpolar constellations?

Here we can use the celestial sphere as a convenient model of the sky (Figure 1-6). A circumpolar constellation is one that does not set or rise. Which constellations are circumpolar depends on our latitude. If we live on Earth's equator, we see all the constellations rising and setting, and there are no circumpolar constellations at all. If we live at Earth's North Pole, all the constellations north of the celestial equator never set, and all the constellations south of the celestial equator never rise. In that case, every constellation is circumpolar. At intermediate latitudes, the circumpolar regions are caps whose angular radius equals the latitude of the observer. If we live in Iceland, the caps are very large, and if we live in Singapore, near the equator, the caps are much smaller.

Check the star charts at the end of this book to locate Ursa Major and Orion. People in Canada count Ursa Major as circumpolar, but people in Mexico see most of this constellation slip below the horizon. From much of the United States, some of the stars of Ursa Major set and some do not. In contrast, Orion rises and sets as seen from nearly everywhere on Earth. Explorers at Earth's poles, however, never see Orion rise or set. How would you improve the definition of a circumpolar constellation to clarify the status of Ursa Major? Would your definition help in the case of Orion?

We now have a model of the sky—the celestial sphere—complete with reference marks and angular measurements to guide us. We can locate constellations, name the stars, and express their brightness numerically. We are now ready to discuss the great cycles of the sun and moon.

1-3 The Annual Cycle of the Sun

The people who built Stonehenge aligned it with the most northerly sunrise of the year. To understand what that means, we must understand how the sun travels around the sky. We know the sun's annual motion is caused by the orbital motion of the Earth revolving around the sun, but the ancient architects of Stonehenge saw only the annual cycle of the sun in the sky.

The Ecliptic

Because Earth orbits around the sun, the sun appears to move against the background of stars. Of course, we don't see the stars during the daytime because the sun lights up Earth's atmosphere, but we can notice where the sun is in the sky by noticing which constellations are in the west after sunset.

In early January, for instance, we find the sun in front of the stars in Sagittarius (Figure 1-10). Each day, Earth moves farther along its orbit, and we find the sun farther eastward. By February it would have moved into the constellation Capricornus, and by March it would be in Aquarius.

Figure 1-10
Earth's orbit around the sun is nearly a perfect circle, shown here tipped at an angle. Because we see the sun from a changing location along Earth's orbit, the sun appears to move eastward against the background of the stars. Earth's circular orbit is thus projected on the sky as the circular path of the sun, the ecliptic.

a

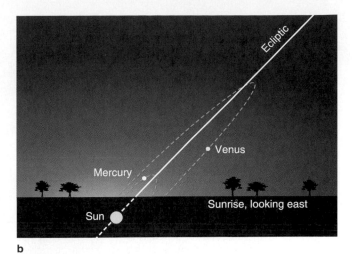

b

Figure 1-11
The orbits of Mercury and Venus lie inside the orbit of Earth, so they always appear near the sun in the sky. They are sometimes visible in the western sky just after sunset (a) or in the eastern sky just before sunrise (b). Because the orbit of Venus is larger, it can move higher above the horizon and is easier to locate than Mercury.

Of course, the sun isn't really "in Aquarius." The sun is only 93 million miles away, and the stars of Aquarius are at least a million times farther away. But in March each year we find the sun in the direction of Aquarius, and we say, "The sun is in Aquarius."

If we watched the sun for a full year, it would appear to circle the sky. A piece of chalk attached to the center of the sun's disk would draw a line called the **ecliptic,** the apparent path of the sun. Another way to define the ecliptic is to say it is the projection of Earth's orbit on the sky. If the celestial sphere were a great screen illuminated by the sun at the center, then the shadow cast by Earth's orbit would be the ecliptic. We can also refer to the ecliptic plane as the plane of Earth's orbit. It is worth defining the ecliptic carefully because it is an important line circling the celestial sphere.

Not only does the sun follow the ecliptic, but the moon and planets also travel eastward along the **zodiac,** a band 18° wide centered on the ecliptic. Mars, Jupiter, and Saturn are bright enough to see with the naked eye, and Uranus is sometimes just bright enough to detect. Neptune and Pluto are always too faint. Mercury and Venus are also bright enough to see, but they never wander far from the sun, and thus we see them in the west in the evening sky or in the east in the dawn sky (Figure 1-11).

You have heard of the signs of the zodiac, which take their names from the 12 main constellations along the ecliptic. The locations of the sun, moon, and planets are the basis of an ancient superstition, astrology. Once an important part of astronomy, astrology is now the exact opposite—astronomy is a science because it depends on evidence, and astrology is a superstition that survives in spite of evidence (Window on Science 1-3).

Because the sun circles the ecliptic in 365.25 days, and a circle includes 360°, the sun travels about 1° eastward each day—about twice its own angular diameter. This stately motion is complicated by the fact that Earth does not rotate perpendicular to the plane of its orbit. Its axis of rotation is tipped 23.5° from perpendicular.

Earth, like a spinning top, holds its axis fixed as it moves around the sun (Figure 1-12). Thus Earth's North Pole always points toward the North Star, Polaris. (Precession changes this so slowly we will not notice any motion in our lifetime.)

Recall that the celestial equator is the projection of Earth's equator on the sky, and the ecliptic is the projection of Earth's orbit on the sky. Because Earth's equator is tipped 23.5° to its orbit, the celestial equator is tipped 23.5° to the ecliptic (Figure 1-13). The ecliptic crosses the celestial equator at two places called equinoxes. The **vernal equinox** is where the sun crosses going northward, and the **autumnal equinox** is where

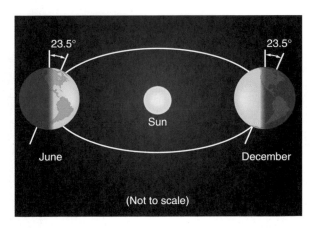

Figure 1-12
Earth's axis of rotation is inclined 23.5° from the perpendicular to its orbit. We have seasons on Earth because the axis remains fixed as Earth orbits the sun. This figure shows Earth at two locations in its orbit: summer solstice (left) and winter solstice (right).

Astronomers have a low opinion of astrology not so much because it is groundless but because it pretends to be a science. It is a pseudoscience, from the prefix *pseudo*, meaning "false." There are many examples of pseudoscience, and it is illuminating to consider the difference between pseudoscience and science.

A pseudoscience is a set of beliefs that appear to be based on scientific ideas but that fail to obey the most basic rules of science. For example, some years ago a fad arose in which people placed objects under pyramids made of paper, plastic, wire, and so on. The claim was that the pyramidal shape would focus cosmic forces on anything inside and so preserve fruit, sharpen razor blades, and do other miraculous things. Many books promoted this idea, but simple experiments showed that any shape would protect a piece of fruit from airborne spores and allow it to dry without rotting. Likewise, any shape would allow oxidation to improve the cutting edge of a razor blade. In short, experimental evidence contradicted the claim. Nevertheless, supporters of the theory declined to abandon or revise their claims. Thus, the fad of pyramid power was a pseudoscience.

One characteristic of a pseudoscience is that it appeals to our needs and desires. Thus, some pseudoscientific claims are self-fulfilling. For example, some people bought pyramidal tents to put over their beds and thus improve their rest. While there is no logical mechanism by which such a tent could affect a sleeper, people slept more soundly because they wanted and expected the claim to be true. Many pseudoscientific claims involve medical cures, ranging from copper bracelets and crystals to focus spiritual power to astonishingly expensive and illegal treatments for cancer. Logic is a stranger to pseudoscience, but human fears and needs are not.

Astrology is a pseudoscience. Over the centuries, astrology has been tested repeatedly, and no correlation between the movement of the heavens and human lives has been found. But it survives, and its supporters disregard any evidence that it doesn't work. Like all pseudosciences, astrology is not open to revision in the face of contradictory evidence. Furthermore, astrology fulfills our human need to believe that there is order and meaning to our lives. It may comfort us to believe that our sweetheart has rejected us because of the motions of the sun, moon, and planets along the zodiac rather than to admit that we behaved badly on our last date. Comfort aside, astrology is a poor basis for life decisions.

Human nature and human needs probably ensure that pseudoscientific beliefs will continue to plague us like emotional viruses propagating from person to person. If we recognize them for what they are, we can more easily guide our lives by rational principles and not assign the stars blame for our failures and credit for our successes.

---

the sun crosses moving southward. The sun crosses the vernal equinox about March 21, and it crosses the autumnal equinox about September 22. The exact dates vary by a day or two because of leap year and other factors.

The most northerly and most southerly points on the ecliptic are called the solstices. The sun crosses the **summer solstice** in the northern sky about June 22, and it crosses the **winter solstice** in the southern sky about December 22.

Note that the equinoxes and solstices are points on the sky, but the same words refer to the times when the sun crosses those points. You might hear someone say, "This year the vernal equinox occurs at 3:02 AM on March 21."

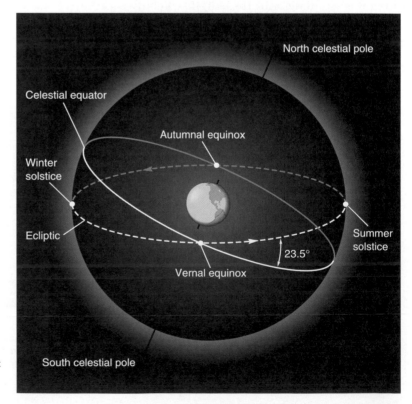

Figure 1-13
The ecliptic crosses the celestial equator at the equinoxes; it reaches its most northerly point at the summer solstice and its most southerly point at the winter solstice.

Now we can begin to understand why the builders of Stonehenge oriented their monument toward the most northerly sunrise of the year. That sunrise marks the summer solstice, and as you can guess from its name, it is related to the cycle of the seasons.

The Seasons

The seasonal temperature at any given latitude on Earth depends on the amount of heat received from the sun and the amount radiated to space. If, on the average, a region gains more heat than it loses, it gets warmer. If it gains less than it loses, it gets cooler. Thus temperature change through the seasons depends on the latitude of the observer and the amount of heat that latitude receives from the sun.

The motion of the sun around the ecliptic tips the heat balance one way in summer and the opposite way in winter. Because the ecliptic is inclined with respect to the celestial equator, the sun spends half the year in the northern celestial hemisphere and half the year in the southern celestial hemisphere (Figure 1-13). When the sun is in the northern celestial hemisphere, the northern half of Earth receives more direct sunlight—and therefore more heat—than the southern half (Figure 1-12). This makes North America, Europe, and Asia warmer.

The seasons are reversed in Earth's southern half (see Figure 1-12). While the sun is in the northern celestial hemisphere warming North America, South America becomes cooler. Southern Chile has warm weather on New Year's Day and cold in July.

To see how the sun can give us more heat in summer, think about the path the sun takes across the sky between sunrise and sunset. Figure 1-14 shows these paths when the sun is at the summer solstice and at the winter solstice as seen by a person living at latitude 40°, a good average latitude for most of the United States. Notice that at the summer solstice, the sun rises in the northeast, moves high across the sky, and sets in the northwest (Figure 1-14a). But at the winter solstice, the sun rises in the southeast, moves low across the sky, and sets in the southwest (Figure 1-14b). Two features of these paths tip the heat balance.

First, the summer sun is above the horizon for more hours of each day than the winter sun. Summer days are long and winter days are short. Because the sun is above our horizon longer in summer, we receive more energy each day.

Second, the sun stands high in the sky at noon on a summer day. It shines almost straight down, as shown by our small shadows. On a winter day, however, the noon sun is low in the southern sky. Each square meter of the ground gains little heat from the winter sun because the sunlight strikes the ground at an angle and spreads out, as shown by our longer shadows (Figure 1-14). These two effects work together to tip the heat balance and produce the seasons.

a

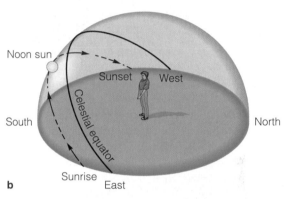

b

Figure 1-14
The seasons for middle northern latitudes typical of the United States. (a) The summer sun rises in the northeast and sets in the northwest. It is above the horizon for more than 12 hours and stands high in the sky at noon. (b) The winter sun rises in the southeast and sets in the southwest. It is above the horizon for less than 12 hours and never stands high in the sky, even at noon. Notice the figure's shadows.

We mark the beginning of the seasons by the position of the sun. Spring begins at the moment the sun crosses the celestial equator going north (the vernal equinox). Summer begins at the moment the sun reaches its most northerly point (the summer solstice), and autumn begins when the sun crosses the celestial equator going south (the autumnal equinox). We mark the official beginning of winter when the sun reaches its most southerly position (the winter solstice).

Of course, the weather does not turn warm the instant spring begins. The ground, air, and oceans are still cool from winter, and they take a while to warm up. Likewise, in the autumn the ground, air, and oceans slowly release the heat stored through the summer. Due to this thermal lag, the average daily temperatures lag behind the solstices by about one month. Although the sun crosses the summer solstice on about June 22, the hottest months at northern latitudes are July and August. The coldest months are January and February, even though the sun passes the winter solstice earlier, about December 22.

The seasons are unrelated to the variation in the Earth–sun distance. Earth's orbit is only slightly elliptical, and its average distance varies by only about 1.7 percent. Earth reaches **perihelion,** its closest point to the sun, about January 4, and it reaches **aphelion,** its farthest point, about July 4. This makes Northern Hemisphere summers a bit cooler and Northern Hemisphere winters a bit warmer than they would be if Earth had a circular orbit, but the effect is very small.

The Milankovitch Hypothesis

Weather is what happens today; climate is the average of what happens over decades and centuries. We know that Earth has gone through ice ages during which glaciers advance and retreat. These ice ages appear to occur roughly every 250 million years, and cycles of glaciation within ice ages occur with a period of about 40,000 years. These climate cycles may be related to astronomical cycles.

In 1920, Yugoslavian meteorologist Milutin Milankovitch proposed what became known as the **Milankovitch hypothesis**—that small changes in Earth's orbit, precession, and inclination affect the heat balance and modify climate. Here we will examine each of these three motions in turn.

First, astronomers know that the shape of Earth's orbit changes slightly, becoming more and then less elliptical with a period of about 93,000 years. At present, the elliptical shape of Earth's orbit carries it about 1.7 percent closer to the sun during Northern Hemisphere winters and about 1.7 percent farther from the sun during Northern Hemisphere summers. This makes the northern climate warmer, and that is critical—most of the land mass where ice could accumulate is in the Northern Hemisphere. If Earth's orbit became more elliptical, northern winters and summers might be warm enough to prevent the accumulation of snow and ice to form glaciers, and Earth's climate would warm.

But a second factor is also at work. Precession causes Earth's axis to wobble with a period of about 26,000 years, and that changes the location of the seasons around Earth's orbit. Perihelion now occurs during northern winters, but in 13,000 years aphelion will occur during northern winters. That is, Earth will be slightly farther from the sun, and northern winters will be colder, perhaps cold enough to allow the accumulation of snow to create glaciers.

The third factor is the inclination of Earth's equator to its orbit. Presently at 23.5°, this inclination varies from 22° to 24° with a period of roughly 41,000 years. If the inclination is large, seasons are more severe, but if the inclination is lower, the seasons are less pronounced.

In 1920, Milankovitch proposed the hypothesis that these three variations cycled against one another to produce variations in Earth's climate that resulted in the ice ages (Figure 1-15). The theory was given little attention until 1976 when evidence was found to test the hypothesis (Window on Science 1-4). Teams studying cores from the seafloor constructed a history of ocean temperature that matched the predictions of the theory.

Since then more evidence has been found that agrees with the theory, but contradictory evidence has also been found. For example, in Nevada a water-filled fault called Devil's Hole contains calcite deposited over the last 500,000 years, and experts can determine the past temperature history from the calcite layers. The timing of that variation seems to contradict the Milankovitch hypothesis, but some scientists believe the Devil's Hole evidence tells us about local climate changes and not about world climate.

Conflicting evidence is common in science. Much of scientific work is focused on resolving such conflicts and trying to reach sound conclusions about nature. Although the Milankovitch hypothesis is now widely accepted, scientists continue to test it by searching for more evidence of Earth's past climate.

REVIEW Critical Inquiry

How does the elliptical shape of Earth's orbit affect the seasons?

In the critical analysis of an idea it can be helpful to exaggerate the importance of a single factor. Doing so not only reveals the effect of that factor, but can also reveal the inner workings of the process itself. In this case, we might try to imagine what life on Earth would be like if Earth's orbit were more elliptical than it really is. Earth now passes perihelion about January 4, and on that date we are slightly closer to the sun than average. If Earth's orbit were more elliptical, we would be significantly closer. This would make winters in the Northern Hemisphere much warmer than they are now; and the weather in the Southern Hemisphere, normally warm in January, would be dreadfully hot. Six months later, about July 4, Earth reaches aphelion, its farthest point from the sun. If it had a more elliptical orbit, Earth would be significantly farther from the sun on this date. Then the Northern Hemisphere would experience a much cooler summer than it now does; and in the Southern Hemisphere, the cool season would be bitter cold.

Of course, Earth's orbit is nearly circular; its distance from the sun varies by only about 3 percent from nearest to farthest. That makes the northern seasons a little milder and the southern seasons a little more extreme, an effect that is hardly noticeable. Now use exaggeration to analyze the true cause of the seasons. Try to imagine what the seasons would be like if Earth's axis were inclined more than 23.5° or less than 23.5° from the perpendicular to its orbit.

Science is based on evidence. Every theory and conclusion must be supported by evidence obtained from experiments or from observation. If a theory is supported by many pieces of evidence but is clearly contradicted by a single experiment or observation, scientists quickly abandon it. For a theory to be true, there can be no contradictory evidence.

Of course, scientists argue about the significance of particular evidence and often disagree on the interpretation of evidence. Some observations may seem significant at first glance, but upon closer examination we may find that the procedure was flawed and so the piece of evidence is not important. Or we might conclude that the observational fact is correct but is being misinterpreted. The observation may not mean what it seems. Some of the most famous disagreements in science, such as those surrounding Galileo and Darwin, have arisen over the interpretation of well-established factual evidence.

Furthermore, scientists are not allowed to be selective in considering evidence. A lawyer in court can call a certain witness and intentionally fail to ask a critical question that would reveal evidence harmful to the lawyer's case. But a scientist may not ignore any known evidence. The difference in their methods is revealing. The lawyer is attempting to prove only one side of the case and rightly may ignore contradictory evidence. The scientist, however, is searching for the truth and so must test any theory against all available evidence. In a sense, the scientist, in dealing with evidence, must act as both the prosecution and the defense.

As you read about any science, look for the evidence in the form of measurements or observations. Every theory or conclusion should have supporting evidence. If you can find and understand the evidence, the science will make sense. All scientists, from astronomers to zoologists, demand evidence. You should, too.

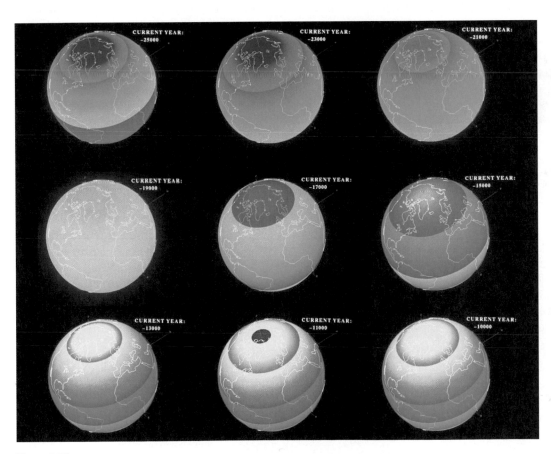

Figure 1-15
Color-coded globes illustrate the importance of the Milankovitch effect. These globes show Earth at the summer solstice from 25,000 years ago (upper left), during the last glaciation, to only 10,000 years ago (lower right), after the end of the glaciation. Red, yellow, and green regions receive more solar energy than average; blue and violet regions receive less. Changes in Earth's rotational axis and orbital shape affect the total amount of solar energy received and thus alter the climate. *(Courtesy Arizona State University, Computer Science and Geography Departments.)*

Earth's rotation and revolution produce the daily cycle and the annual cycle, and the builders of Stonehenge recognized those cycles when they aligned Stonehenge with the summer solstice sunrise. That sunrise marks the most northerly point in the sun's seasonal cycle, a fitting observation for a Stone Age culture that lived so close to the cycles of nature. But Earth's moon reveals more cyclic beauty in the sky and in Stonehenge.

I-4 The Cycles of the Moon

Stonehenge has been marking the summer solstice sunrise for 4000 years, but some experts believe certain stones there mark specific moonrise positions. Could the ancient builders of Stonehenge have marked special moonrises? Why would they do that? Certainly the moon passes through some of the most dramatic cycles in the sky.

The Phases of the Moon

The moon orbits eastward around Earth in 27.322 days. This is called the moon's **sidereal period,** because it refers to the moon's position relative to the stars. The moon takes 27.322 days to circle the sky once and return to the same place among the stars.

The moon rotates on its axis as it circles Earth, keeping the same side toward Earth. Thus we always see the same features on the moon's face. A mountain on the near side of the moon remains facing Earth throughout the moon's circuit around Earth.

The moon moves rapidly across the sky, traveling 13°, about 26 times its apparent diameter, in 24 hours. If you compare the position of the moon to the stars in the background, you will see it move slightly more than its own diameter in an hour. Because the moon's orbit is inclined a bit over 5° to Earth's orbit, the moon never travels farther than about 5° north or south of the ecliptic.

Because Earth's moon, like the planets, does not produce visible light of its own, it is visible only by the sunlight it reflects, and we can see only that portion illuminated by the sun. As the moon moves around the sky, the sun lights up different portions of the side of the moon facing Earth, and so the moon passes through a sequence of phases.

The top panel of Figure 1-16 shows the phases of the moon. We can understand these by looking at the location of the moon relative to the sun. For example, the *new moon* in the second panel of Figure 1-16 is located roughly between Earth and the sun, and the side of the moon facing Earth is dark. We see no moon at new moon. A few days after new moon, it has moved far enough along its orbit to allow the sun to illuminate a small sliver of the side toward us, and we see a thin crescent. Night by night this crescent moon waxes (grows), until, about a week later, we see half of the side toward us illuminated by sunlight and refer to

As seen from Earth

New | Waxing crescent | First quarter | Waxing gibbous | Full | Waning gibbous | Third quarter | Waning crescent

a

Figure 1-16
The phases of the moon are caused by the varying amounts of sunlight that reach the side of the moon facing Earth. At first quarter, for example, only half of the near side is illuminated. At full moon, all of the near side is illuminated. The human figures shown on Earth illustrate our location at different times of day. To set the diagram for a given time, rotate it until the corresponding human figure is at the top of the Earth globe. The moon's orbit will then show the location of the moon above or below our horizon at different phases.

b

it as *first quarter*. The moon continues to wax, becoming gibbous (from the Latin for humpbacked), and then, when it is nearly opposite the sun, the side toward Earth is fully illuminated, and we see a *full moon*.

The second half of the lunar cycle reverses the first half. After reaching full, the moon wanes (shrinks) through gibbous phase to *third quarter,* then through crescent back to new moon. To distinguish between the gibbous and crescent phases of the first and second half of the cycle, we refer to *crescent waxing* and *gibbous waxing* when the moon is growing, and to *gibbous waning* and *crescent waning* when it is shrinking.

The cycle of lunar phases takes 29.53 days, the synodic period of the moon, or about four weeks. Thus new moon, first quarter, full moon, and third quarter occur at roughly one-week intervals. In general, an object's **synodic period** is its orbital period relative to the sun. (*Synodic* comes from the Greek words meaning "together" and "path.") To see why the moon's synodic period is longer than its sidereal period, imagine we begin observing at new moon—that is, when the moon is near the sun in the sky. After 27.322 days, the moon's sidereal period, it has circled the sky and returned to the same place among the stars where it was last new, but the sun has moved about 27° eastward along the ecliptic. The moon needs slightly more than two days to catch up with the sun and reach new moon again. Thus the moon's synodic period is longer than its sidereal period.

You can make a moon-phase dial from Figure 1-16 by covering the lower half of the moon's orbit with a sheet of paper, aligning the edge of the paper to pass through the word "Full" at the left and the word "New" at the right. Push a pin through the edge of the paper at Earth's North Pole to make a pivot, and under the word "Full" write on the paper "Eastern Horizon." Under the word "New" write "Western Horizon." The paper now represents the horizon we see facing south.

You can set your moon-phase dial for a given time by rotating the diagram behind the horizon-paper. Set the dial to sunset by turning the diagram until the human figure labeled "sunset" is standing at the top of the Earth globe; the dial shows, for example, that the full moon at sunset would be at the eastern horizon.

We can now follow the cycle of the moon around the sky, and we begin at sunset. If the moon were new, it would be invisible near the sun on the western horizon. A few days after new moon, we would see the waxing crescent moon moving upward in the sunset sky (Figure 1-17a). A week after new moon, we would find the first quarter moon standing high in the sky at sunset. As the days passed, we would see the moon wax gibbous as it moved eastward reaching full moon at the eastern horizon about two weeks after new moon.

Lunar phases after full moon are not visible in the sunset sky, so we must shift to the sunrise sky to con-

Figure 1-17
(a) We see the waxing crescent moon above the western horizon soon after sunset, and it grows fatter each evening until it reaches first quarter. (b) We find the waning crescent moon above the eastern horizon before dawn, and it grows thinner each morning until it reaches new moon.

tinue. At sunrise, the full moon would be at the western horizon, and it would wane through gibbous each day until it reached third quarter high in the sunrise sky about three weeks after new moon. Each morning we would see the moon farther eastward as it waned through crescent phase (Figure 1-17b) until it reached new moon again at the eastern horizon in the sunrise sky. New moon to new moon takes just over four weeks.

The monthly cycle of the lunar phases is so dramatic it is not surprising to find evidence that the people who built Stonehenge were keeping track of the motion of the moon. Another dramatic lunar cycle is played out at the seashore.

Tides

Isaac Newton understood that Earth's gravity attracts the moon, but he also understood that the moon has gravity of its own that pulls on the Earth, and that produces the advance and retreat of the ocean tides.

Tidal forces arise because of the difference in the force of gravity acting on different parts of a body. The side of Earth facing the moon is about 4000 miles closer to the moon than the center of Earth is, and the moon's gravity pulls on the near side of Earth more strongly than on Earth's center. Though we think of our planet as solid, it is not perfectly rigid, so the moon's gravity draws the rocky surface of the near side up into a bulge a few centimeters high.

The oceans respond to the force of the moon's gravity by flowing into a bulge of water on the side of Earth facing the moon. There is also a bulge on the side away from the moon, which develops because the moon pulls more strongly on Earth's center than on the far side. Thus the moon pulls Earth away from the oceans, which flow into a bulge on the far side.

We can see dramatic evidence of this effect if we watch the ocean shore for a few hours. Though Earth rotates on its axis, the tidal bulges remain fixed with respect to the moon. As the turning Earth carries us into a tidal bulge, the ocean water deepens and the tide crawls up the beach. Later, when Earth carries us out of the bulge, the water becomes shallower and the tide falls. Because there are two bulges on opposite sides of Earth, the tides should rise and fall twice a day on an ideal coast.

In reality, the tide cycle at any given location can be quite complex because of the latitude of the site, shape of the shore, winds, and so on. Tides in the Bay of Fundy (New Brunswick, Canada), for example, occur twice a day and can exceed 40 feet, while the northern coast of the Gulf of Mexico has only one tidal cycle, of roughly 1 foot, each day.

The sun, too, produces tidal bulges on Earth. At new moon and at full moon, the moon and sun produce tidal bulges that add together (Figure 1-18a) and produce extreme tidal changes; high tide is very high, and low tide is very low. Such tides are called **spring tides,** even though they occur at every new and full moon, and not just in the spring. **Neap tides** occur at first- and third-quarter moon, when the moon and sun pull at right angles to each other (Figure 1-18b). Then the tides caused by the sun reduce the tides caused by the moon, and the rise and fall of the ocean is less extreme than usual.

Tidal forces can have surprising effects. The friction of the ocean waters with the seabeds slows Earth's rotation and makes the length of a day grow by 0.0023 second per century. Fossils of marine animals and tidal sediments confirm that only 900 million years ago Earth's day was 18 hours long. In addition, Earth's gravitational field exerts tidal forces on the moon, and, although there are no bodies of water on the moon, friction within the flexing rock has slowed the moon's rotation to the point that it now keeps the same face toward Earth.

Tidal forces can also affect orbital motion. Friction with the ocean beds drags the tidal bulges eastward out of a direct Earth–moon line. These tidal bulges contain a large amount of mass, and their gravitational field pulls the moon forward in its orbit (Figure 1-18c). As a result, the moon's orbit is growing larger by about 4 cm

a

b

c

Figure 1-18
Looking down on Earth from above the North Pole, we can draw the shape of the tidal bulges (exaggerated here for clarity). (a) When the moon and sun pull along the same line, their tidal forces add and the tidal bulges are larger. (b) When the moon and sun pull at right angles, their tidal forces do not add and the tidal bulges are smaller. (c) Earth's rotation drags the tidal bulges ahead of the Earth–moon line (much exaggerated here). The gravitational attraction of these masses of water does not pull directly toward Earth's center, but rather pulls the moon slightly forward in it orbit (green arrow). This forces the orbit to grow larger.

per year, an effect that astronomers can measure by bouncing laser beams off reflectors left on the lunar surface by the Apollo astronauts.

These and other tidal effects are important in many areas of astronomy. In later chapters, we will see how tidal forces can pull gas away from stars, rip galaxies apart, and melt the interiors of satellites orbiting near massive planets. For now, however, we must consider other aspects of the lunar cycle. The stately progression of the lunar phases and the ebb and flow of the ocean tides are commonplace, but occasionally something peculiar happens: Earth's moon darkens and turns copper-red in a lunar eclipse.

Lunar Eclipses

A **lunar eclipse** occurs at full moon when the moon moves through the shadow of Earth. Because the moon shines only by reflected sunlight, we see the moon gradually darken as it enters the shadow.

Earth's shadow consists of two parts. The **umbra** is the region of total shadow. If we were in the umbra of Earth's shadow, we would see no portion of the sun. If we moved into the **penumbra,** however, we would be in partial shadow and would see part of the sun peeking around the edge of Earth. Thus in the penumbra the sunlight is dimmed but not extinguished.

If the orbit of the moon carries it through the umbra, we see a total lunar eclipse (Figure 1-19). As we watch the moon in the sky, it first moves into the penumbra and dims slightly; the deeper it moves into the penumbra, the more it dims. In about an hour, the moon reaches the umbra, and we see the umbral shadow darken part of the moon. It takes about an hour for the moon to enter the umbra completely and become totally eclipsed. Totality, the period of total eclipse, may last as long as 1 hour 45 min-

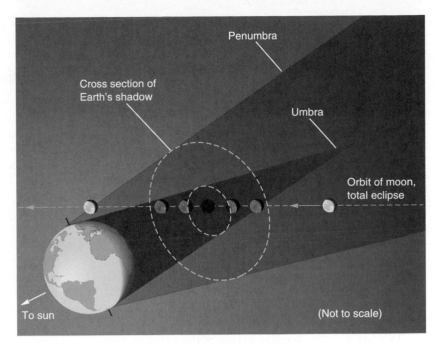

Figure 1-19
During a total lunar eclipse, the moon's orbit carries it through the penumbra and completely into the umbra. Compare the cross section of Earth's shadow with Figure 1-20.

utes, though the timing of the eclipse depends on where the moon crosses the shadow.

When the moon is totally eclipsed, it does not disappear completely (Figure 1-20). Although it receives

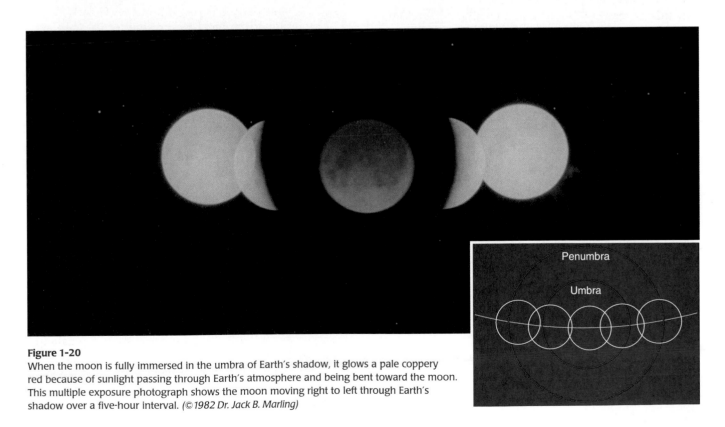

Figure 1-20
When the moon is fully immersed in the umbra of Earth's shadow, it glows a pale coppery red because of sunlight passing through Earth's atmosphere and being bent toward the moon. This multiple exposure photograph shows the moon moving right to left through Earth's shadow over a five-hour interval. (©1982 Dr. Jack B. Marling)

no direct sunlight, it does receive some sunlight refracted (bent) through Earth's atmosphere. If we were on the moon during totality, we would not see any part of the sun because it would be entirely hidden behind Earth. However, we would be able to see Earth's atmosphere illuminated from behind by the sun. The red glow from this "sunset" illuminates the moon during totality and makes it glow coppery red.

If the moon passes a bit too far north or south, it may only partially enter the umbra, and we see a partial lunar eclipse. The part of the moon that remains outside the umbra in the penumbra receives some direct sunlight, and the glare is usually great enough to prevent our seeing the faint coppery glow of the part of the moon in the umbra.

A penumbral lunar eclipse occurs when the moon passes through the penumbra but misses the umbra entirely. Because the penumbra is a region of partial shadow, the moon is only partially dimmed. A penumbral eclipse is not very impressive.

While there are usually no more than one or two lunar eclipses each year, it is not difficult to see one. We need only be on the dark side of Earth when the moon passes through Earth's shadow. That is, the eclipse must occur between sunset and sunrise at our location. Table 1-2 will allow you to determine which upcoming total and partial eclipses will be visible from your location.

Solar Eclipses

We who live on planet Earth can see a phenomenon that is not visible from most planets. It happens that the sun is 400 times larger than the moon and, on the average, 390 times farther away, so the sun and moon have nearly equal angular diameters—0.5°. (See By the Numbers 1-2.) Thus, the moon is just the right size to cover the bright disk of the sun and cause a **solar eclipse.** If the moon covers the entire disk of the sun, we see a total eclipse. If it covers only part of the sun, we see a partial eclipse.

Whether we see a total or partial eclipse depends on whether we are in the umbra or the penumbra of the moon's shadow (Figure 1-22). The umbra of the moon's

	Time* of Mideclipse (GMT)	Length of Totality (Min)	Length of Eclipse (Hr:Min)
Date			
2000 Jan. 21	4:45	76	3:22
2000 July 16	13:57	106	3:56
2001 Jan. 9	20:22	60	3:16
2001 July 5	14:57	Partial	2:38
2003 May 16	3:41	52	3:14
2003 Nov. 9	1:20	22	3:30
2004 May 4	20:32	76	3:22
2004 Oct. 28	3:05	80	3:38
2005 Oct. 17	12:04	Partial	0:56
2006 Sept. 7	18:52	Partial	1:30
2007 Mar. 3	23:22	74	3:40
2007 Aug. 28	10:38	90	3:32
2008 Feb. 21	3:27	50	3:24
2008 Aug. 16	21:11	Partial	3:08
2009 Dec. 31	19:24	Partial	1:00
2010 June 26	11:40	Partial	2:42
2010 Dec. 21	8:18	72	3:28

Table 1-2 Total and Partial Eclipses of the Moon 2000–2010

* Times are Greenwich Mean Time. Subtract 5 hours for Eastern Standard Time, 6 hours for Central Standard Time, 7 hours for Mountain Standard Time, and 8 hours for Pacific Standard Time. From your time zone, lunar eclipses that occur between sunset and sunrise will be visible, and those at midnight will be best placed.

By the Numbers 1-2

The Small-Angle Formula

The angular diameter of an object is related to its linear diameter and its distance by the small-angle formula (Figure 1-21). *Linear diameter* is the distance between an object's opposite sides. The linear diameter of the moon, for instance, is 3476 km. The *angular diameter* of an object is the angle formed by two lines extending from opposite sides of the object and meeting at our eye. Clearly, the farther away an object is, the smaller its angular diameter.

In the small-angle formula, we always express angular diameter in seconds of arc, and we always use the same units for distance and linear diameter:

$$\frac{\text{angular diameter}}{206,265*} = \frac{\text{linear diameter}}{\text{distance}}$$

We can use this formula to find any of these three quantities if we know the other two.

Example: The moon has a linear diameter of 3476 km and is about 384,000 km away. What is its angular diameter? *Solution:* We can leave linear diameter and distance in kilometers, and find the angular diameter in seconds of arc:

$$\frac{\text{angular diameter}}{206,265} = \frac{3467 \text{ km}}{384,000 \text{ km}}$$

The angular diameter is 1870 seconds, or 31 minutes, of arc—about 0.5°.

*The number 206,265 is the number of seconds of arc in a radian. When we divide by 206,265, we convert the angle from seconds of arc into radians.

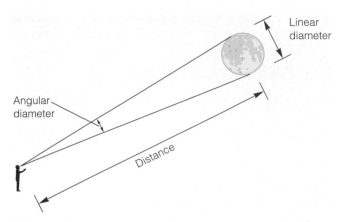

Figure 1-21
The small-angle formula relates angular diameter, linear diameter, and distance. Angular diameter is the angle formed by lines extending from our eye to opposite sides of the object—in this figure, the moon. Linear diameter and distance are typically measured in kilometers or meters.

shadow barely reaches Earth and casts a small circular shadow never larger than 270 km (168 miles) in diameter. If we are standing in that umbral spot, we are in total shadow, unable to see any part of the sun's surface, and the eclipse is total. But if we are located outside the umbra, in the penumbra, we see part of the sun peeking around the edge of the moon, and the eclipse is partial. Of course, if we are outside the penumbra, we see no eclipse at all.

Because of the orbital motion of the moon and the rotation of Earth, the moon's shadow sweeps rapidly across Earth in a long, narrow path of totality. If we want to see a total solar eclipse, we must be in the path of totality. When the umbra of the moon's shadow sweeps over us, we see one of the most dramatic sights in the sky, the totally eclipsed sun.

The eclipse begins as the moon slowly crosses in front of the sun. It takes about an hour for the moon to cover the solar disk, but as the last sliver of sun disappears, dark falls in a few seconds. Automatic street lights come on, drivers of cars turn on their headlights, and birds go to roost. The sky becomes so dark we can even see the brighter stars.

The darkness lasts only a few minutes, because the umbra is never more than 270 km (168 miles) in diameter and sweeps across Earth's surface at over 1600 km/hr (1000 mph). The sun cannot remain totally eclipsed for more than 7.5 minutes, and the average period of totality lasts only 2 or 3 minutes.

When the moon covers the bright surface of the sun, called the **photosphere,** we can see the bright gases of the **chromosphere** just above the photosphere, and the **corona,** the sun's faint outer atmosphere (Figure 1-23). The corona is a low-density, hot gas that glows with a pale white color. Streamers caused by the solar magnetic field streak the corona. The chromosphere is often marked by eruptions on the solar surface called **prominences** and is bright pink. The corona,

a

b

Figure 1-22
(a) Observers in the path of totality see a total solar eclipse when the umbral shadow sweeps over them. Those in the penumbra see a partial eclipse. (b) From inside the umbral shadow, the moon covers the brilliant photosphere, and the fainter corona and prominences are visible. Note the streamers in the corona caused by the sun's magnetic field. *(Daniel Good)*

chromosphere, and prominences are visible only when the brilliant photosphere is covered. As soon as part of the photosphere reappears, the fainter corona, chromosphere, and prominences vanish in the glare, and totality is over. The moon moves on in its orbit, and in an hour the sun is completely visible again.

Just as totality begins or ends, a small part of the photosphere can peek out from behind the moon through a valley at the edge of the lunar disk. Although it is intensely bright, such a small part of the photosphere does not completely drown out the fainter corona, which forms a silvery ring of light with the brilliant spot of photosphere gleaming like a diamond (Figure 1-24). This **diamond ring effect** is one of the most spectacular of astronomical sights, but it is not visible during every solar eclipse. Its occurrence depends on the exact orientation and motion of the moon.

Figure 1-23
This sequence shows the first half of a total solar eclipse. The brilliant surface of the sun is gradually covered by the moon moving from the right. Once all of the bright photosphere is covered, we can see the pink prominences of the chromosphere (d). A much longer exposure is needed to photograph the fainter corona (e). *(Daniel Good)*

Sometimes when the moon crosses in front of the sun, it is too small to fully cover the sun, and we see an **annular eclipse,** a solar eclipse in which a ring (or annulus) of the photosphere is visible around the disk

Figure 1-24
The diamond ring effect can sometimes occur momentarily at the beginning or end of totality if a small segment of the photosphere peeks out through a valley at the edge of the lunar disk. *(NOAO)*

of the moon (Figure 1-25). With a portion of the photosphere visible, the eclipse never becomes total; it never quite gets dark; and we can't see the prominences, chromosphere, and corona. Annular eclipses occur because the moon follows a slightly elliptical orbit around Earth, and thus its angular diameter can vary. When it is at **perigee,** its point of closest approach, it looks significantly larger than when it is at **apogee,** the most distant point in its orbit. Furthermore, Earth's orbit is slightly elliptical, so the Earth–sun distance varies slightly and thus the diameter of the solar disk varies slightly. If the moon is in the farther part of its orbit during totality, its angular diameter will be less than the angular diameter of the sun, and thus we see an annular eclipse.

A list of future total and annular eclipses of the sun is given in Table 1-3. If you plan to observe a solar eclipse, remember that the sun is bright enough to burn your eyes and cause permanent damage if you look at it directly. This is true whether there is an eclipse or not. Solar eclipses can tempt us to look at the sun in spite of its brilliance and thus can tempt us to risk our eyesight. During totality, the brilliant photosphere is hidden and it is safe to look at the eclipse, but the partial phases can be dangerous. See Figure 1-26 for a safe way to observe the partially eclipsed sun.

Predicting Eclipses

Predicting lunar or solar eclipses seems quite complex, and precise predictions require sophisticated calculations. But we can make general eclipse predictions by

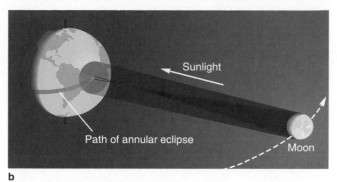

Figure 1-25
(a) The annular eclipse of 1994. A bright ring of photosphere remains visible around the moon, and the corona and prominences are not visible. *(Daniel Good)* (b) An annular eclipse occurs when the moon is in the farther part of its orbit and its umbral shadow does not reach Earth. From Earth, we see an annular eclipse, because the moon's angular diameter is smaller than the angular diameter of the sun.

Table 1-3 Total and Annular Eclipses of the Sun, 2000–2010

Date	Total/ Annular (T/A)	Time of Mideclipse* (GMT)	Maximum Length of Total or Annular Phase (Min:Sec)	Area of Visibility
2001 June 21**	T	12^h	4:56	Atlantic, S. Africa, Madagascar
2001 Dec. 14	A	21^h	3:54	Pacific, Central America
2002 June 10	A	24^h	1:13	Pacific
2002 Dec. 4	T	8^h	2:04	S. Africa, Indian Ocean, Australia
2003 May 31	A	4^h	3:37	Iceland, Arctic
2003 Nov. 23	T	23^h	1:57	Antarctica
2005 Apr. 8	AT	21^h	0:42	Pacific, N. of S. America
2005 Oct. 3	A	11^h	4:32	Atlantic, Spain, Africa
2006 Mar. 29	T	10^h	4:07	Atlantic, Africa, Turkey
2006 Sept. 22	A	12^h	7:09	N.E. of S. America, Atlantic
2008 Feb. 7	A	4^h	2:14	S. Pacific, Antarctica
2008 Aug. 1	T	10^h	2:28	Canada, Arctic, Siberia
2009 Jan. 26	A	8^h	7:56	S. Atlantic, Indian Ocean
2009 July 22	T	3^h	6:40	Asia, Pacific
2010 Jan. 15	A	7^h	11:10	Africa, Indian Ocean
2010 July 11	T	20^h	5:20	Pacific, S. America

The next major total solar eclipse visible from the United States will occur on August 21, 2017.

*Times are in hours Greenwich Mean Time. Subtract 5 hours for Eastern Standard Time, 6 hours for Central Standard Time, 7 hours for Mountain Standard Time, and 8 hours for Pacific Standard Time.

**There are no total or annular eclipses of the sun during 2000.

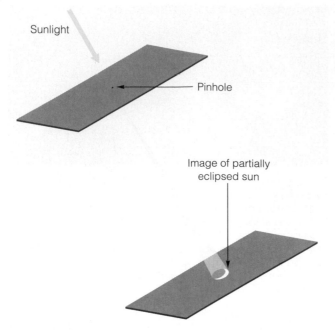

Sunlight

Pinhole

Image of partially
eclipsed sun

Figure 1-26
A safe way to view the partial phases of a solar eclipse. Use a pinhole in a card to project an image of the sun on a second card. The greater the distance between the cards, the larger and fainter the image will be.

thinking about the geometry of an eclipse and the cyclic motions of the sun and moon.

We see a solar eclipse when the moon passes between Earth and the sun, that is, when the lunar phase is new moon. We see a lunar eclipse at full moon. However, we don't see eclipses at every new moon and every full moon. Why not?

Figure 1-27 is a scale drawing of the umbral shadows of Earth and its moon. Notice that they are extremely long and narrow. Earth, moon, and sun must line up almost exactly or the shadows miss their mark and there is no eclipse.

To be eclipsed, the moon must enter Earth's shadow. However, because its orbit is tipped (Figure 1-28), the moon often misses the shadow, passing north or south of it, and no lunar eclipse occurs. Also, in order to pro-

duce a solar eclipse, the moon's shadow must sweep over Earth. The inclination of the moon's orbit, however, means that it often reaches new moon with its shadow passing north or south of Earth, and there is no solar eclipse.

For an eclipse to occur, the moon must reach full or new moon at the same time it passes through the plane of Earth's orbit; otherwise, the shadows miss (Figure 1-28). The points where it passes through the plane of Earth's orbit are called the **nodes** of the moon's orbit, and the line connecting these is called the line of nodes. Twice a year this line of nodes points toward the sun, and for a few weeks eclipses are possible at new moon and full moon. These intervals when eclipses are possible are called eclipse seasons, and they occur about six months apart.

If the moon's orbit were fixed in space, the eclipse seasons would always occur at the same time each year. The moon's orbit precesses slowly, however, because of the gravitational pull of the sun on the moon, and the precession slowly changes the direction of the line of nodes. The line turns westward, making one complete rotation in 18.61 years. As a result, the eclipse seasons occur about three weeks earlier each year. The motion of the line of nodes, combined with the periodicity of the lunar phases, means that every 6585.3 days the eclipse seasons start over and the same pattern of eclipses repeats (6585.3 days equals 18 years 11.3 days or 18 years 10.3 days depending on the number of leap days during the interval). Because this cycle, termed the **Saros cycle,** contains about one-third of a day more than an integer number of days, an eclipse visible in North America will recur after 18 years 11.3 days, about one-third of the way around the world, in this case in the eastern Pacific. Many ancient peoples recognized the Saros cycle from their records of previous eclipses and were able to predict when eclipses would occur, even though they did not understand what the sun and moon were or what alignments produced eclipses.

Could Stonehenge have been used to predict eclipses? Astronomers and archaeologists have found stones and markers there that seem to align with the

New moon

Earth

Full moon

10^6 km

10^6 miles

Figure 1-27
This scale drawing of the umbral shadows of Earth and its moon shows how easy it is for the shadows to miss their mark at full moon and new moon. That is, it is easy for a full moon to occur and not enter Earth's shadow, or for a new moon to occur and not cast its shadow on Earth. The diameters of Earth and its moon are exaggerated by a factor of 2 for clarity.

Figure 1-28
The moon's orbit is tipped about 5° to Earth's orbit. The nodes N and N' are the points where the moon passes through the plane of Earth's orbit. If the line of nodes does not point at the sun, the shadows miss and there are no eclipses at new moon and full moon. At those parts of Earth's orbit where the line of nodes points toward the sun, eclipses are possible at new moon and full moon.

most northern and most southern moonrises. Without knowing what the sun and moon are and without understanding what an eclipse really is, an observer could use those markers to find the eclipse seasons and guess which full moons were in danger of being eclipsed. That we could use Stonehenge to predict eclipses does not mean that ancient people could do the same, but it would not be too surprising if they could. The cycles of the sun and moon are regular and predictable, and the drama of the lunar phases and lunar eclipses has drawn the attention of every culture around the world.

REVIEW   **Critical Inquiry**

What would solar eclipses be like if the moon's orbit were not tipped to the plane of Earth's orbit?

This question asks about a geometry that does not exist, so we must use our imagination. As we struggle to analyze this imaginary situation, we will better understand the geometry of real eclipses. If the moon's orbit were not inclined, then the moon would follow the ecliptic, and it would cross in front of the sun at every new moon. There would be a solar eclipse every month at new moon. Because the ecliptic is inclined 23.5° to the celestial equator, the sun can never be farther than 23.5° from the celestial equator. That means the moon's shadow would sweep over Earth from west to east somewhere in a band exending 23.5° north or south of Earth's equator. The shadow would never sweep over those of us who live too far north of Earth's equator. We would see the moon pass south of the sun at new moon, and we would never see a solar eclipse. If we lived too far south, the moon would pass north of the sun, and, again, we would see no eclipse.

Analyzing a problem with one parameter slightly altered is a good way to study a complicated process. We could repeat this analysis for lunar eclipses, but it will be more interesting to change a different factor. What would lunar eclipses look like if Earth had no atmosphere?

The elegant cycles of the sun and moon produce beautiful phenomena to decorate Earth's sky, they control our environment through the seasons and tides, and they may even influence Earth's climate. It is not surprising that the people who built Stonehenge incorporated into their monument the cycles of the sun and moon.

The cycles in the sky are a rich part of our culture, but those same motions reveal an astonishing fact—Earth is a planet. In the next chapter, we will see how humanity made that discovery.

Summary

Astronomers divide the sky into 88 areas called constellations. Although the constellations originated in Greek and Middle Eastern mythology, the names are Latin. Even the modern constellations, added to fill in the spaces between the ancient figures, have Latin names. The names of stars usually come from ancient Arabic, though modern astronomers often refer to a star by constellation and Greek letters assigned according to brightness within each constellation.

The magnitude system is the astronomer's brightness scale. First-magnitude stars are brighter than 2nd-magnitude stars, which are brighter than 3rd-magnitude stars, and so on. The magnitude we see when we look at a star in the sky is its apparent visual magnitude.

The celestial sphere is a model of the sky, carrying the celestial objects around Earth. Because Earth rotates eastward, the celestial sphere appears to rotate westward on its axis. The northern and southern celestial poles are the pivots on which the sky appears to rotate. The celestial equator, an imaginary line around the sky above Earth's equator, divides the sky in half.

Astronomers often refer to angles "on" the sky as if the stars, sun, moon, and planets were equivalent to spots painted on a plaster ceiling. Then the angle on the sky between two objects is independent of the true distance between the objects in light-years.

The gravitational forces of the moon and sun act on the spinning Earth and cause it to precess like a gyroscope. Earth's axis of rotation sweeps around in a conical motion with a period of 26,000 years, and consequently the celestial poles and celestial equator move slowly against the background of the stars.

Because Earth orbits the sun, the sun appears to move eastward along the ecliptic through the constellations. Because the ecliptic is tipped 23.5° to the celestial equator, the sun spends half the year in the northern celestial hemisphere and half the year in the southern celestial hemisphere, producing the seasons. The seasons are reversed south of Earth's equator.

The motion of Earth changes in ways that can affect the climate. Changes in orbital shape, in precession, and in axial tilt can alter the planet's heat balance and may be responsible for the ice ages and glacial periods.

Because we see the moon by reflected sunlight, its shape appears to change as it orbits Earth. The lunar phases wax from new moon to first quarter to full moon and wane from full moon to third quarter to new moon. A complete cycle of lunar phases takes 29.53 days.

The moon's gravitational field exerts tidal forces on Earth that pull the ocean waters up into two bulges, one on the side of Earth facing the moon and the other on the side away from the moon. As the rotating Earth carries the continents through these bulges of deeper water, the tides ebb and flow. Friction with the seabeds slows Earth's rotation, and the gravitational force the bulges exert on the moon force its orbit to grow larger.

When the moon passes through Earth's shadow, sunlight is cut off and the moon darkens in a lunar eclipse. If the moon only grazes the shadow, the eclipse is partial, or penumbral, and not total.

If the moon passes directly between the sun and Earth, it produces a total solar eclipse. During such an eclipse, the bright photosphere of the sun is covered, and the fainter corona, chromosphere, and prominences become visible. An observer outside the path of totality sees a partial eclipse. If the moon is in the farther part of its orbit, it does not cover the photosphere completely, resulting in an annular eclipse.

New Terms

constellation	winter solstice
asterism	perihelion
magnitude scale	aphelion
apparent visual magnitude	Milankovitch hypothesis
celestial sphere	sidereal period
horizon	synodic period
zenith	spring tide
north celestial pole	neap tide
south celestial pole	lunar eclipse
north, south, east, and west points	umbra
	penumbra
celestial equator	solar eclipse
minute and second of arc	photosphere
scientific model	chromosphere
north circumpolar constellations	corona
	prominence
south circumpolar constellations	diamond ring effect
	annular eclipse
precession	perigee
ecliptic	apogee
zodiac	node
vernal equinox	Saros cycle
autumnal equinox	
summer solstice	

Review Questions

1. What is the difference between an asterism and a constellation? Give some examples.

2. Do people in other parts of the world see the same constellations, asterisms, lunar phases, lunar eclipses, and solar eclipses that you see? Explain.

3. What does the word *apparent* mean in *apparent visual magnitude*?

4. In what ways is the celestial sphere a scientific model?

5. If Earth did not rotate, could we define the celestial poles and celestial equator? the ecliptic?

6. Where would you go on Earth so you could see both north and south celestial poles at the same time? Where would you go to see a celestial pole at your zenith?

7. What is the difference between the daily and annual motion of the sun?

8. What would our seasons be like if Earth were tipped 35° instead of 23.5°? What would they be like if Earth's axis were perpendicular to its orbit?

9. Why are the seasons reversed in the Southern Hemisphere?

10. Where would you look in the sunset or dawn skies to find the full moon, the first quarter moon, the waning crescent moon, the waxing gibbous moon?

11. What phase would Earth be in if you were on the moon when the moon was full? first quarter? waning crescent?

12. How does the moon slow Earth's rotation? How does Earth slow the moon's revolution?

13. Why isn't there an eclipse at every new and at every full moon?

14. Why is the moon red during a total lunar eclipse?

15. Why have most people seen a total lunar eclipse while fewer people have seen a total solar eclipse?

Discussion Questions

1. All cultures around the world named constellations. Why do you suppose this was such a common practice?

2. Do planets orbiting other stars have ecliptics? Could they have seasons? Must they have seasons?

3. Why would it be difficult to see prominences if you were on the moon during a total lunar eclipse?

Problems

1. If light from one star is 40 times more intense than light from another star, what is their difference in magnitudes?

2. If two stars differ by 8.6 magnitudes, what is their intensity ratio?

3. Star A has a magnitude of 2.5; Star B, 5.5; and Star C, 9.5. Which is brightest? Which are visible to the unaided eye? Which pair of stars has an intensity ratio of 16?

4. By what factor is sunlight more intense than moonlight? (*Hint:* See Figure 1-4.)

5. If you are at a latitude of 35 degrees north of Earth's equator, what is the angular distance from the northern horizon up to the north celestial pole? from the southern horizon down to the south celestial pole?

6. If Earth is about 5 billion (5×10^9) years old, how many precessional cycles have occurred?

7. Identify the phases of the moon if on March 21 the moon were located at (a) the vernal equinox, (b) the autumnal equinox, (c) the summer solstice, (d) the winter solstice.

8. The average distance of Mars from the sun is 1.52 AU. Use the small-angle formula to calculate the angular diameter of the sun as seen from Mars.

9. Draw a diagram showing Earth, the moon, and shadows during (a) a total solar eclipse, (b) a total lunar eclipse, (c) a partial lunar eclipse, (d) an annular eclipse.

10. A total solar eclipse was visible from eastern Canada on July 10, 1972. When did this eclipse next occur? When will it next be visible from eastern Canada?

Critical Inquiries for the Web

1. We've all heard of the "North Star" (Polaris, shown in Figure 1-8), but is there a "South Star" for Southern Hemisphere observers? As Earth precesses on its axis, Polaris will cease to be our pole star and others will eventually hold that title. Search for star maps available online and for information on precession, and determine what bright Southern Hemisphere stars might be future pole stars for people living south of the Equator.

2. Would the stars of a familiar constellation like Pegasus or Orion look similar to our view from Earth if we lived on planets orbiting other stars? Find a table of distance data for the bright stars in a familiar constellation, and construct a diagram that visualizes the positions of these stars in space (similar to Figure 1-3).

Exploring *The Sky*

1. As discussed in this chapter, Earth's rotation about its own axis gives us the impression that the whole sky rotates around the **north celestial pole** in a period of one day. This apparent motion of the **celestial sphere** is difficult to notice because it happens so slowly. However, *The Sky* makes it possible to simulate this motion at a pace that is easy to observe by using a feature called "Time Skip." Set the "Time Skip Increment" to 1 minute, and click on "Go Forward" to begin the simulation. View the sky from the four cardinal directions, due north, south, east, and west. (You'll find "Time Skip" under the "Tools" menu as well as a separate tool bar with buttons for the "Go Forward," "Stop," etc.)

2. In the course of a year, the sun appears to move through 12 constellations called the **zodiac**. Using *The Sky*, determine in which constellation of the zodiac the sun appears today (or some other chosen date). How to proceed: Locate the sun on your screen using "Find" (a button on the toolbar or the key stroke "F"). Then turn on the *constellation outlines* as well as the *constellation boundaries* (two buttons on the toolbar or "Reference Lines" under the "Tool" menu).

 Go to the Brooks/Cole Astronomy Resource Center (www.brookscole.com/astronomy) for critical thinking exercises, articles, and additional readings from InfoTrac College Edition, Brooks/Cole's online student library.

The Origin of Modern Astronomy

The passions of astronomy are

no less profound because they are

not noisy.

John Steinbeck
The Short Reign of Pippin IV

Guidepost

In the 16th and 17th centuries, astronomers tried to understand the motions they saw in the sky, motions we discussed in the last chapter. In doing so, they invented a new way of understanding nature, what we now call science. This chapter, then, is really about the birth of science.

To understand the 17 chapters that follow, we must understand how sci-entists use evidence to test hypotheses and build theories to explain nature. Although science has become a powerful force in shaping our society, it remains now as it was four centuries ago, nothing more than a logical way of thinking about nature—a way of understanding what we are and where we are.

The story of modern astronomy begins with the death of the great Polish astronomer Nicolaus Copernicus in May 1543 and the almost simultaneous publication of his model of the universe. That model revolutionized not only astronomy but all science, and it inspired a new consideration of our place in nature. We will trace this story over the 99 years from 1543 and Copernicus's death to 1642, the year that saw the death of another great astronomer, Galileo, and the birth of one of the greatest scientists in history, Isaac Newton.

Two themes run through our story: the place of the Earth in the cosmos and the nature of planetary motion. The debate over Earth's place involved theological questions and eventually brought Galileo Galilei before the Inquisition. The nature of planetary motion involved the most basic physics of motion and eventually led Isaac Newton to understand gravitation and motion. Thus the Copernican model of the universe led to the birth of modern science.

To understand why the Copernican model was so important, we must backtrack to ancient Greece and meet the two great authorities of ancient astronomy, Aristotle and Ptolemy. It was the Greeks who first struggled to understand the place of the Earth and the nature of planetary motion.

Figure 2-1
Aristotle (384–322 BC), honored on this Greek stamp, wrote on such a wide variety of subjects and with such deep insight that he became the great authority on all matters of learning. His opinions on the nature of Earth and the sky were widely accepted for almost two millennia.

2-1 Pre-Copernican Astronomy

Aristotle

The history of astronomy before Copernicus extends back over 3000 years to ancient Babylon and beyond, but it is dominated by a single man, Aristotle (Figure 2-1). He lived from 384 to 322 BC in Greece and taught and wrote on the entire range of human knowledge. Because of his insight, he became the great authority of antiquity.

Much of what Aristotle wrote about scientific subjects was wrong, but he was not a scientist. He was trying to understand the universe by creating a system of formal philosophy. Modern science depends on evidence and hypotheses, but scientific methods had not been invented at the time of Aristotle. Rather, he attempted to use basic observations combined with first principles (ideas that were obviously true) to understand the world and the heavens. His work on astronomy was accepted for almost 2000 years.

Aristotle adopted ideas from his teacher, the great philosopher Plato. Earth was imperfect and changeable, and the heavens were perfect and immutable. The most perfect geometrical figure, said Plato, was the sphere, and the natural motion of a sphere is rotation, so the heavens must be made up of rotating spheres. Thus Aristotle's description of the universe consisted of a perfect celestial realm rotating around a fixed Earth at the center (Figure 2-2).

Figure 2-2
According to Aristotle, Earth is motionless (*Terra immobilis*) at the center of the universe. Earth is surrounded by spheres of water, air, and fire (*ignus*), above which lie spheres carrying the celestial bodies beginning with the moon (*lune*) in the lowest celestial sphere. This woodcut is from C. Cornipolitanus's book *Chronographia* of 1537. *(From the Granger Collection, New York)*

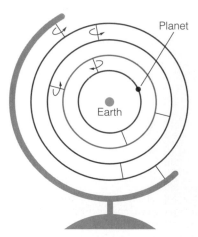

Figure 2-3
In this schematic drawing of nested spheres, the planet is carried around Earth in a complex path by the combined motions of four spheres.

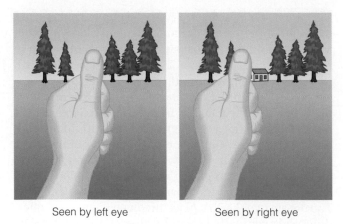

Seen by left eye   Seen by right eye

Figure 2-4
To demonstrate parallax, close one eye and cover a distant object with your thumb held at arm's length. Look with the other eye, and your thumb appears to have shifted position.

Another of Plato's students, Eudoxus of Cnidus (409–356 BC), proposed that each planet was attached to a separate sphere that rotated about an axis connecting it to the next sphere outward (Figure 2-3). By adjusting the rotation rate and angle of the axes of 27 spheres, Eudoxus tried to account for the observed motions of the seven planets—Mercury, Venus, Mars, Jupiter, Saturn, the sun, and the moon. Because the spheres were located in the heavens, they were thought to be made of perfect crystalline material.

Aristotle adopted the idea of nested spheres to carry the planets, but he added more spheres, bringing the total to 55. Above the spheres of the planets was the sphere of fixed stars, and the entire heavenly vault rotated westward around Earth once a day and thus made the sun, moon, and stars rise in the east and set in the west. Thus, Aristotle and the classical astronomers who accepted his philosophy in the centuries following his life thought of the sky as a perfect, heavenly machine not many times larger than Earth itself.

Classical astronomers felt it was obvious that Earth did not move. They could feel no motion, of course, but more important, they saw no parallax in the position of the stars. **Parallax** is the change in the apparent position of an object due to a change in the location of the observer. It is actually an old friend, though you may not have recognized its name, for you use parallax to judge the distance to things. To see how this works, close your right eye and use your thumb, held at arm's length, to cover some distant object, a building perhaps. Now look with your right eye. Your thumb seems to move to the left, uncovering the building (Figure 2-4). This apparent shift is parallax, and your brain uses it to estimate distances to objects around you.

For centuries astronomers reasoned that if Earth moved around the sun, we would view the stars from different places in Earth's orbit at different times of the year. Then parallax would distort the shapes of the constellations; they would look largest when Earth was nearest that part of the sky. Because they did not see this—that is, because they saw no parallax—they concluded that Earth did not move. Actually, they saw no parallax because the stars are much farther away than they supposed, and the parallax is much too small to be visible to the naked eye.

Ptolemy

One reason for the long survival of Aristotle's views was the work of the mathematician Claudius Ptolemy. His nationality and birth date are unknown, but he lived and worked in the Greek settlement at Alexandria in about AD 140, roughly five centuries after Aristotle. There he studied mathematics and astronomy and developed a mathematical model of the universe based on the teachings of Aristotle. Working from Aristotle's general description of the universe based on first principles, Ptolemy tried to devise a mathematical model that would precisely describe the motions of the planets.

The Ptolemaic model, following Aristotle's philosophy, was a **geocentric** (Earth-centered) **universe.** Ptolemy also included the Greek belief that the heavenly bodies, being perfect, moved in **uniform circular motion,** an idea we trace back to Plato. But simple, circular paths centered on Earth do not account for the motions of the planets in the sky. The planets sometimes move faster and sometimes slower, and occasionally they appear to slow to a stop and move backward (westward) for a time in **retrograde motion** (Figure 2-5).

To describe the complicated planetary motions and yet preserve a geocentric model with uniform circular

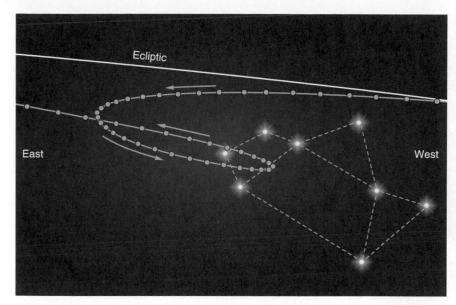

Figure 2-5
The word *planet* comes from the Greek for "wanderer," referring to the motion of the planets against the background of fixed stars. Here the motion of Mars along the ecliptic near the teapot shape of Sagittarius is shown at four-day intervals. Though the planet usually moves eastward, it sometimes slows to a stop and moves westward in retrograde motion.

motion, Ptolemy adopted a system of wheels within wheels. Each planet moved in a small circle called an **epicycle,** and the center of the epicycle moved along a larger circle around Earth called a **deferent.** By adjusting the size of the circles and the rate of their motion, Ptolemy could account for most planetary movement. But as a final adjustment, he placed Earth off center in the deferent circle and specified that the center of the epicycle would appear to move at constant speed only if viewed from a point called the **equant,** located on the other side of the deferent's center (Figure 2-6). Thus, by using a few dozen circles of various sizes rotating at various rates, Ptolemy's system reproduced the motions of the planets (Figure 2-7).

Whereas Aristotle had attempted to describe the motions of the heavens consistent with the first principles of his system of philosophy, Ptolemy was not as concerned with first principles. He wanted accurate mathematical predictions of the positions of the planets. Motion around an epicycle or centered on an equant was not truly Earth-centered nor was it truly uniform circular motion, but it improved the mathematical predictions. Thus, Ptolemy slightly weakened the principles of geocentrism and uniform circular motion in order to make his model more accurate.

About AD 140, Ptolemy included this work in his book *Mathematical Syntaxis,* a book known today as the *Almagest.* The title changed because of one of the great tides in human history. The death of Ptolemy in the middle of the 2nd century marks the end of classical astronomy. The breath of civilization passed to invading Islamic peoples, and they found the great classical works elegant and beautiful. While Europe slumbered through the Middle Ages, Islamic astronomers translated and studied Ptolemy's book, renaming it *Al Magisti* (The Greatest). Beginning in the 12th cen-

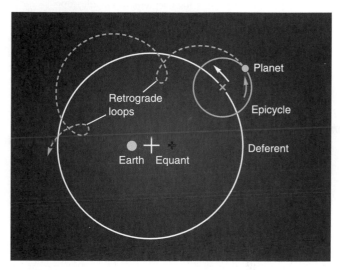

Figure 2-6
To explain the periodic retrograde motion of the planets, ancient astronomers proposed that the planet traveled around a small circle called the epicycle, which revolved around a bigger circle called the deferent. To slightly adjust the speed of the planet, they supposed that Earth was slightly off center and that the center of the epicycle moved such that it appeared to move at a constant rate as seen from the equant point. A planet moving in this way periodically passes through a retrograde loop and appears to move backward (westward) against the background of stars.

tury, European astronomers rediscovered the classical books and translated them into Latin. *Al Magisti* became *Almagest.* Thus when we say the name of Ptolemy's masterpiece, we recall a chapter in the history of astronomy that lasted a millennium.

At first the Ptolemaic system predicted planetary positions well, and Ptolemy probably died thinking his model was a success. But as centuries passed, the model

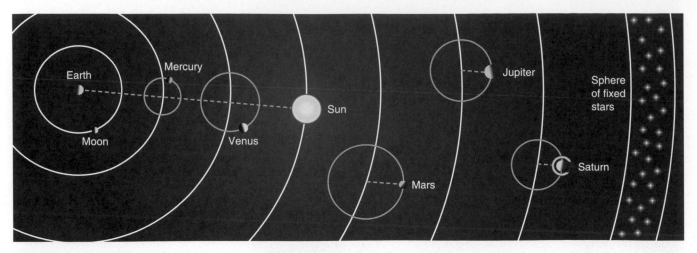

Figure 2-7
The Ptolemaic model of the universe was geocentric (Earth-centered) and based on uniform circular motion of epicycles moving around deferents. The centers of the epicycles of Mercury and Venus must always lie on the Earth–sun line, but Mars, Jupiter, and Saturn are free of this restriction. Notice that this modern illustration shows rings around Saturn and sunlight illuminating the globes of the planets, features the ancient astronomers could not have known before the invention of the telescope.

accumulated errors, like a clock that gains one second a day. Astronomers repeatedly adjusted the system. In the mid-13th century, a team of astronomers supported by King Alfonso X of Castile worked for 10 years revising the Ptolemaic system and publishing the result as the *Alfonsine Tables*. It was the last great adjustment of the Ptolemaic system.

R E V I E W   **Critical Inquiry**

Why did classical astronomers conclude the heavens were made up of spheres?

Classical astronomers did not use evidence and hypotheses the way modern scientists do. Rather, they argued from first principles, and Plato argued that the perfect geometrical figure was a sphere. Then the heavens, which everyone agreed were perfect, must be made up of spheres. The natural motion of a sphere is rotation, and the only perfect motion is uniform motion, so the heavenly spheres were thought to move in uniform circular motion. Thus, classical philosophers argued that the daily motion of the heavens around Earth and the motions of the seven planets (the sun and moon were counted as planets) against the background of the stars had to be produced by the combination of uniformly rotating spheres.

Although ancient astronomers didn't use evidence as modern scientists do, they did observe the world around them. What observations led them to conclude that Earth didn't move?

Pre-Copernican astronomy consists of the elaboration of Aristotle's philosophy of the universe as expressed by Ptolemy in a mathematical model. For nearly 1500 years, astronomers believed in this geocentric universe. Earth was finally set in motion by a Polish church official with a talent for mathematics.

2-2 Copernicus

Nicolaus Copernicus was born in 1473 (Figure 2-8) in what is now Poland. In 1543, the year of his death, the Ptolemaic system, in spite of many revisions, was still a poor predictor of planetary positions. Yet because of the authority of Aristotle, it was the officially accepted theory of the universe.

The Sun-Centered Model

According to the Aristotelian universe, the most perfect region was in the heavens, and the most imperfect at Earth's center. Thus the classical geocentric universe matched the commonly held Christian geometry of heaven and hell, and anyone who criticized Earth's central location was challenging belief in heaven and hell and thus risking at least criticism and perhaps a charge of heresy.

Throughout his life, Copernicus had been associated with the Church. His uncle, by whom he was raised and educated, was an important bishop in Poland, and after studying canon law and medicine in some of the major universities in Europe, Copernicus became a

Figure 2-8
Nicolaus Copernicus (1473–1543) proposed that the sun and not Earth was the center of the universe. These stamps were issued in 1973 to commemorate the 500th anniversary of his birth. Note the sun-centered model on the U.S. stamp and the pages from his book *De Revolutionibus* on the East German (DDR) stamp.

canon of the Church at the age of 24. He served as secretary and personal physician to his powerful uncle for 15 years. When his uncle died, Copernicus went to live in quarters adjoining the cathedral in Frauenburg. Because of this long association with the Church and his fear of persecution, he hesitated to publish his revolutionary ideas in astronomy.

In fact, his ideas were already being discussed long before the publication of his book. His interest in astronomy had begun during his college days, and he apparently doubted the Ptolemaic system even then. About 1507, at the age of 34, he wrote a short pamphlet that discussed the motion of the sky and outlined his hypothesis that the sun, not Earth, was the center of the universe and that Earth rotated on its axis and revolved around the sun. He distributed his summary in handwritten form without a title and, in some cases, anonymously. Perhaps his caution grew from a sense of modesty, but he may also have feared criticism for challenging the Ptolemaic system and the place of the Earth.

De Revolutionibus

Copernicus worked on his book *De Revolutionibus Orbium Coelestium* over a period of many years and was essentially finished by about 1530, yet he hesitated to publish it even though other astronomers knew of his theories. Church officials concerned about the reform of the calendar sought his advice and looked forward to the publication of his book.

One reason he hesitated was the controversial idea of a heliocentric model. This was a time of rebellion in the Church—Martin Luther was speaking harshly about fundamental Church teachings, and others, both scholars and scoundrels, were questioning the authority of the Church. Even matters as abstract as astronomy could stir controversy. Remember, too, that the place of the Earth in astronomical theory was linked to the geometry of heaven and hell, so moving Earth from its central place was a controversial and perhaps heretical idea. Another reason Copernicus may have hesitated to publish was because his work was incomplete. His model could not accurately predict planetary positions. Thus, he did not give permission for the publication of his book until he realized he was dying.

The most important idea in the book was placing the sun at the center of the universe. That single innovation had an astonishing consequence—the retrograde motion of the planets was immediately explained in a straightforward way without the large epicycles that Ptolemy used. In the Copernican system, Earth moves faster along its orbit than the planets that lie farther from the sun. Consequently, Earth periodically overtakes and passes these planets. Imagine that you are in

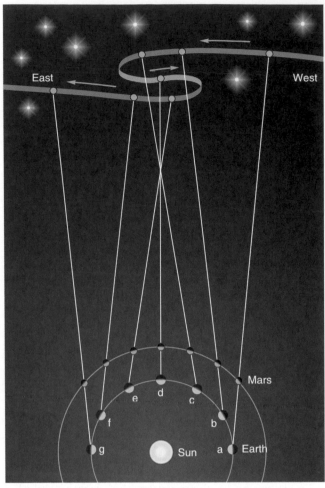

Figure 2-9
The Copernican explanation of retrograde motion. As Earth overtakes Mars (a–c), it appears to slow its eastward motion. As Earth passes Mars (d), it appears to move westward (retrograde). As Earth draws ahead of Mars (e–g), it resumes its eastward motion against the background stars. Compare with Figure 2-5. The positions of Earth and Mars are shown at equal intervals of one month.

a race car, driving rapidly along the inside lane of a circular race track. As you pass slower cars driving in the outer lanes, they fall behind, and if you did not know you were moving, it would seem that the cars in the outer lanes occasionally slowed to a stop and then backed up for a short interval. The same thing happens as Earth passes a planet such as Mars. Although Mars moves steadily along its orbit, as seen from Earth, it appears to slow to a stop and move westward (retrograde) as Earth passes it (Figure 2-9). Because the planetary orbits do not lie in precisely the same plane, a planet does not resume its eastward motion in precisely the same path it followed earlier. Consequently, it describes a loop whose shape depends on the angle between the orbital planes.

Copernicus could explain retrograde motion without epicycles, and that was impressive. The Copernican system was elegant and simple compared to the whirling epicycles and off-center equants of the Ptolemaic system. However, *De Revolutionibus* failed to disprove the geocentric model for one critical reason— the Copernican theory could not predict the positions of the planets any more accurately than the Ptolemaic system could. To understand why it failed this critical test, we must understand Copernicus and his world.

Although Copernicus proposed a revolutionary idea in making the planetary system **heliocentric** (sun-centered), he was a classical astronomer with tremendous respect for the old concept of uniform circular motion. In fact, Copernicus objected strongly to Ptolemy's use of the equant. It seemed arbitrary to Copernicus, a direct violation of the elegance of Aristotle's philosophy of the heavens. Copernicus called equants "monstrous" in that they violated both geocentrism and uniform circular motion. In devising his model, Copernicus returned to a strong belief in uniform circular motion. Although he did not need epicycles to explain retrograde motion, Copernicus discovered that the sun, moon, and planets suffered other smaller variations in their motions that he could not explain with uniform circular motion centered on the sun. Today we recognize those motions as typical of objects following elliptical orbits, but Copernicus held firmly with uniform circular motion, so he had to introduce small epicycles to reproduce these minor variations in the motions of the sun, moon, and planets.

Because Copernicus imposed uniform circular motion on his model, it could not accurately predict the motions of the planets. The *Prutenic Tables* (1551) were based on the Copernican model, and they were not significantly more accurate than the 13th century *Alfonsine Tables* that were based on Ptolemy's model. Both could be in error by as much as 2°, which is four times the angular diameter of the full moon.

The Copernican *model* was inaccurate. It included small epicycles and could not precisely describe the motions of the planets. But the Copernican *hypothesis,* that the universe was heliocentric, was correct. Why that hypothesis gradually won acceptance in spite of the inaccuracy of the epicycles and deferents is a question historians still debate. There are probably a number of reasons, including the revolutionary temper of the times, but the most important factor may be the elegance of the idea. Placing the sun at the center of the universe produced a symmetry among the motions of the planets that was pleasing to the eye as well as to the intellect (Figure 2-10). No longer did Venus and Mercury revolve around empty points located between Earth and the sun. Now they, like the rest of the planets, moved in orbits around the sun.

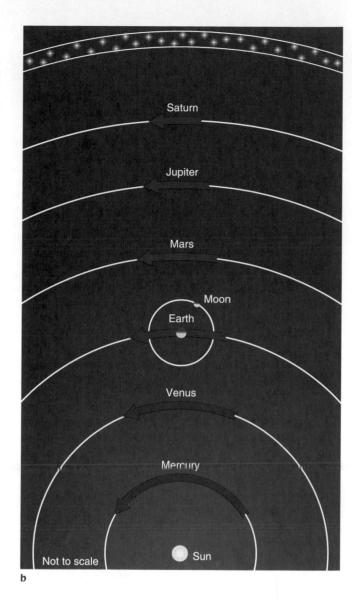

Figure 2-10

(a) The Copernican universe as reproduced in *De Revolutionibus*. Earth and all the known planets revolve in separate circular orbits about the sun (Sol) at the center. The outermost sphere carries the immobile stars of the celestial sphere. Notice the orbit of the moon around Earth. *(Yerkes Observatory photograph)* (b) The model was elegant, not only in the arrangement of the planets but also in their motions. Orbital velocities (red arrows) decreased from Mercury, the fastest, to Saturn, the slowest. Compare the elegance of this model with the complexity of the Ptolemaic model in Figure 2-7.

The most astonishing consequence of the Copernican hypothesis was not what it said about the sun, but what it said about Earth. By placing the sun at the center, Copernicus made Earth move along an orbit like the other planets. By making Earth a planet, Copernicus revolutionized humanity's view of its place in the universe and triggered a controversy that would eventually bring Galileo before the Inquisition, a controversy that continues even today.

Although astronomers throughout Europe read and admired *De Revolutionibus,* they did not usually accept the Copernican hypothesis. The mathematics was elegant, and the astronomical observations and calculations were of tremendous value; but few astronomers believed, at first, that the sun actually was the center of the planetary system and that Earth moved. How the Copernican hypothesis was gradually recognized as correct has been called the Copernican revolution, because it involved not just the adoption of a new idea but a total change in the way astronomers thought about the place of the Earth (Window on Science 2-1).

The Copernican revolution is often cited as the perfect example of a scientific revolution. Over a few decades, astronomers abandoned a way of thinking about the universe that was almost 2000 years old and adopted a new set of ideas and assumptions. The American philosopher of science Thomas Kuhn has referred to a commonly accepted set of scientific ideas and assumptions as a scientific **paradigm.** Thus, the pre-Copernican astronomers had a geocentric paradigm that included uniform circular motion and the perfection of the heavens. That paradigm survived for many centuries until a new generation of astronomers was able to overthrow the old paradigm and establish a new paradigm that included heliocentrism and Earth's motion.

A scientific paradigm is powerful because it shapes our perceptions. It determines what we judge to be important questions and what we judge to be significant evidence. Thus, it is often difficult for us to recognize how our paradigms limit what we can understand. For example, the geocentric paradigm contained problems that seem obvious to us, but because astronomers before Copernicus lived and worked inside that paradigm, they had difficulty seeing the problems. Overthrowing an outdated paradigm is not easy, because we must learn to see nature in an entirely new way. Galileo, Kep-

ler, and Newton saw nature from a new paradigm that would have been almost incomprehensible to astronomers of earlier centuries.

We can find examples of scientific revolutions in many fields, including biology, geology, genetics, and psychology. These scientific revolutions have been difficult and controversial because they have involved the overthrow of accepted paradigms. But that is why scientific revolutions are exciting. They give us not just a new idea or a new theory, but an entirely new insight into how nature works—a new way of seeing the world.

REVIEW Critical Inquiry

Why do we say the Copernican hypothesis was correct but the model was wrong?

The Copernican hypothesis was that the sun and not the Earth was the center of the universe. Given the limited knowledge of the Renaissance astronomers about distant stars and galaxies, that hypothesis was correct. The sun is at the center of our planetary system. That hypothesis produces an elegant model of the universe without large epicycles and equants and explains retrograde motion in a simple way. The Copernican model, however, included not only the heliocentric hypothesis but also uniform circular motion. The model is wrong, because the planets don't really follow circular orbits. Thus, the Copernican model could not make accurate predictions of planetary position even though it placed the sun at the center of the planetary system.

The Copernican hypothesis won converts because it was elegant and could explain retrograde motion. How did its explanation of retrograde motion work, and how was it more elegant than the Ptolemaic explanation?

The central issue in the Copernican hypothesis was the place of the Earth, but that was linked to a second issue, the nature of planetary motion. Thus, our story of the Copernican revolution shifts from Copernicus, a mathematician and theorist, to one of the first great ob-

servational astronomers, a man who measured the daily positions of the sun, moon, and planets.

2-3 Tycho Brahe

The great observational astronomer of our 99-year story was a Danish nobleman, Tycho Brahe (Figure 2-11), born December 14, 1546, only three years after the publication of *De Revolutionibus*. In histories of astronomy, the great astronomers are often referred to by their last name, but Tycho Brahe is usually called Tycho. Were he alive today, he would no doubt object to such familiarity from his obvious inferiors—he was well known for his vanity and lordly manners.

The New Star

During his university days, Tycho was officially studying law, but he made it clear to his family that his real passion was mathematics and astronomy. Yet his life was shaped by two events outside his studies—a duel and a new star in the sky.

While at the university, Tycho became involved in a duel over some supposed insult and received a wound that disfigured his nose. For the rest of his life, he wore false noses made of gold and silver and stuck on with wax (see Figure 2-11). He was already known as a proud example of Danish nobility, and the disfigurement seemed to do little to improve his disposition.

Early in his university days, Tycho showed an interest in astronomy and began measuring the positions of

Figure 2-11
Tycho Brahe (1546–1601), a Danish nobleman, established an observatory at Hveen and measured planetary positions with high accuracy. His artificial nose is suggested in this engraving.

the planets in the sky. In 1563, Jupiter and Saturn passed very near each other in the sky, nearly merging into a single point on the night of August 24. Tycho found that the *Alfonsine Tables* were a full month in error and that the *Prutenic Tables* were in error by a number of days.

Then in 1572 a "new star" (now called Tycho's supernova) appeared in the sky, shining more brightly than Venus. Tycho carefully measured its position and concluded that it displayed no parallax. To understand the significance of this observation, we must note that the rotation of Earth carries us eastward and that the positions of celestial objects near Earth, such as the moon, change slightly through the night because the observer is in different positions. Tycho believed Earth was stationary and that the heavens were rotating westward, but that makes no difference. Objects moving with the heavens near a stationary Earth should still have measurable parallax. That he detected no change in the position of the new star through the night proved it was farther away than the moon. Therefore, he concluded, it was a new star in the heavens. Because Aristotle and Ptolemy held that the heavens were perfect and unchanging, the new star led Tycho to question the Ptolemaic system. He summarized his results in a small book, *De Stella Nova* (*The New Star*), published in 1573.

The book attracted the attention of astronomers throughout Europe, and soon Tycho was summoned to the court of the Danish King Frederik II and offered funds to build an observatory on the island of Hveen just off the Danish coast. Tycho also received a steady source of income as landlord of a coastal district from which he collected rents. (He was not a popular landlord.) On Hveen, Tycho constructed a luxurious home with four towers especially equipped for astronomy, and populated it with servants, assistants, and a dwarf

to act as jester. Soon Hveen was an international center of astronomical study.

Tycho Brahe's Legacy

Tycho's great contribution to the birth of modern astronomy was not theoretical. Because he could measure no parallax for the stars, he concluded that Earth had to be stationary, thus rejecting the Copernican hypothesis. However, he also rejected the Ptolemaic model because of its inaccurate predictions. Instead, he devised a complex model in which Earth was the immobile center of the universe around which the sun and moon moved (Figure 2-12). The remaining planets circled the sun. This model was the same as the Copernican model with Earth stationary instead of the sun. Thus Tycho preserved the central, immobile Earth described in scripture. Although Tycho's model was very popular at first, the Copernican model replaced it within a century.

The true value of Tycho's work was observational. Because he was able to devise new and better instruments, he was able to make highly accurate observations of the positions of the stars, sun, moon, and planets. Tycho had no telescopes—they were not invented until the next century—so his observations were made by the naked eye peering along sights on his large instruments (Figure 2-13). In spite of these limitations, he measured the positions of 777 stars to better than 4 minutes of arc and measured the positions of the

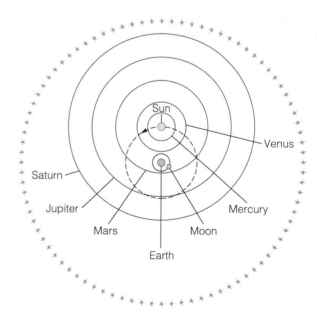

Figure 2-12
Tycho Brahe's model of the universe retained the most important of the classical principles. The stars were in a starry sphere. Earth was fixed and immobile at the center with the sun and moon revolving around Earth. The rest of the planets circled the sun. Thus Tycho's model was very similar to the Copernican model with Earth held stationary.

sun, moon, and planets almost daily for the 20 years he stayed on Hveen.

Unhappily for Tycho, King Frederik II died in 1588, and his young son took the throne. Suddenly Tycho's temper, vanity, and noble presumptions threw him out of favor. In 1596, taking most of his instruments and books of observations, he went to Prague, the capital of Bohemia, and became imperial mathematician to the Holy Roman Emperor Rudolph II. His goal was to revise the *Alfonsine Tables* and publish the revision as a monument to his new patron. It would be called the *Rudolphine Tables.*

Tycho did not intend to base the *Rudolphine Tables* on the Ptolemaic system but rather on his own Tychonic system, proving once and for all the validity of his hypothesis. To assist him, he hired a few mathematicians and astronomers, including one Johannes Kepler. Then in November 1601, Tycho overate at a nobleman's home and collapsed. Before he died, 11 days later, he asked Rudolph II to make Kepler imperial mathematician. Thus the newcomer, Kepler, became Tycho's replacement (though at one-half Tycho's salary).

REVIEW Critical Inquiry

Why did Brahe conclude that the new star of 1572 contradicted the Ptolemaic universe?

According to Aristotle, the heavens were perfect and therefore unchanging. Any change in the heavens had to occur below the moon, which was attached to the lowest of the crystalline spheres. When Tycho saw the new star of 1572 in the sky, he could detect no parallax. If the star lay below the moon's sphere, then it should appear to be east of its average position when it was rising in the east, and it should appear west of its average position when it was setting in the west. That is, it should show parallax. Tycho could measure no parallax at all, which meant the new star had to be at least above the moon's sphere and was probably part of the starry sphere itself. Thus, it was a change in the supposedly unchanging heavenly realm.

Tycho didn't believe in the Ptolemaic universe, and he produced his own model to explain planetary motion. In what ways did his model incorporate ideas from classical astronomy and ideas also found in the Copernican model?

Tycho Brahe tried to resolve the problem of the place of the Earth by assuming the universe was geocentric, but his attempts to understand planetary motion were cut short by his death. The newly appointed imperial mathematician, Johannes Kepler, used Tycho's observations to finally solve the second great problem of the age, the motion of the planets.

2-4 Johannes Kepler
An Astronomer of Humble Origins

No one could have been more different from Tycho Brahe than Johannes Kepler (Figure 2-14). He was born on December 27, 1571, to a poor family in a region now included in southwest Germany. His father was unreliable and shiftless, principally employed as a mercenary soldier fighting for whoever paid enough. He finally failed to return from a military expedition, either because he was killed or because he found circumstances more to his liking elsewhere. Kepler's mother was apparently an unpleasant and unpopular woman. She was accused of witchcraft in later years, and Kepler had to defend her in a trial that dragged on for three years. She was finally acquitted but died the following year.

Kepler was the oldest of six children, and his childhood was no doubt unhappy. The family was not only poor and often lacking a father, but it was also Protestant in a predominantly Catholic region. In addition, Kepler was never healthy, even as a child, so it is surprising that

Figure 2-13
Much of Tycho's success was due to his skill in designing large, accurate instruments such as this one known as the mural quadrant. In this engraving, the figure of Tycho at the center, his dog, and the scene in the background are a mural painted on the wall within the arc of the quadrant. The observer at the extreme right (Tycho) peers through a sight out the loophole in the wall at the upper left and thus measures the object's altitude above the horizon. *(The Granger Collection, New York)*

Figure 2-14
Johannes Kepler (1571–1630) derived three laws of planetary motion from Tycho Brahe's observations of the positions of the planets. This Romanian stamp commemorates the 400th anniversary of Kepler's birth. Ironically, it contains an error—the orientation of the moon.

Figure 2-15
In this diagram based on one drawn by Kepler, the sizes of the celestial spheres carrying the outer three planets, Saturn, Jupiter, and Mars, are determined by spacers (blue) consisting of a cube and a tetrahedron. Inside the sphere of Mars, the remaining regular solids (not shown here) separate the spheres of Earth, Venus, and Mercury. The sun lay at the very center of this Copernican universe based on geometrical spacers.

he did well in the pauper's school he attended, eventually winning a scholarship to the university at Tübingen, where he studied to become a Lutheran pastor.

During his last year of study, Kepler accepted a job in Graz teaching mathematics and astronomy, a job he resented because he knew little about the subjects. Evidently he was not a good teacher either—he had few students his first year, and none at all his second. His superiors put him to work teaching a few introductory courses and preparing an annual almanac that contained astronomical, astrological, and weather predictions. Through good luck, in 1595 some of his weather predictions were fulfilled, and he gained a reputation as an astrologer and seer. Even in later life he earned money from his almanacs.

While still a college student, Kepler had become a believer in the Copernican hypothesis, and at Graz he used his extensive spare time to study astronomy. By 1596, the same year Tycho left Hveen, Kepler was ready to solve the mystery of the universe. That year he published a book called *The Forerunner of Dissertations on the Universe, Containing the Mystery of the Universe.* The book, like nearly all scientific works of that age, was written in Latin, and is now known as *Mysterium Cosmographicum.*

By modern standards, the book contains almost nothing of value. It begins with a long appreciation of Copernicanism and then goes on to speculate on the reasons for the spacing of the planetary orbits. Kepler felt he had found the underlying architecture of the universe in the five regular solids*—the cube, tetrahedron, dodecahedron, icosahedron, and octahedron. Because these five solids were the only regular solids, he supposed they were the "spacers" between the plane-

tary orbits (Figure 2-15). In fact, Kepler concluded that there could be only six planets (Mercury, Venus, Earth, Mars, Jupiter, and Saturn) because there were only five regular solids to act as spacers between their spheres. He advanced astrological, numerological, and even musical arguments for his theory.

The second half of the book is no better than the first, but it has one virtue—as Kepler tried to fit the five solids to the planetary orbits, he demonstrated that he was a talented mathematician and that he was well versed in astronomy. He sent copies to Tycho and to Galileo in Rome, and both recognized his talent in spite of the mystical content of the book.

Joining Tycho

Life was unsettled for Kepler because of the persecution of Protestants in the region, so when Tycho Brahe invited him to Prague in 1600, Kepler went readily, eager to work with the famous Danish astronomer. Tycho's sudden death in 1601 left Kepler in a position to use the observations from Hveen to analyze the motions of the planets and complete the *Rudolphine Tables.* Tycho's family, recognizing that Kepler was a Copernican and guessing that he would not follow the Tychonic system in completing the *Rudolphine Tables,* sued to recover the instruments and books of observations. The legal wrangle went on for years. Tycho's family did get back the instruments Tycho had brought to Prague, but Kepler had the books and he kept them.

*A regular solid is a three-dimensional body, each of whose faces is the same. A cube is a regular solid, each of whose faces is a square.

Even scientists misuse the words *hypothesis*, *theory*, and *law*. We must try to distinguish these terms from one another because they are key elements in science.

A **hypothesis** is a single assertion or conjecture that must be tested. It could be true or false. "All Texans love chili" is a hypothesis. To know whether it is true or false, we need to test it against reality by making observations or performing experiments. Copernicus asserted that the universe was heliocentric; his assertion was a hypothesis subject to testing.

In Chapter 1, we saw that a model (Window on Science 1-1) is a description of some natural phenomenon; it can't be right or wrong. A model is not a conjecture of truth but merely a convenient way to think about a natural phenomenon.

Thus, a model such as the celestial sphere cannot be a hypothesis. Copernicus used his hypothesis to build a model of the universe, and to the extent that the model described the motion of the planets, it was useful.

A **theory** is a system of rules and principles that can be applied to a wide variety of circumstances. A theory may have begun as one or more hypotheses, but it has been tested, expanded, and generalized. Many textbooks refer to the "Copernican theory," but some experts argue that it was not complete in that it lacked a precise description of orbital motion and gravitation, features added by Kepler and Newton. Thus, it is probably more accurate to call it the Copernican hypothesis.

A **natural law** is a theory that almost everyone accepts as true. Such a theory is usually very specific and has been tested many times and applied to many situations. Thus, natural laws are the most fundamental principles of scientific knowledge. Kepler's laws of planetary motion are good examples.

Science is not always consistent about these words. For example, Einstein's theory of relativity is much more precise than Newton's laws of gravity and motion, but tradition dictates the words we use. Darwin's theory of evolution is well tested and widely accepted, but no one refers to it as Darwin's law. The distinctions are subtle and sometimes depend more on custom than on the definition of the words.

Whether Kepler had any legal right to Tycho's records is debatable, but he put them to good use. He began by studying the motion of Mars, trying to deduce from the observations how the planet moved. By 1606, he had solved the mystery, this time correctly. The orbit of Mars (and all planets) was an ellipse with the sun at one focus. Thus he abandoned the 2000-year-old belief in the circular motion of the planets. But the mystery was even more complex. The planets did not move at constant speed along their orbits—they moved faster when close to the sun and slower when farther away. Thus Kepler abandoned both uniform motion and circular motion and finally solved the problem of planetary motion.

Kepler published his results in 1609 in a book called *Astronomia Nova* (*New Astronomy*). Like Copernicus's book, *Astronomia Nova* was written in Latin for other scientists and was highly mathematical. In some ways the book was surprisingly advanced. For instance, Kepler discussed the force that holds the planets in their orbits and came within a paragraph of stating the principle of mutual gravitation.

In spite of the abdication of Rudolph II in 1611, Kepler continued his astronomical studies. He wrote about a supernova that had appeared in 1604 (now known as Kepler's supernova) and about comets, and he authored a textbook about Copernican astronomy. In 1619, he published *Harmonice Mundi* (*The Harmony of the World*) in which he returned to the cosmic mysteries of *Mysterium Cosmographicum*. The only thing of note in *Harmonice Mundi* is his discovery that the radii of the planetary orbits are related to the planet's orbital period. That and his two previous discoveries have become known as the most fundamental rules of orbital motion.

Kepler's Three Laws of Planetary Motion

Although Kepler dabbled in the philosophical arguments of the day, he was a mathematician, and his triumph was the solution of the problem of the motion of the planets. His three rules have been confirmed so often that astronomers have great confidence in them and describe them as natural laws (Window on Science 2-2). They are commonly known as Kepler's Laws (Table 2-1).

Table 2-1 Kepler's Laws of Planetary Motion

  I. The orbits of the planets are ellipses with the sun at one focus.
 II. A line from the planet to the sun sweeps over equal areas in equal intervals of time.
III. A planet's orbital period squared is proportional to its average distance from the sun cubed:

$$P_{\text{y}}^2 = a_{\text{AU}}^3$$

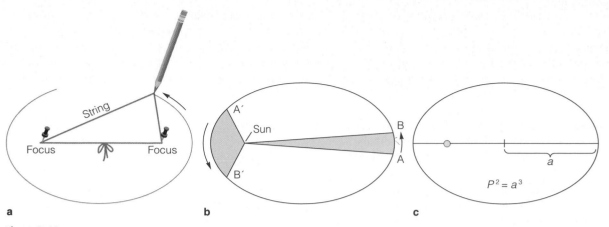

Figure 2-16
(a) Drawing an ellipse with two tacks and a loop of string. (b) Kepler's second law: "A line from a planet to the sun sweeps over equal areas in equal intervals of time." (c) Kepler's third law: The square of the orbital period (*P*) equals the cube of the semimajor axis (*a*).

Kepler's first law states that the orbits of the planets around the sun are ellipses with the sun at one focus. An **ellipse** is defined as a figure drawn around two points called the *foci* in such a way that the distance from one focus to any point on the ellipse and then back to the other focus equals a constant. This makes it very easy to draw ellipses with two thumbtacks and a loop of string. Press the thumbtacks into a board, loop the string about them, and place a pencil in the loop. If you keep the string taut as you move the pencil it traces out an ellipse (Figure 2-16a). The closer together the thumbtacks, the more circular the ellipse. Though Kepler was able to determine the elliptical shape of the planetary orbits, they are nearly circular. Of the planets known to Kepler, Mercury has the most elliptical orbit, but even it deviates only slightly from a circle (Figure 2-17).

Kepler's second law states that a line from the planet to the sun sweeps over equal areas in equal intervals of time (Figure 2-16b). This means that when the planet is closer to the sun and the line connecting it to the sun is shorter, the planet must move more rapidly if the line is to sweep over the same area. Thus the planet in Figure 2-16b would move from point A′ to point B′ in one month, sweeping out the area shown. But when the planet was farther from the sun, one month's motion would carry it from A to B.

Kepler's third law states that a planet's orbital period squared is proportional to its average distance from the sun cubed. Because of the way the planet moves along its orbit, its average distance from the sun is equal to half of the long diameter of the elliptical orbit, the semimajor axis *a* (Figure 2-16c). If we measure this quantity in astronomical units (AU) and the orbital period in years (y), we can summarize the third law as:

$$P_{\text{y}}^2 = a_{\text{AU}}^3$$

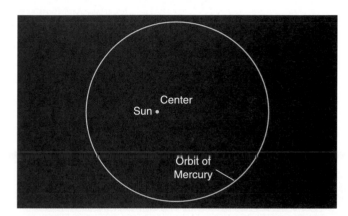

Figure 2-17
The orbits of the planets are nearly circular. Of the planets known to Kepler, Mercury has the most elliptical orbit. Pluto's orbit is only slightly more elliptical.

For example, Jupiter's average distance from the sun is roughly 5.2 AU. The semimajor axis cubed would be about 140.6, so the period must be the square root of 140.6, about 11.8 years.

It is important to notice that Kepler's three laws are empirical. That is, they describe a phenomenon without explaining why it occurs. Kepler derived them from Tycho's extensive observations, not from any fundamental assumption or theory. In fact, Kepler never knew what held the planets in their orbits or why they continued to move around the sun. His books are a fascinating blend of careful observation, mathematical analysis, and mystical theory.

The *Rudolphine Tables*

In spite of Kepler's recurrent involvement with astrology and numerology, he continued to work on the *Rudolphine Tables*. At last, in 1627, they were ready, and he financed their printing himself, dedicating them to the memory of Tycho Brahe. In fact, Tycho's name appears in larger type on the title page than Kepler's own. This is especially surprising when we recall that the tables were based on the heliocentric model of Copernicus and the elliptical orbits of Kepler, and not on the Tychonic system. The reason for Kepler's evident deference was Tycho's family, still powerful and still intent on protecting the memory of Tycho. They even demanded a share of the profits and the right to censor the book before publication, though they changed nothing but a few words on the title page.

The *Rudolphine Tables* was Kepler's masterpiece, the precise model of planetary motion that Copernicus had sought but failed to find. The accuracy of the *Rudolphine Tables* was strong evidence that both Kepler's model for planetary motion and the Copernican hypothesis for the place of the Earth were correct. Copernicus would have been pleased.

Kepler died on November 15, 1630. He had solved the problem of planetary motion, and his *Rudolphine Tables* demonstrated his solution. Although he did not understand the nature of motion, his three rules worked; and in science the only test of a theory is "Does it describe reality?" Kepler's laws have been used for almost four centuries as a true description of orbital motion.

REVIEW Critical Inquiry

What principles of classical astronomy made it difficult for Kepler to understand planetary motion?

Critical analysis is more than careful thinking. We can think carefully about what Kepler did, but until we analyze the paradigm (the network of assumptions and principles) within which he worked, we cannot understand how impressive his achievement was. Kepler lived at a time when every astronomer believed in uniform circular motion. The heavens were assumed perfect, and, according to classical philosophers such as Plato and Aristotle, circular motion at a constant rate is the only perfect motion. Before Kepler could adopt elliptical orbits and variable planetary speeds, he had to step outside that classical paradigm and see nature in a new way. Only then could he recognize how the planets really moved.

This makes us wonder what false paradigms surround us and prevent us from understanding nature better. Kepler was never troubled by geocentrism. From his days in college he accepted that the sun was the center of the universe. Of course, his first law partially violates a strict rule of heliocentrism. How?

Kepler was a Copernican, and he wrote a number of books that proclaimed his belief, but he was never seriously persecuted because he lived in northern Europe, beyond the sway of the Inquisition. Others were not so lucky. The next astronomer in our story was born in southern Europe under the rule of the Inquisition.

2-5 Galileo Galilei

Most people know two facts about Galileo, and both are wrong. Galileo did not invent the telescope, and he was not condemned by the Inquisition for believing Earth moves around the sun. Then why is Galileo so important that in 1979, almost 400 years after his trial, the Vatican reopened his case? As we discuss Galileo, we will discover that his trial concerned not just the place of the Earth and the motion of the planets, but also a new method of understanding nature, a method we call science.

Telescopic Observations

Galileo Galilei (Figure 2-18) was born in Pisa in 1564, and he studied medicine at the university there. His true love, however, was mathematics, and although he had to leave school early for financial reasons, he returned only four years later as a professor of mathematics. Three years after that he became professor of mathematics at the university at Padua, where he remained for 18 years.

During this time, Galileo seems to have adopted the Copernican model, although he admitted in a 1597 letter to Kepler that he did not support Copernicanism publicly because of the criticism such a declaration would bring. It was the telescope that drove Galileo to publicly defend the heliocentric model.

Galileo did not invent the telescope. It was apparently invented around 1608 by lens makers in Holland. Galileo, hearing descriptions in the fall of 1609, was able to build working telescopes in his workshop (Figure 2-19). In fact, Galileo was not the first person to look at the sky through a telescope, but he was the first person to observe the sky systematically and apply his observations to the theoretical problem of the day—the place of the Earth.

What Galileo saw through his telescopes was so amazing he rushed a small book into print. *Sidereus Nuncius* (*The Sidereal Messenger*) reported three major discoveries. First, the moon was not perfect. It had mountains and valleys on its surface, and Galileo used the shadows to calculate the height of the mountains. Aristotle's philosophy held that the moon was perfect,

Figure 2-18
Galileo Galilei (1564–1642), remembered as the great defender of Copernicanism, also made important discoveries in the physics of motion. He is honored here on an Italian 2000-lira note. The reverse side shows one of his telescopes at lower right and a modern observatory above it.

but Galileo showed that it was not only imperfect but was also a world like Earth.

The second discovery reported in the book was that the Milky Way was made up of a myriad of stars too faint to see with the unaided eye. While intriguing, this discovery could not match the third discovery. Galileo's telescope revealed four new "planets" circling Jupiter, planets that we know today as the Galilean moons of Jupiter.

The moons of Jupiter supported the Copernican model over the Ptolemaic model. Critics of Copernicus had said Earth could not move because the moon would be left behind. But Jupiter moved yet kept its satellites, so Galileo's discovery proved that Earth, too, could move and keep its moon. Also, Aristotle's philosophy included the belief that all heavenly motion was centered on Earth. Galileo showed that Jupiter's moons revolve around Jupiter, so there could be other centers of motion.

Soon after *Sidereus Nuncius* was published, Galileo made two additional discoveries. When he observed the sun, he discovered sunspots, raising the suspicion that the sun was less than perfect. Further, by noting the movement of the spots, he concluded that the sun

Figure 2-19
Galileo's telescopic discoveries generated intense interest and controversy. Some critics refused to look through a telescope lest it deceive them. *(Courtesy Yerkes Observatory)*

Chapter 2 The Origin of Modern Astronomy **51**

Figure 2-20
(a) If Venus moved in an epicycle on the Earth–sun line (see Figure 2-7), it would always appear as a crescent.
(b) Galileo's telescope showed that Venus goes through a full set of phases, proving that it must orbit the sun.

was a sphere and rotated on its axis. When he observed Venus, Galileo saw that it was going through phases like those of the moon. In the Ptolemaic model, Venus revolves around an epicycle that is located between Earth and the sun. Thus, it would always be seen as a crescent. But Galileo saw Venus go through a complete set of phases, which proved that it did indeed revolve around the sun (Figure 2-20).

Sidereus Nuncius was very popular and made Galileo famous. He became chief mathematician and philosopher to the Grand Duke of Tuscany in Florence. In 1611, Galileo visited Rome and was treated with great respect. He had long, friendly discussions with the powerful Cardinal Barberini, but he also made enemies. Personally, Galileo was outspoken, forceful, and sometimes tactless. He enjoyed debate, but most of all he enjoyed being right. Thus, in lectures, debates, and letters he offended important people who questioned his telescopic discoveries.

By 1616, Galileo was the center of a storm of controversy. Some critics said he was wrong, and others said he was lying. Some refused to look through a telescope lest it mislead them, and others looked and claimed to see nothing (hardly surprising given the awkwardness of those first telescopes). Pope Paul V decided to end the disruption, so when Galileo visited Rome in 1616 Cardinal Bellarmine interviewed him privately and ordered him to cease debate. There is some controversy today about the nature of Galileo's instructions, but he did not pursue astronomy for some years after the interview. Books relevant to Copernicanism were banned, although *De Revolutionibus,* recognized as an important and useful book in astronomy, was only suspended pending revision. Everyone who owned a copy of the book was required to cross out certain statements and add handwritten corrections stating that the motion of the Earth and the central location of the sun were only theories and not facts.

Dialogo and Trial

In 1621 Pope Paul V died, and his successor, Pope Gregory XV, died in 1623. The next pope was Galileo's friend Cardinal Barberini, who took the name Urban VIII. Galileo rushed to Rome hoping to have the prohibition of 1616 lifted, and although the new pope did not revoke the orders, he did encourage Galileo. Thus, Galileo began to write his great defense of the Copernican model, finally completing it on December 24, 1629. After some delay, the book was approved by both the local censor in Florence and the head censor of the Vatican in Rome. It was printed in February 1632.

Called *Dialogo Dei Due Massimi Sistemi* (*Dialogue Concerning the Two Chief World Systems*), it confronts the ancient astronomy of Aristotle and Ptolemy with the Copernican model. Galileo wrote the book as a debate among three friends. Salviati, a swift-tongued defender of Copernicus, dominates the book; Sagredo is intelligent but largely uninformed. Simplicio is the dismal defender of Ptolemy. In fact, he does not seem very bright.

The publication of *Dialogo* created a storm of controversy, and it was sold out by August 1632, when the Inquisition ordered sales stopped. The book was a clear defense of Copernicus, and, either intentionally or unintentionally, Galileo exposed the pope's authority to ridicule. Urban VIII was fond of arguing that, as God was omnipotent, He could construct the universe in any form while making it appear to us to have a different form, and thus we could not deduce its true nature by mere observation. Galileo placed the pope's argument in the mouth of Simplicio, and Galileo's enemies showed the passage to the pope as an example of Galileo's disrespect. The pope thereupon ordered Galileo to face the Inquisition.

Galileo was interrogated by the Inquisition four times and was threatened with torture. He must have thought often of Giordano Bruno, tried, condemned, and burned at the stake in Rome in 1600. One of Bruno's offenses had been Copernicanism. But the trial did not center on Galileo's belief in Copernicanism. *Dialogo* had been approved by two censors. Rather, the trial centered on the instructions given Galileo in 1616. From his file in the Vatican, his accusers produced a record of the meeting between Galileo and Cardinal Bellarmine that included the statement that Galileo was "not to hold, teach, or defend in any way" the principles of Copernicus. Many historians believe that this document, which was signed neither by Galileo nor by Bellarmine nor by a legal secretary, was a forgery and that Galileo's true instructions were much less restrictive. But Bellarmine was dead, and Galileo had no defense.

The Inquisition condemned him not for heresy but for disobeying the orders given him in 1616. On June 22, 1633, at the age of 70, kneeling before the Inquisition, Galileo read a recantation admitting his errors. Tradition has it that as he rose he whispered "E pur si muove" ("Still it moves"), referring to the Earth. Although he was sentenced to life imprisonment, he was actually confined at his villa for the next 10 years, perhaps through the intervention of the pope. He died there on January 8, 1642, 99 years after the death of Copernicus.

Galileo was not condemned for heresy, nor was the Inquisition interested when he tried to defend Copernicanism. He was tried and condemned on a charge we might call a technicality. Then why is his trial so important that historians have studied it for almost four centuries? Why have some of the world's greatest authors, including Bertolt Brecht, written about Galileo's trial? Why in 1979 did Pope John Paul II create a commission to reexamine the case against Galileo?

To understand the trial, we must recognize that it was the result of a conflict between two ways of understanding our universe. Since the Middle Ages, scholars had taught that the only path to true understanding was through religious faith. St. Augustine (AD 354–430) wrote "Credo ut intelligame," which can be translated as "Believe in order to understand." But Galileo and other scientists of the Renaissance used their own observations to try to understand the universe, and when their observations contradicted Scripture, they assumed their observations were correct. Galileo paraphrased Cardinal Baronius in saying, "The Bible tells us how to go to heaven, not how the heavens go."

The trial of Galileo was not about the place of the Earth. It was not about Copernicanism. It wasn't really about the instructions Galileo received in 1616. It was about the birth of modern science as a rational way to understand our universe (Figure 2-21). The commission appointed by John Paul II in 1979, reporting its conclusions in October 1992, said of Galileo's inquisitors, "This subjective error of judgment, so clear to us today, led them to a disciplinary measure from which Galileo 'had much to suffer.'" Galileo was not found innocent so much as the Inquisition was forgiven for having charged him in the first place.

REVIEW Critical Inquiry

How were Galileo's observations of the moons of Jupiter evidence against the Ptolemaic model?

Galileo presented his arguments in the form of evidence and conclusions, and the moons of Jupiter were key evidence. Ptolemaic astronomers argued from first principles

Figure 2-21
Although he did not invent it, Galileo will always be remembered along with the telescope as on these postage stamps. By depending on direct observation instead of the first principles of philosophy and scripture, Galileo led the way to the invention of modern astronomy and modern science as a way to know about the natural world.

given by Aristotle that the heavens were made up of perfect spheres rotating uniformly around Earth at the center. Then Earth could be the only center of motion. Although Ptolemy's epicycles and equants undermined this belief, it was still critical to classical astronomers. When Galileo saw four moons revolving around Jupiter, he knew that Earth was not the only center of motion in the universe, and thus at least that one aspect of classical astronomy had to be wrong.

Furthermore, Ptolemaic astronomers had argued that Earth could not move or it would leave the moon behind. The fact that the moon continued to circle Earth seemed to prove that Earth was immobile. But Galileo saw moons orbit Jupiter, and everyone agreed that Jupiter moved. Thus, if Jupiter could move and keep its moons, then Earth might move and keep its moon.

Of all of Galileo's telescopic observations, the moons of Jupiter caused the most debate. But the craters on the moon and the phases of Venus were also critical evidence. How did they argue against the Ptolemaic model?

Galileo faced the Inquisition because of a conflict between two ways of knowing and understanding the world. The Church taught faith and understanding through revelation, but the scientists of the age were inventing a new way to understand nature that relied on the evidence of observation and measurement. Thus, the problems of the place of the Earth and the motion of the planets triggered the birth of modern science. The final episode in that adventure involves a man who wondered what makes an apple fall.

2-6 Isaac Newton and Orbital Motion

We date the birth of modern astronomy and of modern science from the 99 years between the deaths of Copernicus and Galileo. The Renaissance is commonly taken to be the period between 1300 and 1600, and thus the 99 years of this story lie at the culmination of the reawakening of learning in all fields (Figure 2-22). Not

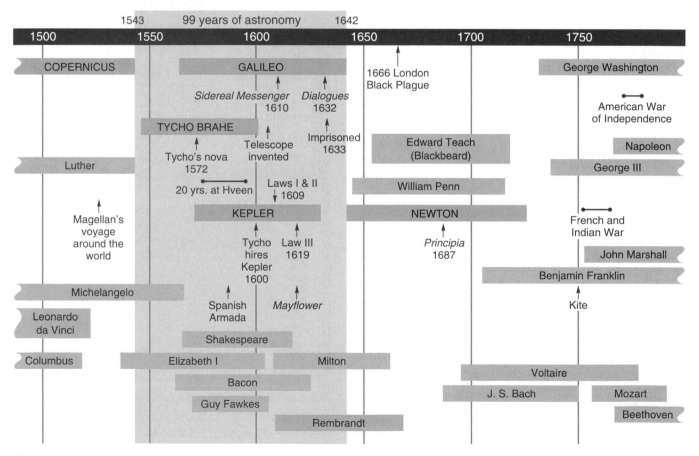

Figure 2-22
The 99 years between the death of Copernicus in 1543 and the birth of Newton in 1642 marked the transition from the ancient astronomy of Ptolemy and Aristotle to the revolutionary theory of Copernicus. This period saw the birth of modern scientific astronomy.

only did the world adopt a new model of the universe, but it also adopted a new way of understanding our place in nature.

The problem of the place of the Earth was resolved by the Copernican Revolution, but the problem of planetary motion was only partly solved by Kepler's laws. For the last 10 years of his life, Galileo studied the nature of motion, especially the accelerated motion of falling bodies. Although he made some important progress, he was not able to relate his discoveries about motion to the heavens. That final step fell to Isaac Newton.

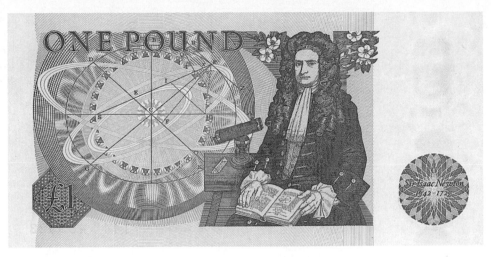

Figure 2-23
Isaac Newton (1642–1727), working from the discoveries of Galileo and Kepler, derived three laws of motion and the principle of mutual gravitation. He and some of his discoveries are honored on this English one-pound note.

Isaac Newton

Galileo died in January 1642. Some 11 months later, on Christmas day 1642,* a baby was born in the English village of Woolsthorpe. His name was Isaac Newton (Figure 2-23), and his life represented the first flower of the seeds planted by the four astronomers of our story.

Newton was a quiet child from a farming family, but his work at school was so impressive that his uncle financed his education at Trinity College, where he studied mathematics and physics. In 1665, the Black Plague swept through England, and the colleges were closed. During 1665 and 1666, Newton spent his time in Woolsthorpe, thinking and studying. It was during these years that he made most of his discoveries in optics, mechanics, and mathematics. Among other things, he studied optics, developed three laws of motion, divined the nature of gravity, and invented differential calculus. The publication of his work in his book *Principia* in 1687 placed science on a firm analytical base.

It is beyond the scope of this book to analyze all of Newton's work, but his laws of motion and gravity had an important impact on the future of astronomy. From his study of the work of Galileo, Kepler, and others, Newton extracted three laws that related the motion of a body to the forces acting on it (Table 2-2). These laws made it possible to predict exactly how a body would move if the forces were known (Window on Science 2-3).

*Because England had not yet reformed its calendar, December 25, 1642, in England was January 4, 1643, in Europe. It is only a small deception to use the English date and thus include Newton's birth in our 99-year history.

In thinking about the motion of the moon, Newton realized that some force had to pull the moon toward Earth's center. With no force to alter its motion, it would continue moving in a straight line and leave Earth forever. It could circle Earth only if Earth attracted it. Newton's insight was to recognize that the force that held the moon in its orbit was the same as the force that made apples fall from trees. The force of gravity was universal—that is, all objects attract all other objects with a force that depends only on their masses and the distance between their centers.

The **mass** of an object is a measure of the amount of matter in the object, usually expressed in kilograms. Mass is not the same as weight. An object's *weight* is the force that Earth's gravity exerts on the object. Thus an object in space far from Earth might have no weight, but it would contain the same amount of matter and would thus have the same mass that it would have on Earth.

Newton also realized that the distance between the masses was important. He recognized that the force of

---

Table 2-2 Newton's Three Laws of Motion

I. A body continues at rest or in uniform motion in a straight line unless acted upon by some force.

II. A body's change of motion is proportional to the force acting on it and is in the direction of the force.

III. When one body exerts a force on a second body, the second body must exert an equal and opposite force back on the first body.

One of the most often used and least often stated principles of science is cause and effect, and we could argue that Newton's second law of motion was the first clear statement of the principle. Ancient philosophers such as Aristotle argued that objects moved because of tendencies. They said that earth and water, and objects made of earth and water, had a natural tendency to move toward the center of the universe. This natural motion had no cause but was inherent in the nature of the objects. But Newton's second law says $F = ma$. If an object changes its motion (a in the equation), then it must be acted on by a force (F in the equation). Any effect (a) must be the result of a cause (F).

The principle of cause and effect goes far beyond motion. The principle of cause and effect gives scientists confidence that every effect has a cause. Hearing loss in certain laboratory rats, color changes in certain chemical dyes, and explosions on certain stars are all effects that must have causes. All of science is focused on understanding the causes of the effects we see. If the universe were not rational, then we could never expect to discover causes. Newton's second law of motion was arguably the first statement that the behavior of the universe depends rationally on causes.

---

gravity had to decrease as the square of the distance between the objects increased. Specifically, if the distance from, say, Earth to the moon were doubled, the gravitational force between them would decrease by a factor of 2^2, or 4. If the distance were tripled, the force would decrease by a factor of 3^2, or 9. This relationship is known as the **inverse square relation.** (We will discuss it in more detail in Chapter 11 where we apply it to the intensity of light.)

With these definitions of mass and the inverse square relation, we can describe Newton's law of gravity in a simple equation:

$$F = -G\,\frac{Mm}{r^2}$$

Here M and m are the masses of two objects, r is the distance between their centers, G is the gravitational constant, and F is the force of gravity acting between the two objects.

Orbital Motion

In developing the three laws of motion and the theory of gravity, Newton resolved the problem Kepler had toyed with—why the planets move along their orbits. In fact, Newton's work explained orbital motion and made it possible to analyze the motions of the heavens with great precision.

To illustrate the principle of orbital motion, imagine that we place a large cannon at the top of a mountain, point the cannon horizontally, and fire it (Figure 2-24). The cannonball falls to Earth some distance from the foot of the mountain. The more gunpowder we use, the faster the ball travels, and the farther from the foot of the mountain it falls. If we use enough powder, the ball travels so fast it never strikes the ground. Our planet's gravity pulls it toward Earth's center, but Earth's surface curves away from it at the same rate at which it falls. Thus we say it is in orbit. If the cannonball is high above the atmosphere, where there is no friction, it will fall around Earth forever.

Simple physics says that an object in motion tends to stay in motion in a straight line unless acted upon by some force. Thus the cannonball in this example trav-

Figure 2-24
A cannon on a high mountain could put its projectile into orbit if it could achieve a high enough velocity. Newton published a similar figure in *Principia*.

When you read about any science, you should notice that scientific theories face in two directions. They look back into the past and explain phenomena we have previously observed. For example, Newton's laws of motion and gravity explained how the planets move. But theories also face forward in that they enable us to make predictions about what we should find as we explore further. Thus, Newton's laws allowed astronomers to calculate the orbits of comets, predict their return, and eventually understand their origin.

Scientific predictions are important in two ways. First, if a theory leads to a prediction and scientists later discover the prediction was true, the theory is confirmed, and scientists gain confidence that it is a true description of nature. But predictions are important in science in a second way. Using an existing theory to make a prediction may lead us into an unexplored avenue of knowledge. Thus, the first theories of genetics made predictions that confirmed the genetic theory of inheritance, but those predictions also created a new understanding of how living creatures evolve.

As you read about any scientific theory, think about both what it can explain and what it can predict.

els in a curve around Earth only because Earth's gravity acts to pull it away from its straight-line motion.

Newton's explanation of orbital motion is only one example of the power of his conception of nature. His laws of motion and gravity were *general* laws that described the motions of all bodies under the action of external forces. In addition, the laws were *productive* because they made possible specific calculations that could be tested by observation. For example, Newton's laws of motion could be used to derive Kepler's third law from the law of gravitation.

Newton's discoveries remade astronomy into an analytical science in which astronomers could measure the positions and motions of celestial bodies, calculate the gravitational forces acting on them, and predict their future motion (Window on Science 2-4).

Were we to trace the history of astronomy after Newton, we would find scientists predicting the motion of comets, the gravitational interaction of the planets, the motions of double stars, and so on. Astronomers built on the discoveries of Newton, just as he had built on the discoveries of Copernicus, Tycho, Kepler, and Galileo. It is the nature of science to build on the discoveries of the past, and Newton was thinking of that when he wrote, "If I have seen farther than other men, it is because I stood upon the shoulders of giants."

tracted the moon, then the moon had to attract the Earth. That is, gravitation had to be *mutual* between any two objects.

When Newton looked beyond the Earth–moon system, he realized that Earth's gravity must attract the sun as well as the moon, and if Earth attracts the sun, then the sun had to attract Earth. Furthermore, if the sun has gravity that acts on Earth, then the same gravity would act on all the planets and even on the distant stars. Step by step, Newton's third law of motion led him to conclude that gravitation had to apply to all masses in the universe. That is, it had to be *universal*.

Newton's explanation of orbital motion was powerful because he explained it in terms of common processes known on Earth's surface. For example, why is it correct to say that the moon is falling?

Newton said he stood on the shoulders of giants, and we too have that vantage point. We understand what we see in the sky because we understand the work of Copernicus, Kepler, and Newton—not just their discoveries, but also their creation of a new way to understand nature. Modern astronomers gaze at the sky with powerful modern tools, the subject of the next chapter.

REVIEW   Critical Inquiry

What do the words *universal* and *mutual* mean in the term *universal mutual gravitation*?

Newton realized that gravity was a force that drew the moon toward the Earth. But his third law of motion said that forces always occur in pairs, so if the Earth at-

Summary

Classical astronomy was based on the writings of the Greek philosopher Aristotle. He taught that Earth was the immobile center of the universe and that the stars and planets were carried around Earth by great crystalline spheres. This model of the universe was given mathematical form about AD 140 in the *Almagest,* the great work of Ptolemy. Ptolemy preserved the classical belief in geocentrism, but he replaced the concentric, Earth-centered heavenly spheres with a system of epicycles, deferents, and equants, and thus tried to create a mathematical model that could accurately predict the positions of the sun, moon, and planets.

In contrast to the geocentric universe of classical astronomy, the universe devised by Copernicus was heliocentric, or sun-centered. One advantage of a heliocentric universe is that retrograde motion, the occasional westward motion of the planets, is easily explained. Copernicus did not publish his book *De Revolutionibus* until 1543, the year he died. The teachings of Aristotle had become part of church doctrine. As a critic of the central place of the Earth, Copernicus was exploring controversial ideas, ideas that some would claim were heretical.

The Danish astronomer Tycho Brahe did not accept the Ptolemaic or the Copernican model but rather developed his own, in which the sun and moon circled Earth and the planets circled the sun. Although his hypothesis was not correct, Tycho made precise observations of planetary positions that later led to a true understanding of planetary motion.

Johannes Kepler, Tycho Brahe's assistant, inherited the Danish astronomer's records in 1601 and used his observations to uncover three laws of planetary motion. Kepler discovered that the planets follow ellipses with the sun at one focus, that they move faster when near the sun, and that a planet's period squared is proportional to its orbital radius cubed.

Galileo Galilei was a great defender of the Copernican hypothesis. Galileo was the first person to use a telescope to observe the heavens and to recognize the significance of what he saw. His discoveries of the phases of Venus, the satellites of Jupiter, the mountains of the moon, and other phenomena helped undermine the Ptolemaic universe. In 1633, Galileo was finally condemned before the Inquisition for refusing to halt his defense of Copernicanism.

Born in 1642, the same year that Galileo died, Isaac Newton used the work of Kepler and Galileo to discover three laws of motion and the theory of universal gravitation. These laws made it possible to understand the orbital motions of the planets as the consequence of the sun's gravity. In addition, Newton's work made it possible to analyze the motion of any celestial body and predict its path in the future.

The 99 years from the death of Copernicus to the birth of Newton marked the birth of modern science. From that time on, science depended on evidence to support theories and relied on the analytic methods first demonstrated by Newton.

New Terms

parallax	paradigm
geocentric universe	hypothesis
uniform circular motion	theory
retrograde motion	natural law
epicycle	ellipse
deferent	mass
equant	inverse square relation
heliocentric universe	

Review Questions

1. Why did Greek astronomers conclude that the heavens were made up of perfect crystalline spheres moving at constant speeds?

2. Why did classical astronomers conclude that Earth had to be motionless?

3. How did the Ptolemaic model explain retrograde motion?

4. In what ways were the models of Ptolemy and Copernicus similar?

5. Why did the Copernican hypothesis win gradual acceptance?

6. Why is it difficult for scientists to replace an old paradigm with a new paradigm?

7. Why did Tycho Brahe expect the new star of 1572 to show parallax? Why was the lack of parallax evidence against the Ptolemaic model?

8. How was Brahe's model of the universe similar to the Ptolemaic model? How did it resemble the Copernican model?

9. Explain how Kepler's laws contradict uniform circular motion.

10. What is the difference between a hypothesis, a theory, and a law?

11. How did the *Alfonsine Tables,* the *Prutenic Tables,* and the *Rudolphine Tables* differ?

12. Review Galileo's telescope discoveries, and explain why they supported the Copernican model and contradicted the Ptolemaic model.

13. Galileo was condemned by the Inquisition, but Kepler, also a Copernican, was not. Why not?

14. Why did Newton conclude that gravitation had to be universal?

15. Explain why we might describe the orbital motion of the moon with the statement, "The moon is falling."

Discussion Questions

1. Why might Copernicus have delayed publication of his book? Discuss as many reasons as possible.

2. What parallels and contrasts do you see between the Vatican's recent reconsideration of Galileo's trial and the controversy over teaching evolution in public schools?

3. Why might Tycho have hesitated to hire Kepler? Why did Tycho appoint Kepler, a commoner and a Copernican, to be his scientific heir?

4. Tycho's description of the universe is usually called the Tychonic theory. What should it be called—a hypothesis, a model, a theory, or a law?

Problems

1. If you lived on Mars, which planets would describe retrograde loops? Which would never be visible as crescent phases?

2. Galileo's telescope showed him that Venus has a large angular diameter (61 seconds of arc) when it is a crescent and a small angular diameter (10 seconds of arc) when it is nearly full. Use the small-angle formula to find the ratio of its maximum distance to its minimum distance. Is this ratio compatible with the Ptolemaic universe shown in Figure 2-7?

3. Galileo's telescopes were not of high quality by modern standards. He was able to see the rings of Saturn, but he never reported seeing features on Mars. Use the small-angle formula to find the angular diameter of Mars when it is closest to Earth. How does that compare with the maximum diameter of Saturn's rings?

4. If a planet had an average distance from the sun of 4 AU, what is its orbital period?

5. If a space probe is sent into an orbit around the sun that brings it as close as 0.5 AU and as far away as 5.5 AU, what will be its orbital period?

6. Pluto orbits the sun with a period of 247.7 years. What is its average distance from the sun?

Critical Inquiries for the Web

1. The trial of Galileo is an important event in the history of science. We now know, and the Church now recognizes, that Galileo's view was correct, but what were the arguments on both sides of the issue as it was unfolding? Research the Internet for documents chronicling the trial, Galileo's observations and publications, and the position of the Church. Use this information to outline a case for and against Galileo in the context of the times in which the trial occurred.

2. It's hard to imagine that an observatory could exist before the invention of the telescope, but Tycho Brahe's observatory at Hveen was a great astronomical center of its day. Search for Web sites on Tycho and his instruments, and describe what an observing session at Hveen might have involved.

Exploring *The Sky*

1. The three laws of Johannes Kepler mathematically describe the motion of the planets around the sun. The following activities help to get a qualitative and visual sense of what these laws mean. In *The Sky*, use a feature called the "3-D Solar System Mode," which gives you a three-dimensional view of the solar system as it might be viewed from a spaceship far from the sun. Set the inner solar system (planets out to Mars) in motion using a time increment of 1 day.

   a. How is Kepler's first law demonstrated in the simulation?

   b. How is Kepler's third law demonstrated in the simulation? Qualitatively, Kepler's third law states that planets further from the sun take longer to make one orbit about the sun.

   c. Use the solar system simulation to estimate the length of the year of Mercury, Earth, and Mars. You should be able to find these values to within a day or two. Compare your results to values given in the table "Properties of Planets" in Appendix A.

   How to proceed: To view the 3-D solar system simulation, click on the "View" menu, and select "3-D Solar System Mode." Or you can simply click on the "3-D Solar System Mode" button on the View Toolbar. Use the magnifier button to zoom in on the inner solar system. Then set the time increment to 1 day on the Time Skip toolbar, and set the system in motion. It is helpful to have a grid in place to give you spatial reference: Press the View menu button, select "Reference Lines," and under "General Lines," select "Ecliptic." Also note that you can change the viewing perspective using the arrow keys on Orientation Toolbar.

2. Mars is one of the planets for which retrograde motion can be readily observed. From our Earth-based viewing platform, planets usually seem to move eastward against the background stars. However, near the time of opposition, when Mars appears opposite from the sun, Mars temporarily seems to move *westward* against the background stars—that is retrograde motion (see Fig 2-9 for a visual explanation). After a few months, the apparent motion of Mars reverts back to its usual *eastward* motion against background stars. To help to clarify this possibly confusing situation, try the following activities:

   a. Use *The Sky* to simulate retrograde motion of Mars between January 1, 1999, and August 1, 1999. Note the initial apparent eastward motion of Mars, then the westward retrograde motion, and then a reversion to its more usual eastward motion against the background stars. Compare the results of the computer simulation to Figure 2-9.

   How to proceed: Click on "Data" of the menu bar, then on "Site Information," and set the date to 1/1/99, the time to 7:00. Then set the time increment for the Time Skip simulation to a sidereal day (23h 56m 04s). Because the sidereal day is the time it takes Earth to rotate once on its axis with respect to the stars, the stars will not move in the simulation—that is, we have effectively "stopped" Earth's rotation! Orient yourself to look southward. Before you begin the actual simulation, make sure that Mars is on your screen.

   b. Find the dates of the beginning and end of retrograde motion.

 Go to the Brooks/Cole Astronomy Resource Center (www.brookscole.com/astronomy) for critical thinking exercises, articles, and additional readings from InfoTrac College Edition, Brooks/Cole's online student library.

Astronomical Tools

He burned his house down for the

fire insurance and spent the proceeds

on a telescope.

Robert Frost
The Star-Splitter

Guidepost

Chapter 1 described the sky as it appears to our unaided eyes, but modern astronomers turn powerful telescopes on the sky. This chapter introduces us to the modern astronomical telescope and its delicate instruments.

The study of the universe is so challenging that astronomers cannot ignore any source of information, and so they use the entire spectrum, from gamma rays to radio waves. This chapter shows how critical it is for astronomers to understand the nature of light.

In each of the 16 chapters that follow, we will study the universe using information gathered by the telescopes and instruments described in this chapter.

Starlight is going to waste. Every night, light from the stars falls on trees, oceans, roofs, and empty parking lots, and it is all wasted. To an astronomer, nothing is so precious as starlight. It is our only link to the sky, and the astronomer's quest is to gather as much starlight as possible and extract from it the secrets of the universe.

The telescope is the emblematic tool of the astronomer because its purpose is to gather and concentrate light for analysis. Nearly all of the interesting objects in the sky are faint sources of light, so modern astronomers are driven to build the largest possible telescopes to gather the maximum amount of light. Thus, our discussion of astronomical tools is concentrated on large telescopes and the specialized tools used to analyze light.

If we wish to gather visible light, a normal telescope will do; but as we will see in this chapter, visible light is only one kind of radiation. We can extract information from other forms of radiation by using specialized telescopes. Radio telescopes, for example, give us an entirely different view of the sky. Some of these specialized telescopes can be used from Earth's surface, but some must go into orbit above Earth's atmosphere. Telescopes that observe X rays, for instance, must be placed in orbit.

In addition to using telescopes to gather light, astronomers need instruments to analyze light. Such instruments measure brightness, record images, or form spectra.

Astronomers no longer study the sky by mapping constellations, charting the phases of the moon, or following the retrograde motion of the planets. Modern astronomers use the most sophisticated telescopes and instruments to analyze starlight. Thus, we begin this chapter with a discussion of the nature of light.

3-1 Radiation: Information from Space

A book on bread baking might begin with a discussion of flour, so our chapter on telescopes begins with a discussion of light—not just visible light, but the entire range of radiation from the sky.

Light as a Wave and a Particle

If you have admired the colors in a soap bubble, you have seen light behave as a wave. But when that light enters the light meter on a camera, it behaves as a particle. How it behaves depends on how we observe it—it is both wave and particle.

We experience waves whenever we hear sound. Sound waves are a mechanical disturbance that travels through the air from source to ear. Sound requires a medium; so on the moon, where there is no air, there can be no sound. In contrast, light is made up of electric and magnetic waves that can travel through empty space. Unlike sound, light does not require a medium and thus can travel through a perfect vacuum.

Because light is made up of both electric and magnetic fields, we refer to it as **electromagnetic radiation.** As we will see later, visible light is only one form of electromagnetic radiation. These oscillating electric and magnetic fields travel through space at about 300,000 km/s (186,000 mi/s). This is commonly referred to as the speed of light c, but it is in fact the speed of all such radiation.

Electromagnetic radiation is a wave phenomenon—that is, it is associated with a periodically repeating disturbance, a wave. We are familiar with waves in water: If we disturb a pool of water, waves spread across the surface. Imagine that we use a meter stick to measure the distance between the successive peaks of a wave. This distance is the **wavelength,** usually represented by the Greek letter lambda (λ) (Figure 3-1).

While light does behave as a wave, it also behaves as a particle. We refer to a "particle of light" as a **photon,** and we can think of a photon as a bundle of waves.

The amount of energy a photon carries depends inversely on its wavelength. That is, shorter wavelength photons carry more energy, and longer wavelength photons carry less. We can express this relationship in a simple formula:

$$E = \frac{hc}{\lambda}$$

Here h is Planck's constant (6.6262×10^{-34} joule s) and c is the speed of light (3×10^8 m/s). The important point in this formula is not the value of the constants but the inverse relationship between the energy E and

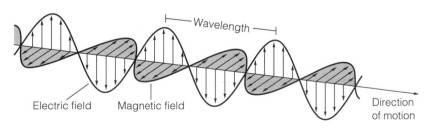

Electric field   Magnetic field   Direction of motion

Figure 3-1
Electric and magnetic fields travel together through space as electromagnetic radiation. The wavelength, commonly represented by the Greek letter lambda (λ), is the distance between successive peaks in the wave.

the wavelength λ. As λ gets smaller, E must get larger. Thus a photon of visible light carries a very small amount of energy, but a photon with a very short wavelength can carry much more energy.

The Electromagnetic Spectrum

A spectrum is an array of electromagnetic radiation in order of wavelength. We are most familiar with the spectrum of visible light—which we can see in rainbows—but the visible spectrum is merely a small segment of the much larger electromagnetic spectrum (Figure 3-2).

The average wavelength of visible light is about 0.0005 mm. We could put 50 light waves end to end across the thickness of a sheet of household plastic wrap. It is too awkward to measure such short distances in millimeters, so we will measure the wavelength of light using the **nanometer (nm),** one billionth of a meter (10^{-9} m). Another unit that astronomers commonly use is called the **angstrom (Å)** (named after the Swedish astronomer Anders Jonas Ångström). One angstrom is 10^{-10} m. The wavelength of visible light ranges between 400 nm and 700 nm or between 4000 Å and 7000 Å.

Radio astronomers often refer to wavelength using meters, centimeters, or millimeters.

Just as we sense the wavelength of sound as pitch, we sense the wavelength of light as color. Light near the short-wavelength end of the visible spectrum (400 nm) looks violet to our eyes, and light near the long-wavelength end (700 nm) looks red (see Figure 3-2).

Beyond the red end of the visible spectrum lies infrared radiation, where wavelengths range from 700 nm to about 1 mm. Our eyes are not sensitive to this radiation, but our skin senses it as heat. A "heat lamp" is nothing more than a bulb that gives off large amounts of infrared radiation.

Beyond the infrared part of the electromagnetic spectrum lie radio waves. The radio radiation used for AM radio transmissions has wavelengths of a few kilometers down to a few hundred meters, while FM, television, military, governmental, and ham radio transmissions have wavelengths that range down to a few tens of centimeters. Microwave transmissions, used for radar and long-distance telephone communications, for instance, have wavelengths from a few centimeters down to about 1 mm.

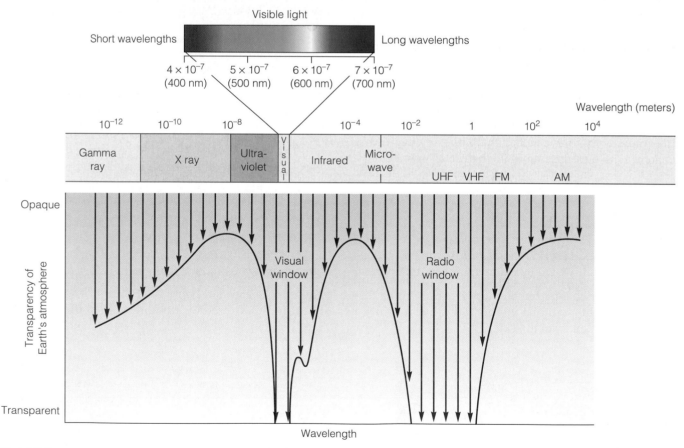

Figure 3-2
The spectrum of visible light, extending from red to violet, is only part of the electromagnetic spectrum. Most radiation is absorbed in Earth's atmosphere, and only radiation in the visual window and the radio window can reach Earth's surface.

You may not think of radio waves in terms of wavelength because radio dials are marked in units of frequency, the number of waves that pass a stationary point in one second. To calculate the wavelength of a radio wave, divide the speed of light by the frequency. Thus, when you tune in your favorite FM station at 89.5 MHz (million cycles per second), you are adjusting your radio to detect radio waves with a wavelength of 335 cm.

The distinction between the wavelength ranges is not sharp. Long-wavelength infrared radiation and the shortest microwave radio waves are the same. Similarly, there is no clear division between the short-wavelength infrared and the long-wavelength part of the visible spectrum. It is all electromagnetic radiation.

At wavelengths shorter than violet, we find ultraviolet radiation, with wavelengths ranging from 400 nm down to about 10 nm. At even shorter wavelengths lie X rays and gamma rays. Again, the boundaries between these wavelength ranges are not clearly defined.

X rays and gamma rays can be dangerous, and even ultraviolet photons have enough energy to do us harm. Small doses produce a suntan; larger doses can cause sunburn, and extreme doses might produce skin cancers. Contrast this to the lower-energy infrared photons. Individually, they have too little energy to affect skin pigment, a fact that explains why you can't get a tan from a heat lamp. Only by concentrating many low-energy photons in a small area, as in a microwave oven, can we transfer significant amounts of energy.

We are interested in electromagnetic radiation because it brings us clues to the nature of stars, planets, and other celestial objects. Only a small part of this radiation, however, in the form of visible light and some radio waves, can get through Earth's atmosphere and reach the surface of Earth; other wavelengths are absorbed. The highest parts of the atmosphere absorb X rays, gamma rays, and some radio waves, and a layer of ozone (O_3) at an altitude of about 30 km absorbs ultraviolet radiation. In addition, water vapor in the lower atmosphere absorbs infrared radiation. The two wavelength regions in which our atmosphere is transparent are called **atmospheric windows.** Obviously, if we wish to study the sky from Earth's surface, we must look out through one of these windows (Figure 3-2).

Now that we know something about electromagnetic radiation, we can study the tools astronomers use to gather and analyze that radiation.

3-2 Optical Telescopes

Astronomers build optical telescopes to gather light and focus it into sharp images. This requires sophisticated optical and mechanical designs, and it leads astronomers to build gigantic telescopes on top of high mountains. To begin, we need to understand the terminology of telescopes, but it is more important to understand how different kinds of telescopes work and why some are better than others.

Two Kinds of Telescopes

Because there are two ways to focus light, there are two kinds of astronomical telescopes. **Refracting telescopes** use a large lens to gather and focus the light, while **reflecting telescopes** use a concave mirror (Figure 3-3). The advantages of the reflecting telescope have made it the preferred design for modern observatories.

Either a lens or a mirror can focus light from a distant star into a sharp image (Figure 3-4). The lens bends (refracts) the light as it passes through the glass and brings it to a focus to form a small inverted image. A concave piece of glass with a reflective surface can also form an image by reflecting the light before it enters the glass. In either case, the **focal length** is the distance from the lens or mirror to the image formed of a distant light source, such as a star (Figure 3-5). Short-focal-length lenses and mirrors must be strongly curved, and long-focal-length lenses and mirrors are less strongly curved. Grinding the proper shape on a lens or mirror is a delicate, time-consuming, and expensive process.

The main lens in a refracting telescope is called the **primary lens,** and the main mirror in a reflecting telescope is called the **primary mirror.** These are also called the **objective lens** or **mirror.** Both kinds of telescopes form a very small, inverted image that is difficult to observe directly, so astronomers use a small lens called

Figure 3-3
(a) A refracting telescope uses a primary lens to focus starlight into an image that is magnified by a lens called an eyepiece. The primary lens has a long focal length, and the eyepiece has a short focal length. (b) A reflecting telescope uses a primary mirror to focus the light by reflection. A small secondary mirror reflects the starlight back down through a hole in the middle of the primary mirror to the eyepiece.

the **eyepiece** to magnify the image and make it convenient to view.

Refracting telescopes suffer from a serious optical distortion that limits their usefulness. When light is refracted through glass, shorter wavelengths bend more than longer wavelengths, and blue light comes to a focus closer to the lens than does red light (Figure 3-6). If we focus the eyepiece on the blue image, the red light is out of focus and we see a red blur around the image. If we focus on the red image, the blue light blurs. The color separation is called **chromatic aberration.** Telescope designers can grind a telescope lens of two components made of different kinds of glass and so bring two different wavelengths to the same focus. This does improve the image, but these **achromatic lenses** are not totally free of chromatic aberration because other wavelengths still blur. Telescopes made with such lenses were popular until the end of the 19th century.

The primary lens of a refracting telescope is very expensive to make because it must be achromatic, and the glass must be pure and flawless because the light passes through the lens. The four surfaces must be ground precisely, and the lens can be supported only along its edge. The largest refracting telescope in the world is at Yerkes Observatory in Wisconsin. Its lens, 1 m (40 inches) in diameter, weighs half a ton, and the glass sags under its own weight. Larger refracting telescopes are prohibitively expensive.

Reflecting telescopes are much less expensive because only the front surface of the mirror must be ground

Figure 3-4
(a) To see how a lens can focus light, we trace three light rays from the flame and base of a candle through a lens, where they are refracted to form an inverted image. (b) A mirror forms an image by reflection from a concave surface. Notice that the light is reflected from the aluminized front surface of the mirror and does not enter the glass. Thus, mirrors do not produce chromatic aberration.

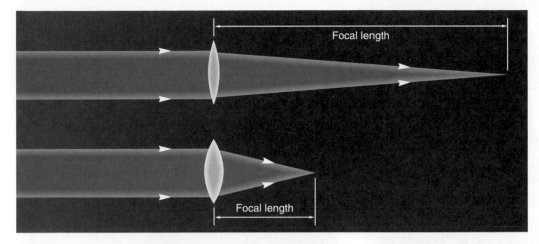

Figure 3-5
The focal length of a lens is the distance from the lens to the point where parallel rays of light come to a focus. The lens at the top has a longer focal length than does the lens at the bottom.

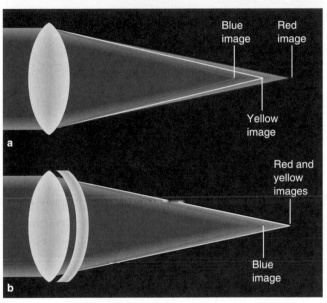

Figure 3-6
(a) A normal lens suffers from chromatic aberration because short wavelengths bend more than long wavelengths. (b) An achromatic lens, made in two parts, can bring any two colors to the same focus, but other colors remain slightly out of focus.

Figure 3-7
The mirror of a reflecting telescope forms the image in the middle of the telescope tube, so astronomers have devised various focal arrangements to bring the light to a convenient place for observing. Observing in a prime focus cage is possible only in the largest telescopes. Most large telescopes use the Cassegrain focus, but many amateur astronomers use the smaller Schmidt-Cassegrain telescopes.

to precise shape. The glass of the mirror need not be perfectly transparent, and the mirror can be supported over its back surface to reduce sagging. Most important, reflecting telescopes do not suffer from chromatic aberration because the light is reflected toward the focus before it can enter the glass. Thus, every large astronomical telescope built since 1900 has been a reflecting telescope.

The only difficulty with a reflecting telescope is looking through it. The mirror forms an image in the middle of the telescope tube where it is inconvenient to place your eye. Isaac Newton solved this problem by using a small mirror to direct the light out the side of the telescope tube (Figure 3-7), and these **Newtonian telescopes** have been popular for smaller instruments.

In the largest telescopes, an astronomer can ride in a cage inside the tube to make observations at the **prime focus.** In the **Cassegrain telescope,** the light is reflected back through a hole in the middle of the primary mirror to a convenient observing location. Many large telescopes are Cassegrains (Figure 3-8). **Schmidt-Cassegrain**

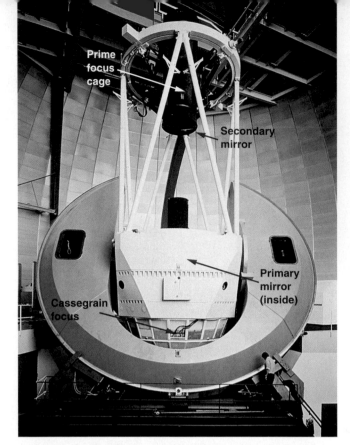

Figure 3-8
The 4-meter Mayall telescope at Kitt Peak National Observatory can be used as a prime focus telescope or, with the secondary mirror in place, as a Cassegrain telescope. Note the human figure at the lower right for scale. (© *Association of Universities for Research in Astronomy, Inc., Kitt Peak National Observatory*)

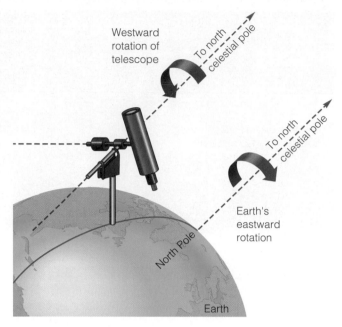

Figure 3-9
Westward motion around the polar axis of an equatorial mounting counters Earth's eastward rotation and keeps the telescope pointed at a given star.

telescopes, using a thin lens added to a Cassegrain telescope, are popular with amateur astronomers. Many other focal arrangements are used for special purposes, such as photography.

In addition to gathering and focusing light, astronomical telescopes must move easily and smoothly to point anywhere in the sky, and that places special restrictions on the machinery that supports the telescope.

Telescope Mountings

A telescope mounting can be more expensive than the telescope itself. The mounting must support the optics, protect against vibration, move accurately to designated objects, and then compensate for Earth's rotation. Because Earth rotates eastward, the telescope mounting must contain a **sidereal drive** (literally, a star drive) that can move the telescope smoothly westward to follow the object being studied. To simplify this motion, most telescope mountings are **equatorial mountings** with one axis of rotation, the **polar axis,** parallel to Earth's axis. Rotation around the polar axis moves the telescope parallel to the celestial equator (Figure 3-9).

Improvements in computer technology are making telescope mountings and observatories much less expensive. Instead of building a large, awkward equatorial mounting inside a large, heavy dome, astronomers now take advantage of the short focal lengths typical of reflecting telescopes. Such telescopes are so short they can be mounted like a cannon; that is, the mounting can move the telescope both in altitude (perpendicular to the horizon) and in azimuth (parallel to the horizon). Such **alt-azimuth mountings** are compact, but they must move simultaneously about both axes at varying rates in order to track stars in different parts of the sky. This complicated motion is made possible by computers that guide the telescope.

Such computer-controlled alt-azimuth mountings are stronger and smaller and do not require buildings as large as those required by equatorial mountings. In fact, many of the newest telescopes are designed so that the telescope is part of the observatory building. Like the turret of a tank, the entire building rotates as the telescope moves.

Nearly all large astronomical telescopes and many smaller ones are operated from control rooms. Thinking of the old days of observing all through the night in the unheated dome beside their telescopes, astronomers often refer to these control rooms as "warm rooms." Comfort, however, is not the main goal. Astronomical telescopes have become so large and so sophisticated that computers are needed to control their motions and record data, and computers, unlike astronomers, must

Have you ever seen a movie in which the hero magnifies a newspaper photo and reads some tiny detail? It isn't really possible, because newspaper photos are made up of tiny dots of ink, and no detail smaller than a single dot will be visible no matter how much you magnify the photo. In fact, all images are made up of elements of some sort, and that means there is a limit to the amount of detail you can see in an image. In an astronomical image, the resolution is often set by "seeing." It is foolish to attempt to see a detail in the image that is smaller than the resolution.

This limitation is true of all measurements in science. A zoologist might be trying to measure the length of a live snake, or a sociologist might be trying to measure the attitudes of people toward drunk driving, but both face limits to the resolution of their measurements. The zoologist might specify that the snake was 43.28932 cm long, and the sociologist might say that 98.2491 percent of people oppose drunk driving, but a critic might point out that it isn't possible to make these measurements that accurately. The resolution of the techniques does not justify the accuracy implied.

Science is based on measurement, and whenever we make a measurement we should ask ourselves how accurate that measurement can be. The accuracy of the measurement is limited by the resolution of the measurement technique, just as the amount of detail in a photograph is limited by the resolution of the photo.

---

be protected from extremes of temperature and humidity. From a control room, astronomers can work through the night controlling the computers that control the telescope.

Some telescopes can be operated from control rooms in remote locations. The two giant Keck telescopes sit at 13,600 feet (4100 m) atop an extinct volcano in Hawaii. Astronomers can observe with the telescopes from sea-level control rooms at the foot of the mountain. Other telescopes can be operated by astronomers from computers in their offices on the other side of the world. A few telescopes are designed so that computers can operate them every night with no human present, and the astronomer merely sends instructions to the computer before dark falls and retrieves the data the next day.

Astronomical telescopes are not only large and heavy, but they require intricate computer control systems. We might wonder what advantages they give us.

The Powers of a Telescope

A telescope can aid our eyes in only three ways. These are called light-gathering power, resolving power, and magnifying power.

Most interesting celestial objects are faint sources of light, so we need a telescope that can gather large amounts of light to produce a bright image. **Light-gathering power** refers to the ability of a telescope to collect light. Catching light in a telescope is like catching rain in a bucket—the bigger the bucket, the more rain it catches (Figure 3-10). This is why astronomers use large telescopes and why they refer to telescopes by their diameters.

The second power, **resolving power,** refers to the ability of the telescope to reveal fine detail. Whenever light is focused into an image, a blurred fringe surrounds the image (Figure 3-11). Because this **diffraction fringe** surrounds every point of light in the image, we can never see any detail smaller than the fringe (Window on Science 3-1). There is nothing we can do to eliminate diffraction fringes; they are produced by the wave

Figure 3-10
Gathering light is like catching rain in a bucket. A large-diameter telescope gathers more light and has a brighter image than does a smaller telescope of the same focal length.

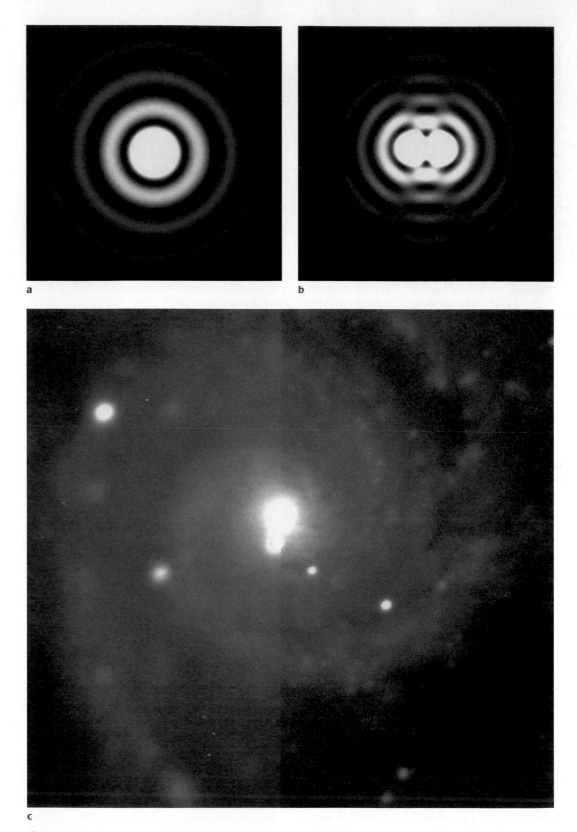

a

b

c

Figure 3-11
Diffraction within the telescope and the stability of Earth's atmosphere limit the detail we can see through a telescope. (a) Stars are so far away that their images are points, but the wave nature of light surrounds each point with diffraction fringes (much magnified in this model). (b) Two stars close to each other have overlapping diffraction fringes and become impossible to detect separately. *(Computer graphics by M. A. Seeds)* (c) Seeing, the distortion caused by Earth's atmosphere, can reduce the detail visible through a telescope. Note that the left half of this galaxy image was recorded under worse seeing conditions than the right half. *(Courtesy William Keel)*

nature of light. If we use a large-diameter telescope, however, the fringes are smaller, and we can see smaller details. Thus the larger the telescope, the better its resolving power.

Two other factors—optical quality and atmospheric conditions—limit the detail we can see. A telescope must contain high-quality optics to achieve its full potential resolving power. Even a large telescope shows us little detail if its optics are marred with imperfections. In addition, when we look through a telescope, we look through miles of turbulent air in Earth's atmosphere, which makes the image dance and blur, a condition called **seeing.** A related phenomenon is the twinkling of a star. The twinkles are caused by turbulence in Earth's atmosphere, and a star near the horizon, where we look through more air, will twinkle more than a star overhead.

On a night when the atmosphere is unsteady, the stars twinkle, the images are blurred, and the seeing is bad. Even under good seeing conditions, the detail visible through a large telescope is limited, not by its diffraction fringes, but by the air through which the observer must look. A telescope performs better on a high mountaintop, where the air is thin and steady, but even there Earth's atmosphere limits the detail the telescope can reveal.

The third and least important power of a telescope is **magnifying power,** the ability to make the image bigger. Because the amount of detail we can see is limited by the seeing conditions and the resolving power, very high magnification does not necessarily show us more detail. You may have noticed this effect in television news footage. Magnifying the surveillance camera image of a bank robber does not allow us to see more; we just see a big fuzzy image instead of a small fuzzy image.

Another reason magnification is not important in an astronomical telescope is that it is easy to change. An astronomer can increase the magnification by switching to a shorter-focal-length eyepiece. But an astronomer can't easily change the light-gathering power or the resolving power of a telescope.

Compare an astronomer's telescope with a biologist's microscope. A microscope is designed primarily to magnify and thus show us things too small to see. An astronomer's telescope solves a different problem. Its primary function is to gather light and show us things too faint to see.

If you visit a department store to shop for a telescope, you will probably find telescopes described according to magnification. One may be labeled an "80-power telescope" and another a "40-power telescope." However, the magnifying power really tells us little about the telescopes. Astronomers identify telescopes by diameter because that determines both light-gathering power and resolving power.

Because astronomical telescopes are so sensitive, we often find observatories located on mountaintops far from cities. The air above high mountaintops is thin, so it is more transparent; the sky is darker; and the stars are brighter. Also, the winds blow smoothly over some mountains, so there is little turbulence to ruin the seeing. In addition, mountaintops located far from cities suffer less from **light pollution,** the brightening of the night sky by light scattered from artificial outdoor lighting. Many of our cities are so brightly lit by street lights and advertising that the night sky glows with scattered light (Figure 3-12), and faint stars and galaxies are invisible. Even some of our largest observatories are threatened by growing light pollution from nearby population centers, and astronomers work actively with city planners to reduce outdoor lighting. As a consequence, astronomers place their largest telescopes on high mountaintops as far as possible from civilization (Figure 3-13).

When we compare telescopes, we should consider their powers. By the Numbers 3-1 (on page 71) shows how to calculate the powers of a telescope. This will be useful if you decide to buy a telescope of your own.

Buying a Telescope

Thinking about how we should shop for a new telescope will not only help us if we decide to buy one but will also illustrate some important points about astronomical telescopes.

Assuming we have a fixed budget, we should buy the highest-quality optics and the largest-diameter telescope we can afford. Of the two things that limit what we see, optical quality is under our control. We can't make the atmosphere less turbulent, but we should buy good optics. If we buy a telescope from a toy store and it has plastic lenses, we shouldn't expect to see very much. Also, we want to maximize the light-gathering power of our telescope, so we want to purchase the largest-diameter telescope we can afford. Given a fixed budget, that means we should buy a reflecting telescope (which includes Schmidt-Cassegrains) rather than a refracting telescope. Not only will we get more diameter per dollar, but our telescope will not suffer from chromatic aberration.

We can safely ignore magnification. Department stores and camera stores may advertise telescopes by quoting their magnification, but it is not an important number. What we can see is fixed by light-gathering power, optical quality, and Earth's atmosphere. Besides, we can change the magnification by changing eyepieces.

Other things being equal, we should choose a telescope with a solid mounting that will hold the telescope steady and allow us to point at objects easily. A sidereal drive would be very useful even on a small

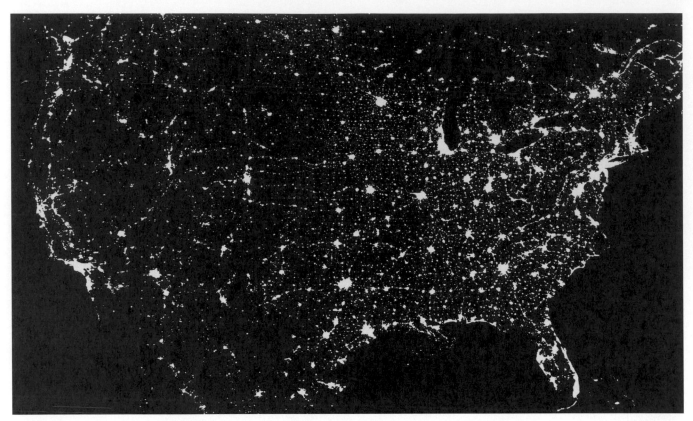

Figure 3-12
This satellite view of the continental United States at night shows the light pollution produced by outdoor lighting. The glow from such lighting makes faint stars and galaxies invisible and forces astronomers to place their telescopes as far from population centers as possible. Even so, some observatories are threatened by spreading light pollution. *(National Geophysical Data Center)*

telescope, and computer-controlled pointing systems are available for a price on many small telescopes. A good telescope on a poor mounting is almost useless.

We might be buying a telescope to put in our backyard, but we must think about the same issues astronomers consider when they design giant telescopes to go on mountaintops. In fact, some of the newest telescopes solve these traditional problems in new ways.

New-Generation Telescopes

For most of this century, astronomers faced a serious limitation in the size of telescopes. A large telescope mirror sags under its own weight and cannot focus the light accurately. To avoid this sag, astronomers made telescope mirrors very thick, but that produced two further problems—weight and cost.

Supporting such a heavy mirror required a massive telescope. The two largest conventional telescopes in the world are the 6-m (236-inch) reflector in the former Soviet Union and the 5-m (200-inch) Hale Telescope on Mount Palomar (Figure 3-14). The 5-m mirror alone weighs 14.5 tons, and its mounting weighs about 530 tons.

Figure 3-13
The southern Gemini telescope, 8 meters in diameter, Is being built on top of an 8895-ft-high mountain in the Chilean Andes. A twin telescope Is being built at 13,882 ft atop the extinct volcano Mauna Kea in Hawaii. The difficulty of building giant telescopes on remote mountaintops is justified by the dark skies, clear air, and the outstanding seeing. *(Gemini 8-m Telescopes Project)*

Light-gathering power is proportional to the area of the telescope objective. A lens or mirror with a large area gathers a large amount of light. Because the area of a circular lens or mirror of diameter D is $\pi(D/2)^2$, we can compare the areas of two telescopes, and therefore their relative light-gathering powers, by comparing the square of their diameters. That is, the ratio of the light-gathering powers of two telescopes A and B is equal to the ratio of their diameters squared:

$$\frac{\text{LGP}_A}{\text{LGP}_B} = \left(\frac{D_A}{D_B}\right)^2$$

Example A: Suppose we compare a 4-cm telescope with a 24-cm telescope. How much more light will the large telescope gather? *Solution:*

$$\frac{\text{LGP}_{24}}{\text{LGP}_4} = \left(\frac{24}{4}\right)^2 = 6^2 = 36 \text{ times more light}$$

Example B: Our eye acts like a telescope with a diameter of about 0.8 cm, the diameter of the pupil. How much more light can we gather if we use a 24-cm telescope? *Solution:*

$$\frac{\text{LGP}_{24}}{\text{LGP}_{\text{eye}}} = \left(\frac{24}{0.8}\right)^2 = (30)^2 = 900 \text{ times more light}$$

The resolving power of a telescope is the angular distance between two stars that are just barely visible through the telescope as two separate images. For optical telescopes, the resolving power α, in seconds of arc, equals 11.6 divided by the diameter of the telescope in centimeters:

$$\alpha - \frac{11.6}{D}$$

Example C: What is the resolving power of a 10-cm telescope? *Solution:* The resolving power is 11.6 divided by 10, or 1.16 seconds of arc:

$$\alpha = \frac{11.6}{10} = 1.16 \text{ seconds of arc}$$

If the lenses are of good quality and if the seeing is good, we should be able to distinguish as separate points of light any pair of stars farther apart than 1.16 seconds of arc. If the stars are any closer together, diffraction fringes blur the stars together into a single image (see Figure 3-11).

The magnification of a telescope equals the focal length of the objective lens or mirror F_o divided by the focal length of the eyepiece F_e.

$$M = \frac{F_o}{F_e}$$

Example D: What is the magnification of a telescope whose focal length is 80 cm used with an eyepiece whose focal length is 0.5 cm? *Solution:* The magnification is 80 divided by 0.5, or 160 times.

a

b

Figure 3-14
Until recent decades, telescope mirrors were single, thick pieces of glass that required massive telescope mountings and large observatory domes. (a) The world's largest traditional telescope, located in the Caucasus Mountains, is the 6-m reflector built by the former Soviet Union. (b) The 5-m (200-inch) Hale Telescope on Mount Palomar contains a mirror weighing 14.5 tons. The telescope dome weighs 1000 tons.

Grinding a conventional telescope to shape is expensive because it requires the slow removal of large amounts of glass. The 5-m mirror, for instance, was begun in 1934 and finished in 1948, and 5 tons of glass were ground away.

Astronomers have been developing new techniques to make large mirrors that weigh and cost less. For example, the Steward Observatory Mirror Laboratory has built a revolving oven under the football stands at the University of Arizona that can produce preshaped mirrors. The oven turns like a merry-go-round, and the molten glass flows outward in the mold to form a concave upper surface. After they slowly cool, the preshaped mirrors can be quickly ground to final shape. The oven is currently able to cast mirrors as large as 8.4 m in diameter.

A mirror of conventional thickness with diameter of 8 m would be too heavy to support in a telescope, so astronomers have devised ways to make the mirrors thinner (Figure 3-15). One technique is to make the mirror in segments. Small segments are less expensive and sag less under their own weight. The largest general-purpose optical telescopes in the world are the twin 10-m Keck I and Keck II Telescopes in Hawaii. Each of these giant telescopes uses 36 hexagonal mirror segments held in alignment by computer-controlled thrusters to form a single mirror (Figure 3-16a,c). The special purpose Hobby-Eberly Telescope in Texas uses 91 hexagonal segments to make up a mirror 11 m in diameter; it is especially designed to focus starlight into a sophisticated spectrograph (Figure 3-16b).

Another way to reduce the weight of telescope mirrors is to make them thin. Thin mirrors sag so easily under their own weight that they are called floppy mir-

a

a

b

c

Figure 3-15
The problem and two solutions: Conventional telescope mirrors (a) are very thick so they will not sag out of shape. To reduce the weight and build bigger mirrors, astronomers have begun making mirrors of segments (b) or as relatively thin disks (c). To hold these mirrors in shape, computer-controlled thrusters (red) are positioned under the mirrors.

b

Figure 3-16
Segmented mirrors can be assembled to create very large diameter telescopes. (a) The hexagonal mirror segments in the Keck I Telescope are visible in this photo. The human figure is crouching in the Cassegrain opening at the center of the mirror array. *(© Russ Underwood. W. M. Keck Observatory)* (b) The 91 hexagonal mirrors of the Hobby-Eberly Telescope in Texas reflect the structure of the interior of the telescope dome. Spanning 11 m, the segmented mirror is specially designed to feed starlight into either of two state-of-the-art spectrographs. *(Photo courtesy Dr. Thomas G. Barnes III, McDonald Observatory, The University of Texas at Austin.)* (c) The operation of a modern telescope requires an impressive array of computers. *(AAO)*

c

a

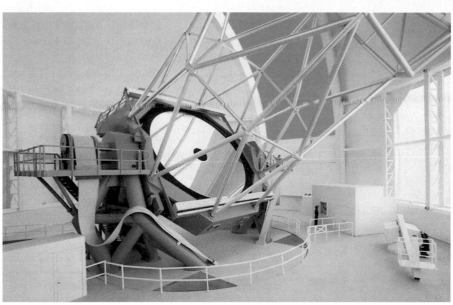

b

Figure 3-17
(a) The four telescopes of the VLT are housed in separate domes, but they can combine their light to work with the light-gathering power of a telescope over 16 m in diameter. *(ESO)* (b) This model of the Gemini Telescope design shows how the thin 8-m mirror will be carried in an alt-azimuth mount under computer control. *(The Gemini 8-m Telescope Project)*

rors, but a computer can control their shape in what astronomers call **active optics.** The New Technology Telescope at the European Southern Observatory in Chile contains a 3.58-m (141-in.) mirror only 24 cm (10 in.) thick. With a computer controlling the mirror shape, the telescope produces very high quality images.

Segmented or floppy, thin telescope mirrors have yet another advantage—they cool quickly. During the day, a telescope mirror warms to air temperature, but after sunset, the air cools quickly. A massive telescope

mirror cannot cool rapidly and uniformly. As some parts of the mirror cool faster than others, the material expands and contracts, producing strains in the mirror that distort the image. Thin telescope mirrors cool more quickly and uniformly as night falls and thus produce better images.

New technology is allowing astronomers to build new, giant telescopes. The Very Large Telescope (VLT), built by the European Southern Observatory, is located high in the remote Andes mountains in northern Chile. The VLT consists of four telescopes (Figure 3-17a) with computer-controlled main mirrors 8.2 m in diameter and only 17.5 cm (6.9 in.) thick. The four telescopes will be able to work together as a single giant telescope or separately. The Gemini project is completing a pair of 8-m thin-mirror telescopes (Figure 3-17b), which are located in Chile and in Hawaii so they can cover the entire sky. Italian and American astronomers are building the Large Binocular Telescope, which will carry a pair of 8.4-m mirrors on a single mounting. Astronomers are now designing truly gigantic telescopes with segmented mirrors 30 m and even 100 m in diameter (larger than a football field). Computer control of the optics makes such huge telescopes worth considering.

R E V I E W Critical Inquiry

Why do astronomers build observatories at the tops of mountains?

Astronomers have joked that the hardest part of building a new observatory is constructing the road to the top of the mountain. It certainly isn't easy to build a large, delicate telescope at the top of a high mountain, but it is worth the effort. A telescope on top of a high mountain is above the thickest part of Earth's atmosphere. There is less air to dim the light, and there is less water vapor to absorb infrared radiation. Even more important, the thin air on a mountaintop causes less disturbance to the image, and thus the seeing is better. A large telescope on Earth's surface has a resolving power much better than the distortion caused by Earth's atmosphere. Thus, it is limited by seeing, not by its own diffraction. It really is worth the trouble to build telescopes atop high mountains.

Astronomers not only build telescopes on mountaintops; they also build gigantic telescopes many meters in diameter. What are the problems and advantages in building such giant telescopes?

Astronomers sometimes refer to a telescope that produces distorted images as a "light bucket." In a sense, all astronomical telescopes are just light buckets, because the light they focus into images tells us very little until it is recorded and analyzed by special instruments attached to the telescopes.

3-3 Special Instruments

Just looking through a telescope doesn't tell us much. To use an astronomical telescope to learn about stars, we must be able to analyze the light the telescope gathers. Special instruments attached to the telescope make that possible.

Imaging Systems

The original imaging device in astronomy was the photographic plate. It could record faint objects in long exposures and could be stored for later analysis, but photographic plates have been almost entirely replaced in astronomy by electronic imaging systems.

Most modern astronomers use **charge-coupled devices (CCDs)** to record images. CCDs are typically about the size of a postage stamp and contain roughly a million microscopic, light-sensitive diodes (Figure 3-18a). These devices can be used like a small photographic plate, but they have dramatic advantages. They can detect both bright and faint objects in a single exposure, are much more sensitive than photographic plates, and can be read directly into computer memory for later analysis.

However an image is recorded, astronomers often reproduce the image in exaggerated ways to bring out subtle details. If an image of a faint object is reproduced as a negative, the sky is white and the stars are dark (Figure 3-18b). This makes the faint parts of the image easier to see. With a computer, astronomers can easily manipulate an image to produce **false-color images** (Figure 3-18c). The colors in such images are merely codes to intensity and are not related to the true colors of the objects.

Measurements of intensity and color were made in the past using a photometer, a highly sensitive light meter attached to the telescope. Today, however, most such measurements are made on CCD images. Because the CCD image is so easily digitized, brightness and color can be measured to high precision.

The Spectrograph

To analyze light in detail, we need to spread the light out according to wavelength into a spectrum, a task performed by a **spectrograph.** We can understand how this works if we reproduce an experiment performed by Isaac Newton in 1666. Boring a hole in his window shutter, Newton admitted a thin beam of sunlight into his bedroom. When he placed a prism in the beam, the sunlight spread into a beautiful spectrum on the far wall. From this Newton concluded that white light was made of a mixture of all the colors.

Newton didn't think in terms of wavelength, but we can use that modern concept to see that the light passing through the prism is bent at an angle that depends on the wavelength. Violet (short wavelengths) bends most, and red (long wavelengths) least. Thus, the white light entering the prism is spread into a spectrum (Figure 3-19a). A typical prism spectrograph contains

Figure 3-18

(a) This large CCD chip contains over 4 million light-sensitive detectors in an array 55.3 mm square. *(Copyright © 1986 Tektronix, Inc. All rights reserved. Reprinted by permission of Tektronix, Inc.)* (b) An electronic image of a giant galaxy (arrow) so distant that its light took 10 billion years to reach us. This image is reproduced as a negative. The sky is light, and the stars and galaxy are black. *(Hyron Spinrad)* (c) This false-color image of the galaxy NGC1232 displays different levels of brightness in different colors. The original CCD image was manipulated by computer. *(California Association for Research in Astronomy)*

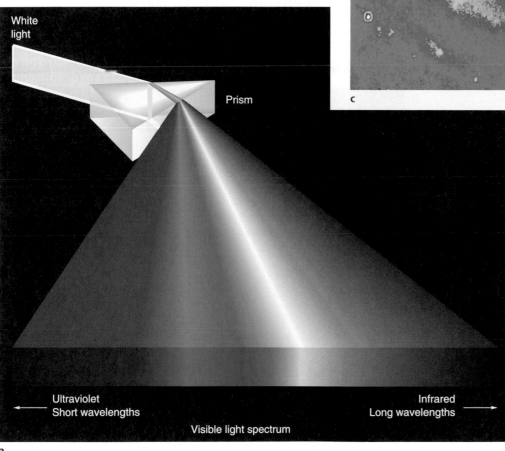

Figure 3-19

(a) A prism bends light by an angle that depends on the wavelength of the light. Short wavelengths bend most, and long wavelengths least. Thus, white light passing through a prism is spread into a spectrum. (b) A stellar spectrum is flanked above and below by comparison spectra containing features of known wavelength. These act as a calibration of wavelengths in the spectrum of the star. Note that this is a negative reproduction with bright features shown dark and dark features shown light. *(Palomar Obs/Caltech)* more than one prism to spread the light farther and lenses to guide the light into the prism and to focus the light onto a photographic plate.

Nearly all modern spectrographs use a grating in place of a prism. A **grating** is a piece of glass with thousands of microscopic parallel lines scribed onto its surface. Different wavelengths of light reflect from the grating at slightly different angles, so white light is spread into a spectrum and can be recorded, often by a CCD camera.

Because astronomers need to measure the precise location of features in a spectrum, most spectrographs can add a **comparison spectrum** above and below the spectrum of the observed object (Figure 3-19b). The features in this comparison spectrum have known wavelengths carefully measured in the laboratory, and thus the astronomer can use the comparison spectrum as a precise calibration of the wavelengths in the spectrum being studied.

Because astronomers understand how light interacts with matter, a spectrum carries a tremendous amount of information (as we will see in the next chapter), and that makes a spectrograph the astronomer's most powerful instrument. An astronomer recently remarked, "We don't know anything about an object 'til we get a spectrum," and that is only a slight exaggeration.

REVIEW   **Critical Inquiry**

What is the difference between light going through a lens and light passing through a prism?

A refracting telescope producing chromatic aberration and a prism dispersing light into a spectrum are two examples of the same thing, but one is bad and one is good. When light passes through the curved surfaces of a lens, different wavelengths are bent by slightly different amounts, and the different colors of light come to focus at different focal lengths. This produces the color fringes in an image called chromatic aberration, and that's bad. But the surfaces of a prism are made to be precisely flat, so all of the light enters the prism at the same angle, and any given wavelength is bent by the same amount wherever it meets the prism. Thus, white light is dispersed into a spectrum. We could call the dispersion of light by a prism "controlled chromatic aberration," and that's good.

CCDs have been very good for astronomy, and they are now widely used. Explain why they are so much more useful than photographic plates.

Figure 3-20
(a) In most radio telescopes, a dish reflector concentrates the radio signal on the antenna. The signal is then amplified and recorded. For all but the shortest radio waves, wire mesh is an adequate reflector (b). *(Courtesy Seth Shostak/SETI Institute)*

So far, our discussion has been limited to visual wavelengths. Now it is time to consider the rest of the electromagnetic spectrum.

3-4 Radio Telescopes

Operation of a Radio Telescope

A radio telescope usually consists of four parts: a dish reflector, an antenna, an amplifier, and a recorder (Figure 3-20). The components, working together, make it possible for astronomers to detect radio radiation from celestial objects.

The dish reflector of a radio telescope, like the mirror of a reflecting telescope, collects and focuses radiation. Because radio waves are much longer than light waves, the dish need not be as smooth as a mirror. Wire mesh works quite well as a reflector of all but the shortest radio waves. In some radio telescopes, the reflector may not even be dish-shaped, or the telescope may contain no reflector at all.

Though a radio telescope's dish may be many meters in diameter, the antenna may be as small as your hand. Like the antenna on a TV set, its only function is to absorb the radio energy and direct it along a cable to an amplifier. After amplification, the signal goes to some kind of recording instrument. Most radio observatories record data on magnetic tape or feed it directly to a computer. However it is recorded, an observation with a radio telescope measures the amount of radio energy coming from a specific point on the sky.

Because humans can't see radio waves, astronomers must convert them into something perceptible.

One way is to measure the strength of the radio signal at various places in the sky and draw a map in which contours mark areas of uniform radio intensity. We might compare such a map to a seating diagram for a baseball stadium in which the contours mark areas in which the seats have the same price (Figure 3-21a). Contour maps are very common in radio astronomy and are often reproduced using false colors (Figure 3-21b).

Limitations of the Radio Telescope

A radio astronomer works under three handicaps: poor resolution, low intensity, and interference. We saw that the resolving power of an optical telescope depends on the diameter of the objective lens or mirror. It also depends on the wavelength of the radiation. At very long wavelengths, like those of radio waves, images become fuzzy because of the large diffraction fringes. As in the case of an optical telescope, the only way to improve the resolving power is to build a bigger telescope. Consequently, radio telescopes must be quite large.

Even so, the resolving power of a radio telescope is not good. A dish 30 m in diameter receiving radiation with a wavelength of 21 cm has a resolving power of about 0.5°. Such a radio telescope would be unable to show us any details in the sky smaller than the moon. Fortunately, radio astronomers can combine two or more radio telescopes to improve the resolving power. Such a linkup of radio telescopes is called a **radio interferometer** and has the resolving power of a radio telescope whose diameter equals the separation of the radio telescopes. For example, the Very Large Array (VLA) radio interferometer (Figure 3-22) consists of 27 radio dishes located in a Y-shaped pattern about 20 km (13 miles)

a

Figure 3-21
(a) A contour map of a baseball stadium shows regions of similar admission prices. The most expensive seats are those behind home plate. (b) A false-color-image radio map of Tycho's supernova remnant, the expanding shell of gas produced by the explosion of a star in 1572. The radio contour map has been color-coded to show intensity. Red is the strongest radio intensity and violet the weakest. *(Courtesy NRAO)*

b

tons of radio energy have such long wavelengths that their individual energies are quite low. In order to get strong signals focused on the antenna, the radio astronomer must build large collecting dishes.

The largest radio dish in the world is 300 m (1000 ft) in diameter. So large a dish can't be supported in the usual way, so it is built into a mountain valley in Arecibo, Puerto Rico (Figure 3-23). A thin metallic surface supported above the valley floor by cables serves as the reflecting dish, and the antenna hangs on cables from three towers built on the three mountains that surround the valley. Although this telescope can look only overhead, the operators can change its

Figure 3-22
(a) A radio interferometer consists of two or more radio telescopes whose signals are combined to give the resolving power of a much larger instrument. (b) The Very Large Array radio interferometer uses up to 27 dishes along the 20-km-long arms of a Y-shaped layout to simulate a radio telescope 40 km in diameter. *(NRAO)*

Figure 3-23
The largest radio telescope in the world is the 300-m (1000-ft) dish suspended in a valley in Arecibo, Puerto Rico. The antenna hangs above the dish on cables stretching from towers. The Arecibo Observatory is part of the National Astronomy and Ionosphere Center, which is operated by Cornell University under contract with the National Science Foundation. *(David Parker/Science Photo Library)*

on each leg. The signals combined by computer simulate a single radio telescope with a diameter of 40 km (25 miles). It can produce radio maps with a resolution better than 1 second of arc. The Very Long Baseline Array (VLBA) consists of matching radio dishes spread from Hawaii to the Virgin Islands and has an effective diameter of about 8000 km. Thus, radio astronomers can use interferometers to compensate for low resolving power.

The second handicap radio astronomers face is the low intensity of the radio signals. We saw earlier that the energy of a photon depends on its wavelength. Phoaim slightly by moving the antenna and by waiting for the rotation of Earth to point the telescope in the proper direction. This may sound clumsy, but the telescope's ability to detect weak radio sources, together with its good resolution, makes it one of the most important radio observatories in the world.

The third handicap the radio astronomer faces is interference. A radio telescope is an extremely sensitive radio receiver listening to radio signals thousands of times weaker than artificial radio and TV transmissions. Such weak signals are easily drowned out by interference. Sources of such interference include everything from poorly designed transmitters in Earth satellites to automobiles with faulty ignition systems. To avoid this kind of interference, radio astronomers locate their telescopes as far from civilization as possible. Hidden deep in mountain valleys, they are able to listen to the sky protected from human-made radio noise.

Advantages of Radio Telescopes

Building large radio telescopes in isolated locations is expensive, but three factors make it all worthwhile. First, and most important, a radio telescope can show us where clouds of cool hydrogen are located between the stars. Because 90 percent of the atoms in the universe are hydrogen, that is important information. Large clouds of cool hydrogen are completely invisible to normal telescopes, because they produce no visible light of their own and reflect too little to be detected on photographs. However, cool hydrogen emits a radio signal at the specific wavelength of 21 cm. (We will see how the hydrogen produces this radiation when we discuss atoms in the next chapter.) The only way we can detect these clouds of gas is with a radio telescope that receives the 21-cm radiation. These hydrogen clouds are the places where stars are born, and that is one reason that radio telescopes are so important.

A second reason: Because radio signals have relatively long wavelengths, they can penetrate the vast clouds of dust that obscure our view at visual wavelengths. Light waves are short, and they interact with tiny dust grains floating in space; thus, the light is scattered and never penetrates the clouds to reach optical telescopes on Earth. However, radio signals from far across the galaxy pass unhindered through the dust, giving us an unobscured view.

Finally, a radio telescope can detect objects that are more luminous at radio wavelengths than at visible wavelengths. This includes everything from the coldest clouds of gas to the hottest stars. Some of the most distant objects in the universe, for instance, are detectable only at radio wavelengths.

REVIEW Critical Inquiry

Why do optical astronomers build big telescopes, while radio astronomers build groups of widely separated smaller telescopes?

Comparison of seemingly similar concepts is a powerful tool in critical analysis because it often reveals subtle differences. In this case, optical astronomers are trying to maximize light-gathering power, but radio astronomers try to maximize resolving power. Because radio waves are so much longer than light waves, a single radio telescope can't see details in the sky much smaller than the moon. By linking radio telescopes miles apart, radio astronomers build a radio interferometer that can simulate a radio telescope miles in diameter and thus increase the resolving power.

This difference between the wavelength of light and radio waves makes a big difference in building the best telescopes. But why don't radio astronomers build their telescopes on mountaintops as optical astronomers do?

Our atmosphere causes trouble for Earth's astronomers in two ways. It distorts images, and it absorbs many wavelengths. The only way to avoid the limitations completely is to send telescopes above the atmosphere, into space.

3-5 Space Astronomy

Ground-based telescopes can operate only at wavelengths in the optical and radio "windows." The rest of the electromagnetic radiation—infrared, ultraviolet, X ray, and gamma ray—never reaches Earth's surface. To observe at these wavelengths, telescopes must fly above the atmosphere in high-flying aircraft, rockets, balloons, and satellites.

Infrared Astronomy

Some infrared radiation does leak through our atmosphere. This radiation enters narrow, partially open atmospheric windows scattered from 1200 nm to about 40,000 nm (1.2 to 40 micrometers). In this range, called the near infrared, much of the radiation is absorbed by water vapor, carbon dioxide, and oxygen molecules in Earth's atmosphere, so it is an advantage to place telescopes on mountains where the air is thin and dry. Two major infrared telescopes, for example, observe from the 4150-m (13,600-ft) summit of Mauna Kea in Hawaii. At this altitude, they are above much of the water vapor, which is the main absorber of infrared.

Figure 3-24
The advantage of infrared observations is illustrated in this mosaic of images of the constellations Vela and Puppis obtained by the IRAS satellite (inset). Three images made at different infrared wavelengths are color-coded and combined. Stars, the hottest objects in the photo, appear as blue dots, while clouds of interstellar dust heated by starlight are color-coded red or yellow. These clouds are colder than 100 K (−280°F) and are undetectable at visible wavelengths. *(NASA)*

The far-infrared range, which includes wavelengths longer than 40 micrometers, can tell us about planets, comets, forming stars, and other cool objects, but these wavelengths are absorbed high in the atmosphere. To observe in the far infrared, telescopes must venture to high altitudes. Remotely operated infrared telescopes suspended under balloons have reached altitudes as high as 41 km (25 miles). For many years, a NASA jet transport carried a 91-cm infrared telescope and a crew of astronomers to altitudes of 12,000 m (40,000 ft) to get above 99 percent of the water vapor in Earth's atmosphere. Now retired from service, that airborne observatory will soon be replaced with the Stratospheric Observatory for Infrared Astronomy (SOFIA), a Boeing 747 carrying a 2.5-m telescope to the fringes of the atmosphere.

The ultimate solution is to place the infrared telescope in space above the atmosphere. In the early 1980s, the Infrared Astronomy Satellite (IRAS) mapped the sky at infrared wavelengths (Figure 3-24). In the middle 1990s, European astronomers launched the Infrared Space Observatory, which carried detectors more sensitive than IRAS. NASA now plans the Space Infrared Telescope Facility (SIRTF) as a major infrared observatory in space.

If a telescope observes in the far infrared, it must be cooled. Everything emits some infrared, and if the telescope is warm it will emit many times more infrared radiation than that coming from a distant object. Imagine trying to look at a dim, moonlit scene through binoculars that are glowing brightly. In the near infrared, only the detector, the element on which the infrared radiation is focused, must be cooled. To observe in the far infrared, however, as IRAS did, the entire telescope must be cooled.

Ultraviolet Astronomy

Some infrared telescopes can observe from mountaintops, but ultraviolet radiation beyond about 290 nm, the far ultraviolet, is completely absorbed by the ozone layer extending from 20 km up to about 40 km in our atmosphere. To get above this layer, ultraviolet telescopes must go into space.

One of the most successful ultraviolet observatories was the International Ultraviolet Explorer (IUE). Launched in 1978, it carried a 45-cm (18-in.) telescope and used TV systems to record spectra from 320 nm to 115 nm. Although IUE was expected to last only a year or two, it survived and was widely used by astronomers from all around the world until, partially shut down because of budget cutbacks, it finally failed in 1996.

The Extreme Ultraviolet Explorer (EUVE) telescope was launched in 1992 to observe at wavelengths from 100 nm to 10 nm. With four telescopes and more modern detectors than IUE, EUVE has surveyed the entire sky and studied specific targets at its shorter wavelength range.

Observations in the infrared reveal cool material, but observations in the ultraviolet tend to show hot, excited regions. Hot stars and hot gas are mapped by such telescopes.

X-Ray Astronomy

Beyond the UV, at wavelengths from 10 nm to 0.01 nm, lie the X rays. These photons can be produced only by high-energy processes, so we see them coming from very hot regions in stars and from violent events such as matter smashing onto a neutron star—a subject we will discuss in Chapter 14. X-ray images can give us information about the heavens that we can get in no other way.

Although early X-ray observations of the sky were made from balloons and small rockets in the 1960s, the age of X-ray astronomy did not really begin until 1970, when an X-ray telescope named Uhuru (Swahili for "freedom") was put into orbit. Uhuru detected nearly 170 separate sources of celestial X rays. In the late 1970s, the three High Energy Astronomy Observatories (HEAO) satellites, carrying more sensitive and more sophisticated equipment, pushed the total to many hundreds. The second HEAO satellite, named the Einstein Observatory, used special optics to produce X-ray images. Other orbiting X-ray telescopes have been designed to make special kinds of observations. Japanese and American astronomers, for example, have collaborated to build the Yohkoh X-ray telescope, which observes the sun (Figure 3-25).

For many years, astronomers have worked to create a major X-ray observatory in space. The Advanced X-Ray Astrophysics Facility (AXAF) was carried into orbit by the Space Shuttle, where it was renamed Chandra in honor of the late Indian-American Nobel Laureate Subrahmanyan Chandrasekhar, who worked on theoretical astrophysics in many areas of astronomy that the X-ray observatory Chandra will study (Figure 3-26).

Focusing X rays is difficult because the high-energy X-ray photons do not reflect off conventional mirrors but rather penetrate into the mirrors. The optics in Chandra are especially designed mirrors in which the X-ray photons graze the surface of the mirrors at very small angles. Under these circumstances, the photons are reflected to form an image. Just as a CCD chip records visible light photons, the detector in Chandra has been designed to absorb and record X-ray photons to produce

Figure 3-25
This X-ray image of the sun was recorded by the Yohkoh solar research spacecraft of the Japanese Institute of Space and Astronautical Science. The image reveals intensely hot regions in the solar atmosphere linked by magnetic fields. *(Courtesy Japan/US Yohkoh Team, NASA and ISAS)*

Figure 3-26
From Earth orbit, Chandra can photograph the sky at X-ray wavelengths. Its images will help astronomers study a wide range of astrophysical problems ranging from the birth of stars out of the gas in space to eruptions in the most distant galaxies. *(Chandra X-ray Center, Smithsonian Astrophysical Observatory)*

an X-ray image. Chandra can detect X-ray-emitting objects 50 times fainter than any previous X-ray telescope and can resolve details 10 times smaller.

The Hubble Space Telescope

Astronomers have dreamed of having a giant telescope in orbit since the 1920s. Such a telescope would not only allow observation at nonvisible wavelengths but would also avoid the turbulence in Earth's atmosphere. That dream was fulfilled on April 24, 1990, when the Hubble Space Telescope (HST) was released into orbit (Figure 3-27a). Named after the man who discovered the expansion of the universe (Chapter 18), it is the largest orbiting telescope ever built. As big as a large bus, the HST carries a 2.4-m (96-in.) mirror. The light can be directed to any one of two spectrographs and two cameras.

Soon after the telescope was launched, NASA engineers and astronomers discovered a defect in the shape of the mirror that prevented it from focusing sharp images. While some astronomers devised computer programs to partially remove the distortion from images, others developed corrective optics. Although the press emphasized the problem, astronomers used the telescope with great success even before astronauts visited the telescope in December 1993 and installed the corrective optics package in a series of five space walks. With

its optics corrected, the telescope has exceeded its original design criteria. It can detect objects so faint it can look 50 times farther into space than can Earth-based telescopes. With its precise optics and its location above Earth's distorting atmosphere, it can resolve details 10 times smaller than can be resolved by any conventional telescope on Earth (Figure 3-27b). Astronauts on the Space Shuttle visit the telescope every few years and install new instruments that make the telescope more sensitive and productive.

HST is controlled from the Space Telescope Science Institute located at Johns Hopkins University in Baltimore, Maryland. Astronomers there plan the telescope's observing schedule to maximize the scientific return while engineers study the operation of the telescope. HST transmits its data to the Institute, where it is analyzed. Because the telescope is a national facility, anyone may propose research projects for the telescope. Competition is fierce, and only the most worthy projects win approval. Nevertheless, even some amateur astronomers have projects approved for the Space Telescope.

In its first years in space, the Hubble Space Telescope made exciting discoveries in stellar formation, evolution, and death. It has revealed young galaxies at great distance, and it has recorded detailed images of planets in our own solar system. Future projects include studies of the age and evolution of the universe. With its new optics and instruments, it will be one of

a

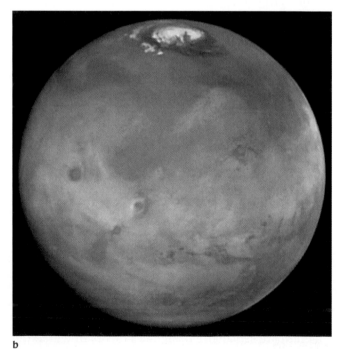

b

Figure 3-27
(a) The Hubble Space Telescope is the largest orbiting telescope ever launched. It was carried into space by the Space Shuttle in 1990 and released into Earth orbit by the robotic arm at upper right. (b) This image of Mars recorded by the Hubble Space Telescope reveals far more detail than is possible from ground-based telescopes. Here thin clouds drift around high volcanoes at the left, and details of the polar cap are visible at the top. *(NASA)*

the most productive tools in astronomy for many years to come.

REVIEW Critical Inquiry

Why can infrared astronomers observe from high mountaintops, while X-ray astronomers must observe from space?

In this analysis, we find similar consequences springing from different causes. Although both infrared and X-ray telescopes have been put into orbit, they differ dramatically. Infrared radiation is absorbed by water vapor in Earth's atmosphere, which is confined to the lower atmospheric layers. If we built an infrared telescope on top of a high mountain, we would be above most of the water vapor and we could collect some infrared radiation from the stars. The longer-wavelength infrared is absorbed much higher in the atmosphere, so telescopes to observe in the far infrared must go into space. X rays are absorbed in the uppermost layers of the atmosphere, and no mountain is high enough to extend above those layers. To observe the stars at X-ray wavelengths, a telescope must go into space.

X-ray and far-infrared telescopes must observe from space, but the Hubble Space Telescope observes in the visual wavelength range. Then why must it observe from orbit?

The tools of the astronomer are designed to gather radiation from the sky and extract information. Perhaps no tool is as important as the spectrograph, because no form of observation is as loaded with information as a spectrum. In the next chapter, we will see how we can harvest the information in a star's spectrum.

three powers of the telescope depend on the telescope's diameter; thus, astronomical telescopes often have large diameters.

Astronomical telescopes are of two types, refractor and reflector. A refractor uses a lens to bend the light and focus it into an image. Because of chromatic aberration, refracting telescopes cannot bring all colors to the same focus, resulting in color fringes around the images. An achromatic lens partially corrects for this, but such lenses are expensive and cannot be made larger than about 1 m in diameter.

Reflecting telescopes use a mirror to focus the light and are less expensive than refracting telescopes of the same diameter. In addition, reflecting telescopes do not suffer from chromatic aberration. Thus most recently built large telescopes are reflectors.

The light gathered by an astronomical telescope can be recorded and analyzed by special instruments attached to the telescope. For many decades, astronomers have used photographic plates to record images at the telescope, but modern electronic systems such as CCDs have now replaced photographic plates in many applications. Spectrographs spread starlight out according to wavelength to form a spectrum.

To observe radio signals from celestial objects, we need a radio telescope, which usually consists of a dish reflector, an antenna, an amplifier, and a recorder. Such an instrument can measure the intensity of radio signals over the sky and construct radio maps. The poor resolution of the radio telescope can be improved by combining it with another radio telescope to make a radio interferometer. Radio telescopes have three important features—they can detect cool hydrogen, they can see through dust clouds in space, and they can detect certain objects invisible at other wavelengths.

Earth's atmosphere admits radiation primarily through two wavelength intervals, or windows—the visual window and the radio window. At other wavelengths, our atmosphere absorbs radiation. To observe in the far infrared, astronomers must fly telescopes high in balloons or aircraft, though they can work at some wavelengths in the near infrared from high mountaintops. To observe in the ultraviolet and X-ray range and some parts of the infrared, they must send their telescopes into space to get above our atmosphere.

Summary

Electromagnetic radiation is an electric and magnetic disturbance that transports energy at the speed of light. The electromagnetic spectrum includes radio waves, infrared radiation, visible light, ultraviolet radiation, X rays, and gamma rays.

We can think of a "particle" of light, or a photon, as a bundle of waves that sometimes acts as a particle and sometimes as a wave. The energy a photon carries depends on its wavelength. The wavelength of visible light, usually measured in nanometers (10^{-9} m), ranges from 400 nm to 700 nm. Infrared and radio photons have longer wavelengths and carry less energy. Ultraviolet, X-ray, and gamma-ray photons have shorter wavelengths and carry more energy.

To obtain data, astronomers use telescopes to gather light, see fine detail, and magnify the image. The first two of these

New Terms

electromagnetic radiation	objective mirror
wavelength	eyepiece
photon	chromatic aberration
nanometer (nm)	achromatic lens
angstrom (Å)	Newtonian telescope
atmospheric window	prime focus
refracting telescope	Cassegrain telescope
reflecting telescope	Schmidt-Cassegrain telescope
focal length	
primary lens	sidereal drive
primary mirror	equatorial mounting
objective lens	polar axis

alt-azimuth mounting

light-gathering power

resolving power

diffraction fringe

seeing

magnifying power

light pollution

active optics

charge-coupled device
  (CCD)

false-color image

spectrograph

grating

comparison spectrum

radio interferometer

Review Questions

1. Why would you not plot sound waves in the electromagnetic spectrum?

2. If you had unlimited funds to build a large telescope, which type would you choose, a refractor or a reflector? Why?

3. Why do nocturnal animals usually have large pupils in their eyes? How is that related to astronomical telescopes?

4. Why do optical astronomers often put their telescopes at the tops of mountains, and radio astronomers often put their telescopes in deep valleys?

5. Optical and radio astronomers both try to build large telescopes but for different reasons. How do these goals differ?

6. What are the advantages of making a telescope mirror thin? What problems does this cause?

7. Small telescopes are often advertised as "200 power" or "magnifies 200 times." As someone knowledgeable about astronomical telescopes, how would you improve such advertisements?

8. An astronomer recently said, "Some people think I should give up photographic plates." Why might she change to something else?

9. What purpose do the colors in a false-color image or false-color radio map serve?

10. How is chromatic aberration related to a prism spectrograph?

11. Why would radio astronomers build identical radio telescopes in many different places around the world?

12. Why do radio telescopes have poor resolving power?

13. Why must telescopes observing in the far infrared be cooled to low temperature?

14. What might we detect with an X-ray telescope that we could not detect with an infrared telescope?

15. If the Hubble Space Telescope observes at visual wavelengths, why must it observe from space?

Discussion Questions

1. Why does the wavelength response of the human eye match so well the visual window of Earth's atmosphere?

2. Basic research in chemistry, physics, biology, and similar sciences is supported in part by industry. How is astronomy different? Who funds the major observatories?

Problems

1. The thickness of the plastic in plastic bags is about 0.001 mm. How many wavelengths of red light is this?

2. Measure the actual wavelength of the wave in Figure 3-1. In what portion of the electromagnetic spectrum would it belong?

3. Compare the light-gathering powers of a 5-m telescope and a 0.5-m telescope.

4. How does the light-gathering power of the largest telescope in the world compare with that of the human eye? (*Hint:* Assume that the pupil of your eye can open to about 0.8 cm.)

5. What is the resolving power of a 25-cm telescope? What do two stars 1.5 seconds of arc apart look like through this telescope?

6. Most of Galileo's telescopes were only about 2 cm in diameter. Should he have been able to resolve the two stars mentioned in Problem 5?

7. How does the resolving power of the 5-m telescope compare with that of the Hubble Space Telescope? Why will the Space Telescope outperform the 5-m telescope?

8. If we build a telescope with a focal length of 1.3 m, what focal length should the eyepiece have to give a magnification of 100 times?

9. Astronauts observing from a space station need a telescope with a light-gathering power 15,000 times that of the human eye, capable of resolving detail as small as 0.1 second of arc, and having a magnifying power of 250. Design a telescope to meet their needs. Could you test your design by observing stars from Earth?

10. A spy satellite orbiting 400 km above Earth is supposedly capable of counting individual people in a crowd. What minimum-diameter telescope must the satellite carry? (*Hint:* Use the small-angle formula.)

Critical Inquiries for the Web

1. How do professional astronomers go about making observations at major astronomical facilities? Visit several observatory Web sites to determine the process an astronomer would go through to secure observing time and make observations at the facility.

2. NASA is in the process of completing a fleet of four space-based "Great Observatories." (The Hubble Space Telescope is one; what are the others?) Examine the current state of these missions by visiting their home pages on the Internet. What advantages would these facilities have over ground-based observatories?

Exploring *The Sky*

1. Although the largest of modern telescopes are usually mounted with computer-controlled **alt-azimuth mountings,** many of the smaller telescopes in existence, including amateur telescopes, use **equatorial mountings.** The polar axis of such mountings must be aligned parallel to Earth's axis of rotation and must be rotated at the sidereal rate to compensate for Earth's rotation. To set up such a telescope, the polar axis of the mounting is set to point to the celestial north pole (CNP), the northern point in the sky about which all stars seem to rotate. In practice, however, it is difficult to determine the exact location of the celestial north pole, and it is often sufficient to point the polar axis to Polaris, the North Star. Use *The Sky* to determine the error in using Polaris instead of the celestial north pole. That is, find the angle between Polaris and the celestial north pole.

How to proceed: Set up *The Sky* to show you a northern view of the sky. To locate Polaris, first find Ursa Major, clicking on the "Constellation Lines" and "Common Names" buttons on the "View" toolbar. Polaris is the last star in the handle of what is called the "Little Dipper." Click on Polaris, and a new window will appear that gives a variety of information about Polaris, including the **declination.** The declination of any object in the sky is one of the coordinates used to define its location. It corresponds to latitude on Earth. The angle between Polaris and the celestial north pole is the difference between the declination of the celestial north pole, which is 90 degrees, and that of Polaris.

 Go to the Brooks/Cole Astronomy Resource Center (www. brookscole.com/astronomy) for critical thinking exercises, articles, and additional readings from InfoTrac College Edition, Brooks/Cole's online student library.

CHAPTER FOUR

Atoms and Starlight

Awake! for Morning in the Bowl

of Night

Has flung the stone that put the

Stars to Flight!

And lo! the Hunter of the East

has caught

The Sultan's Turret in a Noose

of Light.

Edward FitzGerald, translator
The Rubaiyat of Omar Khayyám

Guidepost

This chapter marks the beginning of a new way for us to know about the sky. In Chapter 1 we discussed the sky as we see it, and in Chapter 2 we discussed the theories that astronomers proposed to explain the motions in the sky. To go beyond that, modern astronomers have built sophisticated telescopes, which we discussed in the previous chapter. Now we are ready to begin the analysis of the light that telescopes gather.

In all of the chapters that follow, we will depend on the principles we establish here. Whether we are observing Mars, or a star cluster or a distant galaxy, we can learn about astronomical objects only by analyzing the light our telescopes gather. Thus, what we do in this chapter will serve us through all of the chapters that follow.

Astronomers are in the light business. Because they cannot leave Earth and visit the objects they study, they must analyze the light their telescopes gather. Almost everything we know about heavenly bodies has been puzzled out by astronomers analyzing light (Figure 4-1).

Of course, we don't limit this discussion to the light we can see. Visible light is only a small part of the electromagnetic spectrum, and, as we saw in the previous chapter on telescopes, astronomers use every part of the electromagnetic spectrum as they search out the secrets in the sky. Although we will concentrate on light in our discussion, we must remember that what we say applies as well to the rest of the electromagnetic spectrum.

To deduce the nature of an object from its light, we must understand how light interacts with matter. After all, if light did not interact with matter, we would never know that apples are red. Thus much of our discussion in this chapter has to do with how light interacts with matter. Because atoms interact with light, the atoms leave their traces on the light, and, like detectives studying clues, we can deduce the nature of the atoms from our analysis of the light. In fact, part of the fascination of astronomy is the solution of these mini-mystery stories hidden in starlight.

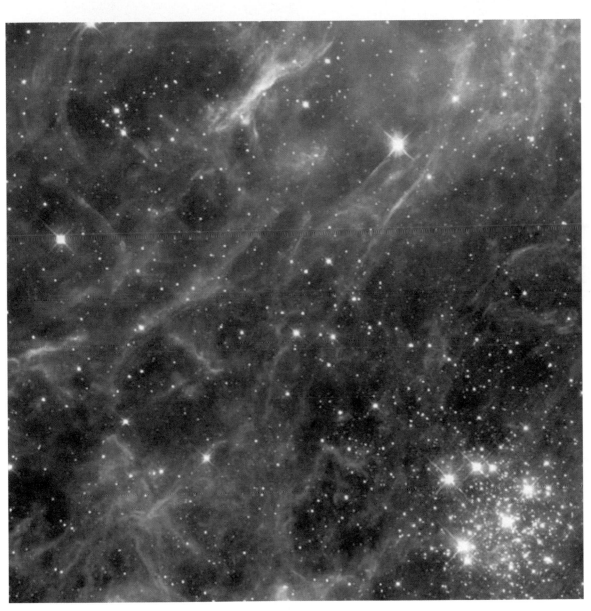

Figure 4-1
The Tarantula Nebula is a region of excited glowing gas within which groups of stars have recently been born. The cluster of stars at the lower right, for example, is about 10 times older than some younger stars in the nebula. All of this information about the nebula and its stars has been discovered by astronomers analyzing the light we receive. *(Hubble Heritage Team)*

Because atoms interact with light, we must also discuss the nature of atoms. We will develop a simple model for an atom and see how it can alter light. Of course, our discussion of atoms is important for a second reason—all material objects such as planets, stars, and people are made from atoms. If our goal is to understand how a planet formed or how a star died, we must understand what atoms are. This chapter links two important ideas—atoms and starlight.

What we see when we look through an astronomical telescope is atoms. Thus we begin with a quick summary of the properties of an atom.

4-1 Atoms

A Model Atom

To think about atoms and how they can interact with light, we must create a working model of an atom. In Chapter 1, we created a working model of the sky, the celestial sphere. We identified and named the important parts and described how they were located and how they interacted. We begin our study of atoms by creating a model of an atom.

In our model atom, we identify a positively charged **nucleus** at the center; this nucleus consists of two kinds of particles. **Protons** carry a positive electrical charge, and **neutrons** have no charge. Thus, the nucleus has a net positive charge.

The nucleus in our model atom is surrounded by a whirling cloud of orbiting **electrons,** low-mass particles with a negative charge. In a normal atom, the number of electrons equals the number of protons, and the positive and negative charges balance to produce a neutral atom. Because protons and neutrons each have a mass 1836 times greater than that of an electron, most of the mass of the atom lies in the nucleus. The hydrogen atom is the simplest of all atoms. The nucleus is a single proton orbited by a single electron, with a total mass of only 1.67×10^{-27} kg, about a trillionth of a trillionth of a gram.

To grasp how small an atom is, we can construct a simple scale model. The nucleus of a hydrogen atom is a proton with a diameter of about 0.0000016 nm, or 1.6×10^{-15} m. If we multiply this by a factor of one trillion (10^{12}) we could represent the nucleus of our model atom with a grape seed about 0.16 cm in diameter. The electron cloud* has a diameter of about 0.4 nm, or 4×10^{-10} m. Multiplying by a trillion magnifies this to about 400 m, or about 4.5 football fields laid end to end (Figure 4-2). When you imagine a grape seed in the midst of 4.5 football fields, orbited by a single magni-

---

*For a representative diameter, we take the size of the atom's second orbit (see Figure 4-3).

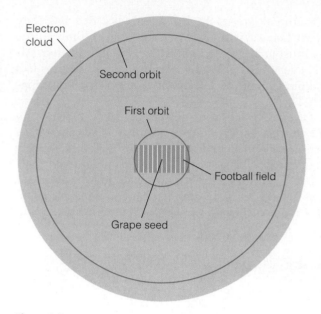

Figure 4-2
Magnifying a hydrogen atom by 10^{12} makes the nucleus the size of a grape seed and the outer electron cloud about 4.5 times bigger than a football field. The electron itself is still too small to see.

fied electron still too small to detect, you can see that an atom is mostly empty space.

Different Kinds of Atoms

There are over a hundred kinds of atoms, called chemical elements. The kind of element an atom represents depends only on the number of protons in the nucleus. For example, carbon has six protons and six neutrons in its nucleus. Adding a proton produces nitrogen, and subtracting a proton produces boron.

We can, however, change the number of neutrons in an atom's nucleus without changing the atom significantly. For instance, if we add a neutron to the carbon nucleus, we still have carbon, but it is slightly heavier than normal carbon. Atoms that have the same number of protons but a different number of neutrons are **isotopes.** Carbon has two stable isotopes. One form contains six protons and six neutrons for a total of 12 particles and is thus called carbon-12. Carbon-13 has six protons and seven neutrons in its nucleus.

Protons and neutrons are bound tightly into the nucleus, but the electrons are held loosely in the electron cloud. Running a comb through your hair creates a static charge by removing a few electrons from their atoms. This process is called **ionization,** and the atom that has lost one or more electrons is an **ion.** The neutral carbon atom, with six protons and six neutrons in its nucleus, has six electrons, which balance the positive charge of the nucleus. If we ionize the atom by re-

Quantum mechanics is the set of rules that describe how atoms and subatomic particles behave. When we think about large objects, such as stars, planets, aircraft carriers, and hummingbirds, we don't have to think about quantum mechanics; but on the atomic scale, particles behave in ways that seem unfamiliar to us.

One of the principles of quantum mechanics specifies that we cannot know simultaneously the exact location and motion of a particle. This is why physicists refer to the electrons in an atom as if they were a cloud of negative charge surrounding the nucleus. Because we can't know the position and motion of the electron, we can't really describe it as a small parti-

cle following an orbit. We can use that image as a model to help our imaginations, but the reality is much more interesting, and describing the electrons as a charge cloud gives us a better and more sophisticated model of an atom.

This raises some serious questions about reality. Is an electron really a particle at all? Quantum mechanics describes particles as waves, and waves as particles. If we can't know simultaneously the position and motion of a specific particle, how can we know how it will react to a collision with a photon or another particle? The answer is that we can't know, and that seems to violate the principle of cause and effect (Window on Science 2-3).

Needless to say, we can't explore these discrepancies here. Scientists and philosophers of science continue to struggle with the meaning of reality on the quantum-mechanical level. Here we should note that the reality we see on the scale of stars and hummingbirds is only part of nature. We have constructed some models to help us think about nature on the scale of atoms, but the truth is much more interesting and much more exciting than anything we see on larger scales. Although we use models of atoms to study stars, there is still much to learn about the atoms themselves.

---

moving one or more electrons, the atom is left with a net positive charge. Under some circumstances, an atom may capture one or more extra electrons, giving it more negative charges than positive. Such a negatively charged atom is also considered an ion.

Atoms form bonds with each other by exchanging or sharing electrons. Two or more atoms bonded together form a **molecule.** Few atoms can form chemical bonds in stars, where the high temperatures produce such violent collisions between atoms that most molecules break up. In planets, however, most atoms are linked together to form molecules. Earth is made of molecules, and so are our bodies. Just as do atoms, molecules interact with light, and thus we can analyze the light we receive from a celestial body to learn its secrets whether it is made of free atoms or atoms linked together in molecules.

Electron Shells

So far we have described the electron cloud only in a general way, but the specific way electrons behave within the cloud is very important in astronomy.

The electrons are bound to the atom by the attraction between their negative charge and the positive charge on the nucleus. This force is known as the **Coulomb force,** after the French physicist Charles-Augustin de Coulomb (1736–1806). If we wish to ionize the atom, we need a certain amount of energy to pull an electron away from its nucleus. This energy is the electron's **binding energy,** the energy that holds it to the atom.

An electron may orbit the nucleus at various distances. If the orbit is small, the electron is close to the nucleus, and a large amount of energy is needed to pull it away. Therefore, its binding energy is large. An electron orbiting farther from the nucleus is held more loosely, and less energy will pull it away. It therefore has less binding energy. The size of an electron's orbit is related to the energy that binds it to the atom.

Nature permits atoms only certain amounts (quanta) of binding energy, and the laws that describe how atoms behave are called the laws of **quantum mechanics** (Window on Science 4-1). Much of our discussion of atoms is based on the laws of quantum mechanics.

Because atoms can have only certain amounts of binding energy, our model atoms can have orbits of only certain sizes, called **permitted orbits** (Figure 4-3). These are like steps in a staircase: You can stand on the number-one step or the number-two step, but not on the number-one-and-one-quarter step. The electron can occupy any permitted orbit but not orbits in between.

The arrangement of permitted orbits depends primarily on the charge of the nucleus, which in turn depends on the number of protons. Thus, each kind of element has its own pattern of permitted orbits (Figure 4-3). Isotopes of the same elements have nearly the same pattern because they have the same number of protons. However, ionized atoms have orbital patterns that differ from their un-ionized forms. Thus the arrangement of permitted orbits differs for every kind of atom and ion.

Figure 4-3
The electrons in atoms may occupy only certain, permitted orbits. Because they have different charges on the nucleus, different elements have different patterns of permitted orbits.

REVIEW   **Critical Inquiry**

How many hydrogen atoms would it take to cross the head of a pin?

This is not a frivolous question. In answering it, we will discover how small atoms really are, and we will see how powerful physics can be as a way to understand nature. First, we will assume that the head of a pin is about 1 mm in diameter. That is 0.001 m. The size of a hydrogen atom is represented by the diameter of the electron cloud, and we will assume that the electron in our atom is in the second orbit. Then the diameter of the electron cloud is about 0.4 nm. Because 1 nm equals 10^{-9} m, we multiply and discover that 0.4 nm equals 4×10^{-10} m. To find out how many atoms would stretch 0.001 m, we divide the diameter of the pinhead by the diameter of an atom. That is, we divide 0.001 m by 4×10^{-10} m, and we get 2.5×10^6. It would thus take 2.5 million hydrogen atoms lined up side by side to cross the head of a pin.

This shows how tiny an atom is and also how powerful basic physics is. A bit of arithmetic gives us a view of nature beyond the capability of our eyes. Now use another bit of arithmetic to calculate how many hydrogen atoms you would need to add up to the mass of a paper clip (1 g).

The energy levels in atoms are familiar territory to astronomers because the electrons in those levels can interact with light. Such interactions fill starlight with clues to the nature of the stars.

4-2 The Interaction of Light and Matter

We begin our study of light and matter by considering the hydrogen atom. Hydrogen is both simple and common. Roughly 90 percent of all atoms in the universe are hydrogen.

The Excitation of Atoms

Each orbit in an atom represents a specific amount of binding energy, so physicists commonly refer to the orbits as **energy levels.** Using this terminology, we can say of the hydrogen atom in Figure 4-4 that the electron is in the lowest permitted energy level, where it is tightly bound to the atom.

We can move the electron from one energy level to another by supplying enough energy to make up the difference between the two energy levels. It is like moving a flowerpot from a low shelf to a high shelf; the greater the distance between the shelves, the more energy we need to raise the pot. The amount of energy needed to move the electron is the energy difference between the two energy levels.

If we move the electron from a low energy level to a higher energy level, we say the atom is an **excited atom.** That is, we have added energy to the atom in moving its electron. If the electron falls back to the lower energy level, that energy is released.

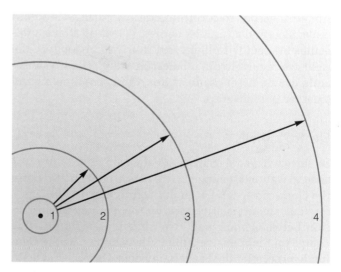

Figure 4-4
A hydrogen atom can absorb one of a number of different wavelength photons and move the electron to one of its higher orbits.

Figure 4-5
An atom can absorb a photon only if it has the correct amount of energy. The excited atom is unstable and within a fraction of a second returns to a lower energy level, reradiating the photon in a random direction.

An atom can become excited by collision. If two atoms collide, one or both may have their electrons knocked into a higher energy level. This happens very commonly in hot gas, where the atoms move rapidly and collide often.

Another way an atom can get the energy to move an electron to a higher energy level is to absorb a photon. Only a photon with exactly the right amount of energy can move the electron from one level to another. If the photon has too much or too little energy, the atom cannot absorb it. Because the energy of a photon depends on its wavelength, only photons of certain wavelengths can be absorbed by a given kind of atom. The atom in Figure 4-4 can absorb any of three wavelengths, moving its electron up to any of three permitted energy levels. Any other wavelength photon has too much or too little energy to be absorbed.

Atoms, like humans, cannot exist in an excited state forever. The excited atom is unstable and must eventually (usually within 10^{-6} to 10^{-9} seconds) give up the energy it has absorbed and return its electron to the lowest energy level. Because the electrons eventually tumble down to this bottom level, physicists call it the **ground state.**

When the electron drops from a higher to a lower energy level, it moves from a loosely bound level to one more tightly bound. The atom then has a surplus of energy—the energy difference between the levels—that it can emit as a photon. Study the sequence of events in Figure 4-5 to see how an atom can absorb and emit photons.

Because only certain energy levels are permitted in an atom, only certain energy differences can occur. Each type of atom or ion has its unique set of energy levels, so each one absorbs and emits photons with a unique set of wavelengths. Thus we can identify the elements in a gas by studying the characteristic wavelengths of light absorbed or emitted.

This process of excitation and emission is a common sight. The gas in a neon sign glows because a high voltage forces electrons to flow through the gas, exciting the atoms by collisions. Almost as soon as an atom is excited, its electron drops back to a lower energy level, emitting the surplus energy as a photon of a certain wavelength. Neon is a popular gas for signs be-cause the pattern of its energy levels makes it emit a rich reddish-orange light when it is excited. So-called neon signs of other colors contain other gases or mixtures of gases instead of pure neon.

Radiation from a Heated Object

To begin our discussion of the interaction of light and matter, we must consider a simple but very important phenomenon. When we heat an object up, it begins to glow. Think of a blacksmith heating a horseshoe in a forge. The hot iron glows bright yellow-orange. What produces this light?

Recall that light is a changing electric and magnetic field. Whenever we produce a changing electric field, we produce electromagnetic radiation. If we run a comb through our hair, we disturb electrons (negatively charged subatomic particles) in both hair and comb, producing static electricity. Because each electron is surrounded by an electric field, any sudden change in the electron's motion gives rise to electromagnetic radiation. Running a comb through your hair while standing near an AM radio produces radio static. This illustrates an important principle: Whenever we change the motion of an electron, we generate electromagnetic waves.

To see what this has to do with a heated object, think of what we mean by hot. When we say an object is hot, we mean that its atoms are vibrating rapidly. The hotter the object is, the more motion among the atoms. The vibrating atoms collide with electrons in the material, and each time the motion of one of these electrons gets disturbed, it emits a photon. Consequently, we should expect a heated object to emit electromagnetic radiation. Such radiation is called **black body radiation** because it refers to light emitted by an ideal object that is a perfect absorber and emitter of radiation. Such an object would look black at room temperature. Black body radiation is quite common. In fact, it is responsible for the light emitted by an incandescent lightbulb, and we recognize the light emitted by the heated horseshoe in the blacksmith's forge as black body radiation. Many objects in astronomy, including the stars, emit radiation as if they were nearly perfect black bodies.

If we graph the energy we get from a heated object versus wavelength, we get a curve like those in Figure 4-6. The curves show that very short wavelength and very long wavelength photons are rare. Most of the emitted photons have some intermediate wavelength. We can understand this because the photons are emitted by collisions, and very violent and very gentle collisions are rare. Most collisions in a heated body have some intermediate energy and produce photons of intermediate wavelength.

The **wavelength of maximum intensity (λ_{max})** depends on the temperature of the object. In a cool object, the average collision among the particles is gentle, and most of the photons have longer wavelengths. Thus λ_{max} is long for cool objects. Hot objects have more violent collisions among the particles, emit more short-wavelength photons, and λ_{max} is shorter. Notice in Figure 4-6 how this changes the color of the object. Cool stars glow a dull red, and hotter objects are yellower. This explains how a blacksmith can judge the temperature of a horseshoe in the forge by its color. It also explains how astronomers can judge the temperature of stars by their color. The light emitted by stars is very similar to black body radiation, so cool stars look red

Figure 4-6
Black body radiation from three bodies at different temperatures demonstrates that a hot body radiates more total energy and that the wavelength of maximum intensity is shorter for hotter objects. The hotter object here will look blue to our eyes, while the cooler object will look red.

By the Numbers 4-1

Black Body Radiation

Black body radiation is described by two simple laws. So many objects in astronomy behave like black bodies that these two laws are powerful tools for the analysis of light from the sky.

Wien's Law shows that the wavelength of maximum in nanometers, λ_{max}, equals 3,000,000 divided by the temperature in degrees Kelvin.

$$\lambda_{max} = \frac{3,000,000}{T}$$

For example, a cool star with a temperature of 3000 K will emit most intensely at a wavelength of 1000 nm, which is in the infrared. Of course, a hot star with a temperature of 30,000 K, for example, will emit most intensely at a wavelength of 100 nm, which is in the ultraviolet part of the spectrum.

The Stefan-Boltzmann Law says that the total energy radiated in 1 second from 1 square meter of an object equals a constant σ times the temperature raised to the fourth power.*

$$E = \sigma T^4 \text{ (J/s/m}^2)$$

Here we express the temperature in degrees Kelvin and the energy is expressed in units called joules. One **joule (J)** is about the energy of an apple falling from a table to the floor. This law shows how strongly the energy radiated depends on temperature. If we doubled an object's temperature, for instance, it would radiate not 2 times, but rather 2^4, or 16, times more energy per second from each square meter of its surface. A small change in temperature can make a big difference to the brightness of a star.

*For the sake of completeness, we note that the constant σ equals 5.67×10^{-8} J/m² s degree⁴.

Figure 4-7
The three types of spectra. A continuous spectrum (top) contains no bright or dark lines, but an absorption spectrum (middle) is interrupted by dark absorption lines. An emission spectrum (bottom) is dark except at certain wavelengths where emission lines occur. Note that the lines in the absorption spectrum of hydrogen (identified by Greek letters) have the same wavelength as the lines in the emission spectrum of hydrogen.

and hot stars look blue. The sun, a medium-temperature star, looks pale yellow.

As we might expect, hot objects emit more radiation than cool objects. We can understand that because the particles in a hot object move faster and collide more often, so they emit more photons. Of course a hot horseshoe emits more energy than a horseshoe at room temperature, and a given area of a hot star emits more energy than the same area of a cool star. These relationships are given as simple formulas in By the Numbers 4-1.

Our study of black body radiation alerts us to two interesting details. First, every wavelength is present in black body radiation. If we spread the light out into a spectrum, we would see a **continuous spectrum** (Figure 4-7) of color with no gaps. Second, although we see no light emitted by objects at room temperature, we must conclude that everything around us is glowing at infrared wavelengths. The wavelength of maximum for room-temperature objects, including our own bodies, lies in the infrared. In fact, some burglar alarms use infrared detectors to pick up the infrared black body radiation emitted by intruders. Even cold objects must emit some radiation. **Absolute zero** is the temperature at which a body's particles contain no motion that can produce photons. In a later chapter we will see that dust in space only a few degrees above absolute zero can be detected by the long wavelength radiation it emits.

To refer to the temperature of astronomical objects such as stars and clouds of dust in space, we will use the Kelvin temperature scale. On this scale, zero degrees Kelvin (written 0 K) is absolute zero (−459.7°F). Nothing can become that cold. Water freezes at 273 K

and boils at 373 K (see Appendix A). The Kelvin scale is important in astronomy because it is defined by absolute zero and thus is directly related to the agitation among the particles in a heated body and to black body radiation.

Now that we understand black body radiation, we can understand how stars can form a spectrum. More important, we can understand how to get information from those spectra.

The Formation of a Spectrum

To see how an astronomical object like a star can produce a spectrum, imagine a cloud of hydrogen gas floating in space with an incandescent lightbulb glowing behind it. The filament in the bulb emits black body radiation and produces a continuous spectrum.

The light from this bulb, however, must pass through the hydrogen gas before it can reach our telescope (Figure 4-8). Most of the photons will pass through the gas unaffected because they have wavelengths the hydrogen atoms cannot absorb, but a few photons will have the right wavelengths. These photons cannot pass through the gas because they are absorbed by the first atom they meet. The atom is excited for a fraction of a second, and the electron then drops back to a lower energy level and a new photon is emitted. The original photon was traveling through the gas toward our telescope, but the new photon is emitted in some random direction. Very few of these new photons leave the cloud in the direction of our telescope, so the light that finally

Figure 4-8
Photons of the proper wavelengths can be absorbed by the gas atoms and re-emitted in random directions. Because most of these particular photons do not reach the telescope, the spectrum is dark at the wavelengths of the missing photons.

enters the telescope has very few photons at the wavelengths the atoms can absorb. When we form a spectrum from this light, photons of these wavelengths are missing and the spectrum has dark lines at the positions these photons would have occupied. These dark lines, like those the Munich optician Fraunhofer saw in the solar spectrum in 1814 and 1815, are called **absorption lines** because the atoms absorbed the photons. A spectrum containing absorption lines is an **absorption spectrum** (also called a **dark-line spectrum**).

What happens to the photons that were absorbed? They bounce from atom to atom, being absorbed and emitted over and over until they escape from the cloud. If, instead of aiming our telescope at the bulb, we swing it to one side so that no light from the bulb enters the telescope, we can photograph a spectrum of the light emitted by the gas atoms (Figure 4-9). In this case, the only photons entering the telescope are photons that were absorbed and re-emitted. A spectrum of this light is almost entirely dark except for the wavelengths corresponding to the photons the gas can absorb and re-emit. Thus we will see a spectrum containing only bright lines on a dark background. These bright lines are called **emission lines,** and a spectrum with emission lines is an **emission spectrum** (also called a **bright-line spectrum**). The spectrum of a neon sign, for instance, is an emission spectrum. Also, the bluish-purple color of mercury-vapor streetlights and the pink-orange color of sodium-vapor

streetlights are produced by the emission lines of these elements.

The properties of the three spectra—continuous, absorption, and emission—can be described by the three rules known as **Kirchhoff's laws.** Law I states that a heated solid, liquid, or dense gas will produce a continuous spectrum. We see this in the filament of a lightbulb. Law II states that a low-density gas excited to emit light will produce an emission spectrum. We see this in neon signs. And Law III states that if light comprising a continuous spectrum passes through a cool, low-density gas, the result will be an absorption spectrum. We see this in stars.

Stellar spectra are absorption spectra formed as light from the bright surface, the photosphere, travels outward through the stellar atmosphere (Figure 4-10). The bright surface of the star is sufficiently dense and hot that collisions between gas atoms and electrons produce a continuous spectrum just as does the fila-

Figure 4-9
Pointing the telescope away from the bulb, we can receive only those photons the atoms can absorb and re-emit, producing emission lines in the spectrum.

Figure 4-10
A star produces an absorption spectrum because its atmosphere absorbs certain wavelength photons in the spectrum.

ment of the lightbulb in Figure 4-8. The lower-density gases in the stellar atmosphere absorb certain wavelength photons, and thus the light that finally reaches our telescope is missing those specific wavelengths, and we see an absorption spectrum.

The absorption and emission lines in the spectra of astronomical objects contain a windfall of information about the gases in stars, in comets, in nebulae in space, and in the atmospheres of planets. We will study the spectra of stars in greater detail in Chapter 11, but we can now review a few of the basic kinds of information we find in astronomical spectra.

4-3 Information from Spectra

A spectrum can give us two kinds of information. It can tell us about the matter that emitted the light we see, and it can also tell us about the motion of that matter.

The Analysis of a Spectrum

Kirchhoff's laws give us a powerful tool in the analysis of a spectrum. If we see that the spectrum is an emission spectrum, we know immediately that we are observing an excited, low-density gas. Nebulae in space such as that shown in Figure 4-1 produce emission spectra, and so do some tails of comets. We will use this clue to discover that some comet tails are made of ionized gas.

If we see an absorption spectrum, we know that we are seeing light that has traveled through a cloud of gas. The spectra of stars are absorption spectra, and so, of course, is the spectrum of the sun.

At visual wavelengths, astronomers almost never see a continuous spectrum, but all planetary objects emit black body radiation. Observing in the infrared, astronomers can measure the wavelength of maximum intensity in the radiation emitted by an object and deduce the temperature of the object. We will see how this can be applied to things such as dust in space and the moons of planets.

Obviously we can find chemical composition from a spectrum. If a star's spectrum contains lines specific to a certain atom, we know that the star must contain that atom. For example, the sun's spectrum contains spectral lines of sodium. Thus the sun must contain sodium. When we discuss stars in detail, we will see that they are mostly hydrogen and helium with only traces of heavier elements.

Deducing the relative amounts of the different atoms in a star is quite difficult, but the mere presence of spectral lines can tell us that certain atoms must be present. The spectrum of Mars, for example, contains absorp-tion lines produced by carbon dioxide gas. Thus we know its atmosphere must contain that gas.

Not only can a spectrum tell us about the object that emitted the light, but it can also tell us about motion.

The Doppler Effect

Astronomers can measure the wavelengths of lines in a star's spectrum and find the velocity of the star. The **Doppler effect** is the apparent change in the wavelength of radiation caused by the motion of the source.

If a star is moving toward Earth, the lines in its spectrum will be shifted slightly toward shorter wavelengths. That is, they are shifted toward the blue end of the spectrum—a **blue shift**. If a star is moving away from Earth, the lines are shifted slightly toward the red end of the spectrum—a **red shift**. These Doppler shifts are small and don't change the color of the star. But it is easy to detect these changes in wavelength in a star's spectrum.

Figure 4-11a shows the Doppler effect in two spectra of the star Arcturus. The lines in the top spectrum are slightly blue shifted because the spectrum was recorded when Earth, following its orbit, was moving toward Arcturus. The lines in the bottom spectrum are red shifted because it was recorded six months later, when Earth was moving away from Arcturus.

The Doppler effect tells us how rapidly the distance between us and the source of light is increasing or decreasing. It does not matter whether we are moving or the star is moving. Only the relative velocity is important. Also, the Doppler shift is only sensitive to the part of the velocity directed away from us or toward us, the **radial velocity (V_r).** We cannot use the Doppler effect to detect any part of the velocity that is perpendicular to our line of sight.

The Doppler effect occurs for any kind of radiation, not just light. Radio astronomers, for instance, observe the Doppler effect when they measure the wavelength of radio waves coming from objects moving toward or away from Earth. Whatever wavelength range we consider—radio, X rays, gamma rays—we will still refer to a lengthening of the wavelength as a red shift and a shortening as a blue shift.

We can even detect the Doppler shift in sound. Sounds with long wavelengths have a lower pitch, and sounds with short wavelengths have a higher pitch. When we hear a truck roar past us on the highway, we hear the sound of the truck drop in pitch as it passes. Its sound is blue shifted while it is approaching and red shifted after it passes.

We can understand how the Doppler shift works by thinking about a fire truck approaching us with a bell clanging once a second. When the bell clangs, the sound travels ahead of the truck to reach our ears. One second later, the bell clangs again, but not at the same place.

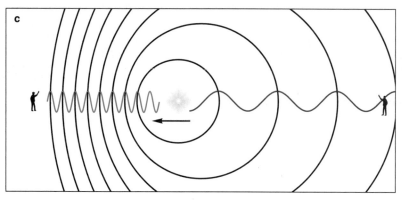

Figure 4-11

The Doppler effect. (a) A blue shift appears in the spectrum of a star approaching Earth (top spectrum). A red shift appears in the spectrum of a star moving away from Earth (bottom spectrum). *(The Observatories of the Carnegie Institution of Washington)* (b) The clanging bell on a moving fire truck produces sounds that move outward (black circles). An observer ahead of the truck hears the clangs closer together, while an observer behind the truck hears them farther apart. (c) A moving source of light emits waves that move outward (black circles). An observer in front of the light source observes a shorter wavelength (a blue shift), and an observer behind the light source observes a longer wavelength (a red shift).

During that one second, the fire truck moved closer to us, so the bell is closer at its second clang. Now the sound has a shorter distance to travel and reaches our ears a little sooner than it would have if the fire truck were not approaching. The third time the bell clangs, it is even closer. By timing the bell, we could observe that the clangs are slightly less than one second apart, all because the fire truck is approaching. If the fire truck were moving away from us, we would hear the clangs sounding more than one second apart, because each

successive clang of the bell occurs farther from us.

Figure 4-11b shows a fire truck moving toward one observer and away from another. The position of the bell at each clang is shown by the black bells with the sound spreading outward as circles. We can see that the clangs are squeezed together ahead of the fire truck and stretched apart behind.

Now we can substitute a source of light waves for the clanging of the bell. If the source is approaching, then each time the source emits the peak of a wave it will be slightly closer to us, and we will observe a shorter wavelength. If it is moving away, we will observe a longer wavelength. This is shown in Figure 4-11c, where the peaks of the waves appear compressed in front of the moving light source and stretched out behind the source.

Of course, how great the change in wavelength is depends on the velocity. Just as a slowly moving car has a smaller Doppler effect in sound than does a high-speed airplane, a slowly moving star has a smaller Doppler effect in light than does a high-speed star. Thus, we can measure the velocity by measuring the amount by which the spectral lines are shifted in wavelength. A large Doppler shift means a high radial velocity. This is a common tool in law enforcement. Police transmit radio signals toward passing autos and measure speed by measuring the wavelength shifts in the reflected signal. By the Numbers 4-2 shows how we could do a Doppler effect calculation.

By the Numbers 4-2

The Doppler Formula

Astronomers can measure radial velocity by using the Doppler effect. The laboratory wavelength λ_0 is the wavelength a certain spectral line would have in our laboratory where the source of the light is not moving. In the spectrum of a star, we see this spectral line shifted by some small amount $\Delta\lambda$. If the wavelength is increased, a red shift, $\Delta\lambda$ is positive, and if the wavelength is decreased, a blue shift, $\Delta\lambda$ is negative. Then the radial velocity of the star is given by the Doppler formula.

$$\frac{V_r}{c} = \frac{\Delta\lambda}{\lambda_0}$$

That is, the radial velocity divided by the speed of light, c, is equal to $\Delta\lambda$ divided by λ_0. In astronomy, radial velocities are almost always given in kilometers per second, so we will express c as 300,000 km/s.

For example, suppose the laboratory wavelength of a certain spectral line is 600.00 nm, and the line is observed in a star's spectrum at a wavelength of 600.10 nm. Then $\Delta\lambda$ is +0.10 nm, and the velocity is 0.10/600 multiplied times the speed of light. The radial velocity equals 50 km/s. Because $\Delta\lambda$ is positive, we know the star is receding from us.

Astronomical spectra are filled to bursting with clues to the nature of the universe. We will use these clues often in the chapters that follow, and we begin in the next chapter where we discuss the formation of our solar system.

Summary

An atom consists of a nucleus surrounded by a cloud of electrons. The nucleus is made up of two kinds of particles: positively charged protons and uncharged neutrons. The number of protons in an atom determines which element it is. Atoms of the same element (that is, having the same number of protons) with different numbers of neutrons are called isotopes.

The motion among the atoms in a solid, liquid, or dense gas causes the emission of black body radiation. The hotter the object, the more radiation it emits. This produces a continuous spectrum, and the energy is most intense at the wavelength of maximum intensity, λ_{max}, which depends on the body's temperature. Hot objects emit mostly short-wavelength radiation, while cool objects emit mostly long-wavelength ra-

diation. This effect gives us clues to the temperatures of stars—hot stars are blue and cool stars are red.

The negatively charged electrons surrounding an atomic nucleus may occupy various permitted orbits. An electron may be excited to a higher orbit during a collision between atoms, or it may move from one orbit to another by absorbing or emitting a photon of the proper energy. If the energy of an absorbed photon is too large, the atom may lose an electron and become ionized.

Because only certain orbits are permitted, only photons of certain wavelengths can be absorbed or emitted. Each kind of atom has its own characteristic set of spectral lines.

If light passes through a low-density gas on its way to our telescope, the gas can absorb photons of certain wavelengths, and we will see dark lines in the spectrum at those positions. Such a spectrum is called an absorption spectrum. If we look at a low-density gas that is excited to emit photons, we see bright lines in the spectrum at those positions. Such a spectrum is called an emission spectrum.

When a source of radiation is approaching us, we observe shorter wavelengths, and when it is receding, we observe longer wavelengths. This Doppler effect makes it possible for the astronomer to measure a star's radial velocity, that part of its velocity directed toward or away from Earth.

New Terms

nucleus

proton

neutron

electron

isotope

ionization

ion

molecule

Coulomb force

binding energy

quantum mechanics

permitted orbit

energy level

excited atom

ground state

black body radiation

wavelength of maximum intensity (λ_{max})

continuous spectrum

absolute zero

joule (J)

absorption line

absorption spectrum (dark-line spectrum)

emission line

emission spectrum (bright-line spectrum)

Kirchhoff's laws

Doppler effect

blue shift

red shift

radial velocity (V_r)

Review Questions

1. Why might we say that atoms are mostly empty space?

2. What is the difference between an isotope and an ion?

3. Why is the binding energy of an electron related to the size of its orbit?

4. Explain why ionized calcium can form absorption lines but ionized hydrogen cannot.

5. Describe two ways an atom can become excited.

6. Why do different atoms have different lines in their spectra?

7. Why does the amount of black body radiation emitted depend on the temperature of the object?

8. Why do hot stars look bluer than cool stars?

9. What kind of spectrum does a neon sign produce?

10. Why does the Doppler effect detect only radial velocity?

11. How can the Doppler effect explain shifts in both light and sound?

Discussion Question

1. In what ways is our model of an atom a scientific model? In what ways is it incorrect?

Problems

1. Human body temperature is about 310 K (98.6°F). At what wavelength do humans radiate the most energy? What kind of radiation do we emit?

2. If a star has a surface temperature of 20,000 K, at what wavelength will it radiate the most energy?

3. Infrared observations of a star show that it is most intense at a wavelength of 2000 nm. What is the temperature of the star's surface?

4. Dust orbiting distant stars may be evidence that planets have formed there. If the dust is brightest at the far infrared wavelength of 100,000 nm, what is the temperature of the dust?

5. If astronomers observe that a liquid flowing out onto the surface of a planet is brightest at a wavelength of 1700 nm, what is its temperature? Do you think it is liquid water or liquid rock?

6. If we triple the temperature of a black body, by what factor will the total energy radiated per second per square meter increase?

7. If one star has a temperature of 6000 K and another star has a temperature of 7000 K, how much more energy per second will the hotter star radiate from each square meter of its surface?

8. In the laboratory, a certain spectral line of hydrogen has a wavelength of 486.10 nm. If the line appears in a comet's spectrum at 486.15 nm, what is the comet's radial velocity?

9. The highest velocity stars an astronomer might observe have velocities of about 400 km/s. What change in wavelength would this cause in the spectral line described in Problem 8?

Critical Inquiries for the Web

1. The name for the element helium has astronomical roots. Search the Internet for information on the discovery of helium. How and when was it discovered, and how did it get its name? Why do you suppose it took so long for helium to be recognized?

2. How was the model of the atom presented in the text you read developed? Search the Web for information on historical models of the atom, and compile a time line of important developments leading to our current understanding. What evidence exists that supports our model?

Exploring *The Sky*

1. Classification of stars of into spectral classes provides invaluable information about the surface temperature of a star. Use *The Sky* to determine the spectral class (or type) and the approximate temperature of the following easily observed bright stars: Antares in Sorpius, Betelgeuse in Orion, Aldebaran in Taurus, Sirius in Canis Major, and Rigel in Orion.

   How to proceed: One way to obtain stellar data is to press the F-key for "Find," then select "Common Star Names," then the star of interest, and finally, click on the "Find" button. If you want actually to view the star and the neighboring sky, click on the "Center & Frame" button instead of the "Find" button. Once you have found the spectral class, use Table 4-1 to find the approximate temperature of the star.

Go to the Brooks/Cole Astronomy Resource Center (www. brookscole.com/astronomy) for critical thinking exercises, articles, and additional readings from InfoTrac College Edition, Brooks/Cole's online student library.

The Origin of the Solar System

What place is this? Where are

we now?

Carl Sandburg
Grass

Guidepost

Preceding chapters have been preparation for our attempt to understand the universe. We begin here with our solar system.

Before we study each of the planets in detail, we must construct a hypothesis for their origin. That will give us a way to organize all of the facts that we will learn about the planets, and it will help us relate the facts to each other.

In addition, we must ask how our solar system originated, because the origin of things is the fundamental question of astronomy. Whatever object we study, as astronomers we must ask, "How did this object form and develop?" In later chapters we will ask this same question about stars, galaxies, and the universe.

The first law of environmental studies is, "Everything has to go somewhere." Perhaps the first law of astronomy should be, "Everything has to have come from somewhere."

Microscopic creatures live in the roots of your eye-lashes. Don't worry. Everyone has them, and they are harmless.* They hatch, fight for survival, mate, lay eggs, and die in the tiny spaces around the roots of our eyelashes without doing us any harm. Some live in renowned places—the eyelashes of a glamorous movie star—but the tiny beasts are not self-aware; they never stop to say, "Where are we?" Humans are more intelligent; we have the ability and the responsibility to wonder where we are in the universe and how we came here.

In this chapter, we begin exploring the universe by studying our solar system: the sun and its family of planets. We must avoid, however, the great pitfall of science. We must avoid memorizing facts instead of understanding nature. Jupiter is 317.83 times more massive than Earth. That is a fact, and it is boring. The excitement of science comes from understanding nature. We want to understand how the sun and planets formed, and that will tell us *why* Jupiter is so massive.

Facts fit together to build understanding, so we begin our study of the solar system by developing a unifying hypothesis for the origin of the sun and planets. As we learn facts about the solar system we can use them to test and improve our hypothesis, and that will increase our understanding of nature. A unifying hypothesis is like a basket into which we can put facts. The facts are easier to remember because they make sense, and the combination of hypothesis and supporting facts represents real knowledge—real understanding—and not just memory.

We have no choice; we must understand our solar system out of self-preservation. We live on a planet with 6 billion other humans, and we are altering our world's environment in unplanned and unknown ways. We could make Earth uninhabitable. The deadly deserts of Mars and sulfuric acid fogs of Venus may help us understand our own more comfortable world.

We must also study planets as astronomers. It seems likely that roughly half of all stars have planetary systems, so there are more planets in the universe than stars. Also, the chemical reactions that give rise to life seem to need the moderate conditions of the surface of planets. To search for life in the universe, we must understand planets.

Above all, we study the solar system because it is our home in the universe. Because we are an intelligent species, we have the right and the responsibility to wonder what we are. Our kind have inhabited this solar system for at least a million years, but only within a lifetime have we begun to understand what a solar system is. Like sleeping passengers on a train, we waken, look out at the passing scenery, and mutter, "What place is this? Where are we now?"

Demodex folliculorum has been found in 97 percent of individuals and is a characteristic of healthy skin.

5-1 The Great Chain of Origins

We are linked through a great chain of beginnings backward through time to a first instant 10 to 20 billion years ago when the universe began. The gradual discovery of the links in that chain is one of the most exciting adventures of the human intellect. In later chapters we will study the origin of the universe, the formation of galaxies, the birth of stars, and the building of the atomic elements. Here we will summarize that story in a few paragraphs so that we can gain a perspective on our place in the cosmos. Then we can propose a hypothesis for the formation of our solar system that places it in its proper place in nature.

The Origin of Matter

Look at your thumb. The matter in your thumb came into existence within minutes of the beginning of the universe. Astronomers have strong evidence that the universe began in an event called the big bang (Chapter 18), and by the time the universe was 3 minutes old, the protons, neutrons, and electrons in your thumb had come into existence. You are made of very old matter.

Although those particles formed quickly, they were not linked together as they are today. Most of the matter was hydrogen, and about 25 percent was helium. Very few heavier atoms were made in the big bang. Although helium is very rare in our bodies, your thumb contains many of those ancient hydrogen atoms unchanged since the universe began.

Within a billion years or so after the big bang, matter began to collect to form galaxies containing billions of stars. Astronomers understand that nuclear reactions inside stars combine low-mass atoms such as hydrogen to make heavier atoms (Chapter 12). Generation after generation of stars cooked the original particles, linking them together to build atoms such as carbon, nitrogen, and oxygen (Chapter 15). Look at your thumb. Even the calcium in the bone and the iron in the blood was assembled inside stars.

The atoms heavier than iron were created by rapid nuclear reactions that can only occur when a massive star explodes (Chapter 13). Gold and silver are rare in our bodies, but iodine is critical in your thyroid gland, and there are no doubt a few of those iodine atoms circulating through your thumb at this very moment, thanks to the violent deaths of massive stars.

Our galaxy contains about 100 billion stars, of which our sun is one. It formed from a cloud of gas and dust about 5 billion years ago, and the atoms in your thumb were part of that cloud. How the sun took shape, how the cloud gave birth to the planets, how those atoms in your thumb found their way into the Earth and into you is the story of this chapter. As we explore the origin of

Many theories in science can be classified as either evolutionary, in that they involve gradual processes, or catastrophic, in that they depend on specific, unlikely events. Scientists have generally preferred evolutionary theories. Nevertheless, catastrophic events do occur.

Something in us prefers catastrophic theories, perhaps because we like to see spectacular violence from a safe distance, which may explain the success of movies that include lots of car crashes and explosions. Also, cataclysmic theories resonate with Old Testament accounts of cata-

strophic events and special acts of creation. Thus, we have an understandable interest in catastrophic theories.

Nevertheless, most scientific theories are evolutionary. Such theories do not depend on unlikely events or special acts of biblical creation, and thus they are open to scientific investigation. For example, geologists much prefer theories of mountain building that are evolutionary, with the mountains being pushed up slowly as centuries pass. All the evidence of erosion and the folding of rock layers show that the process is gradual. Because most such

natural processes are evolutionary, scientists sometimes find it difficult to accept any theory that depends on catastrophic events.

We will see in this and later chapters that catastrophes do occur. Earth, for example, is bombarded by debris from space, and some of those impacts are very large. As you study astronomy or any other natural science, notice that most theories are evolutionary but that we must allow for the possibility of unpredictable catastrophic events.

---

our solar system, keep in mind the great chain of origins that created the atoms. As the geologist Preston Cloud remarked, "Stars have died that we might live."

The Origin of Planets

Over roughly the last two centuries, astronomers have proposed two kinds of theories for the origin of the planets. Catastrophic theories proposed that the planets formed from some improbable cataclysm such as the collision of the sun and another star. Gradualistic theories proposed that the planets formed naturally with the sun. Since the late 1930s, evidence has accumulated to support the gradualistic theories. In fact, nearly all astronomers now believe that planets form naturally as a by-product of star formation (Window on Science 5-1).

Astronomers believe that stars form from the gravitational contraction of clouds of gas and dust. In at least some cases, the contraction may be triggered by the compression of the cloud by the nearby explosion of a dying massive star.

The story of star formation is well understood. As the cloud contracts, the star begins to take shape deep inside, but light from the contracting star is absorbed by its cocoon of gas and dust. Thus, forming stars are not easily visible. The rotation of the cloud will gradually flatten it into a spinning disk around the forming star. As the star grows hotter, it becomes more luminous, and it eventually blows away its surrounding cocoon of gas and dust.

Modern theory proposes that planets form in the rotating disks around young stars (Figure 5-1). Infrared

observations show that some forming stars are cloaked in clouds of gas and dust, and Doppler shifts show that some young stars are blowing away their cocoons of gas

Figure 5-1
(a) Because the solar nebula was rotating, (b) it contracted into a disk, and (c) the planets formed with orbits lying in nearly the same plane.

Beta Pictoris

Size of Pluto's orbit

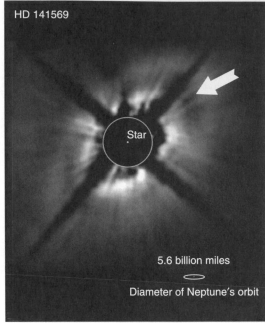

HD 141569

Star

5.6 billion miles

Diameter of Neptune's orbit

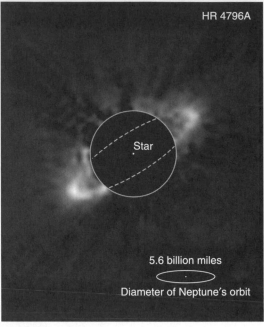

HR 4796A

Star

5.6 billion miles

Diameter of Neptune's orbit

Figure 5-2
The star Beta Pictoris (top) is hidden behind the black bar at center, and the dust disk around it is seen edge-on extending to right and left. *(Al Schultz, CSC/STScI, Sally Heap, GSFC, and NASA)* In the lower two images the stars are hidden behind the round masks with dust disks extending outward. The gap in the ring around HR141569 (arrow) may have been cleared by a planet. The narrowness of the ring around HR4796A suggests it is confined by planets. *(Alycia Weinberger, Eric Becklin, UCLA, Glenn Schneider, University of Arizona, and NASA)*

and dust at speeds up to 200 km/s. Other observations confirm that many young stars are surrounded by disks of gas and dust.

Our own planetary system probably formed in such a disk-shaped cloud around the sun. When the sun became luminous enough, the remaining gas and dust were blown away into space, leaving the planets orbiting the sun. This is known as the **solar nebula theory,** because the planets form from the nebula around the proto-sun (Figure 5-1). If the theory is right, then planets do form as a by-product of star formation, and thus most stars should have planetary systems.

Planets Orbiting Other Stars

If the solar nebula theory is correct, then planets should be common, and we could test the theory by looking for planets orbiting other stars, objects astronomers call **extra-solar planets.** Unfortunately, detecting a planet orbiting another star is a bit like detecting a bug crawling on the bulb in a distant searchlight. Nevertheless, astronomers have found evidence of extra-solar planets.

Infrared astronomers have found very cold, very low-density disks of dust around other stars such as Beta Pictoris. The Beta Pictoris disk is about 100 times the diameter of our solar system and has an inner hole

around the star which may be a place where planets have formed (Figure 5-2). These tenuous disks of cold dust are believed to be debris released from comets or produced by collisions among small bodies such as asteroids and comets. Our own solar system contains this kind of cold dust. Thus the cold, low-density dust disks are not evidence of active planet formation but are evidence that planetary systems have already formed.

Both visible and radio wavelength observations detect dense disks of gas orbiting young stars. For example, at least 50 percent of the stars in the Orion nebula, a region of star formation, are encircled by dense disks of gas and dust. Hubble Space Telescope images of very young stars reveal that many are surrounded by glowing clouds of gas and dust bisected by a dark shadow caused by a dense disk of gas and dust circling the star (Figure 5-3). Thus astronomers have evidence that newborn stars are commonly encircled by dense disks. In contrast with the low-density dust disks, these dense disks appear to be places where planets could be forming now.

In the middle 1990s, astronomers obtained further evidence of extra-solar planets. Although such a planet is too faint to be seen directly, it can be detected from the motion of the star it orbits. The first planet detected

Figure 5-3
Dust disks around young stars are evident in these Hubble Space Telescope infrared images. The stars are so young material is still falling inward and being illuminated by the light from the star. Dark bands across the nebulae (arrows) are caused by dense disks of gas and dust that orbit the stars. Note that the star at lower left is double. *(D. Padgett, IPAC/Caltech; W. Brandner, IPAC; K. Stapelfeldt, JPL; and NASA)*

this way orbits the star 51 Pegasi. As the planet circles the star, the star wobbles slightly. This very small motion of the star is detectable as Doppler shifts in the star's spectrum (Figure 5-4). From the motion of the star and estimates of its mass, astronomers can deduce that the planet has half the mass of Jupiter and orbits only 0.05 AU from the star. Presumably the star has other planets at greater distances, but they are harder to detect.

Astronomers greeted the announcement of a planet orbiting 51 Pegasi with cheers and skepticism (Window on Science 5-2), but more such planets were found. Roughly two dozen are known, including at least three orbiting the star Upsilon Andromidae—the first evidence astronomers have of a true planetary system.

Figure 5-4 ▶
The velocity of the sunlike star 51 Pegasi can be deduced from its spectrum. The star moves with a period of 4.2 days and an amplitude of only 50 meters per second. This wobble in the star is apparently produced by a planet of about half the mass of Jupiter orbiting only 0.05 AU from the star. This was the first planet found to orbit a normal star other than our sun. A number of other such planets have since been detected orbiting other stars. *(Adapted from data by Geoff Marcy and Paul Butler)*

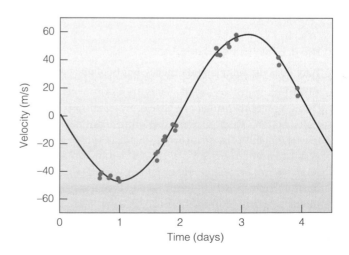

"Scientists are just a bunch of skeptics who don't believe in anything." That is a common compaint about scientists, but it misinterprets the fundamental characteristic of the scientist. Yes, scientists are skeptical about new ideas and discoveries, but scientists do hold strong beliefs about how nature works. Ultimately, scientists try to tell stories about how nature works, and skepticism is the scientists' guide.

Some people think that "telling a story" is telling a fib. The stories that scientists tell are exactly the opposite; perhaps we can call them anti-fibs because they are as true as they can be. The stories are explanations that have been tested over and over to make them accurate descriptions of nature. Skepticism is the tool scientists use to test every aspect of an explanation.

When planets were discovered orbiting 51 Pegasi, astronomers were skeptical—not because they thought the observations were wrong, but because that is how science works. Every observation is tested, and every discovery is confirmed. Only an idea that survives many tests begins to be accepted as a scientific truth.

To nonscientists, this makes scientists seem like irritable skeptics, but among scientists it is not bad manners to say, "Really, how do you know that?" or "Why do you think that?" or "Show me the evidence!"

Scientists try to tell stories about how nature works, stories that are sometimes called theories. A story by John Steinbeck can be a brilliant work of art invented to help us understand some aspect of our lives, and another story, such as a TV script, can be little more than chewing gum for the mind. But scientists' stories are different in a critical way. They are true. To create such stories, scientists cannot create facts to fit their story. Rather every link in a scientific story must be based on evidence and logic. Of course, evidence can be misunderstood, and logical errors happen. To test every aspect of their story, scientists must be continuously skeptical. That is the only way to discover those scientific truths that help us understand how nature works.

Skepticism is not a refusal to hold beliefs. Scientists often believe sincerely in their theories once those theories have been tested over and over. Rather, skepticism is the tool scientists use to find those natural principles worthy of belief.

The planets discovered so far tend to be massive and have short periods because lower-mass planets or longer-period planets are harder to detect. Low-mass planets don't tug on their stars very much, and present-day spectrographs can't detect such small velocity changes. Planets with longer periods are harder to detect because Earth's astronomers have not been making high-precision observations for many years. Thus we should not be surprised that most of the first planets discovered are massive and have short periods.

The discovery of extra-solar planets gives astronomers added confidence in the solar nebula theory. The theory predicts that planets are common, and astronomers are finding them orbiting many stars.

REVIEW   Critical Inquiry

Why does the solar nebula theory imply planets are common?

Often, the implications of a theory are more important in its analysis than its own conjecture about nature. If the solar nebula theory is correct, the planets of our solar system formed from the disk of gas and dust that surrounded the sun as it condensed from a cloud of gas and dust. If that is true, then it should be a common process. Most stars should form with disks of gas and dust around them, and so we would expect them to have planets.

Without knowing any more about the theory, the implication that planets are common suggests a way we could test the theory. If we look at a large number of stars and find lots of planets, the solar nebula theory is confirmed. But if we search and find few planets, it is contradicted. What observational evidence do we have so far on this issue?

So far only a few other stars are known to have planets. We can expect more such worlds to be discovered, and that raises the question of just how planets form. To answer that question, we begin by surveying our own solar system to discover the distinguishing characteristics of planetary systems. Those are the characteristics that our theory must eventually explain.

5-2 A Survey of the Solar System

To test our theories, we must search the present solar system for evidence of its past. In this section, we survey the solar system and compile a list of its most significant characteristics, potential clues to how it formed.

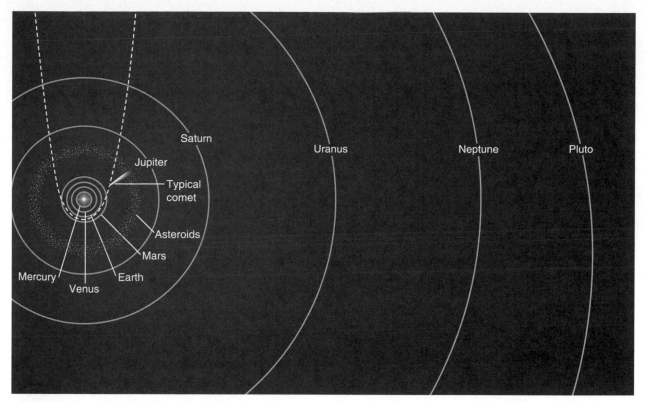

Figure 5-5

The solar system. The radii of the orbits are drawn to scale here. Note the eccentricity of Pluto's orbit. At this scale, only the sun would be visible. The planets are too small to be seen.

We begin with the most general view of the solar system. It is, in fact, almost entirely empty space (Figure 5-5). Imagine that we reduce the solar system until Earth is the size of a grain of table salt, about 0.3 mm (0.01 in.) in diameter. The moon is a speck of pepper about 1 cm (0.4 in.) away, and the sun is the size of a small plum 4 m (13 ft) from Earth. Mercury, Venus, and Mars are grains of salt. Jupiter is an apple seed 20 m (66 ft) from the sun, and Saturn is a smaller seed over 36 m (120 ft) away. Uranus and Neptune are slightly larger than average salt grains, and Pluto, the farthest planet (on average), is a speck of pepper over 150 m (500 ft) from the central plum. We would need a powerful microscope to detect the asteroids and the nuclei of comets. These tiny specks of matter scattered around the sun are the remains of the solar nebula.

Revolution and Rotation

The planets revolve around the sun in orbits that lie close to a common plane. The orbit of Mercury, the planet closest to the sun, is tipped 7° to Earth's orbit, and Pluto's orbit is tipped 17.2°. The rest of the planets' orbital planes are inclined by no more than 3.4°. Thus, the solar system is basically disk-shaped.

The rotation of the sun and planets on their axes also seems related to this disk shape. The sun rotates with its equator inclined only 7.25° to Earth's orbit, and most of the other planets' equators are tipped less than 30°. The rotations of Venus, Uranus, and Pluto are peculiar, however. Venus rotates backward compared with the other planets, and both Uranus and Pluto rotate on their sides (with their equators almost perpendicular to their orbits). We will discuss these planets in detail in Chapters 7 and 8, but later in this chapter we will try to understand how they could have acquired their peculiar rotations.

Apparently, the preferred direction of motion in the solar system—counterclockwise as seen from the north—is also related to its disk shape. All the planets revolve counterclockwise around the sun, and with the exception of Venus, Uranus, and Pluto they rotate counterclockwise on their axes.* Thus, with only a few exceptions, most of which are understood, revolution and rotation in the solar system follow a disk theme.

---

*Recall from Chapter 1 that English distinguishes between the words *revolve* and *rotate*. A planet revolves around the sun but rotates on its axis.

Two Kinds of Planets

Perhaps the most important clue we have to the origin of the solar system is the division of the planets into two categories: terrestrial, or Earthlike, and Jovian, or Jupiter-like (Figure 5-6). The **terrestrial planets** are small, dense, rocky worlds with less atmosphere than the Jovian planets. The terrestrial planets—Mercury, Venus, Earth, and Mars—lie in the inner solar system. By contrast, the **Jovian planets** are large, gaseous, low-density worlds. The Jovian planets—Jupiter, Saturn, Uranus, and Neptune—lie in the outer solar system beyond the asteroids. Note that Pluto does not fit either category very well. It is small like the terrestrial planets but lies far from the sun and has a low density like the Jovian planets.

Although *terrestrial* means "Earthlike," Mercury is more like Earth's moon. It is small, only 40 percent larger than the moon, and its weak gravity has held no atmosphere. Its gray, rocky surface is covered with thousands of overlapping craters, which, like the craters on Earth's moon, are the scars left behind by the impact of large meteorites (Figure 5-7). Such impact craters are very common in the solar system.

Venus is nearly as large as Earth, and its atmosphere is thick with heavy clouds hiding its surface

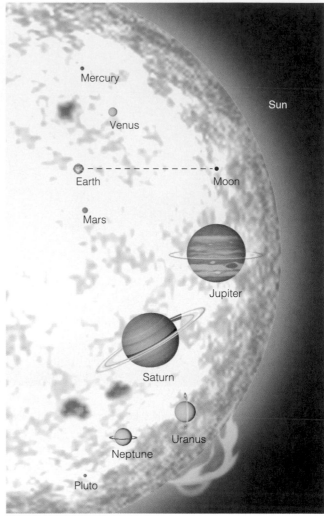

Figure 5-6
The relative sizes of the planets compared with the solar disk.

Figure 5-7
Mercury is about 40 percent larger in diameter than Earth's moon, shown here to the same scale. Like the moon, Mercury lacks an atmosphere and is heavily cratered by the impact of large meteorites. *(Mercury, NASA; moon, Lick Observatory photograph)*

(Figure 5-8). Radio signals can penetrate these clouds and bounce back from its rocky surface to give us a radar image of its terrain. These radar maps reveal that, although Venus is nearly the same size as Earth, it to-tally lacks Earth's oceans of water. Venus is a deadly, dry, desert world marked by impact craters, volcanoes, lava flows, and geological faults (Figure 5-9). In contrast, the atmosphere on Mars is thin, and we can see its surface clearly from space probes. These photos show that Mars, about half the diameter of Earth, is marked by impact craters and volcanoes.

Impact craters seem typical of all but the youngest surfaces in our solar system. The terrestrial planets, including Earth, are marked by such craters, and spacecraft have found craters on the moons of Mars, Jupiter, Saturn, Uranus, and Neptune. (There are no craters on the planets Jupiter, Saturn, Uranus, or Neptune because, as we will see, those planets do not have solid surfaces.) This suggests that craters are a characteristic of every object in the solar system with a surface capable of retaining such features.

The terrestrial planets have densities ranging from about 3 to 5 g/cm^3. Typically, they contain cores of iron and nickel surrounded by a mantle of dense rock. In contrast, the Jovian planets are rich in hydrogen and helium, so their average

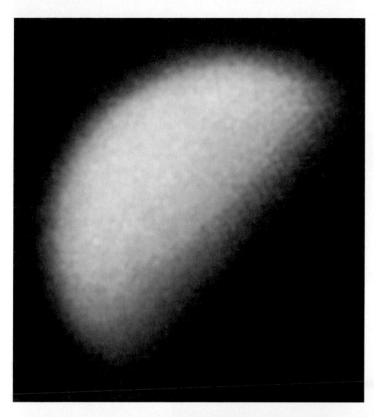

Figure 5-8
Venus is covered by a thick layer of clouds that never allows us to see its surface. In this photo taken through an Earth-based telescope, the planet is visible at quarter phase. Radar mapping from Earth and from satellites orbiting Venus has been able to penetrate the clouds and reveal the planet's rocky terrain. *(NASA)*

Figure 5-9
Earth, Venus, and Mars to scale. Earth (left) is mostly water-covered with large masses of exposed rocky surface and an atmosphere of changing cloud patterns. The surface of Venus is perpetually hidden below thick clouds, so we must depend on radar images to reveal its surface (right), a dry desert marked by impact craters, volcanism, lava flows, and faults. Mars (center) is a smaller world with impact craters and old volcanoes on its rocky surface. *(NASA)*

density is low—less than 1.75 g/cm³. Saturn's very low density, 0.7 g/cm³, is less than the density of water, so the planet would float if we could find a bathtub big enough. Nevertheless, all the Jovian planets are massive. Jupiter is over 318 times as massive as Earth, and Saturn is about 95 Earth masses. Uranus and Neptune are smaller—15 and 17 Earth masses, respectively.

Although photographs of the Jovian planets reveal swirling cloud patterns (Figure 5-10), their atmospheres are not very deep compared to their radii. In fact, if these so-called gas giants were shrunk to a few centimeters in diameter, their gaseous hydrogen and helium atmospheres would be no deeper than the fuzz on a badly worn tennis ball.

Mathematical models predict that the interiors of Jupiter and Saturn are mostly hydrogen in two different states (Figure 5-11). Beneath the cloud belts of these planets are layers where pressure forces the hydrogen into a liquid state. Even deeper, the pressure is so high that the hydrogen atoms can no longer hold their electrons. With the electrons free to move about, the material becomes an excellent electrical conductor and is known as **liquid metallic hydrogen.** The centers of Jupiter and Saturn are believed to be hot cores of heavy elements slightly larger than Earth. These are often called "rocky cores," although the heat and pressure would not allow rock to exist as it does on Earth.

Uranus and Neptune, being smaller, have less pressure inside and consequently cannot force hydrogen into the liquid metallic state. Rather, model calculations suggest they contain heavy-element cores surrounded by deep oceans of liquid water with dissolved minerals. Above that is a very dense, deep atmosphere

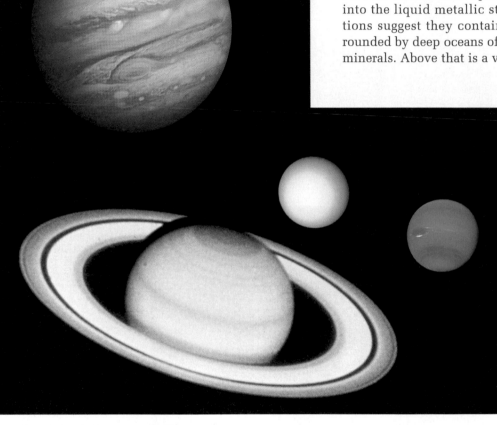

Figure 5-10
Of the Jovian planets (reproduced here to scale), Jupiter and Saturn are the largest. Jupiter is marked by dramatic cloud belts and the Great Red Spot, while Saturn has a bright ring system. Uranus looks featureless, but faint cloud features are visible in computer-enhanced photos. The dark blue color of Neptune and the lighter blue of Uranus are caused by atmospheric methane, which absorbs red photons. *(Adapted from NASA photographs)*

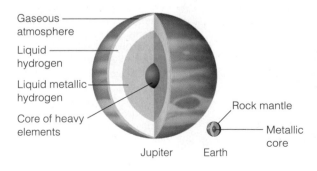

Gaseous atmosphere
Liquid hydrogen
Liquid metallic hydrogen
Core of heavy elements
Jupiter
Rock mantle
Metallic core
Earth

Figure 5-11
Cross sections of Earth and Jupiter illustrate the differences between the interiors of terrestrial and Jovian planets. Earth contains a metallic core with a rock mantle. Jupiter, 11.18 times larger in diameter, is composed mostly of liquid hydrogen. A core composed of heavy elements such as iron, nickel, and silicon lies at the center.

made mostly of hydrogen gas. Although Uranus and Neptune have low densities, hydrogen-rich compositions, and atmospheric features similar to those of Jupiter and Saturn, the two smaller Jovian worlds are clearly different kinds of planets.

Another characteristic of the Jovian planets is their large satellite systems. Jupiter has at least 16 known moons; four of them, the **Galilean satellites** (named after Galileo, who discovered them in 1610), are visible through small telescopes or even binoculars (Figure 5-12a). Saturn, Uranus, and Neptune also have systems of moons.

All of the Jovian planets have ring systems. Saturn has bright rings made of ice particles (Figure 5-12b), and Jupiter has a thin ring of dark, rocky dust. Uranus and Neptune also have thin rings composed of myriads of small particles.

We will examine the satellites and rings of the Jovian planets in later chapters. Here it is sufficient to note that the Jovian planets have large numbers of satellites and that all four have ring systems. In contrast, the terrestrial planets have few satellites and no rings at all. These observations will help us understand how planets were formed in the solar nebula. For now, however, we turn our attention to the smaller members of our solar system.

Space Debris

The sun and planets are not the only remains of the solar nebula. The solar system is littered with three kinds of space debris: asteroids, comets, and meteoroids. Although these objects represent a tiny fraction of the mass of the system, they are a rich source of information about the origin of the planets.

The **asteroids*** are small rocky worlds, most of which orbit the sun in a belt between the orbits of Mars and Jupiter. Roughly 20,000 asteroids have been identified, of which about 1000 follow orbits that bring them into the inner solar system, where they can occasionally collide with a planet. Earth has been struck many times in its history. Some asteroids are located in Jupiter's orbit, some have been found beyond the orbit of Saturn, and a growing number are being discovered among and even beyond the outermost planets.

About 200 asteroids are more than 100 km (60 mi) in diameter, and over 2000 are more than 10 km (6 mi) in diameter. There are probably 500,000 that are larger than 1 km (0.6 mi) and billions that are smaller. Because even the largest are only a few hundred kilometers in diameter, Earth-based telescopes can detect no details on their surfaces.

We do, however, have clear evidence that asteroids are irregularly shaped, cratered worlds. The Galileo

*Also called minor planets.

Figure 5-12
Jupiter and Saturn through a small telescope. (a) Jupiter has at least 16 satellites, but only the four Galilean moons are visible in small telescopes. In this photo, the planet was overexposed to show three of the moons. (b) The icy rings of Saturn are so reflective that they are visible through even a small telescope. Saturn also has a large satellite system. *(Grundy Observatory)*

spacecraft, on its way to Jupiter, flew past two asteroids, Gaspra and Ida, and photographed their surfaces. Both asteroids are irregular lumps of rock pocked by impact craters (Figure 5-13). Furthermore, the Hubble Space Telescope has provided images of some of the larger asteroids and confirms that their surfaces are irregular. We will discuss these observations in detail in Chapter 9, but in our quick survey of the solar system we can note that all evidence suggests the asteroids have suffered many impacts from collisions with other asteroids.

Older theories proposed that the asteroids are the remains of a planet that broke up, but modern astronomers recognize the asteroids as the debris left over by a planet that failed to form at a distance of 2.8 AU from the sun. When we discuss the formation of planets later in this chapter, we must be prepared to explain why material in the solar nebula failed to form a planet at a distance of 2.8 AU.

In contrast to asteroids, the brightest **comets** are impressively beautiful objects (Figure 5-14a). Most comets are faint, however, and are difficult to locate even at their

Figure 5-13
The asteroid Gaspra as photographed by the Galileo spacecraft. The asteroid is 20 by 12 by 11 km and rotates once every 7.04 hours. Its irregular shape and numerous craters testify that it has suffered many impacts. The color here is exaggerated to reveal subtle color variation. Gaspra would look gray to the human eye. *(NASA/Galileo Imaging Team)*

brightest. A comet may take months to sweep through the inner solar system, during which time it appears as a glowing head with an extended tail of gas and dust.

According to the **dirty snowball model** of comets, the tail, which can be greater than 1 AU in length for a bright comet, is produced by the nucleus, a ball of dirty ices (mainly water and carbon dioxide) only a few dozen kilometers in diameter. When the nucleus enters the inner solar system, the sun's radiation begins to vaporize the ices, releasing gas and dust. The pressure of sunlight and the **solar wind** (gas flowing away from the sun) pushes the gas and dust away, forming a long tail. The motion of the nucleus along its orbit, the pressure of sunlight, and the outward flow of the solar wind can create comet tails that are long and straight or gently curved, but in either case the tails of comets always point generally away from the sun (Figure 5-14b).

The nuclei of comets are icy bodies left over from the origin of the planets. Thus, we must conclude that at least some parts of the solar nebula were rich in ices. We will see later in this chapter how important ices were in the formation of the Jovian planets, and we will discuss comets in more detail in Chapter 9.

a

Figure 5-14
(a) A comet may remain visible in the evening or morning sky for weeks as it moves through the inner solar system. Comet West was in the sky during March 1976. (b) A comet in a long, elliptical orbit becomes visible when the sun's heat vaporizes its ices and pushes the gas and dust away in a tail. *(Celestron International)*

Figure 5-15
A meteor is the streak of glowing gases produced by a bit of material falling into Earth's atmosphere. Friction with the air vaporizes the material about 80 km (50 mi) above Earth's surface. *(Daniel Good)*

are produced by meteoroids that weigh less than 1 g. Only rarely is one massive enough and strong enough to survive its plunge and reach Earth's surface.

Thousands of meteorites have been found, and we will discuss their particular forms in Chapter 9. We mention meteorites here for one specific clue they can give us concerning the solar nebula: Meteorites can tell us the age of the solar system.

The Age of the Solar System

If the solar nebula theory is correct, the planets should be about the same age as the sun. The most accurate way to find the age of a celestial body is to bring a sample into the laboratory and determine its age by analyzing the radioactive elements it contains.

When a rock solidifies, it incorporates known percentages of the chemical elements. A few of these elements are radioactive and can decay into another element, called the daughter element. For example, the isotope of uranium ^{238}U can decay into an isotope of lead, ^{206}Pb. The **half life of a radioactive element** is the time it takes for half of the atoms to decay. The half-life of ^{238}U is 4.5 billion years. Thus, the abundance of a radioactive element gradually decreases as it decays (Figure 5-16), and the abundance of the daughter element gradually increases. If we know the abundance of the elements in the original rock, we can measure the present abundance and find the age of the rock. For example, if we studied a rock

Unlike the stately comets, **meteors** flash across the sky in momentary streaks of light (Figure 5-15). They are commonly called "shooting stars." Of course, they are not stars but small bits of rock and metal falling into Earth's atmosphere and bursting into incandescent vapor about 80 km (50 mi) above the ground because of friction with the air. This vapor condenses to form dust, which settles slowly to Earth, adding about 40,000 tons per year to the planet's mass.

Technically, the word *meteor* refers to the streak of light in the sky. In space, before its fiery plunge, the object is called a **meteoroid,** and any part of it that survives its fiery passage to Earth's surface is called a **meteorite.** Most meteoroids are specks of dust, grains of sand, or tiny pebbles. Almost all the meteors we see in the sky

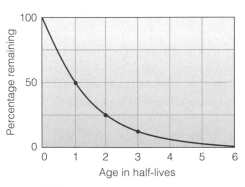

Figure 5-16
The number of nuclei remaining in a radioactive sample decreases such that 50 percent are left after one half-life, 25 percent after two half-lives, 12.5 percent after three half-lives, and so on.

and found that only 50 percent of the ^{238}U remained and the rest had become ^{206}Pb, we would conclude that one half-life must have passed and the rock was 4.5 billion years old.

Uranium isn't the only radioactive element used in radioactive dating. Potassium (^{40}K) decays with a half-life of 1.3 billion years to form calcium (^{40}Ca) and argon (^{40}Ar). Rubidium (^{87}R) decays to strontium (^{87}Sr) with a half-life of 47 billion years. Any of these elements can be used as a radioactive clock to find the age of mineral samples.

Of course, to find a radioactive age, we need a sample in the laboratory, and the only celestial bodies from which we have samples are Earth, the moon, Mars, and meteorites.

The oldest Earth rocks so far discovered and dated are about 3.9 billion years old. That does not mean that Earth formed 3.9 billion years ago. The surface of Earth is active, and the crust is continually destroyed and reformed from material welling up from beneath the crust (see Chapter 6). Thus, the age of these oldest rocks tells us only that Earth is *at least* 3.9 billion years old.

One of the most exciting goals of the Apollo lunar landings was bringing lunar rocks back to Earth's laboratories, where they could be dated. Because the moon's surface is not being recycled like Earth's, some parts of it might have survived unaltered since early in the history of the solar system. Dating the rocks showed the oldest to be 4.48 billion years old. Thus, the solar system must be *at least* 4.48 billion years old.

Although no one has yet been to Mars, a few meteorites found on Earth have been identified by their chemical composition as having come from Mars. Most of these have ages of only 1.3 billion years, but one has an age of 4.5 billion years. Thus, Mars must be at least that old.

Another important source for determining the age of the solar system is meteorites. Radioactive dating of meteorites yields a range of ages, with the oldest about 4.6 billion years old. This figure is widely accepted as the age of the solar system.

One last celestial body deserves mention: the sun. Astronomers estimate the age of the sun to be about 5 billion years, but this is not a radioactive date because they cannot obtain a sample of solar material. Instead, they estimate the sun's age from the radioactive ages of Earth, the moon, and meteorites. Computer models of the sun give only approximate ages, but they generally agree with the age of Earth.

Apparently, all the bodies of the solar system formed at about the same time some 4.6 billion years ago. This completes our survey of the solar system. We have found five significant characteristics (Table 5-1).

Table 5-1 Characteristic Properties of the Solar System

1. Disk shape of the solar system
   Orbits in nearly the same plane
   Common direction of rotation and revolution
2. Two planetary types
   Terrestrial—inner planets; high density
   Jovian—outer planets; low density
3. Planetary ring systems for Jupiter, Saturn, Uranus, and Neptune
4. Space debris—asteroids, comets, and meteors
   Composition
   Orbits
5. Common ages of about 4.6 billion years for Earth, the moon, Mars, meteorites, and the sun

---

We must now try to understand how the formation of the planets in the solar nebula explains these significant characteristics.

REVIEW Critical Inquiry

In what ways is the solar system a disk?

First, the general shape of the solar system is that of a disk. The planets follow orbits that lie in nearly the same plane. The orbit of Mercury is inclined 7° to the plane of Earth's orbit, and the orbit of Pluto is inclined a bit more than 17°. Thus, the planets follow orbits confined to a thin disk with the sun at its center.

Second, the motions of the sun and planets also follow this disk theme. The sun and most of the planets rotate in the same direction, counterclockwise as seen from the north, with their equators near the plane of the solar system. Also, all of the planets revolve around the sun in that same direction. Thus, our solar system seems to prefer motion in the same direction, which further reflects a disk theme.

One of the basic characteristics of our solar system is its disk shape, but another dramatic characteristic is the division of the planets into two groups. What are the distinguishing differences between the terrestrial and Jovian planets?

We have completed our survey of the solar system and gathered the preliminary evidence we can use to evaluate the solar nebula hypothesis. Now it is time to ask how the planets formed from the solar nebula and took on the characteristics we see today.

Scientists often face problems in which they must reconstruct the past. Some of these problems are obvious, such as those faced by an archaeologist excavating the ruins of a burial tomb, but other such problems are less obvious. In each case, success requires the interplay of hypotheses and evidence to re-create a past that no longer exists.

The reconstruction of the past is obvious when we think of an archaeologist, but a biologist studying a centipede is also reconstructing the past. How did this creature come to have a segmented body with so many legs? How did it develop the metabolism that allows it to move quickly and hunt prey? Although the problem might at first seem to be one of mere anatomy, the scientist must reconstruct an environment that no longer exists.

The astronomer's problem is not just to understand what the planets are like but to understand how they got that way. That means planetary astronomers must look at the evidence they can see today and reconstruct a history of the solar system, a past that is quite different from the present. If we had a time machine, it would be a fantastic adventure to go back and watch the planets form. Time machines may be impossible, but we can use the grand interplay of evidence and theory, the distinguishing characteristic of science, to journey back billions of years and reconstruct a past that no longer exists.

5-3 The Story of Planet Building

The challenge for modern planetary astronomers is to compare the characteristics of the solar system with the solar nebula theory and tell the story of how the planets formed. Thus they must re-create the solar system's early past (Window on Science 5-3).

The Chemical Composition of the Solar Nebula

Everything we know about the nature of the solar system and the nature of star formation suggests that the solar nebula was a fragment of an interstellar gas cloud. Such a cloud would have been mostly hydrogen with some helium and tiny traces of the heavier elements.

This is precisely what we see in the chemical composition of the sun. Analysis of the solar spectrum shows that the sun is mostly hydrogen, with 25 percent of its mass being helium and only about 2 percent being heavier elements. Of course, nuclear reactions have fused some hydrogen into helium, but this happens in the sun's core and has not affected its surface composition. Thus, the composition revealed in its spectrum is essentially the composition of the gases from which it formed.

This must have been the composition of the solar nebula, and we can see that composition reflected in the chemical compositions of the planets. One factor, however, can dramatically alter the composition of a planet. We must recognize that planets grow by two processes. Planets begin growing by the sticking together of solid bits of matter. Only when a planet has grown to a mass of about 15 Earth masses does it have enough gravitation to begin capturing gas directly from the solar nebula in a process called **gravitational collapse.** The Jovian planets began forming by the aggregation of bits of rock and ice. Once they accumulated enough mass, they began growing by gravitational collapse. That is, they could capture large amounts of gas, mostly hydrogen and helium, directly from the solar nebula. Jupiter and Saturn grew rapidly and captured the most hydrogen and helium, while Uranus and Neptune grew more slowly and never became massive enough to capture gas as rapidly as Jupiter and Saturn. Thus, the Jovian worlds grew to be low-density, hydrogen-rich planets, with Jupiter and Saturn the largest and least dense.

The terrestrial planets, in contrast, contain very little hydrogen and helium. These small planets have such low masses they were unable to keep the hydrogen and helium from leaking into space. Indeed, because of their small masses, they were probably unable to capture very much of these lightweight gases from the solar nebula. The terrestrial planets are dense worlds because they are composed of the heavier elements from the solar nebula.

Although we can see how the chemical composition of the solar nebula is reflected in the present composition of the sun and planets, a very important question remains: How did gas and dust in the solar nebula come together to form the solid matter of the planets? We must answer that question in two stages. First, we must understand how gas formed small solid particles, and then we must explain how these particles built the planets.

Table 5-2 Observed and Uncompressed Densities

Planet	Observed Density (g/cm³)	Uncompressed Density (g/cm³)
Mercury	5.44	5.4
Venus	5.24	4.2
Earth	5.50	4.2
Mars	3.93	3.3
(Moon)	3.36	3.35

Table 5-3 The Condensation Sequence

Temperature (K)	Condensate	Planet (Estimated Temperature of Formation; K)
1500	Metal oxides	Mercury (1400)
1300	Metallic iron and nickel	
1200	Silicates	
1000	Feldspars	Venus (900)
680	Troilite (FeS)	Earth (600)
		Mars (450)
175	H_2O ice	Jovian (175)
150	Ammonia–water ice	
120	Methane–water ice	
65	Argon–neon ice	Pluto (65)

The Condensation of Solids

The key to understanding the process that converted the nebular gas into solid matter is the variation in density among solar system objects. We have already noted that the four inner planets are high-density, terrestrial bodies, whereas the outermost planets are low-density, giant planets (except for Pluto, which is low in density but not a giant). This division is due to the different ways gases condensed into solids in the inner and outer regions of the solar nebula.

Even among the four terrestrial planets, we find a pattern of subtle differences in density. Merely listing the observed densities of the terrestrial planets does not reveal the pattern, because Earth and Venus, being more massive, have stronger gravity and have squeezed their interiors to higher densities. We must look at the **uncompressed densities**—the densities the planets would have if their gravity did not compress them. These densities (Table 5-2) show that the closer a planet is to the sun, the higher its uncompressed density.

This density variation probably originated when the solar system first formed solid grains. The kind of matter that condensed in a particular region would depend on the temperature of the gas there. In the inner regions, the temperature may have been 1500 K or so. The only materials that could form grains at this temperature are compounds with high melting points, such as metal oxides and pure metals, which are very dense. Farther out in the nebula it was cooler, and silicates (rocky material) could condense. These are less dense than metal oxides and metals. In the cold outer regions, ices of water, methane, and ammonia could condense. These are low-density materials.

The sequence in which the different materials condense from the gas as we move away from the sun is called the **condensation sequence** (Table 5-3). It suggests that the planets, forming at different distances from the sun, accumulated from different kinds of materials. Thus, the inner planets formed from high-density metal oxides and metals, and the outer planets formed from low-density ices.

We must also remember that the solar nebula did not remain the same temperature throughout the formation of the planets but may have grown progressively cooler. Thus, a particular region of the nebula may have begun by producing solid particles of metals and metal oxides but, after cooling, began producing particles of silicates. Allowing for the cooling of the nebula makes our theory much more complex, but it also makes the processes by which the planets formed much more understandable.

The Formation of Planetesimals

In the development of a planet, three groups of processes operate. First, grains of solid matter grow larger, eventually reaching diameters ranging from a few centimeters to kilometers. These objects, called **planetesimals,** are believed to be the bodies that the second group of processes collects into planets. Finally, a third set of processes clears away the solar nebula. The study of planet building is the study of these three groups of processes.

According to the solar nebula theory, planetary development in the solar nebula began with the growth of dust grains. These specks of matter, whatever their composition, grew from microscopic size by two processes: condensation and accretion.

A particle grows by **condensation** when it adds matter one atom at a time from a surrounding gas. Thus, snowflakes grow by condensation in Earth's atmosphere. In the solar nebula, dust grains were continuously bombarded by atoms of gas, and some of these stuck to the grains. A microscopic grain capturing a gas atom increases its mass by a much larger fraction than a gigantic boulder capturing a single atom. Thus, condensation can increase the mass of a small grain rapidly, but as the grain grows larger, condensation becomes less effective.

The second process is **accretion,** the sticking together of solid particles. In building a snowman, we roll a ball of snow across the snowy ground so that it grows by accretion. In the solar nebula, the dust grains were, on the average, no more than a few centimeters apart, so they collided frequently. Their mutual gravitation was too small to hold them to each other, but other effects may have helped. Static electricity generated by their passage through the gas could have held them together, as could compounds of carbon that might have formed a sticky surface on the grains. Ice grains might have stuck together better than some other types. Of course, some collisions might have broken up clumps of grains; on the whole, however, accretion must have increased grain size. If it had not, the planets would not have formed.

There is no clear distinction between a very large grain and a very small planetesimal, but we can consider an object a planetesimal when its diameter becomes a kilometer or so. Objects this size and larger were subject to new processes that tended to concentrate them. One important effect may have been that the growing planetesimals collapsed into the plane of the solar nebula. Dust grains could not fall into the plane because the turbulent motions of the gas kept them stirred up, but the larger objects had more mass, and the gas motions could not have prevented them from settling into the plane of the spinning nebula. This would have concentrated the solid particles into a thin plane about 0.01 AU thick and would have made further planetary growth more rapid.

We will see in later chapters that the collapse of a rotating gas cloud into a disk is a common process. In the solar nebula, the formation of a disk of planetesimals triggered an important process. Computer models show that the rotating disk of particles should have been gravitationally unstable and would have broken up into small clouds (Figure 5-17). This would further concentrate the planetesimals and help them coalesce into objects up to 100 km (60 mi) in diameter. Thus, the theory predicts that the nebula became filled with trillions of solid particles ranging in size from pebbles to tiny planets. As the largest began to exceed 100 km in diameter, new processes began to alter them, and a new stage in planet building began, the growth of protoplanets.

The Growth of Protoplanets

The coalescing of planetesimals eventually formed **protoplanets,** massive objects destined to become planets. As these larger bodies grew, new processes began making them grow faster and altered their physical structure.

Figure 5-17
Gravitational instabilities in the rotating disk of planetesimals may have forced them to collect in clumps, accelerating their growth.

If planetesimals collided at orbital velocities, it is unlikely that they would have stuck together. The average orbital velocity in the solar system is about 30 km/s (67,000 mph). Head-on collisions at this velocity would have pulverized the material. However, the planetesimals were moving in the same direction in the nebular plane and thus avoided head-on collisions. Instead, they merely rubbed shoulders at low relative velocities. Such collisions would have been more likely to fuse them than to shatter them.

In addition, some adhesive effects probably helped. Sticky coatings and electrostatic charges on the surfaces of the smaller planetesimals probably aided formation of larger bodies. Collisions would have fragmented some of the surface rock, but if the planetesimals were large enough, their gravity would have held on to some fragments, forming a layer of soil composed entirely of crushed rock. Such a layer on the larger planetesimals may have been effective in trapping smaller bodies.

The largest planetesimals would grow the fastest, because they had the strongest gravitational field. Not only could they hold on to a cushioning layer to trap fragments, but their stronger gravity could also attract additional material. These planetesimals probably grew quickly to protoplanetary dimensions, sweeping up more and more material. When massive enough, they trapped some of the original nebular gas to form primitive atmospheres. At some point, they crossed the boundary between planetesimals and protoplanets.

To trace the growth of a protoplanet, we can think of the formation of Earth. In its simplest form, our theory of protoplanet growth supposes that all the planetesimals had about the same chemical composition.

The planetesimals accumulated gradually to form a planet-size ball of material that was of homogeneous composition throughout (Figure 5-18). Once the planet formed, heat began to accumulate in its interior from the decay of short-lived radioactive elements, and this heat eventually melted the planet and allowed it to differentiate. **Differentiation** is the separation of material according to density. When the planet melted, the heavy metals such as iron and nickel settled to the core, while the lighter silicates floated to the surface to form a low-density crust.

In this simple form, the theory predicts that Earth's present atmosphere was not its first. The first atmosphere consisted of gases trapped from the solar nebula—mostly hydrogen and helium. They were later driven off by the heat, aided perhaps by outbursts from the infant sun, and new gases that baked from the rocks formed a secondary atmosphere. This creation of a planetary atmosphere from the planet's interior is called **outgassing.**

We can improve this simple theory of planet formation in two ways (Figure 5-18). First, it seems likely that the solar nebula cooled during the formation of the planets, so they did not accumulate from planetesimals of common composition. As planet building began, the first particles to condense in the inner solar system were metals and metal oxides, so the protoplanets may have begun by accreting metallic cores. Later, as the nebula cooled, more silicates could form, and the protoplanets added silicate mantles. A second improvement in the theory proposes that the planets grew so rapidly that the heat released by the in-falling particles, the **heat of formation,** did not have time to escape. This heat rapidly accumulated and melted the protoplanets as they formed. If this is true, then the planets would have differentiated as they formed.

Our improved theory suggests that our present atmosphere was not produced entirely by outgassing. If Earth formed in a molten state, then there was never a time when it had a primitive atmosphere of hydrogen and helium accumulated from the solar nebula. Rather, the gases of the atmosphere were released by the molten rock as the protoplanet grew. Those gases would not have included much water, however, so some astronomers now think that Earth's water and much of its present atmosphere accumulated late in the formation of the planets as Earth swept up volatile-rich planetesimals forming in the cooling solar nebula. Such icy planetesimals may have formed in the outer parts of the solar nebula and been scattered by encounters with the Jovian planets in a bombardment of comets.

The details of the formation of the planets are still being unraveled, but it seems clear that the solar nebula hypothesis can explain the formation of planets

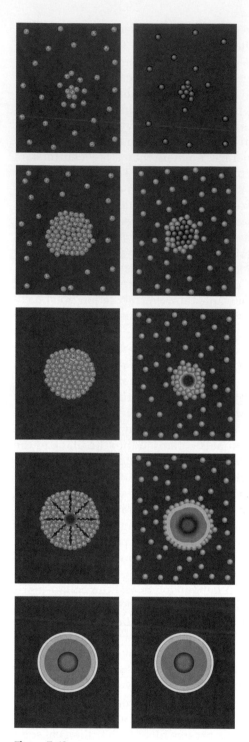

Figure 5-18
At left, a planet grows slowly by the accretion of planetesimals of similar composition. After the protoplanet has formed, the accumulation of heat from radioactive decay in its interior melts it and allows it to differentiate. An improved theory, right, proposes that the solar nebula cooled as the planets formed. The first particles to accumulate were metals, but later particles were silicates. Also, it seems likely that the planets formed fast enough to prevent the loss of the heat of formation, which melted the protoplanets and allowed them to differentiate as they formed.

around the forming sun. But can it explain the distinguishing characteristics of the solar system that we compiled earlier in this chapter?

Explaining the Characteristics of the Solar System

Table 5-1 contains the list of distinguishing characteristics of the solar system. Any theory of the origin of the solar system should explain these characteristics.

The disk shape of the solar system is inherited from the solar nebula. The sun and planets revolve and rotate in the same direction because they formed from the same rotating gas cloud. The orbits of the planets lie in the same plane because the rotating solar nebula collapsed into a disk, and the planets formed in that disk.

The solar nebula hypothesis is evolutionary in that it calls on continuing processes to gradually build the planets. To explain the rotation of Venus, Uranus, and Pluto, however, we must introduce catastrophic events. Apparently, these planets acquired their peculiar rotations when they were struck by very large planetesimals late in their formation. Such impacts may be unusual, but some planets probably experienced them, and an off-center impact by a planetesimal as large as Mars might have made Venus rotate backward. A similar event might have forced Uranus to rotate on its side. In Chapter 8, we will discuss Pluto, and we will see that such a catastrophic event may have been involved in its formation as well.

The second item in Table 5-1, the division of the planets into terrestrial and Jovian worlds, can be understood through the condensation sequence. The terrestrial planets formed in the inner part of the solar nebula, where the temperature was high and most compounds remained gaseous. Only compounds such as the metals and silicates could condense to form solid particles. Planets must begin forming by the sticking together of solid particles because a small blob of matter does not have enough gravitation to capture gas. Thus, the planets that began growing in the inner solar system had to form mainly from metals and silicates. They became the small, dense terrestrial planets.

In contrast, the Jovian planets formed in the outer solar nebula, where the lower temperature allowed the gas to form large amounts of ices, perhaps three times more ices than silicates. Once Jupiter and Saturn reached a mass of roughly 15 Earth masses, they began to grow by gravitational collapse as they captured hydrogen and helium gas directly from the solar nebula. Thus Jupiter and Saturn could have grown very rapidly. Uranus and Neptune never got quite massive enough to grow this way.

Jupiter and Saturn must have reached their present size within 10 million years. Observations show that young stars blow their gas and dust away by then. The other planets, especially the terrestrial planets, could grow by the accretion of planetesimals too large to be blown away, and thus they could have continued growing until the last of the material was swept up. All of the planets must have been complete within 30 to 100 million years.

The asteroids seem to fill a gap between Mars and Jupiter, and we can now understand them as leftovers from a planet that failed to form there. Massive Jupiter, just outside the asteroid belt, gravitationally disrupted the motion of the planetesimals, and they fragmented or were thrown out of the belt and failed to build a planet.

The large satellite systems of the Jovian worlds may contain two kinds of moons. Some moons may have formed in orbit around the forming planet in a miniature of the solar nebula. But some of the smaller moons may be captured planetesimals and asteroids. The large masses of the Jovian planets would have made it easier for them to capture satellites.

In Table 5-1, we noted that all four Jovian worlds have ring systems, and we can understand this by considering the large mass of these worlds and their remote location in the solar system. A large mass makes it easier for a planet to hold onto orbiting ring particles; and being farther from the sun, the ring particles are not as easily swept away by the pressure of sunlight and the solar wind. It is hardly surprising, then, that the terrestrial planets, low mass worlds located near the sun, have no planetary rings.

The last entry in Table 5-1 is the common ages of solar system bodies, and the solar nebula hypothesis has no difficulty explaining that characteristic. If the hypothesis is correct, then the planets formed at the same time as the sun and thus should have roughly the same age. Although we have mineral samples for radioactive dating only from Earth, the moon, Mars, and meteorites, so far the ages seem to agree.

Some of the planets found orbiting near other stars are very massive but orbit close to their stars. At first, that seemed hard to explain with the solar nebula hypothesis, which tells us massive planets should form far from their stars. Detailed studies show that a giant planet forming in a disk filled with many planetesimals might be forced to spiral inward, sweeping up smaller planets until it reached stability close to its star. Also, a few of the giant planets orbiting other stars have highly elliptical orbits, and that could have resulted from interactions between giant planets that formed farther from their star. Thus these newly discovered planets don't tell us the solar nebula theory is wrong, but they warn us that there is more to the story of how planets form.

Our general understanding of the origin of the solar system also explains the origin of asteroids, meteors, and comets. They appear to be the last of the debris left behind by the solar nebula. These objects are such important sources of information about the history of our solar system that we will discuss them in detail in Chapter 9. But for now we ask: What happened to the solar nebula?

Clearing the Nebula

The planets appear to have grown in the solar nebula at the same time that the sun was forming, about 4.6 billion years ago. Indeed, much of the planetary growth may have taken place in darkness because the sun may not have become luminous yet. But once the sun became hot and luminous, it began to clear the nebula of gas and dust and brought planet building to a halt.

Four effects helped clear the nebula. The most important was **radiation pressure.** When the sun became a luminous object, light streaming from its surface pushed against the particles of the solar nebula. Large bits of matter like planetesimals and planets were not affected, but low-mass specks of dust and individual gas atoms were pushed outward and eventually driven from the system.

The second effect that helped clear the nebula was the solar wind, the flow of ionized hydrogen and other atoms away from the sun's upper atmosphere. This flow is a steady breeze that rushes past Earth at about 400 km/s (250 mi/s). When the sun was young, it may have had an even stronger solar wind, and irregular fluctuations in its luminosity, like those observed in some young stars, may have produced surges in the wind that helped push dust and gas out of the nebula.

The third effect for clearing the nebula was the sweeping up of space debris by the planets. All of the old, solid surfaces in the solar system are heavily cratered by meteorite impacts (Figure 5-19). Earth's moon, Mercury, Venus, Mars, and most of the moons in the solar system are covered with craters. A few of these craters have been formed recently by the steady rain of meteorites that falls on all the planets in the solar system, but most of the craters we see appear to have been formed roughly 4 billion years ago in what is called the **heavy bombardment,** as the last of the debris in the solar nebula was swept up by the planets. We will find more evidence of this bombardment in later chapters. Thus, many of the last remaining solid objects in the

a

b

Figure 5-19
Every old, solid surface in the solar system is scarred by craters. (a) Earth's moon is scarred by craters ranging from basins hundreds of kilometers in diameter down to microscopic pits. (b) The surface of Mercury as photographed by a passing spacecraft shows vast numbers of overlapping craters. Throughout the solar system, the heavy bombardment cratered all the solid worlds as the last of the debris in the solar nebula was swept up. *(NASA)*

solar nebula were swept up by the planets soon after they formed.

The fourth effect was the ejection of material from the solar system by close encounters with planets. If a small object such as a planetesimal passes close to a planet, it can gain energy from the planet's gravitational field and be thrown out of the solar system. Ejection is most probable in encounters with massive planets, so the Jovian planets were probably very efficient at ejecting the icy planetesimals that formed in their region of the nebula.

Together, these four effects cleared away the solar nebula. This may have taken considerable time, but eventually the solar system was relatively clear, the planets could no longer gain mass, and planet building ended.

REVIEW **Critical Inquiry**

Why are there two kinds of planets in our solar system?

Planets begin forming from solid bits of matter, not from gas. Consequently, the kind of planet that forms at a given distance from the sun depends on the kind of compounds that can condense out of the gas to form solid particles. In the inner parts of the solar nebula, the temperature was so high that most of the gas could not condense to form solids. Only metals and silicates could form solid grains, and thus the innermost planets grew from this dense material. Much of the mass of the solar nebula consisted of hydrogen, helium, water vapor, and other gases, and they were present in the inner solar nebula but couldn't form solid grains. The small terrestrial planets couldn't grow from these gases, so the terrestrial planets are small and dense.

In the outer solar nebula, the composition of the gas was the same; but it was cold enough for water vapor to condense to form ice grains, and because hydrogen and oxygen are so abundant, there was lots of ice available. Thus, the outer planets grew from solid bits of metal and silicate combined with large amounts of ice. The outer planets grew so rapidly that they became big enough to capture gas directly, and thus they became the hydrogen- and helium-rich Jovian worlds.

The condensation sequence combined with the solar nebula hypothesis gives us a way to understand the formation of the planets. This illustrates the importance of a unifying hypothesis in science. How does the solar nebula theory unify our understanding of planetary orbits and rotation?

The great chain of origins leads from the first instant of the big bang through the birth of galaxies, the formation of stars, the origin of our solar system, and finally to us. One of the most elegant and complex links in that chain is planet building. As we explore the solar system in detail in the following chapters, we must stay alert for further clues to the birth of the planets.

Summary

The solar nebula hypothesis proposes that the solar system began as a contracting cloud of gas and dust that flattened into a rotating disk. The center of this cloud eventually became the sun, and the planets formed in the disk of the nebula.

Observational evidence gives astronomers confidence in this theory. Disks of gas and dust have been found around many young stars, so astronomers suspect that planetary systems are common. Planets orbiting other stars are too faint and too close to their star to image directly, but astronomers have found a number of these planets by observing the motion of the star as the star and planet revolve around each other.

The solar nebula theory explains many of the characteristic properties of the solar system. For example, the solar system has a disk shape. The orbits of the planets lie in nearly the same plane, and they all revolve around the sun in the same direction, counterclockwise as seen from the north. This is also true of many of the satellites of the planets. With only three exceptions, the planets rotate counterclockwise around axes roughly perpendicular to the plane of the solar system. This disk shape and the motion of the planets appear to have originated in the disk-shaped solar nebula.

Another striking feature of the solar system is the division of the planets into two families. The terrestrial planets, which are small and dense, lie in the inner part of the system. The Jovian planets are large, low-density worlds in the outer part of the system. In general, the closer a planet lies to the sun, the higher its uncompressed density.

The solar system is now filled with smaller bodies such as asteroids, comets, and meteors. The asteroids are small, rocky worlds, most of which orbit the sun between Jupiter and Mars. They appear to be material left over from the formation of the solar system.

Another important characteristic of the solar system bodies is their similar ages. Radioactive dating tells us that Earth, the moon, Mars, and meteorites are no older than 4.6 billion years. Thus, it seems our solar system took shape about 4.6 billion years ago.

According to the condensation sequence, the inner part of the nebula was so hot that only high-density minerals could form solid grains. The outer regions, being cooler, condensed to form icy material of lower density. The planets grew from these solid materials, with the denser planets forming in the inner part of the nebula and the lower-density Jovian planets forming farther from the sun.

Planet building began as dust grains grew by condensation and accretion into planetesimals ranging from a kilometer to hundreds of kilometers in diameter. These planetesimals settled into a thin plane around the sun and accumulated into larger bodies, the largest of which grew the fastest and eventually became protoplanets.

Once a planet had formed from a large number of planetesimals, heat from radioactive decay could have melted it and allowed it to differentiate into a dense metallic core and a lower-density silicate crust. In fact, it is possible that the solar nebula cooled as the protoplanets grew so that the first planetesimals were metallic and later additions were silicate. It is also likely that the planets grew rapidly enough that the heat of formation released by the in-falling material melted the planets and allowed them to differentiate as they formed.

The Jovian planets probably grew rapidly from icy materials and became massive enough to attract and hold vast amounts of nebular gas. Heat of formation raised their temperatures very high when they were young, and Jupiter and Saturn still radiate more heat than they absorb from the sun.

Once the sun became a luminous object, it cleared the nebula as its light and solar wind pushed material out of the system. The planets helped by absorbing some planetesimals and ejecting others from the system. Once the solar system was clear of debris, planet building ended.

New Terms

solar nebula theory	half-life
extra-solar planet	gravitational collapse
terrestrial planets	uncompressed density
Jovian planets	condensation sequence
liquid metallic hydrogen	planetesimal
Galilean satellites	condensation
asteroids	accretion
comet	protoplanet
dirty snowball model	differentiation
solar wind	outgassing
meteor	heat of formation
meteoroid	radiation pressure
meteorite	heavy bombardment

Review Questions

1. What produced the hydrogen and helium now present in Jupiter's atmosphere?

2. What produced the iron in Earth's core and the heavier elements like gold and silver in Earth's crust?

3. What evidence do we have that disks of gas and dust are common around young stars?

4. According to the solar nebula theory, why is the sun's equator nearly in the plane of Earth's orbit?

5. Why does the solar nebula theory predict that planetary systems are common?

6. Why do we think the solar system formed about 4.6 billion years ago?

7. If you visited another planetary system, would you be surprised to find planets older than Earth? Why or why not?

8. Why is almost every solid surface in our solar system scarred by craters?

9. What is the difference between condensation and accretion?

10. Why don't terrestrial planets have rings like the Jovian planets?

11. How does the solar nebula theory help us understand the composition and location of asteroids and comets?

12. How does the solar nebula theory explain the dramatic density difference between the terrestrial and Jovian planets?

13. If you visited some other planetary system in the act of building planets, would you expect to see the condensation sequence at work, or was it unique to our solar system?

14. Why do we expect to find that planets are differentiated?

15. What processes cleared the nebula away and ended planet building?

Discussion Questions

1. Discuss the history of the atoms in your thumb. Where did they come from and what processes brought them together?

2. If the solar nebula theory is correct, then there are probably more planets in the universe than stars. Do you agree? Why or why not?

Problems

1. The nearest star is about 4.2 ly distant. If you looked back at the solar system from that distance, what would the maximum angular separation be between Jupiter and the sun? (*Hint:* 1 ly equals 63,000 AU.)

2. Venus is the brightest planet in our sky at a time when it is about 1 AU from the earth. How many times fainter would it look from the nearest star 4.2 ly distant? (*Hints:* 1 ly equals 63,000 AU. Remember the inverse square law.)

3. What is the smallest-diameter crater you can identify in Figure 5-7? (*Hint:* See Table A-13, Properties of the Planets, to find the diameter of Mercury in kilometers.)

4. A sample of a meteorite has been analyzed, and the result shows that out of every 1000 nuclei of ^{40}K originally in the meteorite, only 125 have not decayed. How old is the meteorite? (*Hint:* See Figure 5-16.)

5. In Table 5-2, which object's observed density differs least from its uncompressed density? Why?

6. What composition might we expect for a planet that formed in a region of the solar nebula where the temperature was about 100 K?

7. Suppose that Earth grew to its present size in 100 million years through the accretion of particles averaging 100 g each. On the average, how many particles did Earth capture per second? (*Hint:* See Appendix A, Table A-6 to find Earth's mass.)

8. If you stood on Earth during its formation as described in Problem 7 and watched a region covering 100 m², how many impacts would you expect to see in an hour? (*Hints:* Assume that Earth had its present radius. The surface area of a sphere is $4\pi r^2$.)

9. The velocity of the solar wind is roughly 400 km/s. How long does it take to travel from the sun to Pluto?

Critical Inquiries for the Web

1. How does our solar system compare with the others that have been found? Search the Internet for sites that give information about planetary systems around other stars. What kinds of planets have been detected by these searches so far? Discuss the selection effects (see the Window on Science 16-2, page 367) that must be considered when interpreting this data.

2. How is radioactive dating carried out on meteorites and rocks from surfaces of various bodies in the solar system? Look for Web sites on the details of radioactive dating and summarize the methods used to uncover the abundances of radioactive elements in a particular sample. (*Hint:* Try looking for information on how a particular meteorite—for example, the Martian meteorite ALH84001—was stud-

ied, what age range was determined, and what radioactive elements were used to arrive at the age.)

Exploring *The Sky*

1. Use *The Sky* to observe and describe the shape of the solar system as defined by

   a. The planets
   How to proceed: Enter the **3D Solar System Mode.** Simulate the motion of the terrestrial planets (planets out to Mars) with a time increment of 1 day. Use the up and down arrow keys to view the orbits from different angles. Note the general "flatness" of the inner solar system. Which of the orbits of the terrestrial planets appear to have the greatest inclination to Earth's orbit? Now zoom out to show all the planets, and again observe the motion of the planets from different angles. Again note the relative "flatness" of the outer solar system. Which of the orbits has the largest inclination to the plane of the solar system?

   b. Comets
   How to proceed: Zoom in so the orbit of Jupiter fills the screen, and run the simulation with a time increment of 10 days. Note that aside from showing you the planets, *The Sky* shows a number of pre-installed comets and asteroids. What can you say about the inclination of the orbits of the comets?

 Go to the Brooks/Cole Astronomy Resource Center (www.brookscole.com/astronomy) for critical thinking exercises, articles, and additional readings from InfoTrac College Edition, Brooks/Cole's online student library.

Earth and Its Moon

That's one small step for a man . . .

one giant leap for mankind.

Neil Armstrong
On the moon

Beautiful, beautiful. Magnificent

desolation.

Edwin E. Aldrin, Jr.
On the moon

Guidepost

The preceding chapter explained how planetary systems form, and now we are ready to begin examining our planetary system to see what further evidence we can find to help us understand how planets form and evolve.

We study the Earth and its moon together because they illustrate diverse histories and processes but also because they are the two worlds humans have visited. In later chapters we will study worlds we have not visited and some we can never visit. The experience we gain from the Earth and its moon will help us interpret what we find on these unearthly worlds.

The first two people to visit the moon responded in dramatically different ways. Neil Armstrong responded to the significance of the first human step on the surface of another world, but Buzz Aldrin responded to the moon itself. It *is* desolate, and it *is* magnificent. But it is not unusual. Most planets in the universe probably look like our moon, and astronauts may someday walk on such worlds and compare them with our moon.

The comparison of one planet with another is called **comparative planetology,** and it is one of the best ways to analyze the worlds in our solar system. We will learn much more by comparing planets than we would by studying them individually.

The Earth and its moon are such different worlds that they illustrate almost all of the processes we will find on other rocky planets. Earth is a planet of extremes. Its interior is molten and generates a magnetic field. Its crust is active, with moving sections that push against each other, triggering earthquakes, volcanoes, and mountain building. Almost 75 percent of Earth's surface is covered by liquid water with an average depth of 3 km. Even Earth's atmosphere is extreme. Processes have altered Earth's air from its natural composition to a highly unusual, oxygen-rich sea of gas.

In contrast, the moon may be more typical of rocky worlds. Small, airless, waterless, and scarred by craters, it looks barren and forbidding. Our analysis, however, will reveal that the lunar surface has had a rich and complex history of geological development typical of many of the other worlds we will visit.

6-1 Earth

If Earth is to be the benchmark for our study of other worlds, then we must examine our planet in detail. We will consider its general history, interior, crust, and atmosphere.

Four Stages of Planetary Development

As Earth formed in the inner solar nebula, it passed through four developmental stages (Figure 6-1), stages that other worlds also experienced to varying degrees.

The first stage of planetary evolution is *differentiation,* the separation of material according to density. Earth now has a dense core and a lower-density crust, so we know it differentiated. Some of that differentiation may have occurred very early as the heat of formation released by in-falling matter melted the growing Earth. Some of the differentiation, however, may have occurred later as radioactive decay released heat and further melted Earth, allowing the denser metals to sink to the core.

Figure 6-1
The four stages of planetary development. First: differentiation into core and crust as dense material sank to the center and low-density material floated to the surface. Second: cratering of the newly formed crust by the heavy bombardment of meteorites in the young, debris-filled solar nebula. Third: flooding of lowlands by lava from the interior and later by water from the cooling atmosphere. Fourth: slow surface evolution due to geological activity and erosion.

The second stage, *cratering,* could not begin until a solid surface formed. The heavy bombardment of the early solar system cratered Earth just as it did our moon. As the debris in the solar nebula cleared away, the rate of cratering impacts fell rapidly to its present low rate.

The third stage, *flooding,* began as the decay of radioactive elements heated Earth's interior. Lava welled up through fissures in the crust and flooded the deeper basins. Later, as the atmosphere cooled, water fell as rain and flooded the basins to form the first oceans. Note that on Earth flooding includes both lava and water.

Earth Data File One

A shaded relief map of Earth with the oceans removed reveals the geological features on the continents and under the oceans. *(National Geophysical Data Center)*

Average distance from the sun	1.00 AU (1.495979×10^8 km)
Eccentricity of orbit	0.0167
Maximum distance from the sun	1.0167 AU (1.5210×10^8 km)
Minimum distance from the sun	0.9833 AU (1.4710×10^8 km)
Inclination of orbit to ecliptic	0°
Average orbital velocity	29.79 km/s
Orbital period	1.00 y (365.25 days)
Period of rotation (with respect to the sun)	$24^h00^m00^s$
Inclination of equator to orbit	23°27′
Equatorial diameter	12,756 km
Mass	5.976×10^{24} kg
Average density	5.497 g/cm³
Surface gravity	1.0 Earth gravity
Escape velocity	11.2 km/s
Surface temperature	−50° to 50°C (−60° to 120°F)
Average albedo*	0.39
Oblateness	0.0034

*Albedo is the fraction of light striking an object that the object reflects. The albedo of a perfect reflector is 1.

The fourth stage, *slow surface evolution*, has continued for at least the past 3.5 billion years or more. Earth's surface is constantly changing as sections of crust slide over each other, push up mountains, and shift continents. In addition, moving air and water erode the surface and wear away geological features. Almost all trace of the first billion years of Earth's geology has been destroyed by the active crust and erosion.

Terrestrial planets pass through these four stages, but differences in mass, temperature, and composition emphasize some stages over others and produce surprisingly different worlds.

Earth's Interior

We expect that Earth differentiated, but what evidence do we have to tell us what the interior of Earth is really like?

Earth's mass divided by its volume tells us its average density, 5.50 g/cm³. (See Data File One.) But the density of Earth's rocky crust is only about half that. Clearly, a large part of Earth's interior must be made of material denser than rock.

Although the deepest wells extend only a few kilometers down, Earth scientists have clear proof that Earth did differentiate. Each time an earthquake occurs, seismic waves travel through the interior and register on seismographs all over the world (Figure 6-2). Analysis of these waves shows that Earth's interior is divided

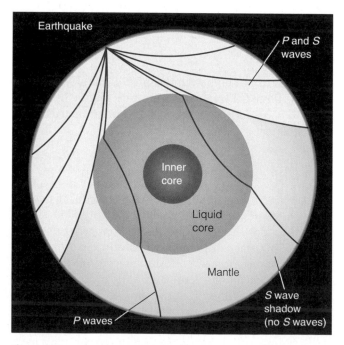

Figure 6-2
P and *S* waves give us clues to Earth's interior. That no direct *S* waves reach the far side shows that Earth's core is liquid. The size of this *S* wave shadow tells us the size of the outer core.

into a metallic core, a dense rocky mantle, and a thin, low-density crust.

The core has a density of 14 g/cm³, denser than lead, and is believed to be composed of iron and nickel at a temperature of roughly 6000 K. In other words, the core of Earth is as hot as the surface of the sun, but the high pressure keeps the metal a solid near the middle of the core and a liquid in its outer parts. Two kinds of seismic waves tell us the outer core is liquid. The **P waves** travel like sound waves, and they can penetrate a liquid; but **S waves** travel as a side-to-side vibration of particles like the jiggle in a bowl of jelly. S waves can't travel through a liquid, so Earth scientists can deduce the size of the liquid core by observing where S waves get through and where they don't (Figure 6-2). The outer boundary of the core lies just over halfway to the surface.

Earth's magnetic field appears to be generated in its molten core by the **dynamo effect,** a process that requires convection in an electrically conducting, rotating body. Earth's core is a very good electrical conductor, and it is stirred by convection, so its rotation can convert some of its energy into a magnetic field. From traces of magnetic field retained by rocks that formed long ago, geologists conclude that Earth's magnetic field reverses itself every million years or so. Periodic reversals, though poorly understood, are a characteristic of the dynamo effect. Thus, the presence of a magnetic field is evidence that part of Earth's core is a liquid metal.

The **mantle** is the layer of dense rock and metal oxides that lies between the molten core and the solid crust. The mantle material is a **plastic,** a material with the properties of a solid but capable of flowing under pressure. Asphalt used in paving roads is a common plastic; it shatters if struck with a sledgehammer, but it bends under the weight of a truck. Just below Earth's crust, where the pressure is less than at great depths, the mantle is most plastic.

Earth's rocky crust is thin and brittle. Only 35 to 60 km thick under the continents, it is even thinner under the oceans—5 to 10 km. Because of its low density (2.5 to 3.5 g/cm³), the crust floats on the denser mantle (3.5 to 5.8 g/cm³) (Figure 6-3). The crust is so brittle it breaks much more easily than does the plastic mantle.

While Earth's hot interior seems far from our everyday experience, it is in fact responsible for most of the scenery we see on the surface.

Plate Tectonics

Earth's surface is active in that it is constantly renewed as large sections of crust drift about. Geologists refer to this activity as **plate tectonics.** *Plate* refers to the sections of crust, and *tectonics* comes from the Greek word for "builder." Plate tectonics is the builder of Earth's surface. It pushes up mountains, destroys old crust, and creates new crust.

The plates move because the mantle is highly plastic just below the crust. Convection currents rise through the mantle and help break the crust into segments, which then slide on the plastic mantle at speeds of a few centimeters per year. The recognition of the importance of plate tectonics is one of the great scientific revolutions of the 20th century.

Direct evidence of plate tectonics lies in the ocean floor. The crust there is thin and composed of dense **basalt,** rock formed by solidified lava. These seafloors are not at all flat but are marked by undersea mountain ranges called **midocean rises,** split down their middle by chasms called **midocean rifts** (see Figure 6-3).

We can understand the midocean rifts when we note that the rocks of the central rift are young and that older rocks with greater accumulations of sediment lie further away. The midocean rifts are sites of seafloor spreading. The Atlantic Ocean floor is an especially good example (Figure 6-4). Two crustal plates spread apart along the rift, and molten rock wells up to solidify and form young basalt. Thus, plate tectonics creates new crust along the midocean rifts. This is dramatically illustrated by measurements of alternating bands of magnetic fields trapped in the rock parallel to the rift (Figure 6-4 inset). These bands evidently form because Earth's magnetic field reverses every million years or so, and the solidifying rock records those changes in alternating strips just as the moving tape in a tape recorder makes a record of the changing magnetic field in the recording head. The detection of these magnetic bands was key evidence in helping geologists understand seafloor spreading.

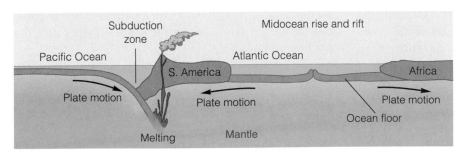

Figure 6-3
A cross section of Earth's crust shows the thick crust of the continents and the thinner crust of the sea floors. The motion of crustal plates allows rising magma to build new crust at the midocean rift, and old sea floor descends under the continents in subduction zones where it triggers mountain building and volcanism.

Figure 6-4

A color-coded relief map of Earth's continents and ocean floors reveals midocean ridges such as that running the length of the Atlantic Ocean. The ocean floor spreads apart along the rift as the crustal plates drift apart. Magma rising into the rift hardens and records the alternate reversals of Earth's magnetic field as parallel bands of fossil magnetism in the seafloor (right). Other ridges mark the Pacific and Indian oceans. (Compare with Figure 6-6.) *(National Geophysical Data Center)*

New crust is created at the midocean rifts, and old crust is destroyed in trenches where a plate slides under another plate. The collision of moving plates often crumples the crust and pushes up mountains. Also, the descending seafloor melts, along with its accumulated sediment, and releases low-density magma that rises to the surface and produces volcanoes. The coastal volcanoes (including Mount Saint Helens) extending from northern California into Alaska are caused by a tectonic plate plunging below the continent. A region where a plate slides downward is called a **subduction zone.**

The seafloor does not always slip below a continent. The floor of the Atlantic Ocean is locked to North and South America and is pushing the continents westward. Tracing this continental drift backward in time, Earth scientists find that North and South America were in contact with Europe and Africa only 200 million years ago. At that point, the land masses on Earth were joined in one great continent, Pangaea (Figure 6-5). About 200 million years ago, Pangaea broke into two land masses, Laurasia in the north and Gondwanaland in the south. These land masses further fragmented and drifted apart to form the continents we know.

Continental drift continues today, driven by moving plates. Continents can split apart in a long, straight, deep depression called a **rift valley.** Africa has split from Arabia, opening a rift valley now filled by the Red Sea. In contrast, continents can collide. The collision of Africa with the east coast of North America, roughly 250 million years ago, folded the crust into the Appalachian Mountains, just as the collision of India with southern Asia is now building the Himalaya Mountains. These **folded mountain chains** are typical of plate collisions. Sections of crust can also slip past each other, as in the case of the Pacific plate carrying part of southern California northward along the San Andreas fault and generating frequent earthquakes.

Plotting the location of earthquakes and volcanism on a world map reveals the edges of the moving plates (Figure 6-6). The Pacific plate is bounded by a ring of earthquake zones known as the "ring of fire." Shallow earthquakes and undersea volcanism also occur along the midocean rises, where new crust forms and the seafloor spreads.

Because Earth's crust is active, all geological features gradually change. The oldest existing portions of the crust, the Canadian Shield and portions of South Africa and Australia, are only about 3.9 billion years old. The constant churning of Earth's surface has wiped away all record of older crust.

Figure 6-5
Continental drift has broken the land mass Pangaea first into Laurasia and Gondwanaland and eventually into the continents we know. (*Adapted from R. S. Dietz and J. C. Holden,* Scientific American *223, April 1970*)

200 million years ago

135 million years ago

65 million years ago

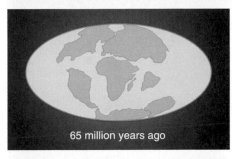

Today

Earth's Atmosphere

Three questions confront us when we consider Earth's atmosphere: How did it form? How has it evolved? How are we changing it? Answering these questions will help us understand other planets as well as our own.

Our planet's first atmosphere, its **primeval atmosphere,** was once thought to contain hydrogen, methane, and ammonia from the solar nebula. Modern studies reveal that the planets formed hot, so carbon dioxide, nitrogen, and water vapor must have been cooked out of Earth. Also, planet building may have ended with the accretion of planetesimals rich in volatiles, including water. Thus the primeval atmosphere must have been rich in carbon dioxide, nitrogen, and water vapor. That primeval atmosphere evolved into the air we breathe today, a **secondary atmosphere.**

Once Earth cooled enough, oceans began to form, and the carbon dioxide began to dissolve in the water. Carbon dioxide is highly soluble in water—which explains the easy manufacture of carbonated beverages. As the oceans removed carbon dioxide from the atmosphere, it reacted with dissolved compounds in the ocean water to form silicon dioxide, limestone, and other mineral sediments. Thus, the oceans transferred the carbon dioxide from the atmosphere to the seafloor and left air rich in nitrogen.

Figure 6-6
Earthquakes, plotted as red dots, outline the edges of Earth's crustal plates (yellow). Volcanism follows the same pattern, and much of the Pacific is ringed by volcanic peaks that extend from Chile through Central America, Alaska, and south through Japan—the "ring of fire." The earthquakes and volcanism in Hawaii are produced where a hot spot in the mantle penetrates upward through the middle of a plate. (*National Geophysical Data Center*)

This removal of carbon dioxide is critical to us because an atmosphere rich in carbon dioxide can trap heat by a process called the **greenhouse effect.** When sunlight shines through the glass roof of a greenhouse, it heats the benches and plants inside (Figure 6-7). The warmed interior radiates heat in the form of infrared radiation, which can't get out through the glass. Heat is trapped in the greenhouse, and the temperature climbs until the glass itself grows warm enough to radiate heat away as fast as sunlight enters.* With a planet, sunlight can enter the atmosphere and warm the surface, but carbon dioxide, being opaque to infrared, would prevent the heat from escaping. Thus, carbon dioxide in a planetary atmosphere can trap energy and warm the surface.

For roughly 4 billion years, Earth's oceans and plant life have been absorbing carbon dioxide and burying it in the form of carbonates such as limestone and in carbon-rich deposits of coal, oil, and natural gas. But in the last century or so, human civilization has begun digging up those fuels, burning them for energy, and releasing the carbon back into the atmosphere as carbon dioxide. This process is steadily increasing the carbon dioxide concentration in our atmosphere and warming our climate. Although the temperature increase is very small now, it is a serious issue for the future. A temperature rise of only a few degrees would begin melting the polar caps and cause a rise in sea levels of as much as 50 meters. Florida, for example, would vanish. When we discuss Venus in the next chapter, we will see an extreme greenhouse effect.

Earth's early atmosphere may have contained some methane (CH_4) and ammonia (NH_3), but ultraviolet photons would have penetrated deep into the atmosphere and broken those molecules up to form carbon, nitrogen, and hydrogen. The hydrogen would escape to space; the carbon would form more carbon dioxide; and the nitrogen would be left in the atmosphere. This no longer happens, because the atmosphere now contains free oxygen, which forms a layer of ozone (O_3) at altitudes of 15 to 30 km.

The ozone layer now protects the lower atmosphere from high-energy photons, but when Earth was young, its atmosphere had no free oxygen. Oxygen is very reactive and quickly forms oxides in the soil, so plant life is needed to keep a steady supply of oxygen in the atmosphere. Although life seems to have begun on Earth within a billion years of its formation, life did not generate free oxygen until the onset of photosynthesis about 3.3 billion years ago. Photosynthesis makes energy for the plant by absorbing carbon dioxide and releasing oxygen. Half a billion years ago, large, shallow seas de-

*A greenhouse also grows warm because the walls prevent the warm air from mixing with the cooler air outside.

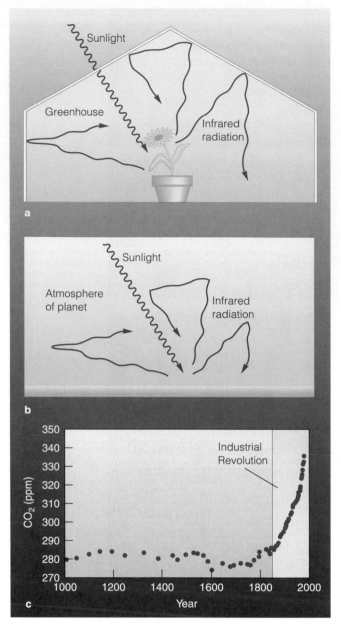

Figure 6-7
The greenhouse effect. (a) Short-wavelength light can enter a greenhouse and heat its contents, but the longer-wavelength infrared radiated by the warm interior cannot get out. (b) The same process can heat a planet's surface if its atmosphere contains CO_2. (c) The concentration of CO_2 in Earth's atmosphere as measured in Antarctic ice cores has remained roughly constant until the Industrial Revolution. *(Adapted from a figure by Etheridge, Steele, Langenfelds, Francey, Barnola, and Morgan.)*

veloped along the continents, and that allowed ocean plants to make oxygen faster than chemical reactions could remove it. Atmospheric oxygen increased rapidly and is still increasing.

Ozone exists because our atmosphere contains oxygen, but chlorofluorocarbons (CFCs), chemicals used in refrigeration, for example, can destroy ozone when they leak into the atmosphere. Since 1970, ozone in the atmosphere has been decreasing and more ultraviolet radiation has been reaching Earth's surface. This poses an immediate health problem, but it is also of interest astronomically. When we discuss Mars, we will see a world with almost no ozone.

REVIEW Critical Inquiry

What evidence do we have that Earth has a molten metallic core?

In science, the critical analysis of ideas must eventually return to evidence. In this case, the evidence is indirect because we can never visit Earth's core. Seismic waves from distant earthquakes pass through Earth, but a certain kind of wave, the S waves, cannot pass through the core. Because the S waves cannot propagate through a liquid, we suspect that Earth's core is a liquid. Earth's magnetic field gives us further evidence of a metallic core. The theory for the generation of magnetic fields, the dynamo effect, requires a liquid, convective, conducting core. If the core were not a liquid, it would not be able to generate a magnetic field. Thus, we have two different kinds of evidence to tell us that our planet has a liquid core.

Every theory must be supported by evidence. What evidence do we have to support the theory of plate tectonics?

As we learn more about other worlds and their atmospheres, we also learn more about our own. Ironically, the next world we will discuss has no atmosphere at all.

6-2 The Moon

Only 12 people have stood on the moon, but planetary astronomers know it well. The photographs, measurements, and samples brought back to Earth paint a picture of an airless, ancient, battered crust.

The View from Earth

A few billion years ago, our moon must have rotated faster, but Earth is four times larger in diameter than the moon (Data File Two), and its tidal forces on the moon are strong. Earth's gravity raised tidal bulges on the moon, and friction in the bulges slowed the moon until it now rotates once each orbit to keep the same side facing Earth in what is called **tidal coupling.** This means that we who live on Earth always see the same side of the moon; we never see the back. The moon's

The Moon Data File Two

Earth   Moon

The moon's diameter is roughly 27 percent that of Earth.

Average distance from Earth	384,400 km (center to center)
Eccentricity of orbit	0.055
Maximum distance from Earth	405,500 km
Minimum distance from Earth	363,300 km
Inclination of orbit to ecliptic	5°9′
Average orbital velocity	1.022 km/s
Orbital period (sidereal)	27.321661 days
Orbital period (synodic)	29.5305882 days
Inclination of equator to orbit	6°41′
Equatorial diameter	3476 km
Mass	7.35×10^{22} kg (0.0123 $M_{\oplus}$*)
Average density	3.36 g/cm^3 (3.35 g/cm^3 uncompressed)
Surface gravity	0.167 Earth gravity
Escape velocity	2.38 km/s (0.21 $V_{\oplus}$*)
Surface temperature	−170° to 130°C (−274° to 266°F)
Average albedo	0.07

*In astronomy, the symbol $\oplus$ refers to Earth; thus $M_{\oplus}$ means the mass of Earth and $V_{\oplus}$ means the velocity of Earth.

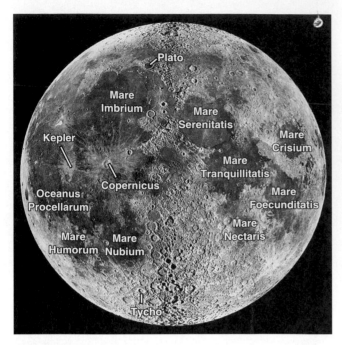

Figure 6-8

The moon is tidally locked to Earth and rotates to keep the same familiar side facing us as it follows its orbit. We never see the back side from Earth. The lunar maria are dark and smooth and have picturesque names such as Mare Imbrium, the Sea of Rains, and Mare Tranquillitatis, the Sea of Tranquility. In fact, there is no water on the moon, and the maria are ancient lava plains. The largest craters are named after famous scientists and philosophers such as Kepler, Copernicus, Tycho, and Plato. *(UCO/Lick Observatory photograph)*

familiar face has shone down on Earth since long before there were humans (Figure 6-8).

The moon is obviously airless. It is small, only a quarter of Earth's diameter (Data File Two), and its gravity is too weak to hold any gases. In Chapter 7, we will discuss in detail how a small planet can lose its atmospheric gases into space, but we can cite evidence for an airless moon. We can see dramatic shadows near the **terminator,** the dividing line between daylight and darkness. There is no air on the moon to scatter light and soften shadows. We also see stars disappear behind the **limb** of the moon—the edge of its disk—without dimming. Clearly, the moon is an airless (and soundless) world.

The most dramatic features on the moon, regions that are visible even to our unaided eyes, are **mare,** Latin for "sea." The dark maria (plural of *mare*) that make up the face of the "man in the moon" were so named because early observers thought the great smooth expanses were seas. Even a small telescope will show that the maria are marked by ridges, faults, smudges, and craters and therefore can't be water. The maria are actually great lowland plains covering 17 percent of the moon's surface.

Not many decades ago, astronomers argued about the nature of the maria. Some said that they were great solidified lava flows, but others argued that they were vast oceans of dust blasted from the lunar mountains by billions of years of meteorites and drawn to the lowlands by the lunar gravity. Science fiction writers wrote fascinating stories about the dust oceans. For example, *A Fall of Moon Dust* by Arthur C. Clarke tells of tourists in a lunar sightseeing boat trapped when their boat sinks in the dust to the bottom of a mare. Although it is a wonderful rescue story, it can never happen. We know now that the maria are great lava flows with only a few centimeters of loose dust on their surfaces.

These lava flows suggest volcanism, but only small traces of past volcanic activity are visible from Earth. No major volcanic peaks are visible, and no active volcanism has been detected. The lava flows that created the maria happened long ago and were much too fluid to build peaks. We can see a few small domes pushed up by lava below the surface, and we can see long, winding channels called **sinuous rilles** (Figure 6-9), often found near the edges of the maria. These channels were evidently cut by flowing lava. In some cases, such a channel may have had a roof of solid rock, forming a lava tube. When the lava drained away, meteorite impacts collapsed the roof to form a sinuous rille. Thus, the view from Earth gives us only hints of volcanic activity associated with the maria.

Besides the maria, the most striking feature of the moon is the millions of impact craters. The lighter parts of the moon's face are heavily cratered highlands. Although they look mountainous at first glance, there are no folded mountain ranges on the moon. Its crust has never wrinkled into folded mountains as Earth's crust has. The mountainous regions of the highlands are actually millions of meteorite craters, one on top of the other (Figure 6-10a). We will discuss the process that forms craters in the next section.

The crater density in different regions of the moon gives us a way of estimating the ages of different features, important information if we are to tell the story of the moon. The highlands must be old to have accumulated so many craters, but the maria contain only a few craters, so they can't be that old (Figure 6-10b). Most of the craters we see on the moon were produced during the first half billion years of the moon's existence by the heavy bombardment at the end of planet formation (Chapter 5). As the debris left over from the solar nebula was gradually swept up, the rate of cratering fell rapidly (Figure 6-11), and craters from the early heavy bombardment were probably totally destroyed by later cratering. The craters we see in the highlands were made by the last of the heavy bombardment about 3.8 billion years ago. The lava flows that fill the maria, which have very few craters, must have formed after the end of the heavy bombardment.

a

b

Figure 6-9
(a) The edge of Mare Imbrium is marked by high mountains and the sinuous rille called Hadley Rille (arrow). Such rilles were caused by lava flowing through a channel. Other rilles, straight faults, and many small craters are visible in the surface of the mare. (b) Apollo 15 landed near the 1200-foot-deep rille (circle in part a) and used a battery-powered rover to visit the rille and nearby mountains and craters. *(NASA)*

A few minutes at a telescope will reveal clear cases of craters that partially cover other craters. Evidently, older craters were partially destroyed by newer craters. We can also find craters that have been partially buried by maria. Evidently, these craters are older than the maria. Such ages are called **relative ages** because we can tell which feature formed first but we can't give either feature an age in years. By comparing the number of craters on a given surface with the rate of cratering as given in Figure 6-11, planetary astronomers can estimate **absolute ages** in years. Of course, an even more accurate way to find the ages of lunar features is to go

b

Figure 6-10
(a) The lunar highlands are the oldest parts of the moon and are heavily battered by craters. (b) The lunar lowlands are filled by solidi-fied lava flows containing few craters and long wrinkle ridges. The lava flows occurred after the end of the heavy bombardment and thus escaped most cratering impacts. *(UCO/Lick Observatory photograph)*

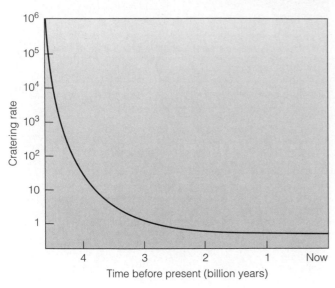

Figure 6-11
The cratering rate was high during the first half-billion years after the formation of the solar system, but as the debris was swept up, the cratering rate fell. The maria formed after the end of the heavy bombardment, and the cratering rate today is very low. Notice that the exponential scale on the vertical axis makes the curve seem less extreme than it really is. By 4 billion years ago the cratering rate had already fallen to 10,000 times less than when the planets formed, and by 3 billion years ago it had fallen to one million times less.

to the moon and bring back rocks that can be dated radioactively. We will discuss these radioactive ages later.

The view from Earth tells us that the moon has an ancient crust that was heavily cratered when the solar system was young. The maria are slightly younger than the highlands, but all parts of the moon are old and scarred by craters. Our goal is to tell the story of the moon (Window on Science 6-1), and to do that we must understand how craters are formed.

Impact Craters

When a meteorite strikes a planet, it delivers tremendous energy and can excavate an impact crater many times its own diameter. Meteorites strike the surface of a planet with velocities that range from 15 to 60 km/s, and even a small meteorite can carry tremendous energy. A meteorite the size of a house would hit with energy equivalent to a large nuclear bomb.

When a meteorite strikes, it penetrates the surface, deforms the rock layers, and converts some of its energy into enough heat to vaporize itself and nearby rock (Figure 6-12). The deformed rock rebounds, and shock waves move upward and outward, driven in part by the heat of the impact. These rebounding shock waves blast away the surface to create a crater. The resulting scar is a circular crater over ten times the diameter of the impacting meteorite and about 10 percent as deep as it is

Figure 6-12
The formation of a crater. An approaching meteorite (a) strikes the lunar surface (b) and penetrates deeply. The meteorite vaporizes, and shock waves fracture the rock and blast the surface away (c) to form a circular crater. (d) Ejected material forms a raised crater rim. The rebound of the rock can form a raised central peak, and slumping along the crater wall can form terraces. The rock below the crater is severely fractured.

wide. Rock blasted out of the crater falls to form a rim that rises slightly above the surrounding surface.

When we look at the moon through even a small telescope, we see many large craters that contain a central mountain and terraces along the inner walls of the crater. The central mountains are produced by the rebound of the fractured rock under the center of the crater. The terraces are produced by slumping as material slides back down into the crater from the inner walls. In even larger craters, we see an inner ring of mountains, and astronomers refer to these features as

Like any technical subject, science includes a mass of details, facts, figures, measurements, and observations. It is easy to be overwhelmed by the flood of details, but one of the most important characteristics of science comes to our rescue. The goal of science is not to discover more details, but to explain the details with a unifying hypothesis or theory. A good theory is like a basket that makes it easier for us to carry a large assortment of details.

This is true of all the sciences. When a psychologist begins studying the way the human eye and brain respond to moving points of light, the data are a sea of detailed measurements and observations. Once the psychologist forms a hypothesis about the way the eye and brain interact, the details fall into place as parts of a log-

ical story. If we understand the hypothesis, the details all fit together and make sense, and thus we can remember the details without blindly memorizing tables of facts and figures. The goal of science is understanding, not memorization.

Scientists are in the storytelling business. The stories are often called hypotheses or theories, but they are really just stories to explain how nature works. The difference between scientific stories and works of fiction lies in the use of facts. Scientific stories are constructed to fit all known facts and are then tested over and over against new facts obtained by observation and experiment.

When we try to tell the story of each planet in our solar system, we pull together all the hypotheses and theories and try to

make them into a logical history of how the planet got to be the way it is. Of course, our stories will be incomplete because astronomers don't understand all the factors in planetary astronomy. Nevertheless, our story of each planet will draw together the known facts and details and attempt to make them a logical whole.

Memorizing a list of facts can give us a false feeling of security, just as when we memorize the names of things without understanding them (Window on Science 1-2). Rather than memorizing facts, we should search for the unifying hypothesis that pulls the details together into a single story. Our goal in studying science is understanding nature, not just remembering facts.

---

multiringed basins. We will find multiringed basins on the moon and other worlds.

Debris blasted out of the crater is called **ejecta;** It falls back to the surface to create an ejecta blanket surrounding the crater. Some ejecta falls back into the crater, but some can be blasted far across the surface. On the moon, we can see the ejecta blankets around young craters because the rough surface scatters sunlight and looks bright. Ejecta shot out along specific directions can fall back to the surface and form bright **rays.** The crater Copernicus is the center of a vast ray system (Figure 6-8), and rays from the crater Tycho extend halfway around the moon. Older craters must have had rays, but exposure to sunlight and a continuous blasting from micrometeorites smaller than grains of sand gradually darken and smooth the lunar surface and obliterate the rays. Thus, the presence of bright, rayed craters shows that at least a few craters have been formed recently. Tycho, for example, may be only 100 million years old.

Impacts on the moon are tremendously powerful. Chunks of ejecta can travel large distances and fall back to the surface to form more craters called **secondary craters.** Some secondary craters are even big enough to form their own secondary craters. Some lunar impacts have been violent enough to blast rocks completely off the moon, and a few such rocks have been found on Earth. They must have been ejected from the moon and fallen to Earth as meteorites (Figure 6-13).

Figure 6-13
The meteorite ALHA81005 was found in Antarctica in January 1982 and later identified as a fragment of the lunar surface. *(NASA, Ursula B. Marvin)*

To learn more about the moon, we need to go there, and between 1969 and 1972, 12 astronauts did just that. They landed at six different sites on the moon as part of the Apollo project. Exploring the lunar lowlands known as the maria and the mountainous highlands (Figure 6-14), they conducted geological studies of the lunar surface. They brought back a treasure of moon

a

b

Figure 6-14
(a) The first three Apollo landings on the moon explored the lunar maria, ancient, flat lava plains powdered by eons of meteorite impacts. Note the level horizon in this photo on Mare Tranquillitatis. (b) Later landings were made in the mountainous lunar highlands, regions even older than the maria. Notice the mountains on the horizon and the powdery surface. The large boulder is ejecta from a large crater beyond the horizon. *(NASA)*

rocks, which paint a picture of the geology of the moon and give us a detailed insight into the complex history of its ancient battered surface.

Moon Rocks

Of the 382 kg (842 lb) of rock samples that the Apollo astronauts carried back to Earth, every one is igneous. That is, they formed by the solidification of molten rock. No sedimentary rocks were found, which is con-sistent with the moon never having had liquid water on its surface. In addition, the rocks were extremely dry. Almost all Earth rocks contain 1 to 2 percent water, either as free water trapped in the rock or as water molecules chemically bonded with certain minerals. But moon rocks contain no water at all.

Rocks from the lunar maria are dark-colored, dense basalts much like the solidified lava produced by the Hawaiian volcanoes. These rocks are rich in heavy elements such as iron, manganese, and titanium, which

a

b

c

Figure 6-15
Moon rocks. (a) A vesicular basalt contains numerous bubbles frozen into the rock when it solidified from a lava flow on the surface. (b) A breccia is formed from two kinds of rocks cemented together. (c) Rocks exposed on the surface of the moon for long periods become pitted with small micrometeorite craters. *(NASA)*

One of the best ways to think about a scientific problem is to follow the energy. According to the principle of cause and effect, every effect must have a cause, and every cause must involve energy (Window on Science 2-3). Energy moves from regions of high concentration to regions of low concentration and in doing so produces changes. For example, we may burn coal to make steam in a power plant, and the steam passes through a turbine and then escapes into the air. In flowing from the burning coal to the atmosphere, the heat spins the turbine and makes electricity.

It is surprising how commonly scientists use energy as a key to understanding nature. A biologist might ask where certain birds get the energy to fly thousands of miles, and an economist might ask where the economic energy to support the creation of new investments comes from. Energy is everywhere, and when it moves, whether it is birds, money, or molten magma, it causes change. Energy is the cause in cause and effect.

We can think of a planet by following the energy. The heat in the interior of a planet may be left over from the formation of the planet, or it may be heat generated by radioactive decay, but it must flow outward toward the cooler surface, where it is radiated into space.

The rich geology of our world is produced by heat flowing out of the interior. That heat stirs the molten interior and generates a magnetic field. It drives plate tectonics and the collision of plates produces great mountain ranges. The separation of the plates causes rift valleys that widen to become ocean basins. Volcanoes erupt spewing new rock onto the crust and vent more gases into the atmosphere. Perhaps the continents themselves have been built up over the eons by colliding plates. Our world is active, rich, and vibrant. A world without heat flowing from its interior is a dead, quiescent world.

When you think about any world, be it a small asteroid or a giant planet, think of it as a source of heat that flows outward through the planet's surface into space. If we can follow that energy flow, we can understand a great deal about the world. A planetary astronomer once said, "The most interesting thing about any planet is how its heat gets out."

give them their dark color. Some of the basalts are **vesicular,** meaning that they contain holes caused by bubbles of gas in the molten rock (Figure 6-15a). Like bubbles in a carbonated beverage, these bubbles do not form while the magma is under pressure. Only when the rock flows out onto the surface, where the pressure is low, do bubbles appear. Thus, the presence of vesicular basalts assures us that these rocks formed in lava flows that reached the surface and did not solidify underground. The ages of the mare basalts can be found from the radioactive atoms they contain (Chapter 5), and these ages range from about 3.1 to 3.8 billion years. Thus, the lava flows that formed these rocks must have happened some time after the formation of the moon.

Rocks from the highlands are light-colored, lower-density **anorthosite.** This rock is rich in calcium, aluminum, and oxygen and is the kind of rock we would expect to form first and float to the top of cooling molten magma. It seems that the anorthosite is the true crustal rock of the moon, while the basalts are lava flows that filled the lowlands. The ages of the anorthosite range from 4.0 to 4.4 billion years, significantly older than the mare basalts.

A large fraction of the lunar rocks are **breccias,** rocks that are made up of fragments of earlier rocks cemented together by heat and pressure (Figure 6-15b). Evidently, meteorite impacts have broken up the rocks and fused them together time after time.

Both the highlands and the lowlands of the moon are covered by a layer of powdered rock fragments called the **regolith.** It is about 10 m deep on the maria but over 100 m deep in certain places in the highlands. About 1 percent of the regolith is meteoric fragments; the rest is the smashed remains of moon rocks that have been ground down to dust by the constant blasting of meteorites. The smallest meteorites, the **micrometeorites,** do the most damage by their constant sandblasting of the lunar surface (Figure 6-15c). Thus, we see that impact cratering dominates the lunar surface and is even responsible for the lunar regolith.

The moon rocks are old, dry, igneous, and badly shattered by impacts. We can use these facts, combined with what we know of lunar features, to reconstruct the moon's past.

The History of the Moon

The ultimate goal in the study of any world is to describe its history, to explain how it evolved from its formation to its present state (Window on Science 6-1). We have seen that the surface of the moon has been severely battered by impact craters, but the maria tell of a different theme in lunar history—the moon's internal heat. Like Earth, the moon formed with heat trapped in its interior, and radioactive decay released more heat. Unlike Earth, the moon is small, and that has sealed its fate. Because it is small, it could not retain its internal heat, it aged quickly, and it is now a dead world (Window on Science 6-2).

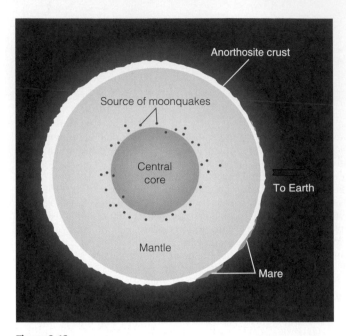

Figure 6-16

A cross section of the moon shows an anorthosite crust thinner on the side facing Earth and a deep mantle of brittle rock. Moonquakes occur deep in the mantle and suggest that the moon has a central core of rock that may be slightly less brittle than the mantle. The low density of the moon and its lack of a magnetic field rule out any iron core more than a few hundred kilometers in diameter.

There is no evidence that the interior of the moon now contains significant heat. The moon has no magnetic field of its own, so it cannot contain a core of molten iron. In fact, the moon's average density is 3.36 g/cm³, so low we can conclude that the moon cannot contain more than a tiny iron core a few hundred kilo-

meters in diameter. The astronauts left seismographs on the lunar surface, which have detected moonquakes about as large as big firecrackers deep in the interior near the boundary between the brittle rock of the mantle and a more plastic rock in a central core (Figure 6-16).

The moon is now a cold, dead, geologically inactive world, but it began as a sea of molten rock. The lack of rocks older than about 4.3 billion years and the chemical composition of the anorthosite have convinced planetary astronomers that the surface of the moon was covered by a sea of magma when it was very young. The energy to melt the rock probably came from the heat of formation (Chapter 5) released by in-falling meteorites.

In the first of the four stages of planetary evolution, the moon cooled and differentiated as the denser minerals sank into the interior and the lower-density minerals floated to the top, like slag in an open-hearth furnace. Given the present chemical composition of the moon, we would expect this slag to cool and form an anorthosite crust within a few hundred million years of the moon's formation.

The second stage of planetary evolution is cratering, and the moon was heavily battered by meteorites. The early heavy bombardment must have blasted the newly formed crust over and over. Most rocks older than 4 billion years are smashed to fragments, and the entire crust of the moon is fractured to a depth of a few kilometers. The largest meteorites excavated vast basins. The mare Oceanus Procellarum may overlie one of the oldest and largest basins on the moon, but later impacts and lava flows have partially obliterated its circular outline. Mare Imbrium was excavated by the impact of a meteorite the size of Rhode Island (roughly 65 km across). The resulting basin was 1300 km across, and the impact smashed the crust into mountain ranges up to 5 km high. Ejecta from this impact covers 16 percent of the lunar surface (Figure 6-17). Apollo 14 landed at Fra Mauro to sample it.

Figure 6-17

The entire surface of the moon has been blasted by meteorite impacts and is fractured to a depth of about 2 km. The uppermost layers of the surface are rock dust produced by meteorite impacts. The large rock visible here is ejecta from a meteorite impact that must have occurred far beyond the horizon. *(NASA)*

The Imbrium Basin is hard to study because it is flooded with later lava flows, but a similar basin, Mare Orientale, lies at the western limb of the moon and has not been flooded. Mare Orientale is a great multiringed basin consisting of three circular mountain ranges that were smashed into the lunar crust by the impact. The Imbrium Basin has similar rings, although they are mostly buried under successive lava flows (Figure 6-18).

Our story of the moon might suggest that it was a violent place during this cratering phase, but large impacts were in fact rare and, except for a continuous rain of smaller meteorites, the moon was, for the most part, a peaceful place even during the heavy bombardment. Centuries might pass between major impacts. Of course, a large impact might have threatened to bury us under

ejecta or jolted us by seismic shocks. We could have felt the Imbrium or Orientale impacts anywhere on the moon, but had we been standing on the side of the moon opposite one of these impacts, we would have been at the focus of seismic waves traveling around the moon from different directions. When the waves met under our feet, the surface would have jerked up and down by as much as 10 m. The places on the moon opposite the Imbrium Basin and the Orientale Basin are both strangely disturbed landscapes called **jumbled terrain.** We will find similar effects of large impacts on other worlds.

By the end of the heavy bombardment, the moon's crust was heavily cratered and fractured, and the moon had cooled significantly. Below the crust, however, the rock was molten, and the decay of radioactive elements added heat. The fractures below the largest basins allowed some of that magma to reach the surface (Figure 6-19), and the resulting lava flows filled the basins with layer after layer of dense magma that solidified to form the maria. Some maria are round, such as Mare Crisium, following the outline of the basin, but others are irregular in shape, showing that the lava overflowed the basin boundaries.

This flooding, the third stage of planetary evolution, did not occur all at once. Studies show that maria consist of successive lava flows one on top of another. In many maria, the weight of the lava pressed the crust

a

Figure 6-18
(a) A topographic map of the moon's nearside shows altitude as color. Blue regions lie about 3 km below the average radius of the moon and red regions about 3 km above. The lunar maria are deep, generally round basins later flooded by lava. In this map from the Clementine spacecraft, north is at the top, and Mare Imbrium is the large round basin at upper left. *(NASA/JPL)* (b) Major impacts produce multiringed basins and break the crust to great depth. Magma can follow these fractures upward to flood the basins. *(Adapted from William Hartmann,* Astronomy: The Cosmic Journey, *Belmont, Calif.: Wadsworth Publishing Company, 1978)*

├─── 200 km ───┤

b

downward and created wrinkle ridges (see Figure 6-10b). In other cases, the lava melted channels or tunnels through the rock, features we see today as sinuous rilles (see Figure 6-9).

Almost all of this flooding occurred on the side of the moon facing Earth. The back side of the moon is heavily cratered, but only a few of the largest craters contain any lava flows (Figure 6-20). Even the multi-ringed basin Mare Orientale, near the west limb, is only slightly flooded. The difference is due to the thickness of the crust. On the side toward Earth, the lunar crust is only about 60 km thick, but on the far side it is more than 100 km thick (Figure 6-16). This asymmetry may have resulted from the immense impact that formed the Oceanus Procellarum Basin and weakened the near-side crust.

The fourth stage of planetary evolution, slow surface evolution, is hardly evident on the moon. It is a small world, and small worlds cool rapidly because they have a large ratio of surface area to volume, just as a small cupcake cools more rapidly than a large cake. The moon has thus lost much of its internal heat. We saw in the case of Earth that geological activity is driven by energy flowing upward from within. But the moon has lost much of its internal energy, and the crust has grown thick and has never divided into moving

Figure 6-19
The evolution of the lunar surface. Mare Imbrium began when a meteorite as large as Rhode Island excavated a great crater basin. Lava welled up to flood the basin, starting about 3.8 billion years ago and lasting until no less than 2 billion years ago. Since then only scattered meteorites have modified Mare Imbrium. *(Don Davis)*

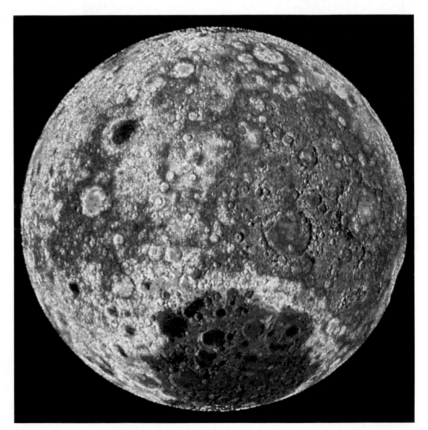

Figure 6-20
Lava flows occurred repeatedly on the lunar nearside but flooded only a few of the small basins on the lunar farside. Nevertheless, the far side of the moon is marked by some very deep basins, including the South Pole–Aitken Basin (purple). Although it is very deep, it has an old, cratered floor and appears to have suffered little flooding. *(NASA/JPL)*

plates. There are no rift valleys or folded mountains on the moon. The moon is too small to retain an atmosphere, so it has never had liquid water on its surface to cause erosion. The only slow surface evolution on the moon is the constant blast of micrometeorites that grinds the surface rocks down to dust. The footprints of the Apollo astronauts will last for millions of years, but they will eventually be erased by the constant flux of meteorites.

On the other hand, the overall terrain on the moon is almost fixed. A billion years from now, plate tectonics will have totally changed the map of Earth, erosion will have worn away the Rocky Mountains, the descent stage of the Apollo 17 lunar lander will be nothing but a peculiar chemical contamination in the regolith at Taurus–Littrow, but the lunar mountains will look much the same.

Our story of the moon's history is as complete as comparative planetology can make it, but it omits one initial detail. Where did Earth get such a large satellite?

The Origin of the Moon

The Apollo missions to the moon told us a great deal, but they failed to answer the most basic question: How did the moon form? The three traditional theories do not work, but a newer hypothesis may solve the problem.

The first of the three traditional theories, the **fission hypothesis,** supposes that the moon formed by the fission of Earth. If the young Earth spun fast enough, tides raised by the sun might make it break into two parts (Figure 6-21a). If this separation occurred after Earth differentiated, the moon would have formed from crust material. Thus, the theory explains the moon's low density.

But the fission theory has problems. No one knows why the young Earth should have spun so fast, nearly ten times faster than today, nor where all of that rotation went after the fission. In addition, the moon's orbit is not in the plane of Earth's equator, as it would be if it had formed by fission.

The second traditional theory is the **co-accretion (or double-planet) hypothesis.** It supposes that Earth and the moon condensed as a double planet from the same cloud of material (Figure 6-21b). However, if they formed from the same material, they should have the same chemical composition and density, which they don't. The moon is very poor in iron and related elements such as titanium, but it also contains almost exactly the same ratios of oxygen isotopes as does Earth's mantle. Co-accretion cannot explain these compositional differences.

The third theory is the **capture hypothesis.** It supposes that the moon formed somewhere else and was later captured by Earth (Figure 6-21c). We might imagine the moon forming inside the orbit of Mercury, where the heat would prevent the condensation of solid metallic grains and only high-melting-point metal oxides could have solidified. A later encounter with Mercury could "kick" the moon out to Earth.

The capture theory is not popular because it requires highly unlikely events involving interactions with Mercury and Earth to move the moon from place to place. Scientists are always suspicious of explanations that require a chain of unlikely coincidences. Also, upon encountering Earth, the moon would have been moving so rapidly that Earth's gravity would have been unable to capture it without ripping the moon to fragments through tidal forces.

Until recently, astronomers were thus left with no acceptable theory to explain the origin of the moon, and they occasionally joked that the moon could not exist. But during the 1980s, planetary astronomers developed a new theory that combines the best aspects of the fission hypothesis and the capture hypothesis.

The **large-impact hypothesis** supposes that the moon formed from debris ejected into a disk around Earth by the impact of a large body (Figure 6-22). The impacting body may have been larger than Mars. Instead of saying that Earth was hit by a large body, we may be more nearly correct in saying that Earth and the moon resulted from the collision and merger of two very large planetesimals. The resulting large body became Earth, and the ejected debris formed the moon.

This hypothesis would explain a number of things. The collision had to occur at a steep angle to eject enough

a

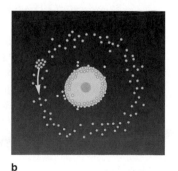

b

c

Figure 6-21
Three traditional theories for the moon's origin. (a) Fission theories suppose that Earth and the moon were once one body and broke apart. (b) Co-accretion theories suppose that the moon formed from material near Earth. (c) Capture theories suggest that the moon formed elsewhere and was captured by Earth. None of these theories explains all the facts.

Figure 6-22
A Mars-sized planetesimal strikes the young, molten Earth, blasting iron-poor mantle material into orbit. In this simulation, the iron core of the planetesimal falls back into Earth within 4 hours, and the moon begins to form within about 10 hours.

matter to make the moon. That is, the objects did not collide head-on. Such a glancing collision would have spun the resulting material rapidly, which would explain the high angular momentum in the Earth–moon system. If the colliding planetesimals had already dif-

ferentiated, the ejected material would be mostly iron-poor mantle and crust. Calculations show that the iron core of the impacting body would have fallen into Earth within 4 hours. This would explain why the moon is so poor in iron and why the abundances of other elements are so similar to rocks from Earth's mantle. Also, the material that eventually became the moon would have remained in a disk long enough for volatile elements, which the moon lacks, to be lost to space.

The moon may be the remains of a giant impact. Until recently, astronomers have been reluctant to consider such catastrophic events, but in the following chapters we will see evidence that a number of the planets have suffered from major impacts. We have already noted the peculiar rotations of Venus, Uranus, and Pluto in Chapter 5. Consequently, we must stay alert to the possibility that the gradual development of the Earth, moon, and the other worlds in our solar system was altered by catastrophic impacts.

REVIEW Critical Inquiry

If the moon was intensely cratered by the heavy bombardment and then formed great lava plains, why didn't the same thing happen on Earth?

In fact, the same thing did happen on Earth. Although the moon has more craters than Earth, the moon and Earth are the same age, and both were battered by meteorites during the heavy bombardment. Some of those impacts on Earth must have been large and dug giant multiringed basins. Lava flows must have welled up through Earth's crust and flooded the lowlands to form great lava plains much like the lunar maria.

Earth, however, is different from the moon in one important respect. It is larger. Because it is a larger world, Earth has more internal heat, and that heat escapes more slowly than the moon's heat did. The moon is now geologically dead, but Earth is very active, with heat flowing outward from the interior to drive plate tectonics. The moving plates long ago erased all evidence of the cratering and lava flows dating from Earth's youth. Earth's surface was almost certainly cratered and flooded by lava flows, as was the moon's, but our planet has continued to evolve, and the moon has not.

Comparative planetology is a powerful tool in that it allows us to see similar processes occurring under different circumstances. For example, suppose you visited a world in our solar system that was even smaller than the moon and so had lost its heat even faster. What kind of surface features would you find, and what kinds of rocks would you pick up? More important, what features and rocks would be missing? We will find such small worlds among the asteroids.

The moon's evolution has been limited by its size. It is too small to have an atmosphere, and it cooled so rapidly it never developed plate motion. In the next chapter, we will study three worlds, one small, one large, and one medium. Mercury, Venus, and Mars will give us further illustrations of the power of cratering and heat flow in shaping planetary surfaces.

The large-impact hypothesis suggests that the Earth–moon system formed in the collision of two large planetesimals. The differentiated bodies formed a single large object that became the higher-density, iron-rich Earth. Ejected mantle material, poor in iron, formed a disk and eventually condensed to form the lower-density, iron-poor moon.

Summary

All terrestrial planets pass through a four-stage history: (1) differentiation; (2) cratering; (3) flooding of the crater basins by lava, water, or both; and (4) slow surface evolution. The importance of each of these processes in the evolution of a planet depends on the planet's mass and temperature.

Earth is the largest terrestrial planet in our solar system, and it has passed through all four stages. Studies show that Earth has differentiated into a metallic core and a silicate crust. Currents in the molten portion of the core produce Earth's magnetic field.

The rocky crust floats on a plastic mantle and is divided into roughly 20 sections that pull apart along midocean rifts, push against each other to build folded mountain chains, and slide over each other at subduction zones where sections of crust are pushed back down into the interior. This process, plate tectonics, continually reshapes Earth's surface.

Earth's first atmosphere, its primary atmosphere, was rich in carbon dioxide and nitrogen, but carbon dioxide dissolved into the ocean waters and became sediments. The atmosphere was left rich in nitrogen, and the evolution of plant life added large amounts of oxygen. Thus the air we breathe is a secondary atmosphere. Today, the greenhouse effect warms Earth's surface, and the ozone layer protects us from ultraviolet radiation.

The moon is, at first glance, dramatically different from Earth. We see two distinct kinds of terrain on the moon. The maria, named after seas, are lowland plains with few craters. They are great lava flows and are younger than the highlands. The highlands are lighter in color and heavily cratered.

Samples returned by the Apollo missions show that the maria are dark basalts, whereas the highlands are light anorthosite. Many samples are breccias, which show how severely the moon's crust has been fractured.

The ages of the mare samples range from about 3.1 to 3.8 billion years, but the highland samples are about 4.0 to 4.4 billion years old. Few samples are older than 4.3 billion years. This suggests that the moon's surface was molten until about 4.4 billion years ago.

As the anorthosite crust solidified, cratering battered it and excavated great basins. Later, molten rock rose through fractures in the crust and flooded the basins to produce the maria. By the time the maria began to form, cratering had declined, so the maria are marked by few craters.

The lunar surface now is evolving very slowly. Because the moon is small, it has cooled rapidly, its crust has grown thick, and it has never divided into moving plates. Now only meteorites alter the surface.

New Terms

comparative planetology	mare
P waves	sinuous rille
S waves	relative age
dynamo effect	absolute age
mantle	multiringed basin
plastic	ejecta
plate tectonics	ray
basalt	secondary crater
midocean rise	vesicular basalt
midocean rift	anorthosite
subduction zone	breccia
rift valley	regolith
folded mountain chain	micrometeorite
primeval atmosphere	jumbled terrain
secondary atmosphere	fission hypothesis
greenhouse effect	co-accretion hypothesis
tidal coupling	capture hypothesis
terminator	large-impact hypothesis
limb	

Review Questions

1. What are the four stages in Earth's development?
2. Why do we expect planets to have differentiated?
3. How do we know that Earth has a molten core?
4. How does plate tectonics create and destroy Earth's crust?
5. Why do we suspect that Earth's primeval atmosphere was rich in carbon dioxide?
6. Why doesn't Earth have as many craters as the moon?
7. How do we find the relative ages of the moon's maria and highlands?
8. Why do we believe that Copernicus is a young crater?
9. Why are there no folded mountain chains on the moon?
10. How does the range of ages of lunar samples suggest that the moon formed with a molten surface?
11. Why are so many lunar samples breccias?
12. What do the vesicular basalts tell us about the evolution of the lunar surface?

13. What evidence would we expect to find on the moon if it had been subjected to plate tectonics?

14. Cite objections to the fission, co-accretion, and capture hypotheses.

15. How does the large-impact hypothesis explain the moon's lack of iron?

12. An astronaut in a space suit is about 1 meter in diameter. Could we have seen the astronauts on the moon with Earth-based telescopes?

13. While two Apollo astronauts went down to the lunar surface, a third stayed behind in the command module orbiting 200 km above the surface. Could the astronauts be seen from the command module?

Discussion Questions

1. If we visited a planet in another solar system and discovered oxygen in its atmosphere, what might we guess about its surface?

2. If liquid water is rare on the surface of planets, then most terrestrial planets must have CO_2-rich atmospheres. Why?

3. Old science-fiction paintings and drawings of colonies on the moon often showed very steep, jagged mountains. Why did the artists assume that the mountains would be more rugged than mountains on Earth? Why are lunar mountains actually less rugged than mountains on Earth?

Problems

1. About what percent of Earth's volume is occupied by its iron core? its rocky crust?

2. If the Atlantic seafloor is spreading at 3 cm/year and is now 6400 km across, how long ago were the continents in contact?

3. The dinosaurs died and mammals arose about 65 million years ago. For what percent of Earth's history have we mammals been a leading life form?

4. If we estimate that our atmosphere is 200 km deep, what percent of Earth's radius is that?

5. What is the ratio of the volume of the moon to the volume of the Earth? Why is that different from the ratio of the masses?

6. Why do small worlds cool faster than large worlds? Compare surface area to volume.

7. The smallest detail visible through Earth-based telescopes is about 1 second of arc in diameter. What size is that on the moon? (*Hint:* Use the small-angle formula.)

8. Midocean rifts and the subduction zones on Earth are 1 km or less across. Why are we sure that such features don't exist on the moon?

9. The Imbrium Basin is about 1300 km in diameter. What percentage of the moon's surface is that?

10. If Earth's moon orbited Jupiter, what would its maximum angular diameter be as seen from Earth? (*Hint:* Use the small-angle formula.)

11. The full moon has an apparent magnitude of −12.5. If it were orbiting Jupiter, how bright would it be? (*Hints:* Sunlight is dimmer at Jupiter's distance, and the moon would be further from Earth. Use the inverse square law.)

Critical Inquiries for the Web

1. The process of plate tectonics has modified Earth's surface significantly over the years. Use Web sites that trace the motion of the plates in the past to determine what the surface of Earth looked like 600 million years ago, 400 million years ago, and 200 million years ago. How did North America appear at these times?

2. What would it be like to walk on the lunar surface? Apollo astronauts visited six different locations on the moon, exploring a variety of lunar terrain. Describe the horizons and general relief of the landing locations of the different missions by exploring Web sites that provide lunar surface photography from the missions. What differences do you see between images from landings in highlands and maria?

Exploring *The Sky*

1. Aside from the sun, the moon is the most prominent object in our sky.

   a. Let *The Sky* show you the moon as it appears at the present time. What kind of features can you identify? What is the phase of the moon?
   How to proceed: Use **Find** to center the moon on the screen. Click on the magnification button to give you a large image of the moon.

   b. Find the following information about the moon at the present time:
   the constellation in which it appears,
   the phase as a percentage of full,
   the angular size,
   its distance from Earth.
   How to proceed: To find the phase, angular size, and distance from Earth, click on the moon to obtain **Object Information.**

   c. Find the date of the next full moon, and find the same information as in part (a).
   How to proceed: To find the date of the next full moon, click on **Tools,** then click on **Moon Phase Calendar.** Then set the date by clicking on **Data,** then **Site Information.**

Go to the Brooks/Cole Astronomy Resource Center (www. brookscole.com/astronomy) for critical thinking exercises, articles, and additional readings from InfoTrac College Edition, Brooks/Cole's online student library.

CHAPTER SEVEN

Mercury, Venus, and Mars

There wasn't a breath in that land

of death . . .

Robert Service
The Cremation of Sam McGee

Guidepost

The previous chapters proposed theories to explain how the solar system formed and how planets like the Earth and the moon have formed and evolved. We are now ready to apply our ideas to other worlds.

Mercury, Venus, and Mars are rocky worlds, so we might expect them to be similar. It is astonishing to see how different they are. Much of this chapter is aimed at understanding how these planets evolved into their present states.

These three worlds cannot tell us much about the origin of the solar system because they have been modified by their own evolution. Nevertheless, we find more hints that the solar system was a dangerous place with major impacts smashing the surfaces of planets, a process we first suspected when we studied Earth's moon.

In the next chapter, we will use similar principles to explore the outer solar system.

Without a space suit, you could live only about 30 seconds on the surface of Mars. The temperature might not be too bad. On a hot summer day at noon, the temperature might reach 20°C (68°F). But the air is 95 percent carbon dioxide, is deadly dry, and contains almost no oxygen. Also, the air pressure at the surface of Mars is only 0.01 that at the surface of Earth. Stepping out onto the surface of Mars without a space suit would be about the same as opening the door of an airplane flying at 30,000 m (100,000 ft) above Earth. (Commercial jets do not generally fly above 12,000 m, or 38,000 ft.)

You would definitely need a space suit on Mercury, where there is virtually no atmosphere at all. If you stood in sunlight the temperature would shoot up to hundreds of degrees Celsius, and if you stepped into the shade the temperature would fall to almost −200°C.

No space suit would save you on Venus. The air pressure is 90 times Earth's, and the atmosphere is mostly carbon dioxide with traces of acids and hot enough to melt lead.

Why are these worlds such rotten places for picnics? In some ways they resemble Earth. They are made of silicates and metals and lie in the inner solar system. We might expect them to be similar. How did they get to be so different from each other and from Earth? Comparative planetology will give us some clues.

When we used comparative planetology to study Earth's moon in the previous chapter, we saw the importance of impact cratering and the outward flow of energy. These same principles will be important in exploring the evolution of Mercury, Venus, and Mars. We will also see the importance of distance from the sun. Our study of these worlds will give new meaning to the word *unearthly*.

7-1 Mercury

The planet Mercury, named for the fleet-footed messenger of the gods, zips around the sun in only 88 days. Because it is so close to the sun, we see Mercury near the horizon at sunset or sunrise (Figure 1-11), and even large telescopes reveal little more than a shimmering globe passing through phases. In the mid-1970s, the Pioneer 10 spacecraft visited Mercury and sent back pictures of a cratered, airless world (Figure 7-1).

Mercury is a third the diameter of Earth (Data File Three), but it has more in common with Earth's moon. Both are cratered, airless, and ruled by tides.

The Motion of Mercury

Earth's moon is tidally locked to Earth, rotating once on its axis as it orbits Earth and thus keeping the same

side facing us. Mercury is also controlled by tidal forces. Each time it circles the sun it rotates 1.5 times (Figure 7-2). This means that a mountain on Mercury directly below the sun at one place in its orbit will point away from the sun one orbit later.

If you flew to Mercury and landed in the middle of the daylight side, the sun would be high overhead and it would be noon. The sun would set 44 Earth days later in the west, and midnight would occur 44 Earth days after sunset. Noon to midnight on Mercury is 88 Earth days, and that is how long Mercury takes to circle the sun once. A full Mercury day is two Mercury years long!

This coupling between rotation and revolution illustrates the power of tides. Although Mercury has no oceans, it has tides. The sun's gravity distorts Mercury from a true sphere, and the friction of these tidal bulges has slowed the rotation until it and the orbital revolution have reached a **resonance** in which the two periods fit together in a simple way—in this case three rotations for every two orbits. We will see many such tidal resonances as we explore the solar system.

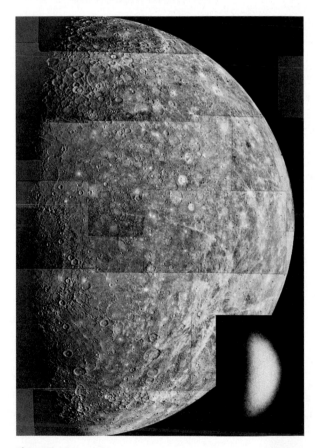

Figure 7-1
Photomosaic of Mercury made by the Mariner 10 spacecraft. Caloris Basin is in shadow at left. Almost no surface detail is visible from Earth (inset). *(NASA; inset courtesy Lowell Observatory)*

Figure 7-2
Mercury's rotation is tidally coupled to our sun, but unlike our moon, which rotates to keep the same face toward Earth, Mercury rotates 1.5 times during each orbit. A mountain facing the sun (a) will face away from the sun after one orbit (e) and will face the sun again after the next orbit.

Mercury Data File Three

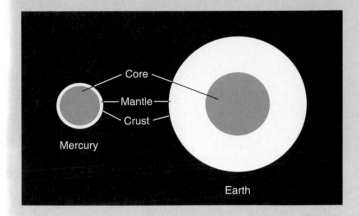

In proportion to its size, Mercury has a very large metallic core.

Average distance from the sun	0.387 AU (5.79×10^7 km)
Eccentricity of orbit	0.2056
Maximum distance from the sun	0.467 AU (6.97×10^7 km)
Minimum distance from the sun	0.306 AU (4.59×10^7 km)
Inclination of orbit to ecliptic	7°00'16"
Average orbital velocity	47.9 km/s
Orbital period	0.24085 y (87.969 days)
Period of rotation	58.646 days (direct)
Inclination of equator to orbit	0°
Equatorial diameter	4878 km (0.382 $D_\oplus$)
Mass	3.31×10^{23} kg (0.0558 $M_\oplus$)
Average density	5.44 g/cm³ (5.4 g/cm³ uncompressed)
Surface gravity	0.38 Earth gravity
Escape velocity	4.3 km/s (0.38 $V_\oplus$)
Surface temperature	−170°C to 430°C (−280°F to 800°F)
Average albedo	0.1
Oblateness	0

a

b

Figure 7-3
(a) Caloris Basin on Mercury lies partly in darkness to the left in this Mariner 10 photo. It is a vast impact basin 1300 km in diameter filled with lava flows and ringed by concentric mountain ranges up to 3 km high. (b) Hero Rupes, a lobate scarp, cuts through craters in smooth planes. The lobate scarps formed when Mercury cooled and its core shrank, wrinkling its crust. *(NASA)*

The Surface of Mercury

Mercury is hot and nearly airless. Daytime temperatures can reach 700 K (427°C) and nighttime temperatures sink to 100 K (−173°C). The temperature extremes occur because Mercury is too small to keep a significant atmosphere. Spectra show it has only traces of hydrogen, helium, oxygen, potassium, and sodium, elements borrowed from the solar wind or cooked out of the crust.

The rocky crust of Mercury is scarred by craters of all sizes, some old and degraded and others young with bright rays. The largest basin on Mercury is called Caloris Basin after the Latin word for "heat," recognition of its location at one of the two "hot poles," which face the sun at alternate perihelions. At the times of the Mariner encounters, the Caloris Basin was half in shadow (Figure 7-3a). Although half cannot be seen, the low angle of illumination is ideal for the study of the lighted half because it produces dramatic shadows.

The Caloris Basin is a gigantic impact basin 1300 km in diameter with concentric mountain rings up to 3 km high. The impact threw ejecta 600 to 800 km across the planet, and the focusing of seismic waves on the far side produced peculiar terrain (called "weird" terrain) that looks much like the jumbled surface of Earth's moon. Like Mare Orientale on the moon, the Caloris Basin is partially filled with lava flows. Some of this lava may be material melted by the energy of the impact, but some may be lava from below the crust that leaked up through cracks. The weight of this lava and the response of the crust have produced deep cracks in the central lava plains. The geophysics of such large, multiringed crater basins is not well understood at present, but Caloris Basin seems to be the same kind of structure as Mare Orientale and the Imbrium Basin on the moon.

When planetary scientists began looking at the Mariner photographs in detail, they discovered something not seen on the moon. Mercury is marked by great curved cliffs called **lobate scarps** (Figure 7-3b). These seem to have formed when the planet cooled and shrank by a few kilometers, wrinkling its crust as a drying apple wrinkles its skin. Some of these scarps are as high as 3 km and reach hundreds of kilometers across the surface. Other faults in Mercury's crust are straight and may have been produced by tidal stresses generated when the sun slowed Mercury's rotation.

Unlike our moon, Mercury does not have dark maria, but Mariner photographs reveal **intercrater plains,** regions that contain few craters larger than 15 km in diameter. Smaller regions called **smooth plains** mostly located around Caloris Basin contain even fewer craters. These appear to be solidified lava flows that covered older terrain. The smooth plains are the youngest and may have formed soon after the Caloris impact. Because of chemical differences, these lava plains on Mercury are not as dark as the maria on the moon, but they show that flooding by lava has occurred on Mercury.

The Interior of Mercury

The mass of Mercury is known from its influence on the passing Mariner 10 spacecraft, and if we divide its mass by its volume, we discover that it is over 60 percent denser than our moon. Spectra show that its surface is rock, so it must have a large core of dense metals such as iron. This poses a mystery.

A world as small as Mercury should lose its internal heat rapidly, and any molten iron should have solidified long ago. If the iron core is not molten, then the dynamo effect cannot generate a magnetic field. Nevertheless, the Mariner 10 spacecraft detected a magnetic field as strong as 0.5 percent of Earth's field. Some models have proposed that the dense core may contain iron mixed with other elements such as sulfur, which would lower the melting point and allow the planet to have some molten core to generate a magnetic field.

The large core, larger in proportion than Earth's iron core (Data File Three), is even larger than we would expect from the condensation sequence (Chapter 5). Mercury condensed near the sun and should have incorporated lots of metals, but not as much as we find in its big core. One hypothesis proposes that Mercury, having differentiated into a metallic core and a rocky mantle, was struck by a giant planetesimal in an impact much like that proposed to explain the origin of Earth's moon. Such an impact could have blasted most of the lower-density mantle into space, leaving behind a large metallic core and whatever rocky material fell back. Thus both Mercury and the moon may be products of giant impacts.

A History of Mercury

The accumulated facts about Mercury tell us little until we have a unifying hypothesis. Like a story, a unifying hypothesis must make sense and bring the known facts together in a logical description of the origin and evolution of Mercury (Window on Science 6-1).

Mercury formed from the solar nebula near the sun, and that left it a small, dense, hot world. It is small because the solar nebula contained little solid material near the sun, and much of that material was metallic. Thus Mercury has a large metallic core and is too small and too hot to retain more than a trace of an atmosphere.

In the first stage, Mercury differentiated to form a metallic core and a rocky mantle. Cratering, the second stage, was intense at first, but as debris in the solar system was swept up, cratering declined. We see evidence of this decline in the lava-flooded intercrater plains and the even younger smooth plains that contain fewer craters. Rare, very large impacts produced large basins such as Caloris Basin, and a planet-shattering impact early in the history of Mercury may explain why its metallic core is so large.

Because Mercury is small, it has lost much of its internal heat. Although the presence of a magnetic field tells us the interior contains some molten material, the core must be mostly solid. Shrinking as it cooled, the planet has wrinkled to form lobate scarps in its crust, but there has been little further geological activity, and, like the moon, Mercury is geologically dead.

Mercury, like Earth's moon, is now an inactive world. To find planets where the heat flowing outward from the interior still drives geological activity, we must look at larger worlds such as Venus and Mars.

7-2 Venus

An astronomer once became annoyed when someone referred to Venus as a planet gone wrong. "No," she said, "Venus is probably a fairly normal planet. It is Earth that is peculiar." To understand how unusual Earth is, we need only compare it with Venus.

In some ways, Venus and Earth are twins, but they are also dramatically different. Venus is nearly 95 percent the diameter of Earth (Data File Four) and has a similar average density, so we might expect it to have a composition similar to Earth's. Yet roughly three dozen spacecraft have visited Venus and never detected a magnetic field, evidence that its core is not a molten conductor like Earth's iron core. Furthermore, Venus rotates backward (retrograde) in 243 Earth days. We might expect Earth's twin world to have a pleasant climate, but the atmosphere is perpetually cloudy (Figure 7-4a) and the surface is dry, hot, and volcanic. How did Earth's twin become so different?

The Atmosphere of Venus

The air on Venus is unbreathable. Spectra and data from spacecraft show that the atmosphere is about 96 percent carbon dioxide and 3.5 percent nitrogen. The

Venus Data File Four

This radar image of Venus strips the cloudy atmosphere away to reveal a world marked by volcanoes and lava flows (bright regions). *(NASA)*

Average distance from the sun	0.7233 AU (1.082×10^8 km)
Eccentricity of orbit	0.0068
Maximum distance from the sun	0.7282 AU (1.089×10^8 km)
Minimum distance from the sun	0.7184 AU (1.075×10^8 km)
Inclination of orbit to ecliptic	3°23′40″
Average orbital velocity	35.03 km/s
Orbital period	0.61515 y (224.68 days)
Period of rotation	243.01 days (retrograde)
Inclination of equator to orbit	177°
Equatorial diameter	12,104 km (0.95 $D_{\oplus}$)
Mass	4.870×10^{24} kg (0.815 $M_{\oplus}$)
Average density	5.24 g/cm³ (4.2 g/cm³ uncompressed)
Surface gravity	0.903 Earth gravity
Escape velocity	10.3 km/s (0.92 $V_{\oplus}$)
Surface temperature	745 K (472°C, or 882°F)
Albedo (cloud tops)	0.76
Oblateness	0

remaining half percent is water vapor, argon, sulfuric acid (H_2SO_4), hydrochloric acid (HCl), and hydrofluoric acid (HF). In fact, the thick clouds that hide the surface are sulfuric acid droplets and microscopic sulfur crystals. Probes falling into the atmosphere reveal that the cloud layers are much higher and much more stable than those on Earth (Figure 7-4b).

These cloud layers are highly stable because the atmospheric circulation on Venus is much more regular than that on Earth. The heated atmosphere at the **subsolar point,** the point on the planet where the sun is directly overhead, rises and spreads out in the upper atmosphere. Convection circulates this gas toward the dark side of the planet and the poles, where it cools and sinks. This circulation produces 300-km/h jet streams in the upper atmosphere, which carry the atmosphere from east to west (the same direction the planet rotates) so rapidly that the entire atmosphere seems to rotate with a period of only four days. As evidence of these circulations, we can see a distinctive Y pattern in ultraviolet images of the cloud pattern (Figure 7-4a).

The details of this atmospheric circulation are not well understood, but it seems that the slow rotation of the planet is an important factor. On Earth, large-scale circulation patterns are broken into smaller cyclonic disturbances by Earth's rapid rotation. Because Venus rotates more slowly, its atmospheric circulation is not broken up into small cyclonic storms but instead is organized as a planetwide wind pattern.

Although the upper atmosphere is cool, the lower atmosphere is quite hot (Figure 7-4b). Instrumented probes that have reached the surface report that the temperature is 745 K (882°F) and the atmospheric pressure is 90 times that of Earth. Earth's atmosphere is 1000 times less dense than water, but on Venus the air is only 10 times less dense than water. If we could survive the unpleasant composition, intense heat, and high pressure, we could strap wings to our arms and fly.

The present atmosphere of Venus is extremely dry, but there is evidence that it once had significant amounts of water. As a Pioneer Venus spacecraft descended through the atmosphere of Venus in 1978, it detected about 150 times more deuterium per hydrogen atom than on Earth. This abundance of deuterium, the heavy isotope of hydrogen, could develop if ultraviolet radiation in sunlight broke water molecules into hydrogen and oxygen. The hydrogen would escape into space, but the heavier deuterium would escape more slowly than normal hydrogen. Thus, the excess of deuterium is evidence that Venus once had more water. Calculations suggest that it may have once had enough water to make a planetwide ocean up to 25 meters deep. (For comparison, the water on Earth would make a uniform planetwide ocean 3000 meters deep.) Without an ozone layer to protect the atmosphere from ultraviolet radiation, the water was broken up and much of the hydro-

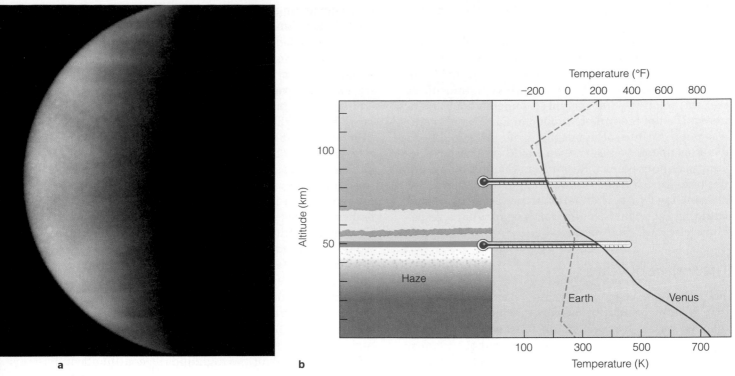

Figure 7-4

(a) The surface of Venus is perpetually hidden by thick cloud layers. This ultraviolet image made by the Hubble Space Telescope shows typical weather patterns, but no part of the surface is visible. *(L. Esposito, University of Colorado, Boulder; and NASA)* (b) The four main cloud layers in the atmosphere of Venus are much higher than cloud layers on Earth. Thermometers inserted into the atmosphere at different levels would register the temperatures indicated by the red line in the graph at the right. The blue dashed line shows the temperatures in Earth's atmosphere. Note the surface temperatures.

gen was lost to space. The oxygen presumably formed oxides in the soil. Thus, Venus is now a deadly dry world with only enough water vapor to make an ocean 0.3 meter deep.

Even if Venus once had lots of water, it was probably too warm for that water to fall from the atmosphere as rain and form oceans, and that lack of oceans is the biggest difference between Earth and Venus. Earth's oceans have made it a living paradise. The lack of oceans on Venus, through the greenhouse effect, has made it a hell.

The Venusian Greenhouse

We saw in Chapter 6 how the greenhouse effect warms Earth. Carbon dioxide (CO_2) is transparent to light but opaque to infrared (heat) radiation. Thus, energy can enter the atmosphere as light and warm the surface, but the surface cannot radiate an equal amount of energy back to space because of the CO_2. Like Earth, Venus has a greenhouse effect, but on Venus the effect is fearsome. Whereas Earth's atmosphere contains only about 0.03 percent CO_2, the atmosphere of Venus contains 96 per-

cent CO_2, and the surface of Venus is nearly twice as hot as a hot kitchen oven.

When Venus was young, it may have been cooler than it is now, but because it formed 30 percent closer to the sun than did Earth, it was warmer than Earth, and that unleashed processes that made it even hotter. As we saw in the previous section, even if Venus once had water, it probably never had large liquid-water oceans. CO_2 is highly soluble in water, but without large oceans to dissolve CO_2 and remove it from the atmosphere, Venus was trapped with a CO_2-rich atmosphere, and the greenhouse effect made the planet warmer. As the planet warmed, any small oceans that did exist evaporated, and Venus lost the ability to cleanse its atmosphere of CO_2. Volcanoes on Earth vent gases that are mostly CO_2 and water vapor, and presumably the volcanoes on Venus did the same. The water vapor was broken up by ultraviolet radiation, and the CO_2 accumulated in the atmosphere. The high temperature baked even more CO_2 out of the surface, and the atmosphere became even less transparent to infrared radiation. Consequently, the temperature rose further. This runaway greenhouse effect has made the surface so hot that even sulfur, chlorine, and fluorine have baked out of the rock and formed sulfuric, hydrochloric, and hydrofluoric acid vapors.

Earth avoided this runaway greenhouse effect because it was farther from the sun and cooler. Thus, it could form and preserve liquid-water oceans to absorb the carbon dioxide, which left a nitrogen atmosphere that was relatively transparent in some parts of the infrared. If all of the carbon dioxide in Earth's sediments was put back into the atmosphere, our air would be as dense as that of Venus, and we would suffer from a severe greenhouse effect. Recall from Chapter 6 how our use of fossil fuels and our destruction of forests is increasing the CO_2 concentration in our atmosphere and warming the planet. Venus warns us of what a greenhouse effect can do.

The Surface of Venus

The clouds of Venus are opaque to visible light, but they are transparent to radio waves, so astronomers can map the surface using radar. The first radar maps, made from Earth, had very low resolution, but the Pioneer Venus spacecraft orbited Venus in 1978 and made maps with a resolution of 25 km. Later the Soviet Venera 15 and 16 spacecraft mapped the north polar region at a resolution of about 2 km. In 1990, the Magellan spacecraft orbited Venus and created a radar map showing features as small as 100 m (Figure 7-5).

Such radar maps can tell us a number of things about the surface. If we transmit a radio signal down through the clouds and measure the time until we hear the echo coming back up through the clouds, we can measure the altitude of the surface. Part of the Magellan data is a detailed altitude map of the surface of Venus. We can also measure the amount of the radio signal that is reflected from each spot on the surface. Rough surfaces such as lava flows are highly reflective (Figure 7-6) and stand out brightly in the radar maps. Other surfaces, because they are smoother or because they have different mineral compositions, are less reflective and appear darker. Thus, part of the Magellan data consists of detailed maps of surface features related to roughness and composition. Many of these Magellan maps are reproduced in a bright orange color to mimic the orange light that filters down through the thick clouds. If we could visit the surface of Venus, the landscape would be bathed in a sunset glow throughout the day (which would last 58 Earth days). Photographs made by robotic landers reveal that the surface rock would look dark gray under normal lighting conditions.

The radar maps reveal that the surface of Venus is approximately 60 percent **rolling plains,** 24 percent highlands, and about 16 percent volcanic peaks. It is tempting to think of the highlands as continents and the rolling plains as sea floors, but that is unwarranted. There is now no liquid water on Venus, and there may never have been any. The geology of the highlands and low rolling plains may arise from entirely different geological conditions than those on the Earth.

By authority of the International Astronomical Union, place names on Venus are female. Thus Ishtar Terra, the northern highlands, is named after Ishtar, the

Figure 7-5
Large numbers of Magellan radar images of the surface of Venus have been combined to form this map showing most of Venus. North and south polar regions are not shown. The hot, dry, rocky surface contains mostly low rolling plains with highlands consisting mainly of Ishtar Terra, Aphrodite Terra, and Lada Terra. Bright regions are rough terrain thought to be associated with volcanism. The nearly featureless region running top to bottom near the center of this map is a region of limited satellite data. (NASA)

a

Figure 7-6
Lava flows are very rough and are good reflectors of radar signals.
(a) This 1000-year-old lava flow near Flagstaff, Arizona, is still rough
enough to be dangerous to hikers. *(M. Seeds)* (b) The slopes of the
volcano Maat Mons are marked by bright regions believed to be rough
lava flows. This radar image has been stretched vertically and transformed into a horizontal view as if we stood on the surface of Venus.
(NASA)

b

Babylonian goddess of love. The only exceptions are
Maxwell Montes, named after the 19th century physicist
James Clark Maxwell, who first described electromagnetic radiation, and Beta Regio and Alpha Regio. They
were discovered in the first radar maps of Venus made
from Earth and were named before the female naming
convention was adopted by the world's astronomers.

Ishtar Terra, larger than the United States, is geologically complex (Figure 7-7). At its eastern edge, the
mountain Maxwell Montes reaches an altitude of 12
km. Mount Everest is only 8.8 km high. The west end
of Ishtar Terra consists of the Lakshmi Planum, a great
plateau rising 4 km above the rolling plains. The plateau

appears to have formed from lava flows from vents such
as Colette and Sacajawea. North and west of Lakshmi
Planum are ranges of wrinkled mountains that suggest
some limited compression of the crust.

Figure 7-7
This radar map of Lakshmi Planum and Maxwell
Montes is based on radar data from both Soviet
and American spacecraft, with ground-based radar
data added (color). Violet and purple regions are
the smoothest, and orange the roughest. Lakshmi
Planum appears to be a volcanic plain produced
by lava flooding from vents such as Collette and
Sacajawea. The folded ranges of mountains around
Lakshmi Planum suggest limited crustal movement.
(USGS)

Volcanism is dramatic on Venus. Beta Regio (Data File Four and Figure 7-5) is a great volcanic uplift containing two peaks on a major fault. Maat Mons (Figure 7-6b) and Sapas Mons (Figure 7-8) are other volcanoes located along fracture zones that presumably opened pathways for magma to rise from below and feed the volcanism. Such faults are widespread on Venus.

Many images created from the Magellan data are reproduced with their vertical scales stretched. This is easy to do with computer-based data (Window on Science 7-1), and it makes the surface features easier to study; however, it makes the mountains look much steeper than they really are. Volcanic peaks on Venus are **shield volcanoes** (Figure 7-9), which form from highly fluid lava and have shallow slopes. Shield volcanoes, such as Mauna Loa in Hawaii, are common on Earth.

The surface of Venus contains roughly 900 impact craters (Figure 7-10), significantly fewer than the moon but many more than Earth. This shows that the surface of Venus is not as rapidly renewed as is Earth's surface where erosion and plate tectonics erase craters rapidly. Earth has only about 150 known impact craters. Counts of craters on Venus show that the average age of the crust is older than Earth's crust but much younger than the moon's.

Much of Venus is marked by circular features that are clearly not impact craters. These features, called **coronae,** are circular bulges bordered by fractures, volcanoes, and lava flows (Figure 7-11). Up to 2100 km in diameter, coronae appear to be caused by rising convection currents of molten magma below the crust. Some experts believe that the largest coronae may be

Figure 7-8
Volcano Sapas Mons is bright with rough lava flows leading away from two volcanic calderas at its peak. The volcano lies in a major fracture zone, which allowed molten rock to reach the surface. Current data cannot tell us whether any of the volcanoes on Venus are still active, but it seems reasonable to assume that some are. The orange color of this radar map mimics the orange light that filters through the thick atmosphere like a daylong sunset glow. *(NASA)*

a

Figure 7-9
(a) A computer-generated image based on Magellan radar data shows volcanoes Gula Mons (left) and Sif Mons (right) with their associated lava flows. The vertical scale has been stretched to make surface details stand out. The true profile of the volcanoes resembles that of normal shield volcanoes (inset). *(NASA)* (b) In this computer model, details of the true shape of the mountain are dramatically revealed by a vertical magnification of 10 times (top) but are invisible when the model is reproduced with no vertical stretching (bottom). *(Model by M. A. Seeds)*

Window on Science 7-1

Graphical Data Manipulation

Because so much modern scientific data is stored inside computers, it is common to see data manipulated to exaggerate certain characteristics and suppress others. In some fields, this would be dishonest, but the goals of science make such manipulation valuable. By purposely distorting data, we may reveal things we would not have seen otherwise.

We have seen one form of data manipulation in false-color images (Figures 3-18c and 3-21). Astronomers commonly change the colors in images and maps to reveal detail invisible to our eyes. Photos of galaxies and radio maps of clouds of gas are only two examples. As long as we remember that we are looking at false-color images, we can learn a great deal with this tool.

Geologists, too, manipulate data. Geological maps have been color coded for over a century, and now computers allow

faster and more precise use of false color in maps and images. But geological data can also be manipulated to reveal other features. A relief map of the United States would look as smooth as glass if the mapmaker did not exaggerate the height of mountain ranges, and having such data in a computer makes it easy to "stretch" the heights to make them more visible. Astronomers often do the same with radar maps of Venus and surface maps of Mars.

Such data distortion would be inadmissible in a court of law. In fact, photographic evidence is generally inadmissible in court because it is so easy to fake a photograph and show the defendant in almost any incriminating situation. Faking a photograph in journalism is at least disreputable, and in some cases it is libelous. In fact, modern computer graphics make it possible to put Humphrey Bogart into a modern TV commercial with no trace of

fakery, so photographic manipulation is unlikely ever to become acceptable in court or in journalism.

In science, however, our quest is not for a first impression but for detailed analysis. Data can be manipulated and distorted any way that is useful as long as the form of the manipulation is specified. Then viewers are free to form their own opinions as to the meaning of the data. Of course, this assumes that the viewer is alert and takes into consideration the nature of any data manipulation. Thus, when we look at a relief map of the United States or a radar image of a mountain on Venus, it is our responsibility to take into account any data manipulation. if we don't, we might think that the Rocky Mountains are 100 miles high and that mountains on Venus are 20 times steeper than mountains on Earth.

sites where new crust is created and spreads or where old crust sinks and is destroyed.

Although some ridges, such as those around Lakshmi Planum, seem to be the result of horizontal motion in the crust, plate tectonics is not important on Venus. We see no spreading centers like Earth's mid-ocean rifts, and we see no subduction trenches where

b

Figure 7-10
Impact craters are dramatic in the Magellan data because the material ejected from the crater creates rough terrain that is bright in reflected radar imaging. The crater Howe in the foreground is 37 km in diameter. A small city would easily fit inside this crater. Note the craters in the background. *(NASA)*

Figure 7-11
Evidence of subsurface magma on Venus. Coronae are circular bulges bordered by circular and radial faults apparently caused by rising currents of magma below the crust. Light gray areas are lava flows, and volcanic cones are visible at lower left. These two coronae, Bahet (left) and Onatah (right), may be related. The vertical dark bar is caused by missing data. *(NASA)*

crust descends. More important, Venus lacks long chains of folded mountains such as the Rockies and the Andes. The surface of Venus may be too hot and flexible to form the rigid plates necessary for plate tectonics. Rather, Venus appears to be dominated by rising currents of molten rock below its crust. Some of its volcanic regions are probably still active.

Seven different Soviet landers reached the surface of Venus and analyzed the soil. The rock appears to be dark gray basalts much like those that squeeze out of the midocean rifts and form the seafloors on Earth. Thus the chemical composition agrees with the radar images; the history of Venus has been dominated by volcanism.

A History of Venus

The four-stage history of Venus is not well understood. It formed only slightly closer to the sun than did Earth, yet Venus is dramatically different.

The planet rotates backward (retrograde) with a period of 243 days. Perhaps this backward rotation is the result of the impact of a very large planetesimal late in the accretion of the planet. The rotation turns nearly the same face of Venus toward Earth each time the planets pass each other, so tidal forces may have produced a resonance in the rotation of Venus.

Presumably a planet as large as Venus has differentiated, but the lack of a magnetic field suggests that its interior is not molten iron, as is Earth's. Some astronomers have wondered if the hot, dense atmosphere has softened the crust and allowed occasional outbursts of volcanism to cool the core. If that happened, the core might be solid. We do not know because there are no seismographs on the surface of Venus.

Impact craters on Venus hint at an interesting story. Small craters are lacking because the thick atmosphere breaks up and vaporizes smaller meteorites. The existing craters seem uniformly scattered over its surface and mostly undamaged by geological processes. There are no ancient craters and no young regions free of craters. It seems as if the entire surface of Venus was wiped smooth by lava flows roughly 800 million years ago.

Flooding by lava has certainly been important on Venus, but it does not appear to have ever had major oceans of water on its surface. Thus the CO_2 in the atmosphere could not dissolve in water and be transferred into the crust. The resulting greenhouse effect increased the temperature, making it even less likely that water could survive on the surface and driving the atmosphere into a runaway greenhouse condition. Had Earth been just 5 percent closer to the sun, it might never have formed oceans and cleansed its atmosphere of CO_2.

The last stage of planetary evolution is slow surface evolution. On Earth, that means erosion and plate tectonics, but with no water, erosion is not important on Venus. Because we see no long, folded mountains, no rift valleys outlining plates, and no long features like Earth's midocean ridges and subduction zones, it also seems very unlikely that plate tectonics is important on Venus.

Fully 70 percent of the heat from Earth's interior flows outward through the midocean rifts. But Venus lacks crustal rifts, and its numerous volcanoes cannot carry much of the heat out of the interior. Rather, Venus seems to get rid of its interior heat through large currents of hot magma that rise beneath the crust. We see coronae above such currents. Planetary scientists now speculate that cooler magma may sink back into the interior and cause minor horizontal motions in the crust that could account for the folded mountains near Ishtar Terra.

True plate tectonics is not important on Venus. For one thing, the crust is very dry and is consequently about 12 percent less dense than Earth's crust. This low-density crust is more buoyant than Earth's crust and thus resists being pushed into the interior. Also, the crust is so hot it is halfway to the melting point of rock. Such hot rock is not very stiff, so it cannot form the rigid plates typical of plate tectonics on Earth.

Planetary astronomers studying Venus are developing a startling hypothesis. Venus may be dominated, not by the slow plate motion that shapes the geology of the Earth, but by catastrophic episodes in which most of its hot, pliable crust sinks rapidly into the mantle and new crust forms from lava flows. The resurfacing that wiped away all older crust and erased craters roughly 800 million years ago may be the most recent of these overturnings. Such a process would allow Venus to lose large amounts of internal heat and might account for a solid core. If this is true, then it may arise because the thick atmosphere keeps the crust hot and flexible.

The overturning of an entire planetary crust would release large amounts of carbon dioxide, but that would be negligible compared to the thick atmosphere Venus already has. However, mathematical models suggest that other greenhouse gases might alter the climate for half a billion years and cause changes in surface temperature as large as 100 K. Clearly Venus has much to teach us about the evolution of planetary surfaces and atmospheres.

Our study of Venus illustrates two principles of comparative planetology. First, large planets cool slowly and remain geologically active much longer than small planets. The moon and Mercury are dead, but Earth and Venus are active. Second, even a small difference in the temperature of a world can have a dramatic effect if it affects the stability of liquid water. Earth has water oceans and Venus does not, and that has made the difference.

REVIEW   Critical Inquiry

What evidence do we have that plate tectonics does not occur on Venus?

When we talked about Earth in Chapter 6, we identified certain geological features that are characteristic of plate tectonics: rift valleys, subduction regions, and folded mountain chains. We don't see these features on Venus even in the highest-resolution radar maps made by the Magellan spacecraft. There are no rift valleys or subduction zones outlining plates, and there are no major chains of folded mountains. Rather, we see evidence of volcanism and lava flows, and the small regions of folded mountains suggest only limited horizontal motion in the crust. Rather than being dominated by the motion of rigid crustal plates, Venus may have a crust made more flexible by the heat, a crust dominated by rising plumes of molten rock that strain the crust to produce coronae and that break through to form volcanoes and lava flows.

At first glance, comparative planetology suggested that Earth and Venus should be sister worlds, but it seems they can be no more than distant cousins. We can blame the thick atmosphere of Venus for altering its geology, but that raises another question: Why doesn't Earth have a similar atmosphere?

Venus is certainly an unearthly world, but the familiar principles of comparative planetology help us understand it. As we turn our attention to Mars, we will see the same principles in action, but we will discover that Mars is a world that lacks what Venus has in abundance—internal heat and air.

7-3 Mars

For terrestrial worlds, Mercury and the moon are small and Venus and Earth are large. Mars, in contrast, is medium sized (Data File Five). It is twice the diameter of the moon but only 53 percent the diameter of Earth. Its small size has allowed it to cool faster than Earth, and much of its atmosphere has leaked away. Its present carbon dioxide atmosphere is only 1 percent as dense as Earth's.

In some ways, Mars is much like Earth. A day on Mars is 24 hours 40 minutes long, and its year lasts 1.88 earth years. Just as Earth is tipped 23.5°, Mars is tipped 25°, and thus it has seasons with white polar caps growing in the hemisphere experiencing winter and shrinking in the hemisphere experiencing summer (Figure 7-12). The atmosphere and geology of Mars make a striking comparison with those of Earth.

Mars Data File Five

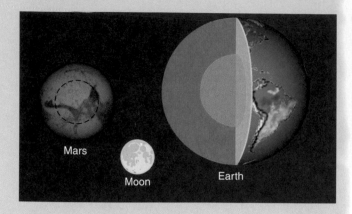

Mars is a medium-size world smaller than Earth but larger than Earth's moon. Earth's core is hot and molten, while the moon is cold and may not have a molten core at all. The nature of the interior of Mars is unknown.

Average distance from the sun	1.5237 AU (2.279×10^8 km)
Eccentricity of orbit	0.0934
Maximum distance from the sun	1.6660 AU (2.492×10^8 km)
Minimum distance from the sun	1.3814 AU (2.066×10^8 km)
Inclination of orbit to ecliptic	1°51′09″
Average orbital velocity	24.13 km/s
Orbital period	1.8808 y (686.95 days)
Period of rotation	$24^h37^m22.6^s$
Inclination of equator to orbit	25°19′
Equatorial diameter	6796 km (0.53 $D_\oplus$)
Mass	0.6424×10^{24} kg (0.1075 $M_\oplus$)
Average density	3.94 g/cm³ (3.3 g/cm³ uncompressed)
Surface gravity	0.379 Earth gravity
Escape velocity	5.0 km/s (0.45 $V_\oplus$)
Surface temperature	−140° to 20°C (−220° to 68°F)
Average albedo	0.16
Oblateness	0.009

Figure 7-12
This Hubble Space Telescope view of Mars shows the northern polar cap at the top, impact craters, and dark surface features. The spot near the left edge is the volcano Ascraeus Mons, 25 km (16 mi) high, poking up through morning clouds of water-ice crystals in the thin CO_2 atmosphere. *(Philip James, University of Toledo; Steven Lee, University of Colorado, Boulder; and NASA)*

The Atmosphere of Mars

If we visited Mars, our first concern, even before we opened the door of our spaceship, would be the atmosphere. Is it breathable? Even for the astronomer observing safely from Earth, the atmosphere of Mars is a major concern. The gases that cloak Mars are critical to our understanding of the history of the red planet.

The air on Mars is 95 percent carbon dioxide, with a few percent each of nitrogen and argon. It contains almost no water vapor or oxygen, and its density at the surface of the planet—only 1 percent that of Earth's atmosphere—does not provide enough pressure to prevent liquid water from boiling into vapor. If you stepped outdoors on Mars without a space suit, your own body heat would make your blood boil.

The thin air on Mars interacts with its dusty surface in dramatic ways. Photographs made on the surface of Mars reveal a dusty, red desert (Figure 7-13), and analysis of the soil shows that it is rich in reddish iron oxides. Much of the oxygen that we humans would prefer in the atmosphere is actually locked in chemical compounds with iron in the soil. Although the atmosphere is thin, it can kick this dust up into dust storms that can occasionally envelop the entire planet. Mars is truly a desert world.

To understand Mars, we must ask why its atmosphere is so thin and dry and why the surface is rich in

Figure 7-13
The reddish, dusty soil of Mars is dramatically evident in this photo made in 1997 by the Mars Pathfinder spacecraft. Its robotic rover is shown here testing the composition of the large rock (arrow) dubbed Yogi. The soil is rich in iron oxides, which accounts for its color. *(NASA)*

oxides. The answers to these questions lie in the origin and evolution of the Martian atmosphere.

Presumably, the gases in the Martian atmosphere were mostly outgassed from its interior when it was younger. Volcanism on terrestrial planets typically releases carbon dioxide, water vapor, and other gases. Being small, Mars has cooled and may be less actively releasing gases from its interior.

To understand how gases can escape from a planet, we must think about the velocity we need to leave the surface of a world, the **escape velocity.** A massive planet such as Earth has a high escape velocity, 11.2 km/s (25,000 mph). To throw a rock off of Earth into space, we would have to throw it at least that fast. The moon's escape velocity is only 2.38 km/s, and Mars has an escape velocity of 5.0 km/s. An atom of gas can easily escape from the moon, but escaping from Mars is more difficult because the escape velocity is higher.

The temperature of a planet's atmosphere is also important. If the gas is hot, its molecules have a higher average velocity and are more likely to exceed escape velocity. Thus, a planet that is near the sun and very hot is less likely to retain an atmosphere than a more distant, cooler planet. The velocity of a gas molecule, however, also depends on the mass of the molecule. On average, a low-mass molecule travels faster than a massive molecule. Thus, a planet loses its lowest-mass gases more easily because those molecules travel fastest.

We can see this principle of comparative planetology if we plot a diagram such as that in Figure 7-14. The points show the escape velocity versus temperature of the larger objects in our solar system. The temperature we use is the temperature of the gas that is in a position to escape. For the moon, which has essentially no atmosphere, this is the temperature of the sunlit surface. For Mars, we use the temperature at the top of the atmosphere. The lines in Figure 7-14 show the typical velocity of the fastest-traveling examples of various molecules. At any given temperature, some water molecules, for example, travel faster than others, and it is the highest-velocity molecules that escape from a planet. The diagram shows that the Jovian planets have escape velocities so high that very few molecules can escape. Earth and Venus can't hold hydrogen, and Mars can hold only the more massive molecules. Earth's moon is too small to keep any of the common gases. We will refer to this diagram again as we study the atmospheres of other worlds.

Over the 4.5 billion years since Mars formed, it has lost some of the lower-mass gases. Water molecules are massive enough for Mars to keep, but ultraviolet radiation can break up the water molecules. On Earth, the ozone layer protects water vapor from ultraviolet radiation, but Mars never had an oxygen-rich atmosphere, so it never had an ozone layer. Ultraviolet photons from the sun can penetrate deep into the atmosphere and break up molecules such as water. The hydrogen escapes, and the oxygen, a very reactive element, forms oxides in the soil—the oxides that make Mars the red planet. Thus, molecules too massive to leak into space can be lost if they break into lower-mass fragments.

The argon in the Martian atmosphere is evidence of a denser blanket of air in the past. Argon atoms are massive, almost as massive as a CO_2 molecule, and would not be lost easily. In addition, argon is inert and would not form compounds in the soil. The 1.6 percent argon in the atmosphere of Mars is evidently left over

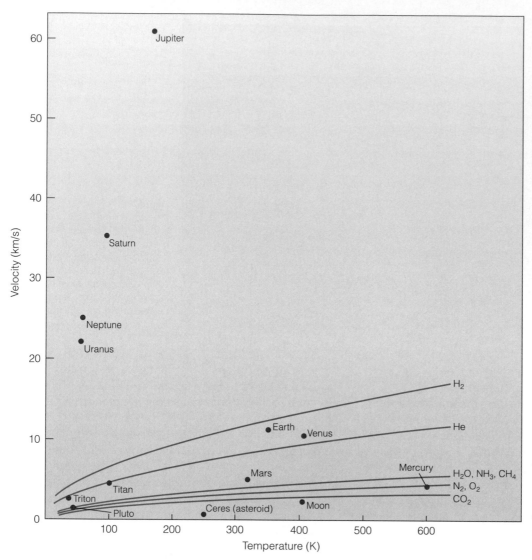

Figure 7-14
The loss of planetary gases. Dots represent the escape velocity and temperature of various solar-system bodies. The lines represent the typical highest velocities of molecules of various masses. The Jovian planets have high escape velocities and can hold on to even the lowest-mass molecules. Mars can hold only the more massive molecules, and Earth's moon has such a low escape velocity that even the most massive molecules can escape.

from an ancient atmosphere that was 10 to 100 times denser than the present Martian air.

Finally, we must consider the interaction of the solar wind with the atmosphere of Mars. This is not an important process for Earth because Earth has a magnetic field that deflects the solar wind. Because Mars has no significant magnetic field, the solar wind interacts directly with the Martian atmosphere, and detailed calculations show that significant amounts of CO_2 could have been carried away by the solar wind over the history of the planet. This process would have been most efficient long ago, however, when the sun was more active and the solar wind was stronger. We will see evidence later in this chapter that Mars may once have had a stronger magnetic field when it still retained significant internal heat. Such a magnetic field could have protected its atmosphere from the solar wind.

Although we remain uncertain as to how much of an atmosphere Mars has had in its past and how much it has lost, it is a good example for our study of com-

parative planetology. When we look at Mars, we see what can happen to the atmosphere of a medium-size world. Like its atmosphere, the geology of Mars is typical of intermediate-size worlds.

The Rocky Surface of Mars

Until 1965, Mars was a planet of mystery, and science fiction writers peopled it with little green men building canals to bring water from its polar caps. That year, the Mariner 4 spacecraft flew past and sent back pictures of a cratered desert world with no canals, no green people, and no water. Since then many spacecraft have visited Mars, and as the 21st century begins, an armada of robotic probes is exploring the red planet.

All of Mars is divided into two parts. The southern highlands are heavily cratered, making them moonlike, while the northern lowlands are smooth and have few craters (Figure 7-15). This tells astronomers that the southern hemisphere is over 3 billion years old and pre-

Geologists are fond of saying "The present is the key to the past." By that they mean that we can learn about Earth's history by looking at the present condition of Earth's surface. The position and composition of various rock layers in the Grand Canyon, for example, tell us that the western United States was once at the floor of an ocean. This principle of geology was astonishing when it was first formulated in the late 1700s, and it continues to be relevant today as we try to understand other worlds, such as Venus and Mars.

In the late 18th century, naturalists recognized that the present gave them clues to Earth's history. This was astonishing because most people assumed that Earth had no history. That is, they assumed either that Earth had been created in its present state as described in the Old Testament or that Earth was eternal. In either case, people commonly assumed that the hills and mountains they saw around them had always existed more or less as they were. The 18th-century naturalists began to see evidence that the hills and mountains were not eternal but were the result of past processes and were slowly changing. That gave birth to the idea that Earth had a history.

As the naturalists of the 18th century made the first attempts to thoughtfully and logically explain the nature of the earth by looking at the evidence, they were inventing modern geology as a way of understanding Earth. What Copernicus, Kepler, and Newton did for the heavens in the 1600s, the first geologists did for the Earth beginning in the late 1700s. Of course, the invention of geology as the study of the Earth led directly to our modern attempts to understand the geology of other worlds.

Geologists and astronomers share a common goal: They are attempting to reconstruct the past (Window on Science 5-3). Whether we study Earth, Venus, or Mars, we are looking at the present evidence and trying to reconstruct the past history of the planet by drawing on observations and logic to test each step in the story. How did Venus get to be covered with lava, and how did Mars lose its atmosphere? The final goal of planetary astronomy is to draw together all of the available evidence (the present) to tell the story (the past) of how the planet got to be the way it is. Those first geologists of the late 1700s would be fascinated by the stories planetary astronomers tell today.

serves craters from long ago (Window on Science 7-2). The northern hemisphere must have been resurfaced by a smoothing process at least a billion years ago. Volcanism and lava flows have been suggested as the smoothing agent, and volcanism is certainly important in the northern hemisphere. Many large volcanoes have been identified on Mars, and most lie in the northern hemisphere.

Figure 7-15
Color-coded Mars maps show elevation. The northern lowlands lie about 4 km below the southern highlands. Volcanoes are very high (white), and the impact basins Hellas and Argyre are low. Note the depth of the canyon Valles Marineris. *(NASA)*

The surface of Mars contains craters and volcanoes, and that is as we might expect from its size. Mars is larger than Earth's moon but smaller than Earth. Unlike the moon, Mars was large enough to retain some internal heat, which explains its prolonged geological activity. It is smaller than Earth, however, so it has not been as active and has not erased all of its craters. In terms of comparative planetology, Mars is a medium-sized rocky planet with thin air and both craters and volcanoes.

Martian volcanoes are shield volcanoes, which have shallow slopes showing that the lava flowed easily. Nevertheless, the largest volcano, Olympus Mons, is a vast structure (Figure 7-16). Its base is 600 km (370 mi) in diameter, and it towers 25 km (16 mi) above the surface. In contrast, the largest volcano on Earth, Mauna Loa in Hawaii, rises only 10 km (6 mi) above its base on the Pacific Ocean floor, and its base is only 225 km (140 mi) in diameter. The weight of Mauna Loa is too great to be supported by the seafloor. It has pushed the seafloor down, just as a bowling ball resting on a bed depresses the surface. This depression is visible as an undersea moat ringing Mauna Loa, but no depression is visible ringing Olympus Mons, even though it is 2.5 times higher than Mauna Loa. Evidently the crust of Mars is thicker than Earth's.

Other evidence shows that the Martian crust is thinner than the moon's. The various Mariner spacecraft discovered a vast valley at least 4000 km (2500 mi) long and, at its widest, 600 km (400 mi) wide. At its deepest point, it is over 6 km (4 mi) deep, making it over four times deeper than the Grand Canyon. On Earth it would stretch farther than from New York to Los Angeles. The valley, now named Valles Marineris (Figure 7-17), is apparently a block of crust that has dropped downward along parallel faults. We see no such features on the moon, so we can conclude that the crust of Mars is thinner than the moon's crust. This is

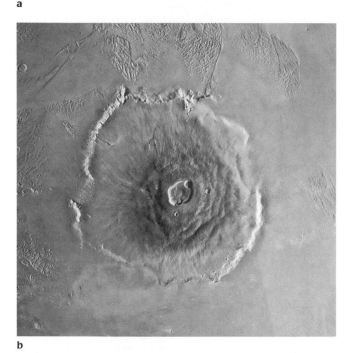

Figure 7-16
(a) Olympus Mons, the tremendous shield volcano on Mars, is much larger than Mauna Loa, the largest volcano on Earth. Mauna Loa has sunk into the crust, producing a moat around its base. (b) Olympus Mons has no moat, suggesting that the Martian crust is stronger than Earth's. *(NASA)*

Figure 7-17
This mosaic of photographs taken by the orbiting Viking spacecraft shows the reddish surface of Mars marked by circular impact craters and larger, darker volcanoes (left). The canyon system Valles Marineris stretches across the planet for a distance greater than the distance from New York to Los Angeles. *(NASA/USGS, courtesy Alfred S. McEwen)*

what we would expect for a planet that is larger than the moon but smaller than Earth.

Valles Marineris begins in an area of volcanism called the Tharsis bulge and may be related to the stresses created in the crust by this volcanic activity. Olympus Mons is only one of a number of volcanoes in the Tharsis region. The area bulges about 10 km (6 mi) higher than the average surface of Mars and has few craters, showing that it is geologically young (Figure 7-18). Some of the volcanoes in this region, such as Olympus Mons, appear to be less than a billion years old, but others may be very old volcanoes whose flanks have been buried by lava from younger volcanoes. The Tharsis bulge itself may be almost as old as the surface of the planet, but recent volcanism has wiped out surface features from more than about 1 billion years ago.

A similar uplifted volcanic plain, the Elysium region, lies almost halfway around the planet. It is more heavily cratered and appears to be slightly older than the Tharsis region. The origin of these uplifted regions is not well understood, but they are probably connected to rising currents of magma in the mantle. On Venus, we see these regions as coronae, but the thick crust on Mars may prevent such scars from forming.

Volcanoes form where the rising magma breaks through the crust, but volcanoes on Earth do not grow as large as they do on Mars, evidently because Earth's crust is in motion. On Earth, a rising current of magma, like that below the Hawaiian islands, breaks through the crust and forms a volcano. The motion of the crustal plate, however, moves the accumulated lava aside, and the next eruption produces a new volcano (Figure 7-19). Thus, the Hawaiian–Emperor island chain extends 3800 km (2300 mi) northwest across the Pacific seafloor because the crust is moving (see Figure 6-4). On Mars, the magma seems to have erupted over and over in the same place and built massive volcanoes such as Olympus Mons, the largest volcano in our solar system. The vast size of Olympus Mons is evidence that the crust of Mars is not separated into moving plates.

We have no evidence that Martian volcanoes are still active, although that is a possibility. Craters in the youngest lava flows on Olympus Mons show that the volcano may have been active as recently as a few hundred million years ago. The Elysium volcanoes appear to be inactive.

Water on Mars

There is no trace of liquid water on Mars now, but two kinds of erosion suggest that water once flowed over the surface (Figure 7-20). That is exciting because it

Mars

Figure 7-18
Earth, Venus, and Mars are shown in these maps, with the highest regions shaded beige and the lowest blue. Note the large volcanic highland, the Tharsis bulge, at left center on the Mars map and the Elysium volcanic uplift at extreme right. Two large impact basins on Mars, Argyre and Hellas, are visible at lower left and lower right. The Mars map is also reproduced to scale. *(NASA)*

Mars to scale

Venus

Earth

a

b

Figure 7-19

(a) On Earth, a rising current of magma breaks through the crust to form a chain of volcanoes because the moving plate carries older volcanic peaks away. The Hawaiian–Emperor island chain formed in this way, and most of the older peaks have now been worn down below sea level. (b) On Mars, volcanoes such as Olympus Mons have grown to great size, suggesting that the crust is not moving. Successive eruptions occur at the same place and add layer after layer to the bulk of the volcano.

suggests that Mars has had a more Earthlike history than Mercury or Venus and because it holds out hope that life may have gotten a foothold on Mars.

One type of erosional feature on Mars resembles the traces left behind by violent floods that scour the surface in a matter of hours or days (Figure 7-20a). Located in the equatorial regions, some of these **outflow channels** are so large they must have been formed by floods 10,000 times greater than the flow of the Mississippi. Perhaps this water was frozen in the soil and melted by volcanic heating. Indeed, counting the craters on top of the outflow channels shows the floods happened about 3 billion years ago, about the same age as the volcanic regions in the northern hemisphere.

In contrast to the flood features, branching **runoff channels** look like dry river beds, and photos made by the Mars Global Surveyor spacecraft show that the water flowed through the river beds for long periods of time, perhaps for millions of years. These channels are located in the old, cratered region of the southern hemisphere, and they have about the same age, about 4 billion years. Evidently, the atmosphere was denser then, and liquid water could survive on the surface.

The water that formed the runoff channels does not seem to have fallen as rain or snow. There are very few small tributary streambeds, as there would be if pre-

cipitation drained from a large area. Rather the water appears to have been frozen in the soil as permafrost that melted and seeped out through springs to drain away down the runoff channels. Some regions of the Martian crust appear to be jumbled valleys that have collapsed, perhaps because subsurface ice melted and drained away.

It seems that the climate of Mars has changed, and some features suggest that it changes in cyclic ways. The polar caps consist of frozen carbon dioxide (dry ice) with frozen water beneath. When spring begins in a hemisphere, the polar cap there begins to shrink as the carbon dioxide turns directly into gas. The water, however, remains behind in a smaller, permanent polar cap. The region around this permanent cap is marked by layered terrain, evidently deposited by wind-borne dust that accumulates on the ice cap (Figure 7-21). As the cap shrinks, the material accumulates in layers. Each layer is a few dozen meters thick and may take about 10,000 years to accumulate. The variation in the layering suggests that some periodic change in climate, perhaps due to orbital changes, may affect the frequency and intensity of planetwide dust storms and thus alter the rate at which material is deposited. Recall that Earth's climate may go through cycles caused by variations in its orbital motion. (See Chapter 1.)

a   b   c

Figure 7-20
Evidence of water on Mars long ago. (a) This image shows signs of erosion due to a sudden flood that skirted elevations and flowed through depressions. Craters on top of these flood features show that the floods occurred when Mars was young. *(NASA)* (b) The steady flow of streams on ancient Mars is suggested by meandering canyons. (c) A detailed image of the central canyon recorded by the Mars Global Surveyor spacecraft reveals signs that the water flowed for a long time. Note the narrow flow channel near the top of the image. *(b and c, Malin Space Science Systems/NASA)*

a   b

c

Figure 7-21
(a) The north polar cap of Mars is made of many regions of ice separated by narrow valleys free of ice. *(NASA)* (b) An image from the Mars Global Surveyor spacecraft in one of the valleys of the polar cap reveals superimposed layers of sediment as thin as 10 m (33 feet). *(Malin Space Science Systems/NASA)* (c) In places, different sets of layers are superimposed, suggesting climate change. *(Adapted from a diagram by J. A. Cutts, K. R. Balasius, G. A. Briggs, M. H. Carr, R. Greeley, and H. Masursky)*

a

b

Figure 7-22

(a) The impact crater Galle at right, known for obvious reasons as the Happy Face Crater, is 210 km in diameter. Such large impacts may occasionally eject fragments of the crust into space. A few fragments from Mars have fallen to Earth as meteorites. *(Malin Space Science System/NASA)* (b) Meteorite ALH84001, found in Antarctica, has been identified as an ancient rock from Mars. Minerals found in the meteorite were deposited in water and thus suggest that the Martian crust was once richer in water than it is now. *(NASA)*

Studies of the North Polar Cap on Mars show that it contains both solid CO_2 and frozen water. In the spring, the CO_2 vaporizes, but the water remains permanently frozen. The amount of frozen water is much less than was first expected. From the size and extent of outflow channels and runoff channels, astronomers know about how much water was present on Mars when it was young. Where that water has gone is a mystery. Evidently only a small amount is frozen in the polar caps.

One astonishing bit of evidence of water on Mars is the analysis of rock samples from the planet. Of course, no astronaut has ever visited Mars and brought back a rock, but over the history of the solar system occasional impacts by asteroids have blasted giant craters on Mars and ejected bits of rock into space. A few of those bits of rock have fallen to Earth as meteorites, and nearly a dozen have been found and identified (Figure 7-22). These rocks are basalts, as we might expect from a planet so heavily covered by lava flows, and chemical analysis shows that the magma from which the rocks solidified must have contained about 1.4 percent water. If all the lava flows on Mars contained that much water, it would be enough to create a planetwide ocean 200 m deep. It isn't clear how much of that water has been lost into space and how much has been frozen in the soil, but estimates suggest that what remains is only a few percent of what once flowed over the surface of Mars.

Nevertheless, there is still water hidden on Mars. Precision measurements of color using Hubble Space Telescope images of Mars can locate minerals that formed in the presence of water (Figure 7-23). These minerals tell us that water was once present and may still be locked in the crust as water-bearing minerals or as ice.

Astronomers are interested in the history of the water on Mars for two reasons. First, it is dramatically clear that liquid water has been present on the surface of Mars and that is it no longer stable there. Understanding how Mars changed so dramatically will help astronomers better understand planets in general and the Earth in particular. A second reason for being interested in the history of water on Mars is life. If water was present, even if it was frozen in the soil and only reached the surface when volcanism heated it and forced it up through hot springs, then there is a chance that life began on Mars just as it did on Earth.

Figure 7-23
Water-related minerals on Mars can be identified through precision measurements of color. On the left, a Hubble Space Telescope image of Mars has been carefully calibrated to reproduce the colors we would see with the unaided eye. The blue shade along the edges of the disk is atmospheric haze and wispy clouds. On the right an infrared image of the same side of Mars is reproduced in false colors to show the location of reddish water-bearing minerals. The large red region, Mare Acidalium, was the site of massive flooding early in Martian history. Mars Pathfinder landed at its southern edge (Figure 7-13). *(Jim Bell, Cornell; Justin Maki, JPL; Mike Wolff, Space Sciences Institute; and NASA)*

A History of Mars

The four-stage history of Mars is a case of arrested development. Because of its small size, its interior cooled and geological activity did not last very long. Also because of its small size, it gradually lost much of its atmosphere, and its surface fell into a deep freeze.

In the first stage of planetary evolution, the planet seems to have differentiated into a core, mantle, and crust. There are no obvious traces of plate tectonics such as folded mountain chains, but the Mars Global Surveyor spacecraft measured residual magnetic fields in the crust of Mars and found them in alternating parallel bands. The existence of any residual field means that Mars must have had a magnetic field long ago. Otherwise there would be no residual field trapped in the rocks of the crust. The presence of the field in parallel bands resembles the alternating bands formed in Earth's crust by plate tectonics. (See Figure 6-4.) This is far from conclusive, but it does tell us that Mars once had a molten conducting core.

During the second stage of planetary evolution, cratering, the crust of Mars was battered during the heavy bombardment as the last of the debris in the young solar system was swept up. The old southern hemisphere survives from this age about 4 billion years ago.

The early atmosphere was thicker, and liquid water carved the erosional features we see today. Some astronomers believe there was once a liquid-water ocean that filled the northern hemisphere lowlands. They argue that the region is smooth because it is old seafloor. Other astronomers argue that the northern hemisphere is smooth because it was flooded by lava flows and that liquid water oceans never existed on Mars. In any case, as the atmosphere leaked into space, water became trapped in the soil as permafrost, and Mars became a frozen world.

The crust of Mars is now too thick to be active. The planet has lost much of its internal heat, and now it lacks a molten core as evidenced by the lack of a planet-wide magnetic field. Although the crust was stressed by rising magma that produced the Tharsis and Elysium volcanic rises and the great canyon Valles Marineris, it is too thick for plate tectonics. Volcanism may still occur on Mars, but the crust has grown too thick for much geological activity.

The fourth stage in the history of Mars has been one of slow decline. Small periodic changes in orbit and rotation may cause cycles in the climate that release some carbon dioxide into the atmosphere, but floods and rivers no longer flow over the surface of Mars. Winds kick up vast dust storms that deposit the dust that builds the layered terrain near the polar caps.

Our history of Mars is far from complete, but it does illustrate how the intermediate size of Mars has influenced its atmosphere and geology. Neither small nor large, Mars is a medium world.

The Moons of Mars

Unlike Mercury and Venus, Mars has moons, two small worlds that orbit close to the planet (Figure 7-24). Phobos (28 × 20 km) and Deimos (12 km) are both tidally locked to Mars, keeping the same side facing the planet as they follow their orbits.

Much about the moons suggests they are captured asteroids. They are small, rocky, heavily cratered, and dark gray with albedos of only 0.06. Their densities are only about 2 g/cm³. All these characteristics are asteroid-like (Chapter 9).

Both satellites have been photographed by spacecraft, and those photos show that the satellites are heavily cratered. Such cratering could have occurred while the moons were either still in the asteroid belt or in orbit around Mars. In any case, the heavy battering has broken the satellites into irregular chunks of rock, and they cannot pull themselves into smooth spheres because their gravity is too weak to overcome the structural strength of the rock. We will discover that low-mass moons are typically irregular in shape, whereas more massive moons are more spherical.

Photos of Phobos reveal a unique set of narrow, parallel grooves (Figure 7-24a). Averaging 150 m wide and 25 m deep, the grooves run from Stickney, the largest crater, to an oddly featureless region on the opposite side of the satellite. One theory suggests that the grooves are deep fractures produced by the impact that formed the crater. The featureless region opposite Stickney may be similar to the jumbled terrain on Earth's moon and the weird terrain on Mercury. All were produced by the focusing of seismic waves from a major impact on the far side of the body. High-resolution photographs show that the grooves are lines of pits, suggesting that the pulverized rock material on the surface has drained into the fractures or that gas, liberated by the heat of impact, escaped through the fractures and blew away the dusty soil.

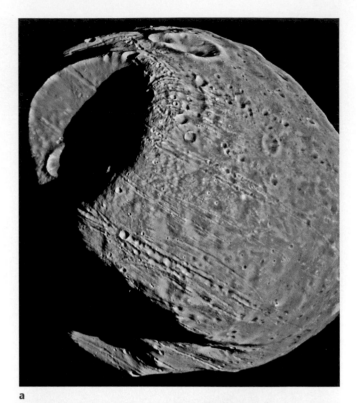

a

b

Figure 7-24
The moons of Mars are small, rocky worlds too small to pull themselves into spherical shape. They may be captured asteroids, or they may have formed with Mars. (a) Phobos, only 28 km (17 mi) in its largest dimension, is irregular in shape, heavily cratered, and marked by grooves that might be related to the impact that formed Stickney, the largest crater. *(Damon Simonelli and Joseph Veverka, Cornell University/NASA)* (b) Deimos, smaller than Phobos, looks more uniform because of dusty soil covering the smaller features. *(NASA)*

Deimos not only has no grooves, it also looks smoother because of a thicker layer of dust on its surface (Figure 7-24b). This material partially fills craters and covers minor surface irregularities. It seems certain that Deimos experienced collisions in its past, so fractures may be hidden below the debris.

The debris on the surfaces of the moons raises an interesting question. How can the weak gravity of small bodies hold any fragments from meteorite impacts? Escape velocity on Phobos is only 12 m/s. An athletic astronaut could almost jump into space. Certainly, most fragments from impacts should escape, but some could fall back and accumulate on the surface.

Deimos, smaller than Phobos, has a smaller escape velocity. But it has more debris on its surface because it is farther from Mars. Phobos is close enough to Mars that most ejecta from impacts on Phobos will be drawn into Mars. Deimos, being farther from Mars, is able to keep a larger fraction of its ejecta.

Deimos and Phobos illustrate three principles. First, some satellites are probably captured asteroids. Second, small satellites tend to be irregular in shape and heavily cratered. Third, tidal forces can have dramatic effects on moons that are close to their planets. We will find such tidal effects when we visit Jupiter's moons (Chapter 8).

REVIEW Critical Inquiry

Why doesn't Mars have coronae like those on Venus?

The coronae on Venus are believed to have been caused by rising currents of molten magma in the mantle pushing upward under the crust and then withdrawing to leave the circular scars called coronae. Earth, Venus, and Mars have had significant amounts of internal heat, and there is plenty of evidence that they have had rising convection currents of magma under their crusts. Of course, we wouldn't expect to see coronae on Earth because its surface is rapidly modified by erosion and plate tectonics. Furthermore, the mantle convection on Earth seems to produce plate tectonics rather than coronae. Mars, however, is a smaller world and must have cooled faster. We see no evidence of plate tectonics, and we do see giant volcanoes that suggest rising plumes of magma erupting up through the crust at the same point over and over. Perhaps we see no coronae on Mars because the crust of Mars rapidly grew too thick to deform easily over a rising plume. On the other hand, perhaps we could think of the entire Tharsis bulge as a single giant corona.

Planetary astronomers haven't explored enough planets yet to see all the fascinating combinations nature has in store for us. Perhaps after we have visited a few hundred terrestrial planets, we will recognize the geology of Mars as typical of medium-size worlds. Of course, Mars is not medium in terms of its location. Of the terrestrial planets, Mars is the farthest from the sun. How has that affected the evolution of its atmosphere?

Venus seems totally inhospitable, but humans may someday visit Mercury and will probably colonize Mars. In the next chapter, we explore the worlds of the outer solar system, worlds so totally unlike Earth that we will need entirely new principles of comparative planetology to describe them.

Summary

Mercury is a small world, only about 40 percent larger than Earth's moon, and like the moon, it has lost any permanent atmosphere it might have had, and its interior has cooled rapidly. Debris from the formation of the solar system cratered Mercury's surface as it did the moon's, and lava flows buried parts of that terrain under the intercrater plains. The Caloris Basin, formed near the end of cratering, is the largest crater basin. Soon after the Caloris impact, lava flows created the smooth plains. These lava plains have about the same color as the intercrater plains, so the contrast between plains and highlands is not as obvious on Mercury as it is on the moon.

The interior of Mercury must contain a large metallic core to account for the planet's high density. A slight shrinkage in the diameter of the planet at the time of the cooling of the core may have led to wrinkling of the crust. This would account for the lobate scarps that mark all of the photographed parts of Mercury.

Venus is Earth's twin in size and density, but the planet has evolved along divergent lines because it is slightly closer to the sun. The higher temperature evaporated any early oceans and prevented the absorption of carbon dioxide from the atmosphere. The accumulating carbon dioxide created a greenhouse effect that produces a surface temperature of 745 K (882°F).

The crust of Venus, mapped by radar, is marked by low rolling plains and highlands with some very high volcanoes. Spacecraft that have reached the surface have found dark basalts. Coronae appear to be produced by rising currents of molten rock below the crust. Where the magma breaks through, it builds volcanoes and floods the surface with lava flows. The surface contains more impact craters than Earth but fewer than the moon, as astronomers suspect the entire planet has been resurfaced by volcanism within the last billion years.

No magnetic field has been detected around Venus, and that suggests that the core is not molten iron generating a magnetic field through the dynamo effect as in Earth's core. The occasional resurfacing of the planet by volcanism may have cooled the interior so it is now solid. Another mystery is the retrograde rotation. This may have been produced by a major impact during formation.

Mars is smaller than Earth and Venus but larger than Mercury. Its southern hemisphere is old and cratered; the northern hemisphere has been resurfaced by lava flows or water. The lack of folded mountains and the vast size of the volcanoes suggest that plate motion does not occur.

Mars once had liquid water on its surface, producing riverlike runoff channels and later floodlike outflow channels. The atmosphere must have been thicker in the past to allow liquid water, but the climate changed as gas leaked into space and water froze in permafrost. Layered terrain near the polar caps shows the climate varies in cycles, but much of the water may be lost forever.

Deimos and Phobos, the two small moons of Mars, may be captured asteroids. Like most small moons, they are irregular in shape and heavily cratered. Tides have locked them to Mars so they keep the same side facing the planet as they follow their orbits.

New Terms

resonance

lobate scarp

intercrater plain

smooth plain

subsolar point

rolling plain

shield volcano

corona

escape velocity

outflow channel

runoff channel

Review Questions

1. How does the tidal coupling between Mercury and the sun differ from that between the moon and Earth?

2. Why does Mercury have lobate scarps and Earth, its moon, Venus, and Mars do not?

3. What evidence do we have about the interior of Mercury?

4. Why would we expect Venus and Earth to be similar? Why do they differ?

5. What evidence do we have that Venus and Mars once had more water than at present? Where did that water come from, and where did it go?

6. How do we know there is no plate tectonics on Venus? How has the crust of Venus evolved?

7. Why doesn't Mars have mountain ranges like those on Earth? Why doesn't Earth have large volcanoes like those on Mars?

8. Explain why Venus and Mars both have carbon dioxide–rich atmospheres. How did Earth avoid such a fate?

9. What evidence do we have that the climate on Mars has changed?

10. How can we estimate the ages of the erosional features on Mars? How did they form?

11. Why are Phobos and Deimos nonspherical? Why is Earth's moon not quite spherical?

Discussion Questions

1. From your knowledge of comparative planetology, describe the view astronauts would have if they landed on the surface of Mercury.

2. From what you know of Earth, Venus, and Mars, do you expect the volcanoes on Venus and on Mars to be active or extinct? Why?

3. If humans colonize Mars, the biggest problem may be finding water and oxygen. With plenty of solar energy beating down through the thin atmosphere, how might colonists extract water and oxygen from the Martian environment?

Problems

1. If we transmitted radio signals to Mercury when it was closest to Earth and waited to hear the radar echo, how long would we wait?

2. Calculate the orbital velocity of Mercury in its orbit around the sun in kilometers per second.

3. Suppose we sent a spacecraft to land on Mercury and it transmitted radio signals to us at a wavelength of 10 cm. If we observed Mercury at its greatest angular distance west of the sun, to what wavelength would we have to tune our radio telescope to detect the signals? (*Hint:* Use the Doppler effect.)

4. What is the maximum angular diameter of Venus as seen from Earth? (*Hint:* Use the small-angle formula.)

5. How long did it take for radio signals from the Magellan spacecraft orbiting Venus to reach Earth?

6. If you send a radio signal down toward the surface of Venus and you hear an echo from a certain spot 0.000133 second before the main echo, how high is the spot above the average surface of Venus?

7. The smallest feature visible through an Earth-based telescope has an angular diameter of about 1 second of arc. If a crater on Mars is just visible when Mars is at its closest to Earth, how big is the crater? (*Hint:* Use the small-angle formula.)

8. What is the maximum angular diameter of Phobos as seen from the surface of Mars? as seen from Earth?

9. Deimos is about 12 km in diameter and has a density of 2 g/cm^3. What is its mass? (*Hint:* The volume of a sphere is $\frac{4}{3}\pi r^3$.)

Critical Inquiries for the Web

1. Who decides how planetary features are named? Surface features on Venus are (mostly) named after female figures from history and mythology, while figures from the arts and music are used to name features on Mercury. Look for information on planetary nomenclature and summarize the way different types of features on Venus are assigned names.

2. "Martians" have fascinated humans for the last century or more. There are many online sources that chronicle the representation of life on Mars throughout history and in literature. Read about the Martians as represented by a particular literary work or nonfiction account and discuss to what extent it is (or is not) based on realistic views of the nature of Mars both in terms of our current understanding and the views of that period.

Exploring *The Sky*

1. Mercury, Venus, and Mars exhibit phases similar to the moon's phases.

   a. Use *The Sky* to sketch the appearance of Mercury, Venus, and Mars at the present time. Also give the phase as a percentage of full.
   How to proceed: Use **Find** to give you a highly magnified view of a particular planet. Click on the planet to obtain **Object Information.**

   b. Explain the observed phase on the basis of geometrical relationship between the sun, Earth, and the planet as shown by the **3D Solar System Mode.**

Go to the Brooks/Cole Astronomy Resource Center (www. brookscole.com/astronomy) for critical thinking exercises, articles, and additional readings from InfoTrac College Edition, Brooks/Cole's online student library.

Worlds of the Outer Solar System

A good many things go around

in the dark besides Santa Claus.

Herbert Hoover

Guidepost

As we begin this chapter, we leave behind the psychological security of planetary surfaces. We can imagine standing on a terrestrial world, but the Jovian worlds have no surfaces. Thus, we face a new challenge—to use comparative planetology to study worlds so unearthly we cannot imagine being there.

One reason we find the moon and Mars of interest is that we might go there someday. Humans may become the first Martians. But the worlds of the outer solar system seem much less useful, and that gives us a chance to think about the cultural value of science.

Among the outer worlds, we see strong evidence that the smaller bodies that fall through the solar system impact planets and satellites. The next chapter will allow us to study these small bodies in detail and will give us new evidence that our solar system formed from a solar nebula.

a

b

Figure 8-1

(a) Jupiter's atmosphere is marked by complex circulation patterns, including the Great Red Spot at lower right. The planet has a large system of moons and a faint ring, not visible here. (b) In its longest dimension, the Great Red Spot is about twice the diameter of Earth. This infrared image from the Galileo spacecraft shows lower clouds as blue; higher thin hazes as pink; and high, thick clouds as white. The Red Spot is a region of rising gases forced to rotate like a giant hurricane by Jupiter's rapid rotation. *(NASA)*

The sulfuric acid fogs of Venus seem totally alien, but compared with the planets of the outer solar system, Venus is a tropical oasis. Of the five planets beyond the asteroid belt, four have no solid surface, and one, Pluto, is so far from the sun that most of its atmosphere lies frozen on its surface. As we begin our study of these strange worlds, we will discover new principles of comparative planetology.

The four Jovian planets—Jupiter, Saturn, Uranus, and Neptune—are strikingly different from the terrestrial planets. Their thick atmospheres of hydrogen, helium, methane, and ammonia are filled with clouds of ammonia, methane, and water-ice crystals; and their interiors are mostly liquid. They seem to be worlds dominated by cloud patterns (Figure 8-1).

The Jovian worlds can be studied from Earth, but much of what we know has been radioed back to Earth from space probes. Pioneer 10 and Pioneer 11 explored Jupiter and Saturn in the mid-1970s, but their instruments were not highly sensitive. The Voyager 1 and Voyager 2 spacecraft were launched in 1977 on a mission to visit all four Jovian worlds, concluding when Voyager 2 flew past Neptune in 1989. In December 1995, the Galileo spacecraft reached Jupiter. While an instrumented probe dropped into Jupiter's atmosphere, the main Galileo spacecraft went into orbit to begin a long-term study of the planet and its moons. Throughout this discussion of the Jovian worlds, you will find images and data returned by these robotic explorers (Window on Science 8-1).

Today's astronomers have a tremendous library of photos, measurements, and facts about the Jovian planets; and although no spacecraft has visited Pluto, Earth's astronomers are beginning to understand its role. Our task is to discover the relationships that explain how these worlds got to be the way they are. We begin with the closest and largest Jovian planet, named after the ruler of the Roman gods.

8-1 Jupiter

Jupiter is the most massive of the Jovian planets, containing 71 percent of all the planetary matter in our solar system. This high mass accentuates some processes that are less obvious or nearly absent on the other Jovian worlds. Just as we used Earth as the basis for comparison in our study of the terrestrial planets, we will examine Jupiter in detail so we can use it as a standard in our comparative study of the other Jovian planets.

Sending spacecraft to another world is expensive, and it may seem pointless when that world seems totally hostile to human life. What practical value is there in sending a space probe to Jupiter? To resolve that question, we need to consider the distinction between science, technology, and engineering.

Science is nothing more than the logical study of nature, and the goal of science is understanding. Although much scientific knowledge proves to have tremendous practical value, the only goal of science is a better understanding of how nature works. Technology, in contrast, is the practical application of scientific knowledge to solve a specific problem. People trying to find a faster way to paint automobiles might use all the tools and techniques of science, but if their goal has some practical outcome, we should more properly call it technology rather than science. Engineering is the most practical form of technology. An engineer is likely to use well-understood technology to find a practical solution to a problem.

Of course, there are situations in which science and technology blur together. For example, humanity has a practical and tragic need to solve the AIDS problem; we need a cure and a method of prevention. Unfortunately, we don't understand the AIDS virus itself or viruses in general well enough to design a simple solution, and thus much of the work involves going back to basic science and trying to better understand how viruses interact with the human body. Is this technology or science? It is hard to decide.

We might describe science that has no known practical value as basic science or basic research. Our exploration of worlds such as Jupiter would be called basic science, and it is easy to argue that basic science is not worth the effort and expense because it has no known practical use. Of course, the problem is that we have no way of knowing what knowledge will be of use until we acquire that knowledge. In the middle of the 19th century, Queen Victoria is supposed to have asked physicist Michael Faraday what good his experiments with electricity and magnetism were. He answered, "Madam, what good is a baby?" Of course, Faraday's experiments were the beginning of the electronic age. Many of the practical uses of scientific knowledge that fill our world—transistors, vaccines, plastics—began as basic research. Basic scientific research provides the raw materials that technology and engineering use to solve problems.

Basic scientific research has yet one more important use that is so valuable it seems an insult to refer to it as merely practical. Science is the study of nature, and as we learn more about how nature works, we learn more about what our existence in this universe means for us. The seemingly impractical knowledge we gain from space probes to other worlds tells us about our planet and our own role in the scheme of nature. Science tells us where we are and what we are, and that knowledge is beyond value.

Jupiter's Interior

Jupiter is mostly hydrogen. How do astronomers know this? From its density. Jupiter is only a third denser than water (Data File Six) and only a fourth the density of Earth. Theoretical models based on the density and the properties of matter show that Jupiter must be composed mostly of hydrogen and helium. The pressure inside compresses the hydrogen into a liquid. Heavier elements such as iron, nickel, silicon, and so on would settle toward the center, and the models tell us that this so-called rocky core can't be much larger than Earth.

From Jupiter's brightness in the infrared, astronomers conclude that heat is flowing out of the interior. In fact, it is radiating twice the energy it receives from the sun. It must be hot inside with heat left over from its formation and with heat from radioactive decay in the core.

With a temperature four times higher than the surface of the sun, the material in the core would flash into vapor were it not for the immense pressure. Clearly the core is not rocklike. It is rocky only in its chemical composition.

Although Jupiter is mostly liquid hydrogen, there is no surface on which we might sail a Jovian boat. The base of the atmosphere is so hot and the pressure is so high that there is no real distinction between liquid and gas, and if we jumped into Jupiter carrying a rubber raft we would be disappointed. As we fell deeper into the atmosphere we would notice the density around us increasing until we realized we were sinking through a liquid. We would never splash into a liquid surface.

Inside Jupiter, the pressure is so high that liquid hydrogen is not stable, and it is compressed into liquid metallic hydrogen—a very good conductor of electricity. This vast globe of conducting liquid, stirred by convection currents caused by heat flowing outward and spun by the planet's rapid rotation, drives the dynamo effect and generates a powerful magnetic field. Jupiter's magnetic field, over 10 times stronger than Earth's, is evidence that it has a large liquid metallic interior.

One dramatic consequence of this magnetic field is **aurora.** Just as in the case of Earth, interactions between Jupiter's magnetic field and the solar wind generate powerful electric currents that flow down into the atmosphere in two rings around the planet's magnetic poles. These are visible at ultraviolet wavelengths as rings of aurora larger in diameter than Earth (Figure 8-2a).

A planet's magnetic field deflects the solar wind and dominates a volume of space around the planet called the **magnetosphere.** Jupiter's magnetosphere is 100 times larger than Earth's (Figure 8-2b). If we could see it in the sky, it would look six times larger than our full moon.

The magnetic field around Jupiter traps charged particles from the solar wind in large, doughnut-shaped radiation belts that surround the planet just as Earth is surrounded by its Van Allen radiation belts. Because Jupiter's magnetic field is stronger, its trapped radiation is a billion times more intense than Earth's. Spacecraft that have flown through these regions received over 400 times the radiation that would have killed a human.

However unearthly Jupiter may be, it obeys the same rules of nature that govern Earth. Thus, it has a magnetic field and aurora just as Earth does. Jupiter's atmosphere, however, seems even more unearthly than its interior.

Jupiter's Atmosphere

The atmosphere on Jupiter is about 78 percent hydrogen, and the rest is mostly helium with small traces of water vapor, methane, ammonia, and similar molecules. The clouds we see are layers of ammonia and ammonia hydrosulfide crystals (Figure 8-3). Through gaps in these clouds, astronomers can see deeper, warmer layers of water-droplet clouds. Below that the atmosphere merges with the liquid interior.

Jupiter's atmosphere is dominated by **belt–zone circulation,** which circles the planet parallel to its equator (Figure 8-1a). Even a small telescope can reveal the dark belts and lighter zones. The colors of the belts and zones are believed to arise from molecules in the clouds produced by sunlight or lightning interacting with ammonia and other compounds.

Jupiter's belt–zone circulation is related to the high- and low-pressure areas we see on Earth's weather maps. Zones are high-pressure regions of rising gas, and belts are low-pressure regions where gas sinks.

On Earth, the temperature difference between the equator and poles drives a wavelike wind pattern that organizes such high- and low-pressure regions into cyclonic circulations familiar from weather maps (Figure 8-4). On Jupiter, the equator and poles appear to be about the same temperature, perhaps because of heat rising from the interior. Thus, there is no temperature difference to drive the wave circulation, and Jupiter's rapid rotation draws the high- and low-pressure regions into bands that circle the planet. On Earth, high- and low-pressure regions are bounded by winds induced by the wave circulation. On Jupiter, the same circulation appears as high-speed winds that blow around

Jupiter **Data File Six**

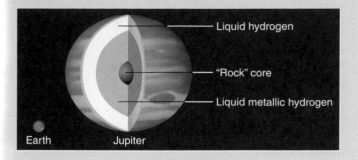

A theoretical model of the interior of Jupiter. Note the oblateness of the planet.

Average distance from the sun	5.2028 AU (7.783 × 10^8 km)
Eccentricity of orbit	0.0484
Maximum distance from the sun	5.455 AU (8.160 × 10^8 km)
Minimum distance from the sun	4.951 AU (7.406 × 10^8 km)
Inclination of orbit to ecliptic	1°18′29″
Average orbital velocity	13.06 km/s
Orbital period	11.867 y (4334.3 days)
Period of rotation	9^h55^m30^s
Inclination of equator to orbit	3°5′
Equatorial diameter	142,900 km (11.20 $D_\oplus$)
Mass	1.899×10^{27} kg (317.83 $M_\oplus$)
Average density	1.34 g/cm^3
Gravity at base of clouds	2.54 Earth gravities
Escape velocity	61 km/s (5.4 $V_\oplus$)
Temperature at cloud tops	−130°C (−200°F)
Albedo	0.51
Oblateness	0.0637

a

b

Figure 8-2

(a) Aurora on Jupiter is confined to a ring around the north magnetic pole and the south magnetic pole, as shown in these ultraviolet images. Earthly aurora follow the same pattern. The small comet-shaped spots are caused by powerful electrical currents flowing from Jupiter's moon Io. *(John Clarke, University of Michigan, and NASA)* (b) Jupiter's large conducting core and rapid rotation create a powerful magnetic field that holds back the solar wind and dominates a region called the magnetosphere. High-energy particles trapped in the magnetic field form giant radiation belts.

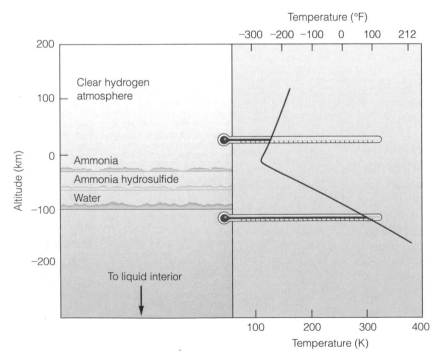

Figure 8-3

The three principal cloud layers on Jupiter are shown at left. Thermometers inserted into the Jovian atmosphere would register temperatures as in the graph at right. Note that the upper atmosphere is clear, cold hydrogen, but that the temperature rises rapidly below the cloud belts. Deep below the clouds, the atmosphere merges gradually with the liquid hydrogen interior.

a

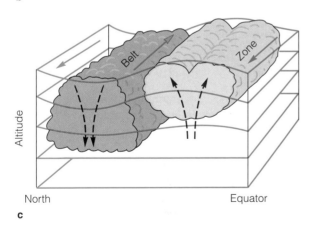

b

c

Figure 8-4

(a) In Earth's atmosphere, a wave circulation separates low- and high-pressure regions. (b) On Jupiter, low- and high-pressure regions take the form of dark belts of sinking gas and bright zones of rising gas. (c) Zones are bright because the rising gas forms clouds high in the atmosphere, where sunlight is strong. On both worlds, winds circulate in the same way—counterclockwise around low-pressure regions in the northern hemisphere and clockwise in the southern hemisphere.

the planet at the boundaries of the belts and zones. Above about 45° latitude, the belt–zone pattern becomes unstable, and the circulation decays to turbulent whirlwinds.

Mixed among the belts and zones are light and dark spots, a few larger than Earth. Some appear and disappear in a few days, while others last for years. The largest is the Great Red Spot, a reddish oval that has been in one of the southern zones for at least 300 years (see Figure 8-1b). Voyager and Galileo spacecraft photographs show that the Great Red Spot is a vast cyclonic storm where winds between adjacent belt and zone meet.

Astronomers around the world watched the atmosphere of Jupiter in the summer of 1994 as fragments of a comet slammed into the planet at high velocity. Over a period of six days, 20-some objects 0.5 km or so in diameter hit with the energy of millions of megatons of TNT, each impact creating a fireball of hot gases and leaving behind dark smudges that remained visible for months afterward (Figure 8-5). Planetary scientists continue trying to model the complex physics of the impacts, but two things are becoming clear. First, the violent impacts had little effect on the long-term circulation patterns in Jupiter's atmosphere. And second, the fragments of the comet were very fragile and didn't penetrate down to the cloud layers before they exploded. In the long run, such impacts on Jupiter probably occur roughly once every century or two. Major impacts on Earth occur less often because Earth is smaller, but they are inevitable. All of the planets in the solar system are occasionally struck by these objects.

One of the interesting features of Jupiter does not lie within its cloudy atmosphere, but orbits above its equator. Jupiter, like the other three Jovian planets, has a ring system.

Jupiter's Ghostly Rings

Jupiter has rings, but they were not discovered from Earth. The main ring was photographed by the Voyager 1 spacecraft in 1979. The ring is less than 30 km thick and extends out to 1.81 planetary radii. Less than 100 times as opaque as Saturn's rings, the main ring around Jupiter is composed of reddish dust. In 1998, the Galileo spacecraft found even more tenuous rings, the gossamer rings, extending over twice as far from Jupiter.

We have strong evidence that Jupiter's rings are formed of very small particles. The photo in Figure 8-6 was taken by the Galileo spacecraft looking back with the sun hidden behind Jupiter. The ring, illuminated from behind, glows brightly in what astronomers call **forward scattering.** This effect can only occur when particles are very small. Objects the size of sand grains, snowflakes, baseballs, or larger do not scatter light forward. Thus we can be sure that Jupiter's rings are made up of microscopic dust specks.

a

b

Figure 8-5

(a) Over six days in July 1994, fragments of a disrupted comet struck Jupiter and produced violent fireballs as they exploded in the atmosphere. This infrared photograph shows the glowing sites of impacts against the darker globe of Jupiter. *(University of Hawaii)* (b) At visible wavelengths, the impact sites show as dark smudges that were slowly dispersed by circulation in the atmosphere. The largest scar at extreme right is the size of Earth. *(NASA)*

The main ring lies inside the **Roche limit,** the distance inside which a moon cannot hold itself together by its gravity alone. It would be pulled apart by the planet's tidal forces. Any moon that orbits inside the Roche limit must be made of physically strong rock. (For a moon and planet of similar density, the Roche limit is about 2.44 times the radius of the planet.)

Because the rings lie inside the Roche limit, we might suspect the rings are material that could not form a moon so close to Jupiter, but the rings can't be very old. The pressure of sunlight and magnetic forces rapidly push such small particles into Jupiter or out into space. Also, the radiation near Jupiter grinds the particles down to nothing in a century or so. Thus, it is clear the rings are not old leftovers. They must be constantly supplied with new particles.

Observations with the Galileo spacecraft show that the main ring is densest at its outer edge where the small moon Adrastea orbits and that another small moon, Metis, orbits inside the main ring. Evidently, meteorite impacts on the moons blast dust into space, and tidal forces inside the Roche limit prevent the moons from pulling the dust back. Electromagnetic effects in the main ring cause some particles to rise above and below the ring to form a faint halo, visible in Figure 8-6. The gossamer rings are most dense at the orbits of two other small moons, Amalthea and Thebe, again giving us evidence that the dust is being blasted into space by impacts on the moons.

Jupiter's Family of Satellites

Jupiter has well over a dozen moons, but most are small, only a few tens of kilometers in diameter, and some of these have retrograde (backward) orbits. This evidence

Figure 8-6

This false-color image of Jupiter's ring was recorded while the Galileo spacecraft passed through Jupiter's shadow. The ring is bright when illuminated from behind, a property called forward scattering, which tells us the ring particles are microscopic. A mist of small particles is visible in a halo extending above and below the rings. Jupiter's upper atmosphere is visible at right, and the ring appears to end where it enters Jupiter's shadow. *(NASA)*

Figure 8-7
The Galilean moons of Jupiter, clockwise from upper left: Io, Europa, Callisto, and Ganymede. Callisto and Ganymede are 50 percent water, and Europa contains an icy crust over a water mantle. Io is dry and sulfurous. Earth's moon is shown at center for scale. *(NASA; inset UCO/Lick Observatory Photo)*

suggests that many of Jupiter's smaller moons are captured asteroids.

In contrast, the four largest moons are called Galilean moons (Figure 8-7) after their discoverer, Galileo. These four moons are clearly related to each other and probably formed with Jupiter. We know these worlds surprisingly well because the two Voyager spacecraft and the Galileo spacecraft have studied them in detail.

The outermost Galilean moon, Callisto, is half again as big as Earth's moon, but it has a low density of only

1.79 g/cm³. This means it must consist roughly of 50:50 rock and ice. Observations of its gravitational field by the Galileo spacecraft reveal that it has a dense core and a lower-density exterior, so it presumably differentiated. Also, it interacts with Jupiter's magnetic field in such a way that astronomers suspect it has a mineral-rich ocean of liquid water 100 km below its icy crust. Photos of its surface, however, show thousands of impact craters with little sign of geological activity (Figure 8-8).

a

b

Figure 8-8
(a) Callisto's surface is heavily cratered, dirty ice, where the youngest craters look bright because they have dug into cleaner ice. The bull's-eye impact feature is called Valhalla and is roughly 3000 km in diameter. (b) A Galileo image of Callisto's surface shows the dirty icy crust and impact features. *(NASA)*

Figure 8-9

(a) The icy surface of Ganymede is divided into old, dark cratered terrain and bands of bright grooves (see Figure 8-7). In this Voyager image, the dark terrain is marked by a few bright craters where more recent impacts have broken through the dirty surface ice to clean ice below. (b) A Galileo spacecraft image shows how a system of icy grooves covered an older region at bottom while an even more recent system of grooves at top covered the previous grooves. The grooves suggest water flooding up through cracks. The age of the grooved terrain is uncertain. *(NASA)*

a

b

Next inward from Callisto is Ganymede, with a density of 1.9 g/cm³, suggestive of a mix of rock and ice. The Galileo spacecraft detected a weak magnetic field, which suggests a differentiated metallic core. The icy crust is marked by old, cratered, dark areas, and younger, brighter regions of **grooved terrain** believed to be systems of faults in the brittle crust. Sets of grooves overlap other sets of grooves, and that suggests extended episodes of geological activity (Figure 8-9).

The density of the next moon inward, Europa, is 3.03 g/cm³, high enough to suggest the moon is mostly rock with a thin icy crust. The visible surface is very clean ice, contains very few craters, and has long scars suggestive of cracks in the icy crust (Figure 8-10). The icy crust we see may overlay an ocean of liquid water or watery slush. The surface is estimated to be only about 10 million years old—very young for a geological surface. Thus Europa is apparently still an active world.

Io, the innermost of the four moons, is also the most active. Its density of 3.55 g/cm³, combined with gravity measurements by the Galileo spacecraft, shows that it has a large metallic core and a rocky, sulfur-rich crust. Spectra show no trace of water. Spacecraft photos reveal active volcanoes venting gases and sulfur-rich ash high above the surface (Figure 8-11). A new crater on Io would be buried under the sulfur ash at the rate of a few millimeters per year. That burying explains why photos show no impact craters on Io.

The activity we see in the Galilean moons must be driven by energy flowing outward. The violently active volcanism of Io is apparently caused by **tidal heating**. Io is too small to have remained hot from its formation, but its orbit is slightly elliptical; and as it moves closer and then farther from Jupiter, the planetary gravitational field flexes the moon with tides, and friction heats its interior. That heat flowing outward causes the volcanism. Europa is not as active as Io, but it too must have a heat source, presumably tidal heating. Ganymede is no longer active, but when it was younger, it had internal heat that broke the crust to produce the grooved terrain. Callisto, the most distant of the four moons, appears never to have been geologically active.

Tidal heating has probably affected the Galilean moons, but we must also suspect they formed together in a disk-shaped nebula around the proto-Jupiter. Heat from Jupiter would have heated the inner part of that disk, and the outer parts would have remained colder. Thus, we can understand the densities and compositions of the four Galilean moons by the condensation sequence (Chapter 5) acting in a miniature solar nebula around Jupiter.

A History of Jupiter

Can we put all of the evidence together and tell the story of Jupiter? Creating such a logical argument of evidence and hypotheses is the ultimate goal of planetary astronomy.

a

a

b

b

Figure 8-10

(a) Long lines on the surface of Europa appear to be cracks in its icy crust where water from below has refrozen. Note the bright surface and lack of craters—signs of youth. (b) This image from the Galileo spacecraft shows ice blocks on Europa that have moved like a disturbed puzzle. This resembles icebergs floating in pack ice on Earth. *(NASA)*

Figure 8-11

(a) These two photos recorded by the Galileo spacecraft show that a new volcanic vent became active on Io during the summer of 1997. The larger feature is the ejecta around the older volcano Pele, but the new dark ejecta sheet is Pillan Patera, covering a region about as big as Arizona. (b) A computer-enhanced image of Io reveals the volcano Loki spewing ash high above the surface where it falls back to blanket the surface. *(NASA)*

Science is an expensive enterprise, and that raises the question of payment. Some science has direct technological applications, but some basic science is of no immediate practical value. Who pays the bill?

In Window on Science 8-1, we saw that technology is of immediate practical use; thus, business funds much of this kind of scientific research. Auto manufacturers need inexpensive, durable, quick-drying paint for their cars, and they find it worth the cost to hire chemists to study the way paint dries. Many industries have large research budgets, and some industries, such as pharmaceutical manufacturers, depend exclusively on scientific research to discover and develop new products.

If a field of research has immediate potential to help society, it is likely that government will supply funds. Much of the research on public health is funded by government institutions such as the National Institutes of Health and the National Science Foundation. The practical benefit of finding new ways to prevent disease, for example, is well worth the tax dollars.

Basic science, however, has no immediate practical use. That doesn't mean it is useless, but it does mean that the practical-minded stockholders of a company will not approve major investments in such research. Digging up dinosaur bones, for instance, is very poorly funded because no industry can make a profit from the discovery of a new dinosaur. Astronomy is another field of science that has few direct applications, and thus very little astronomical research is funded by industry.

The value of basic research is twofold. Discoveries that have no known practical use today may be critically important years from now. Thus, society needs to continue basic research to protect its own future. But basic research, such as studying Saturn's rings or digging up dinosaur bones, is also of cultural value because it tells us what we are. Each of us benefits in intangible ways from such research, and thus society needs to fund basic research for the same reason it funds art galleries and national parks—to make our lives richer and more fulfilling.

Because there is no immediate financial return from this kind of research, it falls to government institutions and private foundations to pay the bill. The Keck Foundation has built two giant telescopes with no expectation of financial return, and the National Science Foundation has funded thousands of astronomy research projects for the benefit of society. Debates rage as to how much money is enough and how much is too much, but ultimately, funding basic scientific research is a public responsibility that society must balance against other needs. There isn't anyone else to pick up the tab.

Jupiter formed far enough from the sun to incorporate large numbers of icy planetesimals, and it must have grown rapidly. Once it was 15 times more massive than Earth, it could grow by gravitational collapse (Chapter 5), the capture of gas directly from the solar nebula. Thus, it grew rich in hydrogen and helium from the solar nebula. Its present composition is quite sunlike and resembles the composition of the solar nebula. The location of Jupiter's point in Figure 7-14 shows that its gravity is strong enough to hold onto all of its gases.

The large family of moons may be mostly captured asteroids, and Jupiter may still encounter a wandering asteroid or comet now and then. Some of these are deflected, some captured into orbit, and some, like the comet that struck Jupiter in 1994, actually fall into the planet. Dust blasted off of the inner moons by micrometeorites settles into the equatorial plane to form Jupiter's rings.

The four Galilean moons seem to have formed in a disk of gas and dust around the forming planet. They have different compositions, much like planets in our solar system, but they have also been affected to different degrees by tidal heating.

REVIEW Critical Inquiry

Why is Jupiter so big?

We can analyze this question by constructing a logical argument that relates the formation of Jupiter to the solar nebula theory. Jupiter is rich in hydrogen and helium, but Earth is relatively poor in these elements. While the solar nebula existed, Earth grew by the accretion of solid, rocky planetesimals, but it never became massive enough to capture gas directly from the solar nebula. That is, it never grew by gravitational collapse. Jupiter, however, grew so rapidly from icy planetesimals in the outer solar nebula that it was eventually able to grow by gravitational collapse. By the time the solar nebula cleared away and ended planet building, Jupiter had captured large amounts of hydrogen and helium and was quite massive.

Often the present nature of a world can be traced back to the way it formed. Can you create a logical argument to explain the nature of the Galilean moons?

We have studied Jupiter in detail because it is the basis of comparison for the other Jovian worlds. Now we can turn our attention to the most beautiful Jovian world, Saturn.

8-2 Saturn

Saturn has played second fiddle to its own rings since Galileo first viewed it through a telescope in 1610. The rings are dramatic, strikingly beautiful, and easily seen through even a small telescope; but Saturn itself, only slightly smaller than Jupiter, is a fascinating planet (Data File Seven).

Although Saturn lies roughly 10 AU from the sun, we know a surprising amount about it. The beautiful rings are easily visible through the telescopes of modern amateur astronomers, and large Earth-based telescopes have explored the planet's atmosphere, rings, and moons. The two Voyager spacecraft flew past Saturn in 1979, transmitting back to Earth detailed measurements and images, and NASA is sending the Cassini spacecraft to orbit Saturn early in the 21st century (Window on Science 8-2). That mission will tell us even more about the Saturn system.

Saturn the Planet

Seen from Earth, Saturn shows only faint evidence of belt–zone circulation, but the Voyager photos show that belts and zones are present and that the associated winds blow up to three times faster than on Jupiter. The belts and zones on Saturn are less visible because they occur deeper in the cold atmosphere below a layer of methane haze (Figure 8-12).

Saturn is less dense than water (it would float), and that suggests that it is, like Jupiter, rich in hydrogen and helium. In fact, its density is so low that it must have a relatively small core of heavy elements.

The shape of a Jovian planet can tell us about the interior. All of the Jovian planets, being mostly liquid and rotating rapidly, are slightly flattened. A planet's **oblateness** is the fraction by which its equatorial diameter exceeds its polar diameter. As photographs show, Saturn is the most oblate of the planets, and that evidence tells us that it is mostly liquid. A world with a large rocky core and mantle would not be flattened much by rotation, but an all-liquid planet would flatten significantly. Thus the oblateness of a Jovian planet, combined with its average density, can help astronomers model the interior.

Models of Saturn predict that it must have a small core of heavy elements and less liquid metallic hydrogen

Saturn Data File Seven

Saturn's rings of ice particles are bright and beautiful, but its belt–zone circulation is subdued. A moon and its shadow lie just above center. *(NASA)*

Average distance from the sun	9.5388 AU (14.27 × 10^8 km)
Eccentricity of orbit	0.0560
Maximum distance from the sun	10.07 AU (15.07 × 10^8 km)
Minimum distance from the sun	9.003 AU (13.47 × 10^8 km)
Inclination of orbit to ecliptic	2°29′17″
Average orbital velocity	9.64 km/s
Orbital period	29.461 y (10,760 days)
Period of rotation	10^{h}39^{m}25^s
Inclination of equator to orbit	26°24′
Equatorial diameter	129,660 km (9.42 $D_\oplus$)
Mass	5.69 × 10^{26} kg (95.147 $M_\oplus$)
Average density	0.69 g/cm^3
Gravity at base of clouds	1.16 Earth gravities
Escape velocity	35.6 km/s (3.2 $V_\oplus$)
Temperature at cloud tops	−180°C (−292°F)
Albedo	0.61
Oblateness	0.102

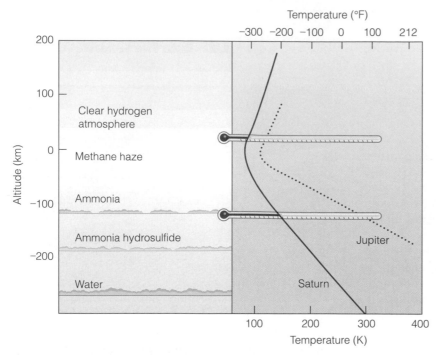

Figure 8-12
Because Saturn is farther from the sun, its atmosphere is colder than Jupiter's (dotted line). The cloud layers on Saturn form at the same temperature as do the cloud layers on Jupiter, but that puts them deeper in Saturn's hazy atmosphere, where they are not as easy to see as the clouds on Jupiter. (See Figure 8-3.)

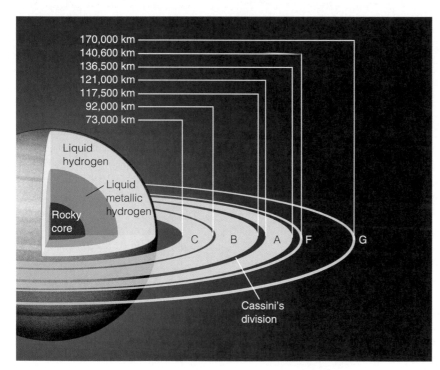

Figure 8-13
Models of the interior of Saturn suggest that it has a core of heavy elements that is smaller than Jupiter's core and that it contains less liquid metallic hydrogen because the pressure inside Saturn is lower. The major rings of Saturn are identified by letters of the alphabet. The gap called Cassini's division is a bit over one-fourth the diameter of Earth.

than Jupiter because Saturn's internal pressure is lower. Perhaps this is why Saturn's magnetic field is 20 times weaker than Jupiter's. Like Jupiter, Saturn has a hot interior and radiates more energy than it receives from the sun (Figure 8-13).

Saturn may look a bit bland, but it makes up for that in the splendor of its rings.

The Rings of Saturn

Saturn's rings contain billions of icy particles ranging from specks of dust to chunks the size of houses. These particles are confined in a thin layer perhaps only meters thick in the equatorial plane of the planet. A few gaps, such as Cassini's division, are visible from Earth (Figure 8-13), but the true complexity of the rings was not evident until the Voyager 1 spacecraft flew past Saturn in

Figure 8-14
The rings of Saturn are made up of ice particles orbiting in the planet's equatorial plane. The transparency of the rings is evident in this image where they cross in front of the planet at left. Generally pale and colorless, the rings are marked by many gaps and ringlets that are produced by dynamic effects. Notice the tenuous C ring near the planet and the narrow F ring beyond the outer edge of the main rings. Look carefully to see ringlets inside Cassini's division. *(NASA)*

1980. The Voyager images showed that the rings were made up of hundreds of narrow ringlets, some as narrow as 2 km. There were even ringlets in Cassini's division.

From Earth, astronomers had identified three main rings, A, B, and C, and had hints of a few other ring features. The Voyager images revealed a dusty E ring extending far from the planet and a threadlike F ring lying just outside the bright A ring (Figure 8-14). This F ring is confined in a narrow strand by two small moons that orbit just inside and just outside the ring. As a particle drifts away from the F ring, the moons shepherd it back into the ring, and thus the moons are known as **shepherd satellites.** A similar G ring and the sharp outer edge of the A ring are believed to be confined by shepherd satellites.

Some gaps in the rings seem to be caused by resonances with satellites. A particle in Cassini's division, for example, orbits Saturn twice in the time the moon Mimas takes to orbit once. Thus, on every other orbit, the ring particle overtakes Mimas at the same place in the particle's orbit, and the gravitational tugs of Mimas gradually pull the particle into a slightly elliptical orbit. Such an orbit is dangerous because the particle crosses the paths of millions of other particles. Our unfortunate particle is almost certain to suffer a collision and have its orbit altered dramatically. Thus, particles cannot remain in orbit at a resonance with a moon, and gaps in the rings mark such spots.

But not all gaps are caused by resonances. Encke's division in the A ring is caused by a small satellite in the gap that keeps the gap clear of particles, and the hundreds of other ringlets cannot all be produced by resonances. Small moons in the rings and waves, such as spiral density waves, moving through the rings are believed to explain many of the gaps.

Saturn's icy rings can't be debris left over from the formation of the planet. The planet would have formed hot and would have vaporized ices and driven away the gas. Also, micrometeorites would tend to darken the icy particles over periods of 100 million years. These facts suggest that the material in Saturn's rings is debris from collisions between the heads of comets and icy moons. Such collisions must occur every few tens of millions of years and occasionally supply the rings with fresh ice. Saturn's moons certainly show signs of impacts.

The Moons of Saturn

Saturn has 18 known moons, many of which are small and all of which are composed of mixtures of rock and ice. These moons are faint points of light as seen from Earth, but astronomers know some of them well because they were photographed by the two Voyager spacecraft in 1980 and 1981.

The largest of Saturn's moons, Titan (Figure 8-15), is a few percent larger than the planet Mercury. Its density suggests that it must contain a rocky core under a thick mantle of ice, but its surface is not visible through its cloudy atmosphere. We might not expect such a small world to have an atmosphere, but Titan is so cold its gas molecules do not travel fast enough to escape (see Figure 7-14). The gas is about 90 percent nitrogen with nearly 10 percent argon and a small amount of methane. Sunlight acting on the methane and nitrogen produces the thick orange smog of organic molecules,* which may drift down through the atmosphere and collect in an organic goo on the surface.

Organic goo on the surface of Titan sounds exciting because it could harbor primitive life, but the surface

---

*Organic molecules are common in living things on Earth but do not have to be derived from living things. One chemist defines an organic molecule as "any molecule with a carbon backbone."

Figure 8-15

(a) The surface of Titan is hidden by its smoggy atmosphere (with a circulation pattern near the north pole). The extent of the atmosphere is visible in photos of the night side (inset). (b) Hubble Space Telescope can detect infrared radiation from the surface. Here images show the side of Titan facing Saturn (left) and the side facing forward in its orbit (right). The surface is not a uniform ethane and methane ocean but consists of at least some solid surface with differing geology. *(NASA)*

of Titan is very cold at −179°C (−290°F). Also, models of the atmosphere suggest that the surface may be partly covered by oceans of liquid methane and ethane. Oddly, Titan is cooled by an inverse greenhouse effect as the organic haze blocks much of the incoming sunlight but allows infrared radiation to escape to space. Clearly, Titan is a peculiar world.

The remaining moons of Saturn are small and icy, have no atmospheres, and are heavily cratered. Most have ancient surfaces. Tethys, for example, is less than a third the diameter of Earth's moon, and its heavily cratered crust seems quite old (Figure 8-16a). However, a valley 3 km deep trails three-fourths of the way around the satellite. Such cracks are found on a few other satellites and appear to have formed long ago, when the interiors of the icy moons froze and expanded.

Enceladus, a bit smaller than Tethys, nevertheless shows signs of recent activity (Figure 8-16c). Some parts of its surface contain 1000 times fewer craters than other regions, showing that these lightly cratered regions must be younger. At some point in its history, this moon had internal heat that produced geological activity and resurfaced some areas of its crust. The energy source is unknown, but tidal heating is likely.

Like nearly all moons in the solar system, Saturn's moons are tidally locked to their planet, rotating to keep the same side facing the planet. The leading side of these moons, the side facing forward in the orbit, is sometimes modified by debris. Iapetus, for example, has a cratered trailing side about as dark as dingy snow, but its leading side is as dark as fresh asphalt (Figure 8-16b). One hypothesis is that the dark material is carbon-rich dust from meteorite impacts on the next moon out, Phoebe.

The History of Saturn

Can we put all of the evidence and hypotheses together and describe the history of Saturn? Doing so is a real test of our understanding.

Saturn formed in the outer solar nebula, where ice particles were stable and may have contained more trapped gases. The protoplanet grew rapidly and became massive enough to attract hydrogen and helium by gravitational collapse. The heavier elements sank to the middle to form a small core, and the hydrogen formed a liquid mantle containing liquid metallic hydrogen. The outward flow of heat from the interior is believed to drive convection inside the planet that produces its magnetic field. Because Saturn is smaller than Jupiter, the internal pressure is less, the planet contains less liquid metallic hyrogen, and its magnetic field is weaker.

The rings can't be primordial. That is, they can't be material left over from the formation of the planet. Such ices would have been vaporized and driven away by the heat of the protoplanet. Rather, we can suppose that the ring material is debris from occasional collisions between comets and Saturn's icy moons. Such giant impacts are not unheard of in the solar system. Jupiter was hit by a comet in 1994.

Some of Saturn's moons are probably captured asteroids that wandered too close, but some of the moons probably formed with Saturn. Many have ancient surfaces. The giant moon Titan may have formed with Saturn, or it may be a very large icy planetesimal captured into orbit around Saturn. We will see more evidence for this hypothesis when we explore farther from the sun.

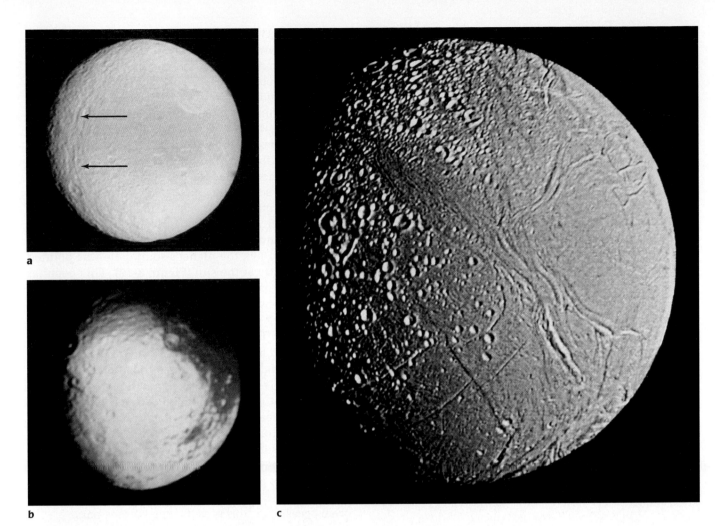

Figure 8-16

(a) Heavily cratered Tethys appears to be a dead world, although it is marked by a deep valley (arrows). (b) The leading side of Iapetus (facing upper right) is much darker than the trailing side. (c) Some portions of Enceladus have fewer craters. These areas apparently have been resurfaced recently, which suggests that Enceladus is an active moon. *(NASA)*

REVIEW Critical Inquiry

Why do the belts and zones on Saturn look so bland?

In a Jovian planet, the light-colored zones form in high-pressure regions where rising gas cools and condenses to form icy crystals of ammonia, which we see as bright clouds. Saturn is twice as far from the sun as Jupiter, so sunlight is weaker and the atmosphere is colder. The rising gas currents don't have to rise as high to reach temperatures cold enough to form clouds. Because the clouds form deeper in the hazy atmosphere, they are not as brightly illuminated by sunlight and look dimmer. A layer of methane haze above the clouds makes the belts and zones look even less distinct.

We have used some simple physics to construct a logical argument that explains the hazy cloud features on Saturn. Can you explain why Saturn's rings have gaps and ringlets?

In any contest, Saturn would win the ribbon for most beautiful planet. But beyond Saturn's orbit are three more planets that could compete for the ribbon for most striking personality.

8-3 Uranus

Uranus is a Jovian planet, but it is quite different from Jupiter and Saturn. Its atmosphere, interior, rings, and satellites are profitable subjects for comparative planetology.

Uranus the Planet

Uranus is only a third the diameter of Jupiter, only a twentieth as massive, and, being four times farther from the sun, its atmosphere is over 100 degrees colder than Jupiter's (Data File Eight).

Uranus Data File Eight

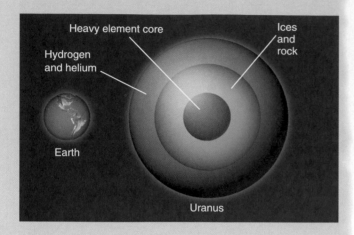

About four times the diameter of Earth, Uranus is believed to contain a liquid interior and dense core.

Average distance from the sun	19.18 AU (28.69 × 10^8 km)
Eccentricity of orbit	0.0461
Maximum distance from the sun	20.1 AU (30.0 × 10^8 km)
Minimum distance from the sun	18.3 AU (27.4 × 10^8 km)
Inclination of orbit to ecliptic	0°46′23″
Average orbital velocity	6.81 km/s
Orbital period	84.013 y (30,685 days)
Period of rotation	17^{h}14^m
Inclination of equator to orbit	97°55′
Equatorial diameter	51,118 km (4.01 $D_\oplus$)
Mass	8.69 × 10^{25} kg (14.54 $M_\oplus$)
Average density	1.29 g/cm^3
Gravity	0.919 Earth gravity
Escape velocity	22 km/s (1.96 $V_\oplus$)
Temperature above cloud tops	−220°C (−364°F)
Albedo	0.35

Because Uranus is smaller than Jupiter, its internal pressure is lower, and it does not contain liquid metallic hydrogen. Models of Uranus based in part on its density and oblateness suggest that it has a small core of heavy elements and a deep mantle of partly frozen water containing rocky material and dissolved ammonia and methane. Circulation in this electrically conducting mantle may generate the planet's peculiar magnetic field, which is highly inclined to its axis of rotation. Above this mantle lies the deep hydrogen and helium of the planet's atmosphere.

Uranus rotates on its side, with its equator inclined 98° to its orbit, and consequently each of the four seasons on Uranus lasts over 20 years. When Voyager 2 flew past in 1986, the planet's south pole was pointed almost directly at the sun (Figure 8-17). This peculiar rotation may have been produced when Uranus collided with a very large planetesimal late in its formation.

Voyager 2 photos show a nearly featureless blue-green world (Figure 8-18). The atmosphere is mostly hydrogen and helium (12 percent), but traces of methane absorb red light and thus make the atmosphere look blue. There is no belt–zone circulation visible in the Voyager photographs, although extreme computer enhancement does reveal a few clouds and bands around the south pole.

Studies subsequent to the Voyager 2 visit show that the atmosphere is complex. Infrared images show different layers of haze (Figure 8-18b), and occasional cloud features have been detected. Uranus may suffer from occasional storms as its long seasons pass.

Atmospheric circulation on a Jovian world may be driven by heat flowing out of the interior, but Uranus has little heat flow. It has cooled and now radiates about the same amount of energy that it receives from the sun. This may account for its limited atmospheric activity.

The Rings of Uranus

The rings of Uranus were discovered in 1977 when astronomers watched the planet occult—cross in front of—a distant star. Although the rings are not easily visible from ground-based telescopes, the astronomers detected them when the star dimmed momentarily as each ring passed in front of it. Voyager 2 captured images of the rings in detail when it visited Uranus in 1986 (Figure 8-19).

The rings of Uranus are tenuous and dark. They are 100 to 1000 times more substantial than Jupiter's ring, but they total less mass than the thin material in Cassini's division—the largest "gap" in Saturn's rings. In addition, the rings of Uranus are composed of very dark material as black as lumps of coal.

Eleven narrow rings have been found. The widest, the ε ring, is slightly elliptical and varies from 12 to 60

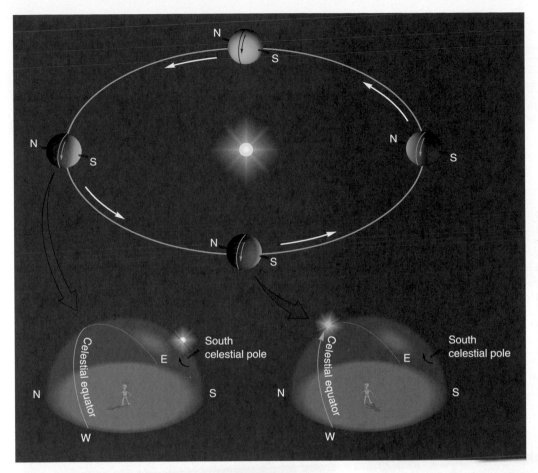

Figure 8-17
Uranus rotates on an axis that is tipped 97.9° from the perpendicular to its orbit, so its seasons are peculiar. When one of its poles is pointed nearly at the sun (a solstice), a citizen of Uranus would see the sun near a celestial pole, and it would never rise or set. As it orbits the sun, the planet maintains the direction of its axis in space, and thus the sun moves from pole to pole. At the time of an equinox on Uranus, the sun would be on the celestial equator and would rise and set with each rotation of the planet. Compare with Figure 1-14.

a

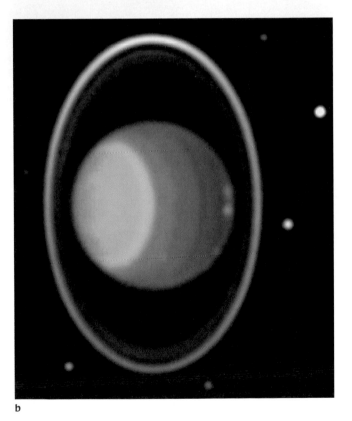

b

Figure 8-18
(a) The planet Uranus shows almost no atmospheric features in this visible light image made by Voyager 2 in 1986. (b) Three infrared images from the Hubble Space Telecope have been combined in this false-color image to reveal layers in the atmosphere. High haze is violet near the edge of the disk, lower haze is yellow, and the clear deep atmosphere is blue. The rings have been digitally enhanced to make them visible. The ring particles are actually as dark as charcoal. *(Erich Karkoschka, University of Arizona Lunar and Planetary Lab; NASA)*

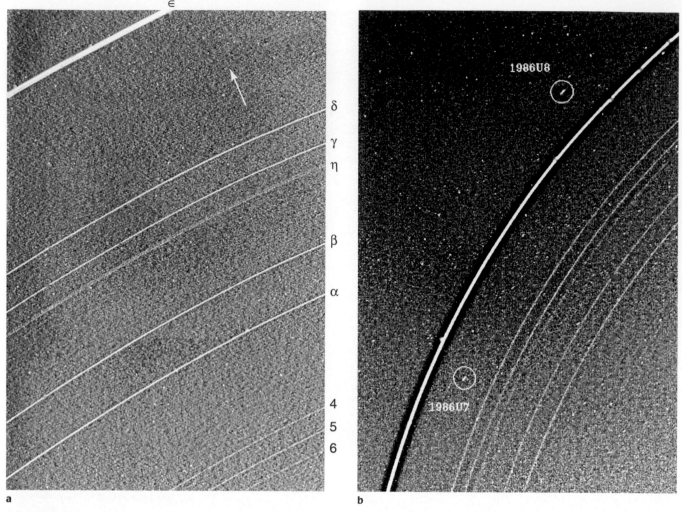

Figure 8-19
(a) The narrow rings of Uranus are identified by Greek letter or by number. Voyager images reveal another faint ring (arrow) halfway from the ε ring to the next innermost (δ) ring. Tip this page and view the faint ring along its length. (Hints of other faint rings can be located by using this edge-on viewing trick.) (b) Two shepherd satellites confine the ε ring into a narrow band. Because the satellites have slightly elliptical orbits, the ring is elliptical and slightly wider where it is farther from the planet. *(NASA)*

km in width. Planetary rings should gradually spread into thin sheets, so the narrowness of these rings suggests that they are confined by shepherd moons. Voyager 2 photos show two moons less than 25 km in diameter shepherding the ε ring (Figure 8-19b), and the other rings are presumably confined by moons too small to detect.

The black color of the ring particles seems to be caused by the radiation trapped in Uranus's magnetosphere. The radiation can convert traces of methane ice into black organic polymers. The smallest moons are also blackened by this radiation.

When Voyager 2 looked back through the rings from their dark side, it saw very little forward-scattered light, and this means the rings contain little dust. Particles must be boulder- to house-sized. Collisions between

ring particles must occasionally chip off dust particles, but the tenuous high atmosphere of Uranus slows these particles and they fall into the planet.

The rings of Uranus could not have survived from the formation of the planets. They must have been produced within the last few million years by debris that got trapped in the spaces between the moons' orbits. Astronomers now think that debris may be produced when comets strike the larger satellites.

The Uranian Moons

Five large moons orbit Uranus, and Voyager 2 discovered 10 more small moons. Others have been found from Earth, and the count now stands at 20 moons orbiting Uranus.

a   b

Figure 8-20

(a) Ariel, a satellite of Uranus, apparently had an active past, as shown by the broad, shallow valleys on its surface. (b) The face of Miranda is marred by ovoids, which may have formed as convection currents circulating through the icy mantle. Note the 5-km-high cliff at the lower edge of the moon. *(NASA)*

The five major satellites were photographed by Voyager 2, and we can analyze their surface features. Oberon and Titania, the two outer moons, are about 45 percent the diameter of Earth's moon, and they have old, cratered surfaces with faults hundreds of kilometers long. Some regions of these icy moons have been resurfaced by liquid water "lava" that covered old craters and froze. In contrast, Umbrial, the next moon inward, is about a third the diameter of Earth's moon and shows no sign of activity on its ancient, cratered surface. Ariel, the fourth moon inward, is slightly smaller than Umbrial, but its cratered crust is marked by broad, smooth-floored valleys that may have been cut by flowing ice (Figure 8-20a).

Miranda, the innermost moon, is only 14 percent the diameter of Earth's moon, but its surface is marked by oval patterns of grooves called **ovoids** (Figure 8-20b). These are believed to have been caused by internal heat that produced convection in the icy mantle. Rising currents of ice have deformed the crust and created the ovoids. From craters on the ovoids, astronomers conclude that the entire surface is old and the moon is no longer active. Perhaps it was warmed by tidal heating long ago.

A History of Uranus

The challenge of comparative planetology is to tell the story of a planet, and Uranus is the most difficult planet to explain. It is not only far away and hard to study, but it is peculiar in a number of ways.

It must have formed from the outer solar nebula, but it must have grown more slowly than Jupiter or Saturn, perhaps because the nebula was less dense beyond Saturn. Uranus captured some gas from the nebula but it did not grow as rich in hydrogen as Jupiter or Saturn. It is now a mostly liquid world.

Perhaps as it was forming it was struck a glancing blow by a large planetesimal and given its highly inclined rotation. These large impacts seem to have been common when the solar system was young. We have seen evidence for giant impacts on Earth's moon, Mercury, Mars, and the satellites of Jupiter and Saturn. Also, we suspect that impacts between the icy heads of comets and moons may produce the debris that replenishes planetary rings.

The highly inclined magnetic field of Uranus may be produced by convection in its electrically conducting mantle. With very little heat flowing out of the interior, this convection must be limited. The lack of interior heat in Uranus is not well understood, but some astronomers suspect that the impact that gave Uranus its odd rotation also stirred it so thoroughly that it lost much of its heat.

REVIEW **Critical Inquiry**

Why are the rings of Uranus so narrow?

Unlike the rings of Jupiter and Saturn, the rings of Uranus are quite narrow, like hoops of wire. We would expect collisions among ring particles to gradually spread the rings out into thin sheets, so something must be confining the narrow rings. In fact, two small moons have been found orbiting just inside and outside the ε ring. If a ring particle drifts away from the ring, the corresponding moon's gravity will boost it back into the ring. Moons too small to detect are thought to shepherd the other rings. Thus, the rings of Uranus resemble Saturn's narrow F ring.

This explains the narrowness of the rings, but isn't it odd that moons just happen to be in the right place to keep the rings narrow? How did that happen?

Uranus seems to be a bland, featureless planet, but it hides some peculiar secrets. Its neighbor, Neptune, is also a peculiar world.

8-4 Neptune

Seen from Earth, Neptune has a faint blue tint, but no details are visible. In 1989, the Voyager 2 spacecraft flew past Neptune and revealed some of its secrets.

Planet Neptune

Only 4 percent smaller in diameter than Uranus, Neptune has a similar interior. A small core of heavy elements lies within a slushy mantle of water below a hydrogen-rich atmosphere (Data File Nine). Yet Neptune looks quite different; it is dramatically blue and has active cloud formations.

The dark-blue tint to the atmosphere is understandable because Neptune contains 3 percent methane compared to only 2 percent in Uranus. The methane absorbs red photons better than blue, giving Neptune a blue

a

b

Figure 8-21

(a) A 1989 Voyager 2 image of Neptune revealed an anticyclonic disturbance about the size of Earth, which was dubbed the Great Dark Spot. High-altitude winds were deflected around the spot and formed high clouds of methane-ice crystals. *(NASA)* (b) A 1996 Hubble Space Telescope image of Neptune shows how cloud patterns have changed since Voyager 2 flew past in 1989. The Great Dark Spot has disappeared, but new spots have been detected. The highest methane clouds appear white to the human eye but appear pink in this image due to their infrared emission. *(Lawrence Sromovsky, University of Wisconsin, Madison; NASA)* tint. Also, the atmosphere scatters shorter-wavelength (blue) photons better than red photons, just as does Earth's atmosphere, and that fact helps make Neptune look blue. Finally, Uranus and Neptune are quite cold, and their temperatures may prevent the formation of the molecules that give the clouds their colors on Jupiter and Saturn.

Atmospheric circulation on Neptune is much more dramatic than on Uranus. When Voyager 2 flew by Neptune in 1989, the largest feature was the Great Dark Spot (Figure 8-21a). Roughly the size of Earth, the spot seemed to be an atmospheric circulation much like Jupiter's Great Red Spot. Smaller spots were visible in Neptune's atmosphere, and photos showed they were circulating like hurricanes. More recently, the Hubble Space Telescope has photographed Neptune and found that the Great Dark Spot is gone and new cloud formations have appeared (Figure 8-21b). Evidently, the weather on Neptune is changeable.

The atmospheric activity on Neptune is apparently driven by heat flowing from the interior. The heat causes convection in the atmosphere, which the rapid rotation of the planet converts into high-speed winds, high-level white clouds of methane ice crystals, and rotating storms that we see as spots. Neptune has more activity than Uranus because it has more heat flowing out of its interior.

Like Uranus, Neptune has a highly inclined magnetic field that must be linked to circulation in the interior. In both cases, we suspect that ammonia dissolved in the liquid water mantle makes it a good electrical conductor and that convection in the water, coupled with the rotation of the planet, drives the dynamo effect and generates the magnetic field.

Neptune's Rings

Voyager 2 photos revealed that Neptune has two narrow rings with radii of about 2 and 2.5 planetary radii with a broad, faint ring between (Figure 8-22a). Small moons shepherd the narrow rings and keep the ring particles from spreading.

The narrow rings are brightest when illuminated from behind—forward scattering—so they must contain large amounts of dust. The broader rings, however, seem to contain less dust. Because the ring particles are not very reflective, they are probably rocky rather than icy.

The outer narrow ring contains short arcs where the ring is denser. The ring arcs are difficult to explain because collisions among particles should smooth out the rings in relatively short times. However, a mathematical model of the rings shows that the small, inner moon Galatea creates waves in the ring that confine the particles in certain small arcs. Furthermore, the model

Neptune Data File Nine

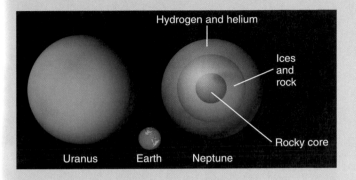

Neptune is a bit smaller than Uranus and four times larger in diameter than Earth. Its interior contains a rocky core with a deep mantle of slushy ices mixed with rock.

Average distance from the sun	30.0611 AU (44.971 × 10^8 km)
Eccentricity of orbit	0.0100
Maximum distance from the sun	30.4 AU (45.4 × 10^8 km)
Minimum distance from the sun	29.8 AU (44.52 × 10^8 km)
Inclination of orbit to ecliptic	1°46′27″
Average orbital velocity	5.43 km/s
Orbital period	164.793 y (60,189 days)
Period of rotation	16^{h}3^m
Inclination of equator to orbit	28°48′
Equatorial diameter	49,500 km (3.93 $D_\oplus$)
Mass	1.030 × 10^{26} kg (17.23 $M_\oplus$)
Average density	1.66 g/cm^3
Gravity	1.19 Earth gravities
Escape velocity	25 km/s (2.2 $V_\oplus$)
Temperature at cloud tops	−216°C (−357°F)
Albedo	0.35
Oblateness	0.027

a

b

Figure 8-22

(a) The brilliant disk of Neptune is hidden behind the central black bar in this Voyager 2 photo of the rings. Although the rings look uniform here, they actually contain short arcs where the ring particles are denser. *(NASA)* (b) This computer-enhanced image shows the arcs in Neptune's rings. A mathematical model reveals that the arcs are created by the gravitational influence of Neptune's moon Galatea. Arrows point to fainter arcs whose existence was predicted by the model. *(NASA image courtesy of Carolyn C. Porco)*

predicts the location of other fainter arcs, and computer-enhanced images confirm their existence (Figure 8-22b) and thus give us added confidence in the model.

The Moons of Neptune

Two moons are visible from Earth, and both have peculiar orbits. Nereid is a small moon in a large, elliptical orbit taking nearly an earth year to circle Neptune once.

Triton orbits Neptune in only 5.877 days, but it orbits backward—clockwise as seen from the north. These odd orbits suggest that the system was disturbed long ago in an interaction with some other body, such as a massive planetesimal.

Voyager 2 photographs revealed six more moons orbiting among the rings. They are all small, dark, rocky bodies. No additional moons were found beyond Triton, and some astronomers suggest that Triton in its retrograde orbit would have consumed such moons.

Although Triton is only 78 percent the radius of Earth's moon, it has an atmosphere of nitrogen and methane about 10^5 times less dense than Earth's. Triton can keep this gas because it is very cold so far from the sun—only 37 K ($-393°$F) (see Figure 7-14). In fact, a significant part of Triton is ice, and deposits of nitrogen frost are visible at the southern pole (Figure 8-23a), which has been turned toward sunlight for the last 30 years. This nitrogen appears to be vaporizing in the sunlight and is probably refreezing in the darkness at the north pole.

Many features on Triton suggest it has had an active past. It has few craters on its surface, but it does have long faults that appear to have formed when the icy crust broke. Some approximately round basins about 400 km in diameter appear to have been flooded time after time by liquids from the interior (Figure 8-23b). Most exciting of all are the dark smudges visible in the southern polar cap (Figure 8-23a). Analysis of the photos reveals that these are deposits produced when liquid nitrogen in the crust, warmed by the sun, erupts through vents and spews up to 8 km high into the atmosphere. Methane in the gas is converted by sunlight into dark deposits that fall to the surface, leaving the black smudges. These geysers make Triton one of only three worlds in our solar system observed to be active—Earth, Io, and Triton.

By counting craters on Triton, astronomers conclude that the surface has been active as recently as a million years ago and may still be active. The energy source for this volcanism could come from radioactive decay. This moon is two-thirds rock, and although such a small world would not be able to generate sufficient radioactive decay to keep molten rock flowing to its surface, frigid Triton may suffer from water–ammonia volcanism. A mixture of water and ammonia could melt at very low temperatures and resurface parts of the moon.

The History of Neptune

Our history of Neptune is quite short because many of its characteristics are understandable in terms of comparative planetology. On the other hand, its peculiarities are very difficult to understand because we have little data.

a

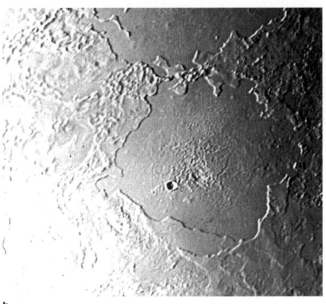

b

Figure 8-23
(a) Triton's southern polar cap is formed of nitrogen frost. Note the absence of craters and the dark smudges caused by nitrogen geysers. (b) Such round basins on Triton appear to have been repeatedly flooded by liquid from the interior. *(NASA)*

Neptune must have formed much as Uranus did, growing slowly in the outer solar nebula and never becoming massive enough to trap large amounts of hydrogen and helium. It developed a core of heavy elements, a mantle of dirty water, and a deep hydrogen-rich atmo-

sphere. It has an internal heat source provided, perhaps, by radioactive decay. The heat flowing to the surface generates its magnetic field and creates atmospheric circulation.

The moons of Neptune suggest some cataclysmic encounter long ago that put Nereid into a long-period elliptical orbit and put Triton into a retrograde orbit. We have seen evidence of major impacts throughout the solar system, so such interactions may have been fairly common. Certainly, impacts on the satellites could provide the debris that we find trapped among the smaller moons to form the rings.

REVIEW **Critical Inquiry**

Why is Neptune blue?

To analyze this question, we must be careful not to be misled by the words we use. When we look *at* something, we really turn our eyes toward it and receive light *from* the object. The light we receive from Neptune is sunlight that is scattered from various layers of Neptune and journeys to our eyes. Because sunlight contains a distribution of photons of all visible wavelengths, it looks white to us; but sunlight entering Neptune's atmosphere must pass through hydrogen gas that contains a small amount of methane, which is a good absorber of longer wavelengths. As a result, red photons are more likely to be absorbed than blue photons, and that makes the light bluer. Furthermore, when the light is scattered in deeper layers, the shorter-wavelength photons are most likely to be scattered, and thus the light that finally emerges from the atmosphere and reaches our telescopes is poor in longer wavelengths. It looks blue.

This discussion shows how a careful, step-by-step analysis of a natural process can help us better understand how nature works. For example, can you explain why the clouds on Neptune look white?

The Voyager 2 spacecraft left the solar system after visiting Neptune. It will not pass near any other worlds unless, millions of years from now, it enters some distant solar system. Nevertheless, we have one more planet to analyze.

8-5 Pluto

Pluto is a small icy world clearly different from the Jovian worlds yet also different from the terrestrial planets of rock and ice. Since its discovery in 1930, it has been a mystery on the edge of the solar system.

Pluto Data File Ten

The Hubble Space Telescope is just able to reveal bright and dark regions on Pluto. Some of the bright regions may be deposits of bright methane ice. *(Alan Stern, Southwest Research Institute, Marc Buie, Lowell Observatory, NASA, ESA)*

Average distance from the sun	39.44 AU (59.00 × 10⁸ km)
Eccentricity of orbit	0.2484
Maximum distance from the sun	49.24 AU (73.66 × 10⁸ km)
Minimum distance from the sun	29.64 AU (44.34 × 10⁸ km)
Inclination of orbit to ecliptic	17°9′3″
Average orbital velocity	4.73 km/s
Orbital period	247.7 y (90,465 days)
Period of rotation	$6^d9^h21^m$
Inclination of equator to orbit	122°
Equatorial diameter	2370 km (0.19 $D_\oplus$)
Mass	1.2×10^{22} kg (0.002 $M_\oplus$)
Average density	2.0 g/cm³
Surface gravity	0.06 Earth gravity
Escape velocity	1.2 km/s (0.11 $V_\oplus$)
Surface temperature	−230°C (−382°F)
Albedo	0.5

Pluto as a Planet

Pluto is very difficult to observe from Earth. Only a bit larger than 0.1 second of arc in diameter, the little planet is only 65 percent the diameter of Earth's moon and shows little surface detail. The best photos by the Hubble Space Telescope reveal large areas of light and dark terrain (Data File Ten), a mystery to be explored when spacecraft finally visit Pluto in the 21st century.

Most planetary orbits in our solar system are nearly circular; but Pluto's is quite elliptical (see Figure I-7). On the average, it is the most distant planet, but it can come closer to the sun than Neptune. In fact, starting January 21, 1979, Pluto moved closer to the sun than Neptune did, not returning outside Neptune's orbit until March 14, 1999. The planets will never collide, however, because Pluto's orbit is inclined 17°.

Orbiting so far from the sun, Pluto is cold enough to freeze most compounds we think of as gases, and spectroscopic observations have found evidence of ices of nitrogen, methane, carbon monoxide, and ethane. It also has a thin atmosphere of gaseous methane mixed with either nitrogen or carbon monoxide. There is no direct evidence of water ice on Pluto, but most astronomers believe there must be water frozen into the crust.

In 1978, Pluto's moon Charon was discovered in a highly inclined orbit with a period of 6.387 days and an average distance of about 19,640 km (Figure 8-24). The moon was named after the mythological ferryman who transports souls across the river Styx into the underworld.

From the size of the orbit and its period, astronomers could calculate the mass of Pluto, about 0.002 earth masses. Given its diameter of about 2370 km, this places its density at roughly 2 g/cm³, typical for a mixture of rock and ices. Charon appears to be about half the size of Pluto and significantly less dense at 1.8 g/cm³.

The History of Planet Pluto

Is Pluto really a planet? It is so dramatically different from its Jovian neighbors that it doesn't seem to fit into the outer solar system, and some have argued that it isn't a planet at all. In fact, this depends almost entirely on our arbitrary definition of the word "planet." Although the news media find this debate interesting, astronomers are more interested in the history of Pluto than in the definition of a word. A better question is, "How did Pluto form?"

An old theory proposed that Pluto was an escaped moon of Neptune. Pluto is much like Neptune's moon Triton. It is roughly the same density, and it is 85 percent the diameter of Triton; it could easily fit in with the icy moons of the Jovian planets. The disturbed orbits of Triton and Nereid could be the result of a complex interaction with a passing massive planetesimal

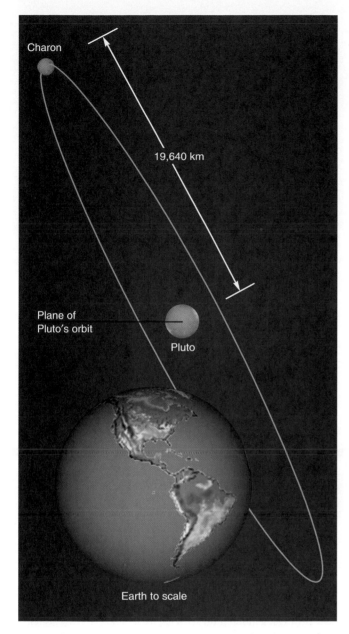

Figure 8-24
The circular orbit of Charon is seen here at an angle. The orbit is only a few times bigger than Earth and is tipped 118° to the plane of Pluto's orbit.

Charon

19,640 km

Plane of
Pluto's orbit

Pluto

Earth to scale

turned up over 100 icy bodies with hints that there are a vast number of more distant bodies. (We will discuss this belt of icy planetesimals in detail in Chapter 9.) If this is true, then Pluto is just a large, icy planetesimal that never became part of a larger planet.

The moon Charon is presumably another of these icy worlds. A collision long ago between Charon and Pluto could have tilted the axis of rotation of Pluto and trapped Charon in an elliptical orbit around Pluto. Tidal forces between Pluto and Charon could then have forced the orbit of Charon to become circular as it is today. If that happened, then we might hope to see dramatic evidence of tidal heating in Pluto and in Charon when spacecraft finally reach Pluto.

At present no spacecraft is heading for Pluto. Plans have been proposed, but no funds have been set aside. It may be a long time before we see the surface of these cold worlds, but our study of comparative planetology helps our imagination go where our cameras cannot.

REVIEW Critical Inquiry

What evidence do we have that cataclysmic impacts have occurred in our solar system?

We would expect objects that formed together from the solar nebula to rotate and revolve in the plane of the solar system, so the high inclination of Pluto and the orbit of Charon suggest a past collision or, at least, a close gravitational interaction with a passing body. And the Pluto-Charon system is not the only evidence for major impacts in the past. The peculiar orbits of Neptune's moons Triton and Nereid, the fractured and cratered state of the Jovian satellites, and the peculiar rotation of Uranus all hint that impacts and encounters with large planetesimals or the heads of comets have been important in the history of these worlds. Furthermore, the existence of planetary rings suggests that impacts have scattered small particles and replenished the ring systems. Even in the inner solar system, we see the backward rotation of Venus, the density of Mercury, and the formation of Earth's moon as possible consequences of major impacts when the solar system was young.

Certainly, we do not imagine that planets bounce around like billiard balls, but we must also recognize that the planets are not totally isolated now and may have been victims (or beneficiaries) of large impacts as they formed long ago. For example, how do the size and composition of Pluto and Charon compared with the moons of the outer Jovian worlds suggest an unexpected population of icy worlds in the outer solar system?

in which Pluto might have been ejected from an orbit around Neptune. This theory has been abandoned, however, because Pluto has a moon, and it doesn't seem possible for it to have kept a moon if it was once a moon itself. Also, a newer interpretation seems more probable.

Modern astronomers have begun thinking of Pluto as the largest of a family of small icy bodies that were once common in the outer solar system. Searches for other icy worlds beyond the orbit of Neptune have

Like clues at the scene of a crime, the evidence we find in the outer solar system alerts us that all of the planets in our solar system have been and still are sitting ducks for impacts by the smaller objects that hurtle along their orbits around the sun. What are these objects and what can they tell us about the origin and evolution of our solar system? We explore that in the next chapter.

Summary

The Jovian planets are so massive that they have been able to capture and retain large amounts of hydrogen and helium. This excess of light elements gives the Jovian planets densities much lower than those of the terrestrial planets. In their interiors, Jupiter and Saturn have such high pressures that the hydrogen becomes a liquid, and deep inside, it becomes liquid metallic hydrogen, a very good electrical conductor. Circulation in this rotating fluid drives the dynamo effect and produces the planets' magnetic fields, which trap particles from the solar wind to form radiation belts.

The atmosphere of Jupiter is marked by high- and low-pressure regions in the form of belts and zones that encircle the planet. Chemical reactions in the clouds are believed to produce compounds that color the belts. The Great Red Spot on Jupiter is believed to be a cyclonic disturbance much like an earthly hurricane.

Jupiter's ring is a tenuous sheet of dust particles orbiting in the planet's equatorial plane. Because certain processes rapidly remove dust from the ring, astronomers believe it must be continuously resupplied by new dust chipped from small satellites.

Jupiter's smaller moons may be captured asteroids, but its four largest, the Galilean moons, clearly formed with the planet. The outermost moon, Callisto, is an icy, dead world, and the next moon inward, Ganymede, shows grooves produced by ancient breaks in its icy crust. Europa is less rich in ice, but it does have an icy crust free of craters. Thus, its surface must be active. Io is very dry and has active volcanos powered by tidal heating.

Saturn's cold atmosphere has belt–zone circulation, though it is partially hidden below methane haze. Saturn's rings are made of icy particles shepherded by small satellites and broken into ringlets by spiral density waves. The rings can't be as old as the planet; they may need an occasional supply of fresh ice from impacts on Saturn's moons. Most of Saturn's moons are small, icy, dead worlds, but at least one, Enceladus, shows signs of an active surface.

Uranus and Neptune are smaller than Jupiter and Saturn and thus have lower internal pressures. They contain no liquid hydrogen, but they do have mantles of dirty water in which the dynamo effect can generate highly inclined magnetic fields.

Uranus rotates on its side, perhaps because it collided with a large planetesimal late in its formation. Its atmosphere was featureless when Voyager 2 flew past, but it appears to have active weather with clouds and storms as its long seasons pass. Uranus's narrow rings of dark boulders appear to be confined by shepherd satellites. Although it has many small moons, only five are large enough to be easily visible from Earth. Voyager 2 images revealed traces of past impacts. The innermost moon, Miranda, is marked by ovoids that suggest past convection in its icy mantle and crust.

Heat is flowing outward from the interior of Neptune, and that drives strong weather patterns in its atmosphere which produce hurricane-like storms and high methane clouds. Neptune's rings are thin hoops with the outermost ring containing denser arcs caused by a resonance with one of its smaller moons. Triton, its largest moon, is an icy world where methane–water volcanism may still be active.

Pluto and its moon Charon are small and made of a mixture of ice and rock. The origin of Pluto and Charon is unknown, but they are similar in many ways to Triton, and astronomers now speculate that such icy bodies may once have been common in the outer solar system.

New Terms

aurora	grooved terrain
magnetosphere	tidal heating
belt–zone circulation	oblateness
forward scattering	shepherd satellite
Roche limit	ovoid

Review Questions

1. Why is Jupiter so much richer in hydrogen and helium than Earth?

2. How can Jupiter have a liquid interior and not have a liquid surface?

3. How does the dynamo effect account for the magnetic fields of Jupiter and Saturn?

4. Why are the belts and zones on Saturn less distinct than those on Jupiter?

5. Why do we conclude that neither Jupiter's ring nor Saturn's rings can be left over from the formation of the planets?

6. How can a moon produce a gap in a planetary ring system?

7. Explain how geological activity on Jupiter's moons varies with distance from the planet.

8. What makes Saturn's F ring and the rings of Uranus and Neptune so narrow?

9. Why is the atmospheric activity of Uranus less than that of Saturn and Neptune?

10. Why do we suspect that Enceladus has been geologically active more recently than some other moons?

11. What are the seasons on Uranus like?

12. Why are Uranus and Neptune blue?

13. What evidence do we have that Triton has been geologically active recently?

14. How do astronomers account for the origin of Pluto?

15. What evidence do we have that catastrophic impacts have occurred in the solar system's past?

Discussion Questions

1. Some astronomers argue that Jupiter and Saturn are unusual, while other astronomers argue that all solar systems should contain one or two such giant planets. What do you think? Support your argument with evidence.

2. Why don't the terrestrial planets have rings? If you were to search for a ring among the terrestrial planets, where would you look first?

Problems

1. Seen from Earth, what is the maximum angular diameter of Jupiter? of Ganymede? of Jupiter's magnetosphere?

2. From the radius of Jupiter and its period of rotation on its axis, compute the velocity of its equator in meters per second.

3. The highest-speed winds on Jupiter are in the equatorial jet stream and travel about 150 m/s. What is this in miles per hour? How long do these winds take to circle Jupiter?

4. What is the angular diameter of Jupiter as seen from Io? How does that compare with Earth's moon in our sky?

5. Saturn's moon Enceladus has a radius of 250 km and a mass of 8.4×10^{19} kilograms. What is its density in grams per cubic centimeter? What does this tell us about its composition? (*Hint:* The volume of a sphere is $\frac{4}{3}\pi r^3$.)

6. Particles at the outer edge of Saturn's A ring travel at about 16.7 km/s. How long does it take for them to orbit Saturn?

7. Compared to the center of Saturn, what is the Doppler shift in nanometers for the particles in Problem 6 if they are observed at a wavelength of 500 nm?

8. If the ε ring around Uranus is 50 km wide and the orbital velocity of Uranus is 6.81 km/s, how long a blink should we expect to see when the ring crosses in front of a star?

9. If Pluto has a surface temperature of 50 K, at what wavelength will it radiate the most energy? (*Hint:* See By the Numbers 4-1.)

10. How long did radio commands take to travel from Earth to Voyager as it passed Neptune?

Critical Inquiries for the Web

1. If you lived on the surface of Pluto and looked into the sky to observe Charon, what phases would you see? (*Hint:* Be sure to consider your location on the planet when answering this question.)

2. What factors caused Voyager 2 to see a bland atmosphere when it encountered Uranus in 1986? Given these circumstances, would images of the Uranian atmosphere taken by a space probe arriving at Uranus in 2006 be similar to those taken in 1986, or would there be significant differences?

3. Imagine what you'd think if you had been the first person ever to see Saturn through a telescope. When Galileo first observed Saturn in 1610, he did not recognize that it was a ringed planet. It was many years later before the strange apparition of Saturn was finally attributed to a ring structure. Search the Web for information on historical observations of Saturn, summarize the observations of Galileo and others, and determine who was first to recognize what he saw as a ring. Why do you suppose it took so long to understand that Saturn is a ringed planet?

Exploring *The Sky*

1. Galileo was the first known person to see Jupiter's four brightest moons (often referred to as "Galilean moons"). Use *The Sky* to show you the appearance of Jupiter and these moons tonight as they would be seen through a small telescope.

   a. Make a properly scaled sketch of this Jovian system, and identify each moon by name. Find the angular distance between Jupiter and the moon that appears furthest from Jupiter.

   How to proceed: Center Jupiter on your screen using the **Find** command to open the Find window. Select Jupiter, and click on **Center & Frame**. To select the proper magnification that will enable you to see the Jovian system, use the Zoom buttons to adjust the diameter of Jupiter to roughly 1 cm. Click on each moon to find its name. To find the angular distance between Jupiter and a moon, click on Jupiter, then on the moon, and read the angular separation in the Object Information box (activated by clicking on an object).

   b. Make a sketch of the Jovian system two hours later—note the change in positions of the moons due their rotation about Jupiter.

   How to proceed: To change the time, click on **Data,** then on **Site Information,** and advance the time by two hours. Jupiter at that point will disappear from your screen, so you'll have to recenter it using the **Center & Frame** button you are already familiar with.

 Go to the Brooks/Cole Astronomy Resource Center (www. brookscole.com/astronomy) for critical thinking exercises, articles, and additional readings from InfoTrac College Edition, Brooks/Cole's online student library.

When they shall cry "PEACE,

PEACE" then cometh sudden

destruction!

COMET'S CHAOS?—What

terrible events will the Comet bring?

From a religious pamphlet predicting
the end of the world because of the
appearance of Comet Kohoutek, 1973

CHAPTER NINE

Meteorites, Asteroids, and Comets

Guidepost

In Chapter 5 we began our study of planetary astronomy by asking how our solar system formed. In the three chapters that followed, we surveyed the planets, but we found only limited insight into the origin of the solar system. The planets are big, and they have evolved as heat has flowed out of their interiors. In this chapter we have our best look at unevolved matter left over from the solar nebula. These small bodies are, in fact, the last remains of the nebula that give birth to the planets.

This chapter is unique in that it covers small bodies. In past chapters we have used the principles of comparative planetology to study large objects—the planets. In this chapter, we see that some principles also apply to the smaller bodies, but we also see that we need some new tools to think about the tiniest worlds in the solar system.

Figure 9-1
Comet Hyakutake swept through the inner solar system in 1996 and was dramatic in the northern sky. Seen here from Kitt Peak National Observatory, the comet passed close to the north celestial pole (behind the observatory dome in this photo). Notice the Big Dipper below and to the left of the head of the comet. *(Courtesy Tod Lauer)*

In 1910, Comet Halley was spectacular, and on the night of May 19, Earth actually passed through the tail of the comet—and millions of people panicked. The spectrographic discovery of cyanide gas in the tails of comets led many to believe that life on Earth would end. Householders in Chicago stuffed rags around doors and windows to keep out the gas, and bottled oxygen was sold out. Con artists in Texas sold comet pills and inhalers to ward off the noxious fumes. An Oklahoma newspaper reported (in what was apparently a hoax) that a religious sect tried to sacrifice a virgin to the comet.

Throughout history, bright comets have been seen as portents of doom. Even in our own time, the appearance of bright comets has generated predictions of the end of the world. Comet Kohoutek (1973), Comet Halley (1986), Comet Hyakutake (1996), and Comet Hale-Bopp (1997) caused concern among the superstitious. A bright comet moving slowly through the night sky is so out of the ordinary (Figure 9-1) that we should not be surprised if it generates some instinctive alarm.

In fact, comets are graceful and beautiful visitors to our skies. Astronomers think of comets as messengers from the age of planet building. By studying comets, we learn about the conditions in the solar nebula from which planets formed. But comets are only the icy remains of the solar nebula; the asteroids are the rocky debris left over from planet building. Because we cannot easily visit comets and asteroids, we begin our discussion with the fragments of those bodies that fall into our atmosphere—the meteorites.

9-1 Meteorites

We discussed meteorites in Chapter 5 when we discussed the age of the solar system. There we saw that the solar system is filled with small particles called meteoroids, which can fall into Earth's atmosphere at speeds of 10 to 40 km/s. Friction with the air heats the meteoroids to glowing, and we see them vaporize as

a

b

Figure 9-2
(a) An iron meteorite sliced open, polished, and etched with acid reveals a Widmanstätten pattern. Such patterns are evidence that the iron cooled from a molten state no faster than a few degrees per million years. (b) A stony meteorite sliced open and polished reveals small, spherical inclusions called chondrules. *(North Museum, Franklin and Marshall College)*

meteors streaking across the night sky. If a meteoroid is big enough and strong enough, it can survive its plunge through the atmosphere and reach Earth's surface. Once the object strikes Earth's surface, we refer to it as a meteorite. The largest of these objects can blast out craters on Earth's surface (see Figure 6-12), but such impacts are rare. The vast majority of meteorites are too small to form craters. We find these meteorites all over Earth, and their value is in what they can tell us about the origin of the planets.

Inside Meteorites

Meteorites can be divided into two broad categories: *Iron* meteorites are solid chunks of iron and nickel. *Stony* meteorites are silicate masses that resemble Earth rocks.

When iron meteorites are sliced open, polished, and etched with nitric acid, they reveal regular bands called **Widmanstätten patterns** (Figure 9-2a). The patterns arise from crystals of nickel–iron alloys that have grown very large, indicating that the meteorite cooled no faster than a few degrees per million years. Explaining how iron meteorites could have cooled so slowly will be a major step in analyzing their history.

In contrast to iron meteorites, most stony meteorites appear never to have been heated hot enough to melt. In fact, we can classify stony meteorites into three types, according to the degree to which they have been heated. **Chondrites** are stony meteorites that contain **chondrules,** rounded bits of glassy rock ranging in size from tiny beads to objects as large as a pea (Figure 9-2b). The origin of chondrules is unknown, but they appear

to be very old; and because melting would have destroyed them, their presence in chondrites indicates that these meteorites never melted.

Nevertheless, chondrites have been heated slightly. The rock in which the chondrules are imbedded is also old, but it contains no volatiles. The condensing solar nebula should have incorporated carbon compounds and water into the forming solids, but no such material is present in the chondrites. They have been heated enough to drive off these volatile compounds but not enough to melt the chondrules.

The **carbonaceous chondrites** contain both chondrules and volatile compounds. Had they been heated, even slightly, the volatiles would have evaporated. Thus the carbonaceous chondrites are the least altered remains of the solar nebula.

Stony meteorites of the third type are called **achondrites** because they contain no chondrules. They also lack volatiles and appear to have been subjected to intense heat that melted chondrules and drove off volatiles, leaving behind rock with compositions similar to Earth's lavas.

The Origin of Meteors and Meteorites

The mineral composition of the meteorites gives us a clue to the origins of these objects. Somehow they have survived almost unchanged since the age of planet building in the solar nebula.

We can find evidence of the origin of meteors through one of the most pleasant observations in astronomy. We can observe a **meteor shower,** a display of meteors that are clearly related in a common origin. For

We can enjoy a meteor shower as Mother Nature's fireworks, but our enjoyment is much greater once we begin to understand what causes meteors and why meteors in a shower follow a pattern. When we know that meteor showers help us understand the origin of our world, the evening display of shooting stars is even more exciting. Science typically increases our enjoyment of the natural world by revealing the significance of things we might otherwise enjoy only in a casual way.

Everyone likes flowers, for example. We enjoy them in a casual way, admiring their colors and fragrances, their graceful petals and massed blooms. But botanists know that nature has carefully designed flowers to attract insects and spread pollen. The bright colors attract insects, and the shapes of the flowers provide little runways so the insect will find it easy to land. Some color patterns even guide the insect in like landing lights at an airport, and many flowers, such as orchids and snapdragons, force the insect to crawl inside in just the right way to exchange pollen and fertilize the flower. The nectar is nothing more than bug bait. Once we begin to understand what flowers are for, a visit to a garden becomes not only an adventure of color and fragrance, but also an adventure in understanding the beauty of design and function, an adventure in meaning as well as an adventure of the senses.

And our understanding of a natural phenomenon helps us understand and enjoy related phenomena. For example, some flowers attract flies for pollination, and such blossoms smell like rotting meat. A few flowers depend on bats, and they open their blossoms at night. Many plants, such as pine trees, depend on the wind to spread their pollen, and those plants do not have colorful flowers at all; but flowers that depend on hummingbirds have long trumpet-shaped blossoms that just fit the hummingbirds' beaks.

The more science tells us about nature, the more enjoyable the natural world is. We can enjoy a meteor shower as nothing more than a visual treat, but the more we know about it, the more interesting and exciting it becomes. The natural world is filled with meaning, and science, as a way of discovering and understanding that meaning, gives us new opportunities to enjoy the world around us.

---

example, the Perseid meteor shower occurs each year in August (Table 9-1), and on that night we might see as many as 40 meteors an hour if we stretched out on a lawn chair at a dark site and watched the sky. Like many natural phenomena, a meteor shower is more enjoyable when we know more about it (Window on Science 9-1).

On any clear, moonless night of the year we might see 5 to 15 meteors per hour, but these are not related to each other. During a meteor shower, we see meteors that are related to each other in that they seem to come from a single spot on the sky. The Perseid shower, for example, appears to come from a spot in the constellation Perseus. These showers are known to occur when Earth passes near the orbit of a comet, and thus the meteors in meteor showers must be caused by dust and debris released by the icy head of the comet. The meteors appear to come from a single place in the sky because they are produced by particles traveling along parallel paths through space. Like railroad tracks approaching from a point on the horizon, the meteors

Table 9-1 Meteor Showers

| Shower | Dates | Hourly Rate | Radiant* | | Associated Comet |
			R. A.	Dec.	
Quadrantids	Jan. 2–4	30	15^h24^m	50°	
Lyrids	April 20–22	8	18^h4^m	33°	1861 I
η Aquarids	May 2–7	10	22^h24^m	0°	Halley?
δ Aquarids	July 26–31	15	22^h36^m	−10°	
Perseids	Aug. 10–14	40	3^h4^m	58°	1982 III
Orionids	Oct. 18–23	15	6^h20^m	15°	Halley?
Taurids	Nov. 1–7	8	3^h40^m	17°	Encke
Leonids	Nov. 14–19	6	10^h12^m	22°	1866 I Temp
Geminids	Dec. 10–13	50	7^h28^m	32°	

*R. A. and Dec. give the celestial coordinates (right ascension and declination) of the radiant of each shower.

appear to approach from a point in space (Figure 9-3). The orbits of comets are filled with such debris. For example, the telescope aboard the Infrared Astronomy Satellite detected the dusty orbits of a number of comets glowing in the far infrared because of the sun-warmed dust scattered along the orbits.

Studies of meteors show that most of the meteors we see on any given night (whether or not there is a shower) are produced by tiny bits of debris from comets. These specks of matter are so small and so weak they vaporize completely in the atmosphere and never reach the ground, but from their motions astronomers can deduce their orbits, and those orbits match the orbits of comets. Thus nearly all of the meteors come from comets.

Meteorites that reach Earth's surface are structurally much stronger than comet dust. They appear to have had a more complex history, and the different kinds of meteorites suggest that they are fragments of larger bodies that grew hot enough to melt and differentiate to form a molten iron core and a rocky mantle and crust (Figure 9-4).

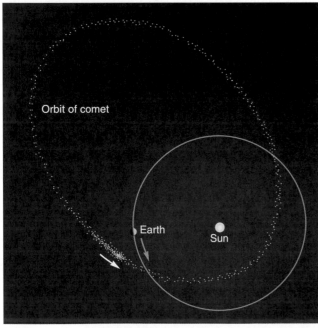

Figure 9-3
(a) Meteors in a meteor shower enter Earth's atmosphere along parallel paths, but perspective makes them appear to diverge from a radiant point in the sky. (b) Similarly, parallel railroad tracks appear to diverge from a point on the horizon. (c) Meteors in a shower are debris left behind as a comet's icy nucleus vaporizes. The rocky and metallic bits of matter spread along the comet's orbit. If Earth passes through such material, we see a meteor shower.

Figure 9-4
If a planetesimal were large enough, it could melt and differentiate (a) into an iron-rich core surrounded by layers of different silicate compositions (b). Impacts (c, d, e) could break up these layers and produce various kinds of meteorites. Presumably, these remaining planetesimals are the asteroids. *(Adapted from a diagram by C. R. Chapman)*

If an object were 100 km in diameter, then the outer layers of rock would insulate the iron core. It would lose heat slowly, and the iron would cool so gradually that large crystals could grow. A sample of such material sliced, polished, and etched would show Widmanstätten patterns. Thus the material in iron meteorites could have formed in the cores of such bodies.

Collisions could break up such an object, and fragments from the core would look like the iron meteorites. Fragments of the object's mantle, which was hot enough to melt chondrules, would make achondrites. Also, some achondrites appear to be pieces of lavas that flowed out onto the surfaces of larger bodies. Fragments from a crust that has never been hot enough to melt the chondrules might look very much like the meteorites called chondrites.

The carbonaceous chondrites may have formed in bodies that were smaller and cooler. Planetesimals farther from the sun would have been cooler and could have retained volatiles more easily. They could not have been large, or heating and differentiation would have altered them.

These theories trace the origin of meteorites to planetesimal-like parent bodies, but they produce a mystery. The small meteorites in the solar system cannot be fragments of the planetesimals that formed the planets because small meteorites would have been swept up by the planets in only a billion years or less. They could not have survived for 4.6 billion years.

In fact, when astronomers study the orbits of objects seen to fall to Earth as meteorites, those orbits lead back into the asteroid belt. Thus we have good evidence to believe that the meteorites now in museums all over the world must have broken off of asteroids within the last billion years. Although nearly all meteors are pieces of comets, the meteorites are pieces of asteroids.

REVIEW Critical Inquiry

How can we say that meteors come from comets, but meteorites come from asteroids?

A selection effect can determine what we notice when we observe nature, and a strong selection effect prevents us from finding meteorites that originated in comets. Cometary particles are physically weak, and they vaporize in our atmosphere easily. Thus, very few ever reach the ground, and we are unlikely to find them. Furthermore, even if a cometary particle reached the ground, it would be so fragile that it would weather away rapidly, and, again, we would be unlikely to find it. Asteroidal particles, however, are made from rock and metal and so are stronger. They are more likely to survive their plunge through the atmosphere and more likely to survive erosion on the ground. Meteors from the asteroid belt are rare. Almost all of the meteors we see come from comets, but not a single meteorite is known to be cometary.

The meteorites are valuable because they give us hints about the process of planet building in the solar nebula. What evidence do we find in the meteorites that tells us they were once part of larger bodies broken up by impacts?

The vast majority of meteorites must have originated in the asteroids. Our next goal is to try to understand the nature and origin of these tiny rocky worlds.

9-2 Asteroids

Space pirates lurk in the asteroid belt in old-time pulp science fiction, but astronomers have found that there isn't much in the asteroid belt for a pirate to stand on. Thousands of asteroids are known, but most are quite small; and given its vast size, the asteroid belt between Mars and Jupiter is mostly empty.

Nevertheless, we must consider the nature of the asteroids. In Chapter 5, we identified the asteroids as a characteristic of the solar system and concluded that they were the last remains of material that was unable to form a planet between Mars and Jupiter. We are now ready to examine these small worlds in more detail.

The Geology of Asteroids

Asteroids are distant objects too small to study in detail with Earth-based telescopes. Yet astronomers have learned a surprising amount about these little worlds, and spacecraft have given us a few close-ups.

From the infrared radiation emitted by asteroids, astronomers can calculate their sizes. Ceres, the largest, is about 30 percent the diameter of our moon, and Pallas, next largest, is only 15 percent the diameter of the moon. Most are much smaller.

Astronomers long suspected that asteroids were not spherical, because the brightness of a typical asteroid varies over periods of hours. Apparently asteroids are irregular in shape, and, as they rotate, they reflect varying amounts of sunlight. Only the largest asteroids have enough gravity to overcome the strength of the rock they are made from and pull themselves into spherical shape.

The irregular shape of asteroids has been confirmed by photographs taken by spacecraft flying past asteroids (Figure 9-5a). All of the asteroids photographed so far are irregular chunks of rock battered by craters. From the gravitational effects of an asteroid on the motion of a passing spacecraft, astronomers can compute the masses of the asteroid, and, dividing by its volume, find its density. This has been done for only a few asteroids, but the results are surprising. Although asteroids are clearly rocky, they have low densities. Mathilde, for example, has a density of just 1.3 g/cm³. It can't be solid rock, but must be a collection of fragments only loosely held together by their mutual gravity.

In fact, some asteroids appear to be double objects. Radar observations of the asteroid Toutatis reveal that it is made up of two objects (Figure 9-5b). Whether they orbit each other, are in contact with each other, or are fused into a single object is unknown. The asteroid Castalia is also a double asteroid.

Some asteroids have moons of their own. When the Galileo spacecraft on its way to Jupiter flew past the asteroid Ida, it took photos that revealed that the 52-km-long asteroid is orbited by a small moon, Dactyl, only 1.5 km in diameter. Photographs made from Earth reveal that the asteroid Eugenia has a small moon orbiting around it. Fluctuations in brightness suggest that the asteroid Dionysus has a moon, and other asteroids are suspected of having moons.

Irregular shapes, craters, fragmented interiors, double asteroids, and asteroids with moons all serve to assure us that the asteroids have collided with each other often throughout their history. It should not surprise us that the inner solar system is filled with meteoroids from the asteroid belt. Many of the meteoroids are debris from impacts among asteroids.

Not all asteroids lie in the asteroid belt. Some 2000 objects larger than 1 km are believed to follow orbits that cross Earth's orbit. Although a number of searches are under way, only about 200 of these asteroids have been found. For example, LONEOS, Lowell Observatory Near Earth Object Search, will be able to survey the entire sky visible from Lowell Observatory three times a month and should be able to locate 1000 near-Earth asteroids over the next 10 years. A number of other projects are searching for these asteroids, not only because we want to understand the asteroids better, but because such asteroids must collide with the Earth occasionally. Although such collisions occur very rarely, a single impact could cause planetwide devastation. We will discuss such catastrophic impacts on Earth later in this chapter.

The color and spectra of asteroids help us understand their composition. From their bright, reddish colors, astronomers classify some asteroids as S types (Figure 9-6). These may be silicates mixed with metals, or

a

b

Figure 9-5
Asteroids Mathilde, Gaspra, and Ida (left to right) are reproduced here to scale from separate photographs taken by spacecraft. Mathilde is 59 km in diameter. The brightness of the three images has been adjusted. Mathilde is actually much darker than the other two asteroids. Ida, at the right, has a small moon not shown in this mosaic. *(Johns Hopkins Applied Physics Lab and NASA)* (b) A radar image of asteroid Toutatis reveals it is two cratered bodies in contact. *(Steven J. Ostro)*

they may resemble chondrite meteorites. C-type asteroids are very dark—about as bright as a lump of coal. They appear to be carbonaceous. M-type asteroids, bright but not red, appear to be mostly iron–nickel alloys. S types are common in the inner belt, and C types are common in the outer belt. That distribution is a clue to the origin of the asteroids.

Some asteroids have clearly evolved since they formed. Vesta, with a diameter of 525 km, is too small to image from Earth's surface, but Hubble Space Telescope images show that its surface is patchy (Figure 9-7), and spectra of these regions suggest solidified lavas. Furthermore, the best images reveal a giant crater 87 percent of Vesta's diameter. A family of small asteroids appear to be fragments from Vesta, and a certain class of meteorites found on Earth have been identified spectroscopically as being from Vesta. Those meteorites are basalts typical of lava flows (Figure 9-7). Thus there is strong evidence that Vesta has been at least partly

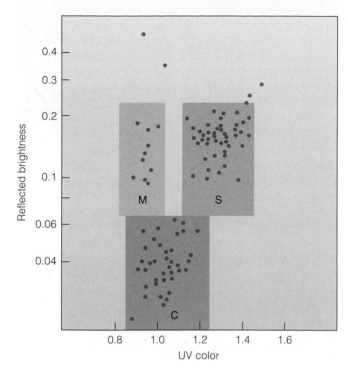

Figure 9-6

The three principal types of asteroids. In a diagram of albedo versus color (ultraviolet minus visual), the most reflective asteroids lie near the top, and the reddest at the right. The S types, which are mixed silicates and metals or perhaps chondrites, are clearly redder than the M types, which seem to be metallic. The darkest asteroids, the C types, are believed to be similar to carbonaceous chondrites. *(Diagram adapted from a figure by B. Zellner)*

resurfaced by lava flowing up from its interior. We must consider the evolution of asteroids as we think about their origin.

The Origin of the Asteroids

An old theory proposed that asteroids are the remains of a planet that exploded. Planet-shattering death rays may make for exciting science fiction movies, but in reality planets do not explode. The gravitational field of a planet holds the mass tightly, and disrupting the planet would take tremendous energy. Shattering Earth, for example, would take all of the energy generated by the sun over a period of two weeks. In addition, the total mass of the asteroids is only about one-fifth the mass of the moon, hardly enough to be the remains of a planet.

Most astronomers now believe that the asteroids are the remains of material that was unable to form a planet at 2.8 AU because of the gravitational influence of Jupiter, the next planet outward. If this is true, then the asteroids are the remains of planetesimals fragmented by collisions with one another. This would explain why the C-type asteroids, which appear to be carbonaceous, are more common in the outer asteroid belt. It is cooler there, and the condensation sequence (see Chapter 5) predicts that carbonaceous material would form there more easily than in the inner belt.

Now we can understand the compositions of the meteorites. They are fragments of planetesimals, the largest of which developed molten cores, differentiated,

Figure 9-7

A Hubble Space Telescope image of the asteroid Vesta (upper left) shows a complex surface with a large crater near the south pole. From 78 such images, astronomers constructed a model of the asteroid showing the impact crater and its central peak (upper right). Detailed studies of the model produce the elevation map (bottom) showing that the crater is up to 13 km deep. (Inset) A meteorite identified spectroscopically as a fragment of Vesta. Chemical and crystalline structure show that it was once part of a molten lava flow. Presumably it was blasted off of Vesta by a large impact. *(Ben Zellner, Georgia Southern University; Peter Thomas, Cornell University; NASA; inset: R. Kempton, New England Meteoritical Services)* may have suffered lava flows on their surfaces, and then cooled slowly. The largest asteroids we see today may be nearly unbroken planetesimals, but the rest are just the fragments produced by 4.6 billion years of collisions.

REVIEW   Critical Inquiry

What is the evidence that asteroids have been fragmented?

First, we might note that the solar nebula theory of the formation of the solar system predicts that planetesimals collided and either stuck together or fragmented. This is suggestive, but it is not evidence. A theory can never be used as evidence to support some other theory or hypothesis. Evidence means observations or the results of experiments, so we must turn to observations of asteroids. Spacecraft photos of asteroids such as Ida, Gaspra, and Mathilde reveal irregularly shaped worlds scarred by impact craters, and the radar images obtained so far of asteroids also show irregularly shaped bodies that appear to be the fragments of larger bodies. In fact, radar images of Castalia and Toutatis show what may be pairs of bodies in contact, and the Galileo image of Ida reveals its small satellite, Dactyl. Furthermore, meteorites appear to come from the asteroid belt, and a few have been linked to specific asteroids such as Vesta. All of this evidence suggests that the asteroids have suffered violent collisions in their history.

The impact fragmentation of asteroids has been important, but it has not erased all traces of the original planetesimals from which the asteroids formed. What evidence do we have to tell us what those planetesimals were like?

The asteroids are only the last crumbs left behind by failed planet building near Jupiter. Farther out in the solar system are other remains—ice shards from the age of planet formation.

9-3 Comets

Of all the fossils left behind by the solar nebula, comets are the most beautiful. Asteroids are dark, rocky worlds, and meteors are flitting specks of fire, but the comets move with the grace and beauty of a great ship at sea (Figure 9-8). Comet Hale-Bopp, for example, was visible for weeks in 1996 and 1997. Scientifically, comets are interesting because observations of the tail and head of a comet tell us about the small icy nucleus and about the ancient solar nebula from which we inherit our wealth of comets.

a

b

Figure 9-8
(a) Comet Ikeya-Seki was beautiful in the dawn sky in 1965. The long, graceful tail sprang from a nucleus of ice only a dozen kilometers in diameter. *(Stephen M. Larson)* (b) Comet Halley, like all comets, looked white to the eye, but computer enhancement and false-color graphics are often used to reveal subtle detail. This image from 1910 has been digitally enhanced to reveal subtle differences in intensity. *(NOAO)*

a

b

Figure 9-9
(a) The parts of a comet include an extended coma, or head, of gas and dust surrounding an icy nucleus only a few kilometers in diameter. Gas is carried away by the solar wind in a wispy gas tail, but dust moves more slowly and forms a curving dust tail. (b) Comet Mrkos in 1957 had pronounced gas and dust tails, which changed dramatically from night to night. *(Caltech)*

Properties of Comets

The head, or **coma,** of a comet is a vast cloud of gas and dust up to 1,000,000 km in diameter—nearly as large as the sun (Figure 9-9). The gas in the coma is made up of H_2O, CO_2, CO, H, OH, O, S, C, and so on. Mixed with the gases of the coma are tiny specks of dust, most smaller than the particles in cigarette smoke. The dirty snowball model, described in Chapter 5, proposes that all of the features of the coma are created by a central lump of dirty ices only a dozen kilometers in diameter. As the ices vaporize, dust is released and larger molecules of gas are broken into the smaller fragments we see in the coma.

The tail of a comet springs from the coma and typically extends 10^7–10^8 km through space, with the longest reaching 1 AU (1.5×10^8 km). Seen from Earth, the tail of a large comet can span 30° across the sky.

A comet can have two kinds of tails (Figure 9-10). A type I tail, also called a *gas tail*, is straight and wispy and looks like a narrow trail of smoke. Its spectrum contains emission lines of ionized gases excited by the

Figure 9-10
Comet Hale-Bopp was brilliant in the sky in early 1997. With an unusually large nucleus estimated to be 40 km in diameter, it produced large amounts of gas and dust. As the gas became ionized, it was carried away rapidly by the solar wind to form the straight, blue type I tail. The dust was less affected by the solar wind. It was pushed outward by the pressure of sunlight and produced the broad, curving type II tail. *(Dean Ketelsen)*

ultraviolet radiation from the sun and blown outward by the moving gas and the embedded magnetic field in the solar wind.

The ionized gases in the gas tail interact with the magnetic field in the solar wind to give the tail its long, thin shape. When a space probe named ICE plunged through the tail of a comet, it found the magnetic field draped over the nucleus like seaweed over a fishhook. Disturbances in the solar wind (caused, perhaps, by activity on the sun) can even produce kinks in the comet's tail.

The type II tail, also known as a *dust tail,* is immune to magnetic fields. Its spectrum is identical to sunlight's, which means it is made up of solid bits of dust that reflect sunlight. The dust is evidently the dirt in the dirty snowball released as the ices vaporize. The Infrared Astronomy Satellite found that this dust fills the solar system, and it can be collected by spacecraft and high-flying aircraft (Figure 9-11).

The dust in the dust tail is not affected by magnetic fields but is pushed outward by the pressure of sunlight. It thus forms a long, curving tail with a much more uniform appearance than a gas tail. The dust tail curves because the dust, once released by the comet, is pushed outward by sunlight, and, obeying Kepler's laws, its orbital velocity decreases. As the dust particles fall behind the head of the comet, the dust tail curves. A dust tail lacks the wisps and twists of a gas tail, because it is not affected by kinks and shifts in the magnetic field of the solar wind. A comet recedes from the sun with its gas and dust tails leading the way.

The Geology of Comet Nuclei

The nucleus of a comet is quite small and cannot be resolved by any telescopes now in use. Nevertheless, astronomers are beginning to understand the geology of these peculiar worlds.

Of course, the nuclei of comets must contain water, but they must also contain compounds such as ammonia and carbon dioxide trapped in the ice. These ices sublime—change from a solid directly into a vapor—to produce the observed tails. Telescopes in space can detect vast clouds of hydrogen gas surrounding the heads of active comets (Figure 9-12a)—hydrogen atoms freed when molecules of water are broken up by sunlight. All of the gases that comets release must come from the volatile ices in the nuclei.

Comet Halley visited the inner solar system in 1985 and 1986, and five spacecraft flew past the nucleus. The photos sent back show a dark, irregularly shaped body venting gas and dust from active regions on its sunward side (Figure 9-12b). The surface is so dark it reflects only 4 percent of the light that hits it; a sooty

a

b

Figure 9-11
(a) Comet Halley is a typical comet in that much of the light we receive from its tail is reflected by microscopic dust particles released from the head of the comet. *(NOAO)* (b) NASA aircraft flying at an altitude of 18 km (59,000 ft) can collect dust falling into Earth's atmosphere from space. This dust speck is only 5 by 6 micrometers. Comets continually release such dust throughout the inner solar system, and it can be detected by infrared telescopes in Earth orbit. *(NASA)*

a

b

c

Figure 9-12

(a) An ultraviolet image of Comet Hale-Bopp reveals a vast cloud of hydrogen gas surrounding the visible comet (inset). Concentric bands in the cloud are produced by the digitizing process. At lower right is the sun to scale. *(SOHO/SWAN Consortium, ESA, and NASA)* (b) This image of the nucleus of Comet Halley taken by the Glotto spacecraft shows jets of dust and gas venting from the dark nucleus. *(©1986 Max-Planck Institute)* (c) In the inner coma of Comet Halley, jets from the nucleus are bent back to form the tail. The black dot is produced by the imaging process. The actual nucleus is too small to detect. *(Steven Larson)*

lump of coal reflects 6 percent. The dark color of this dust suggests the composition of the meteorites called carbonaceous chondrites.

From the gravitational effects the nucleus had on the spacecraft, astronomers can find the mass of the nucleus and thus its density, 0.1 to 0.25 g/cm^3, much less than the density of ice. From these observations we can conclude that the nucleus is not a solid ball of ice but must be a fluffy mixture of ices with a dark, dusty crust.

Photographs of the comas of comets often show jets springing from the nucleus and being swept back by the pressure of sunlight and by the solar wind to form the tail (Figure 9-12c). Studies of the motions of these jets as the nucleus rotates and the photographs of the nucleus of Comet Halley reveal that the jets originate from active regions that may be faults or vents. As the

rotation of a cometary nucleus carries an active region into sunlight, it begins venting gas and dust, and as it rotates into darkness it shuts down.

The irregular shape of the nucleus of Comet Halley suggests that the crust is a fluffy layer of dust left behind by the vaporizing ices. Most of the ice is hidden below this dark crust, and, as it sublimes from ice to vapor, the crust collapses into the resulting voids. Some regions of cometary nuclei seem extra rich in highly volatile ices, so the nuclei are evidently not uniform mixtures.

The nucleus of Comet Halley was observed to be 16 by 8 by 7 km, and the nucleus of Comet Hale-Bopp is believed to be about 40 km in diameter. Each passage near the sun costs such an object many millions of tons of ices, so the nuclei slowly loose their ices until there is nothing left but dust and rock.

The Origin of Comets

Family relationships among the comets give us clues to their origin. Most comets have long, elliptical orbits with periods greater than 200 years. These are known as long-period comets. Their orbits are randomly inclined with comets falling into the inner solar system from all directions. As many circle the sun clockwise as counterclockwise.

In contrast, about 100 or so of the 600 well-studied comets have orbits with periods less than 200 years. These short-period comets follow orbits that lie within 30° of the plane of the solar system, and most revolve around the sun counterclockwise—the same direction the planets orbit. Comet Halley, with a period of 76 years, is a short-period comet.

Comets cannot survive long before the heat of the sun vaporizes their ices and reduces them to inactive bodies of rock and dust. A comet may last only 100 to 1000 orbits around the sun. The comets we see in our skies can't have survived 4.6 billion years since the formation of the solar system, so there must be a continuous supply of new comets. Where do they come from?

In the 1950s, Dutch astronomer Jan Oort proposed that the long-period comets are objects that fall in from the **Oort cloud,** a spherical cloud of icy bodies believed to extend from 10,000 to 100,000 AU from the sun (Figure 9-13). Astronomers estimate that the cloud contains several trillion icy bodies. Far from the sun, they are very cold, lack comas and tails, and are invisible. The gravitational influence of occasional passing stars could perturb a few of these objects to fall into the inner solar system, where the heat of the sun warms their ices and transforms them into comets. Because the Oort cloud is spherical, these long-period comets fall inward from random directions.

The short-period comets, however, require a different explanation. In 1951, Dutch-American astronomer Gerard P. Kuiper proposed that the formation of the solar system should have left behind a belt of small, icy planetesimals beyond the Jovian planets and in the plane of the solar system. Known today as the **Kuiper belt,** this disk extends from 30 AU to as much as 100 AU from the sun. The entire Kuiper belt would be hidden in Figure 9-13 behind the yellow dot symbolizing the solar system. The short-period comets are visitors from this Kuiper belt.

Although the Kuiper belt was originally proposed as a theory, astronomers are now able to observe some of the Kuiper belt objects (Figure 9-14). Searches for small icy bodies in the outer solar system have so far identified over 100 objects that are members of the Kuiper belt. Also, the physical and chemical similari-

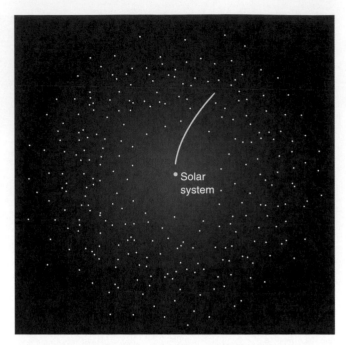

Figure 9-13
The long-period comets appear to originate in the Oort cloud, a spherical volume of icy bodies between 10,000 and 100,000 AU from the sun. Objects that fall into the inner solar system from this cloud arrive from all directions.

ties among the icy moons of the outer solar system such as Triton and Charon suggest these are large Kuiper belt objects that passed too close to planets and were captured as moons. In fact, astronomers commonly think of Pluto not as a tiny planet but as the largest of the remaining Kuiper belt objects. Working at the extreme limit of its capability, the Hubble Space Telescope has produced statistical evidence of a population of Kuiper belt objects roughly the size of the nucleus of Comet Halley.

We can find more evidence that our solar system has a Kuiper belt by looking at other stars. A number of nearby stars such as Beta Pictoris (Chapter 5) are surrounded by disks of dust believed to be released by comets in the equivalent of Kuiper belts around these stars. Thus the detection of the dust implies the presence of Kuiper belts.

If comets are icy planetesimals from the Oort cloud and the Kuiper belt, how did those planetesimals form? The Kuiper-belt objects appear to have formed as icy planetesimals in the outer solar nebula not much farther from the sun than the outer planets. But the objects in the Oort cloud lie much farther from the sun, and they can't have formed out there. The solar nebula would have been too tenuous at such great distances. Also, we would expect objects that formed from the nebula to be confined to a disk and not distributed in a sphere. Rather, astronomers think the objects now in the Oort cloud formed in the outer solar system near

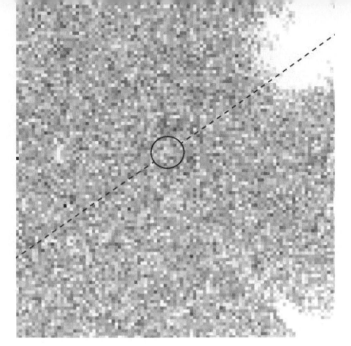

Figure 9-14

A possible Kuiper-belt object lies in the circle in this exposure with the Hubble Space Telescope. The object, probably an icy body a few kilometers in diameter, was found following an orbit (dashed line) typical of a Kuiper-belt object. Hubble Space Telescope and ground-based observations have located dozens of such objects. *(A. Cochran, University of Texas, and NASA)*

the present orbits of Saturn and Uranus. As the Jovian planets grew more massive, they swept up some of these planetesimals and ejected others to form the Oort cloud. If this idea is true, then even the long-period comets are icy planetesimals.

Comet Impacts on Earth

For centuries, superstitious people have associated comets with doom, and that seems silly. Comets are graceful visitors from the icy fringes of the solar system. Of course, comets must hit planets now and then, so we might wonder just how dangerous such impacts would be.

Earthlings watched in awe during the summer of 1994 as the fragmented head of a comet (Figure 9-15a) slammed into Jupiter and produced impacts equaling millions of megatons of TNT (Figure 8-5). Such impacts on Jupiter probably occur every century or so, and we might expect similar impacts on Earth, smaller and with less gravitational power, to occur much less often. Nevertheless, these impacts do occur, and we can find chains of craters on the moon that seem to have been formed by fragmented comets (Figure 9-15b).

a

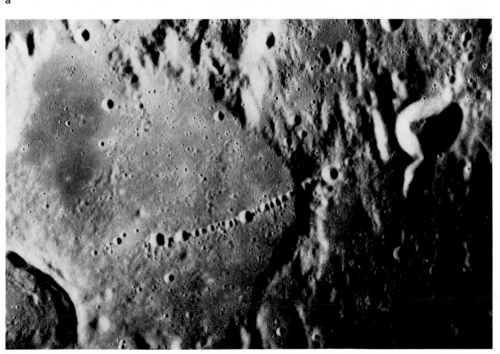

b

Figure 9-15

(a) The heart of Comet Shoemaker-Levy 9 came within the Roche limit of Jupiter in July 1992, and tidal forces pulled it apart. At least 21 fragments slowly spread apart as they looped out into space and eventually returned to hit Jupiter two years later. (See Figure 8-5.) (b) This 40-km-long chain of craters on the moon may have been formed by fragments from the nucleus of a comet ripped to pieces by Earth's tidal forces. Unlike the impacts that occurred on Jupiter, the impacts that made these craters probably happened within a few seconds. *(NASA)*

Figure 9-16
(a) The Barringer Meteorite Crater (near Flagstaff, Arizona) is nearly a mile in diameter and was formed about 50,000 years ago by the impact of an iron meteorite roughly 90 m in diameter. It hit with energy equivalent to that of a 3-megaton hydrogen bomb. Notice the raised and deformed rock strata all around the crater. For scale, locate the brick building on the far rim at right. *(M. A. Seeds)* (b) Like all larger impact features, the Barringer Meteorite Crater has a raised rim and scattered ejecta. *(USGS)*

Small meteorite impacts occur quite often, and a building is damaged by a falling meteorite every few years. Larger impacts are less common, and truly large impacts are rare. Earth is marked by many dozens of meteorite craters that illustrate the power of an impacting object (Figure 9-16 and Figure 6-12). A large impact could have devastating consequences.

Recent studies of sediments laid down all over the world 65 million years ago at the time of the extinction of the dinosaurs have found overabundances of the element iridium—common in meteorites but rare in Earth's crust. This finding suggests that the impact of a large meteorite or comet may have altered the atmosphere and climate on Earth so dramatically that the dinosaurs and over 75 percent of the species then on Earth became extinct.

Mathematical models and observations of the impact of the comet fragments on Jupiter in 1994 have combined to create a plausible scenario of the events following a major impact on Earth. Of course, creatures living near the site of the impact would probably die in the initial shock, and an impact at sea would create tsunamis (tidal waves) many hundreds of meters high that would devastate coastal regions for many kilome-

ters inland halfway around the world. But the worst effects would begin after the initial explosion. On land or sea, a major impact would excavate large amounts of pulverized rock, heat it to high temperatures, and loft it high above the atmosphere. As this material fell back, Earth's atmosphere would be turned into a glowing oven of red-hot rock streaming through the air as a rain of meteors, and the heat would trigger massive forest fires around the world. Soot from such fires has been detected in the layers of clay laid down at the end of the Cretaceous. Once the fire storms cooled, the remaining dust in the atmosphere would block sunlight and produce deep darkness for a year or more. At the same time, large amounts of carbon dioxide locked in limestone deposits would be released into the atmosphere, where it would produce acid rain. All of these consequences make it surprising that any life could have survived such an impact.

Geologists have located a crater at least 150 km in diameter centered near the village of Chicxulub in the northern Yucatán (Figure 9-17). Although the crater is totally covered by sediments, mineral samples show that it contains shocked quartz typical of impact sites

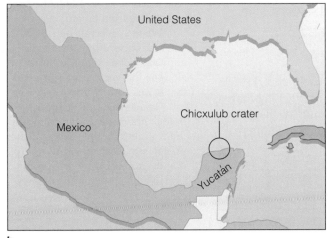

Figure 9-17
(a) The theory that the impact of one or more comets altered Earth's climate and drove dinosaurs to extinction has become so popular it appeared on this Hungarian stamp. The spacecraft shown (ICE) flew through the tail of Comet Giacobini-Zinner in 1985. Note the dead dinosaurs in the background. (b) The giant impact scar buried in Earth's crust near the village of Chicxulub in the northern Yucatán was formed about 65 million years ago. Such a large impact would have altered Earth's climate dramatically, and thus the Chicxulub impact is suspected of being the trigger that ended the Cretaceous period and destroyed the dinosaurs.

a

b

and that it is the right age. A gigantic impact formed the crater about 65 million years ago, just when the dinosaurs and many other species died out, and many Earth scientists now believe that this is the scar of the impact that ended the Cretaceous.

Could such an impact happen again? Earth gets hit by small meteorites every day and by larger objects less often. Impacts by large asteroids may happen many millions of years apart but they continue to happen. In mid-March 1998, newspaper headlines announced, "Mile-Wide Asteroid to Hit Earth in October 2028." The frightening story was true except that the news media did not mention the uncertainty in the orbit. Within days, astronomers found the asteroid on old photographs, recalculated the orbit with the new data, and concluded that the asteroid, known as $1997XF_{11}$, would miss Earth by 600,000 miles. There will be no impact by this asteroid in 2028, but there are plenty more asteroids that haven't been discovered. It is just a matter of time.

A solar system is a dangerous place to put an inhabited planet. Comets, asteroids, and meteoroids con- stantly rain down on the planets, and Earth gets hit often. Chicxulub isn't the only large impact scar on the Earth. About 150 are known, including giant craters buried under sediment in Iowa, and another underlying most of Chesapeake Bay.

REVIEW **Critical Inquiry**

How do comets help us understand the formation of the planets?

According to the solar nebula hypothesis, the planets formed from planetesimals that accreted in a disk-shaped nebula around the forming sun. The planetesimals that formed in the inner solar nebula were warm and could not incorporate much ice. The asteroids may be the last remains of such bodies. In the outer solar nebula, it was colder, and the planetesimals would have contained large amounts of ices. Many of these planetesimals were destroyed when they fell together to make the planets, but some may have survived. The icy bodies of the Oort cloud and the Kuiper belt may be the last surviving icy planetesimals in our solar system. When these bodies fall into the inner solar system, we see them as comets, and the gases they release tell us that many of them are rich in volatile materials such as water and carbon dioxide. These are the ices we would expect to find in the icy planetesimals. Furthermore, comets are rich in dust, and the planetesimals must have included large amounts of dust frozen into the ices when they formed. Thus, the nuclei of comets seem to be frozen samples of the ancient solar nebula.

Nearly all of the mass of a comet is in the nucleus, but the light we see comes from the coma and the tail. What kind of spectra do we get from comets, and what does that tell us about the process that converts a dirty iceberg into a comet?

Our solar system is an elegant planetary system, but we must remember that it is small. The universe contains billions of stars more or less like our sun. We begin our study of the rest of the universe in the next chapter.

Summary

The solar nebula theory proposes that the solar system formed from a disk of gas and dust around the forming sun and that most of that nebula was blown away when the sun became luminous. The asteroids and comets appear to be rocky and icy debris left behind.

Meteorites can be classified according to their mineral content. Iron meteorites are mostly iron and nickel, and when sliced open, polished, and etched with acid, they show Widmanstätten patterns. These suggest the metal cooled from a molten state very slowly.

Stony meteorites can be classified as chondrites if they contain small, glassy particles called chondrules. If such a meteorite is rich in volatiles and carbon, it is called a carbonaceous chondrite. An achondrite is a stony meteorite that contains no chondrules. Achondrites appear to have been melted after they formed, but chondrites were never melted. Carbonaceous chondrites were never even warm enough to drive off volatiles.

This evidence suggests that the different kinds of meteorites were formed in bodies roughly 100 km in diameter. Such objects could have melted, differentiated, and cooled very slowly, and when they were later broken by collisions, their fragments would resemble the different kinds of meteorites. Also, we may suppose that the carbonaceous chondrites were formed farther from the sun, where it was cooler. This evidence suggests that meteorites are fragments of the asteroids.

Most of the meteors we see, however, are small, weak objects that never reach the ground. These bits of matter appear to be debris from comets, and we see meteor showers when Earth crosses the debris-laden orbit of a comet.

Most asteroids lie in the asteroid belt between Mars and Jupiter. Infrared studies show that asteroids in the outer belt are darker and redder than others. These may resemble the volatile-rich carbonaceous chondrites. Apparently, the asteroids are the remains of planetesimals that were unable to form a planet between Mars and Jupiter. The strong gravitational influence of Jupiter could have prevented the material from accumulating into a large body.

A comet is produced by a lump of dirty ices only a few dozen kilometers in diameter. In a long, elliptical orbit, the icy body stays frozen until the object draws close to the sun. Then the ices vaporize and release the imbedded dust and debris. The gas is caught in the solar wind and blown outward to form a type I, or gas, tail. The pressure of sunlight blows the dust away to form a type II, or dust, tail. The coma of a comet can be up to 1,000,000 km in diameter, and it contains jets issuing from the nucleus.

When spacecraft flew past Comet Halley in 1986, astronomers discovered that the nucleus was coated by a dark crust and that jets of vapor and dust were venting from active regions on the sunlit side. The low density of the nucleus showed that it was a fluffy mixture of ices and silicate dust.

Comets are believed to have formed as icy planetesimals in the outer solar system, and some of these objects remain as the Kuiper belt, the source of short-period comets. Some icy planetesimals were swept up by the formation of the Jovian planets and others were ejected to form the Oort cloud, the source of the long-period comets.

New Terms

Widmanstätten pattern	meteor shower
chondrite	coma
chondrule	Oort cloud
carbonaceous chondrite	Kuiper belt
achondrite	

Review Questions

1. What do Widmanstätten patterns tell us about the history of iron meteorites?

2. What do chondrules tell us about the history of chondrites?

3. Why are there no chondrules in achondritic meteorites?

4. Why do astronomers refer to carbonaceous chondrites as "unmodified"?

5. How do observations of meteor showers reveal one of the sources of meteoroids?

6. How can most meteors be cometary if most, perhaps all, meteorites are asteroidal?

7. Why do we think the asteroids were never part of a planet?

8. What evidence do we have that the asteroids are fragmented?

9. What evidence do we have that some asteroids have differentiated?

10. What evidence do we have that some asteroids have had active surfaces?

11. How is the composition of meteorites related to the formation and evolution of asteroids?

12. What is the difference between a type I tail and a type II tail?

13. What evidence do we have to support the dirty snowball model of cometary nuclei?

14. Why do short-period comets tend to have orbits near the plane of the solar system?

15. How did the bodies in the Kuiper belt and the Oort cloud form?

Discussion Questions

1. Futurists suggest we may someday mine the asteroids for materials to build and supply space colonies. What kinds of materials could we get from asteroids? (*Hint:* What are S-, M-, and C-type asteroids made of?)

2. If cometary nuclei were heated by internal radioactive decay rather than by solar heat, how would comets differ from what we observe?

3. From what you know now, do you think the government should spend money to locate near-Earth asteroids? How serious is the risk?

Problems

1. Large meteorites are hardly slowed by Earth's atmosphere. Assuming that the atmosphere is 100 km thick and that a large meteorite falls vertically toward the ground, how long does it spend in the atmosphere? (*Hint:* How fast do meteorites travel?)

2. If a single asteroid 1 km in diameter were fragmented, how many asteroids 1 m in diameter could it make? (*Hint:* The volume of a sphere is $\frac{4}{3}\pi r^3$.)

3. What is the orbital velocity of a meteoroid orbiting the sun at the distance of Earth? (*Hint:* Use Kepler's third law.)

4. If a half million asteroids, each 1 km in diameter, were assembled to make a planet, how large would it be? (*Hint:* The volume of a sphere is $\frac{4}{3}\pi r^3$.)

5. The asteroid Vesta has a mass of 2×10^{20} kg and a radius of about 250 km. What is its density in g/cm^3? (*Hint:* The volume of a sphere is $\frac{4}{3}\pi r^3$.)

6. What is the maximum angular diameter of Ceres, the largest asteroid, as seen from Earth? Could any Earth-based telescope detect surface features?

7. At what distances from the sun would you expect to find an asteroid orbiting with exactly half of Jupiter's orbital period? exactly one third? Such locations in the asteroid belt contain very few asteroids because of these resonances with Jupiter.

8. If the velocity of the solar wind is about 400 km/s and the visible tail of a comet is 100 million km long, how long does it take an atom to travel from the nucleus of a comet to the end of the visible tail?

9. If a comet comes very close to the sun and its orbit takes it out as far as the orbit of Pluto, what is its orbital period? (*Hint:* Use Kepler's third law.)

10. The mass of a comet's nucleus is about 10^{14} kg. If the Oort cloud contains 2 trillion (2×10^{12}) cometary nuclei, what is the mass of the cloud in kilograms? in Earth masses? (*Hint:* Mass of Earth = 6×10^{24} kg.)

Critical Inquiries for the Web

1. Some nights are better for looking for meteors than others (see the Table of Meteor Showers—Table 9-1). We know these showers are associated with comets, but how are these associations made? There are several sites on the Internet that provide information on meteor showers, including historical data and information on parent comets. Pick a shower whose parent comet is known, and summarize how we came to know that the meteors and the comet are related.

2. The chances are small that you will be killed by an asteroid impact, but if there are objects out there that astronomers are not aware of whose orbits intersect Earth, we could be in for a surprise one day. Look for information on the Spacewatch project and other investigations into near-Earth asteroids. How many such objects have been discovered? What is the record for closest known passage of an asteroid to Earth?

Exploring *The Sky*

1. Many people have seen meteors and comets. However, few have seen asteroids because they tend to be rather faint and usually require binoculars or telescopes to see. This is your opportunity to see an asteroid—at least a virtual asteroid! Use *The Sky* to locate the largest of the asteroids, Ceres, and determine:

   a. The constellation in which to find Ceres at the present time.

   b. Its present distance from the sun (between which two planetary orbits is it located?).

   c. Its present distance from Earth.

   d. Its magnitude (if it's brighter than 6th magnitude, it may be possible to see it with the unaided eye).

 Go to the Brooks/Cole Astronomy Resource Center (www. brookscole.com/astronomy) for critical thinking exercises, articles, and additional readings from InfoTrac College Edition, Brooks/Cole's online student library.

CHAPTER TEN

The Sun—
Our Star

Guidepost

We have explored our solar system in space and in time, but we have neglected the central object, the sun. As we examine our sun, we learn not only about the object that gave birth to the planets, but we begin to learn about stars.

In this chapter, we will study the surface of the sun as a source of light and heat and as a typical star. We will study the inside of the sun in Chapter 12 where we consider the structure of stars in general.

As you read this chapter, notice the interplay of evidence and hypothesis. Scientists know things about nature by using evidence to build confidence in theories and hypotheses.

In this chapter we take our first step toward understanding the stars.

A wit once remarked that solar astronomers would know a lot more about the sun if it were farther away. This comment contains a grain of truth; the sun is only a humdrum star, and there are billions like it in the sky, but the sun is the only one close enough to show surface detail. Solar astronomers can see so much detail in the swirling currents of gas and arching bridges of magnetic force that present theories seem inadequate to describe it. Yet the sun is not a complicated object. It is just a star.

In their general properties, stars are very simple. They are great balls of hot gas held together by their own gravity. Their gravity would make them collapse into small, dense bodies were they not so hot. The tremendously hot gas inside stars has such a high pressure that the stars would surely explode were it not for their own confining gravity. Like soap bubbles, stars are simple structures balanced between opposing forces that individually would destroy them. Thus, we study the sun as a close-up example of a star.

Another reason to study the sun is that life on Earth depends critically on the sun. Should the sun's energy output vary by even a small amount, life on Earth might vanish. In addition, we get nearly all our energy from the sun—oil and coal are merely stored sunlight—and our pleasant climate is maintained by energy from the sun. In fact, the sun's atmosphere of very thin gas reaches out past Earth's orbit, and thus any change in the sun, such as an eruption or a magnetic storm, can have a direct effect on Earth.

Finally, we study the sun because it is beautiful. Our analysis of sunlight will reveal that the sun is both powerful and delicate. Thus, we study the sun not only because it is *a star*, not only because it is *our star*, but because it is the sun.

10-1 The Solar Atmosphere

Although the sun is more than 100 times larger in diameter than Earth and contains more than 300,000 times more matter, it is gaseous throughout (Data File Eleven). When we look at the sun, we see only the surface, what astronomers call the atmosphere. We have seen these layers, the photosphere, chromosphere, and corona, revealed during total solar eclipses (Chapter 1), and we are now ready to explore them in detail.

Heat Flow in the Sun

Simple logic tells us that energy in the form of heat is flowing outward from the sun's interior. Observations of the solar spectrum tell us that the temperature of the sun's surface is about 5800 K. At that temperature every square centimeter of the sun's surface must be radiating more energy than a 6000-watt light bulb. With all

The Sun Data File Eleven

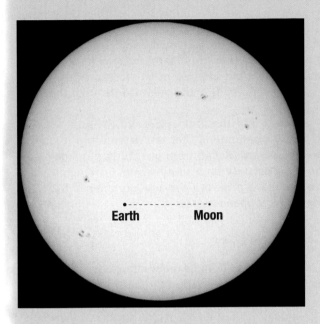

An image of the sun in visible light shows a few sunspots. The Earth–moon system is added for scale. *(Daniel Good)*

Average distance from Earth	1.00 AU (1.495979×10^8 km)
Maximum distance from Earth	1.0167 AU (1.5210×10^8 km)
Minimum distance from Earth	0.9833 AU (1.4710×10^8 km)
Average angular diameter seen from Earth	0.53° (32 minutes of arc)
Period of rotation	25 days at equator
Radius	6.9599×10^5 km
Mass	1.989×10^{30} kg
Average density	1.409 g/cm³
Escape velocity at surface	617.7 km/s
Luminosity	3.826×10^{26} J/s
Surface temperature	5800 K
Central temperature	15×10^6 K
Spectral type	G2 V
Apparent visual magnitude	−26.74
Absolute visual magnitude	4.83

that energy radiating into space, the sun's surface would cool rapidly if energy did not flow outward from the interior to keep the surface hot.

Not until the 1930s did astronomers understand how the sun makes its energy. Nuclear reactions occur in the core of the sun and generate tremendous heat, and that heat flowing outward keeps the surface hot. We will discuss these nuclear reactions in detail later in this chapter, but here we must recognize the importance of the energy flowing outward through the sun's surface.

As we study the atmosphere of the sun, we will find many phenomena that are driven by the energy flowing outward. Recall from our study of planets how geological activity is driven by heat flowing outward from a planet's interior (Window on Science 6-2). The same thing happens to stars.

The Photosphere

The sun is gaseous from its outer atmosphere right down to its center, so it has no solid or liquid surface. The surface we see, the photosphere, is a layer of gas roughly 100 km deep from which we receive most of the sun's light. If the sun were shrunk to the size of a bowling ball, the photosphere would be thinner than a layer of tissue paper wrapped around the ball (Figure 10-1).

Below the photosphere, the gas is hotter and denser and therefore radiates plenty of light, but that light cannot escape from the sun because of the outer layers of gas. Above the photosphere, the gas is less dense and cannot radiate as much light. The photosphere is the layer in the sun's atmosphere that is dense enough to emit plenty of light but not so dense that the light cannot escape.

Figure 10-1
A cross section at the edge of the sun shows the relative thickness of the photosphere and chromosphere. Earth is shown for scale. On this scale, the disk of the sun would be more than 1.5 m (5 feet) in diameter.

Although the photosphere appears to be substantial, it is really a very low density gas. Even in the deepest and densest layers visible, the photosphere is 3400 times less dense than the air we breath. With a fantastically efficient insulation, we could fly an airplane deep into the sun. Only toward the center, where the density is very high, would our airplane meet serious resistance.

The spectrum of the sun is an absorption spectrum, and Kirchhoff's third law tells us we must be looking at light passing through a thin gas. In this case, the deepest and densest layers of the photosphere emit a continuous spectrum, and as that light passes through the higher layers, photons are absorbed, producing dark lines in the solar spectrum. This is a simple model to explain a process that is actually elegant and complex, but our simple model of the formation of the solar absorption spectrum is sufficient to account for what we see when we observe the photosphere.

Good photographs of the photosphere show that it is not uniform but is mottled by a pattern of dark-edged bright cells called **granulation** (Figure 10-2a). Each granule is about the size of Texas and lasts for only 10 to 20 minutes before fading away to be replaced by new granules. Spectra of these granules show that the centers are a few hundred degrees hotter than the edges, and Doppler shifts reveal that the centers are rising and the edges sinking at speeds of about 1 km/s.

From this evidence, astronomers recognize the granulation as the surface effects of convection just below the photosphere (Figure 10-2b). **Convection** occurs when hot fluid rises and cool fluid sinks, as when, for example, a convection current of hot gas rises above a candle flame. You can create convection in a liquid by adding a bit of nondairy creamer to an unstirred cup of hot coffee. The cool creamer sinks, warms, rises, cools, sinks again, and so on, creating small convection currents. The tops of these currents look much like solar granules. Rising currents of hot gas heat small regions of the photosphere, which, being slightly hotter, emit more black body radiation and look brighter. The cool sinking gas of the edges emits less light and thus looks darker. Thus the granulation reveals that energy flows upward from below by convection and heats the photosphere.

The Chromosphere

Above the photosphere lies the chromosphere. The depth of the layers in the sun's atmosphere are not precisely defined, but we can think of the chromosphere as being on average no deeper than Earth's diameter (see Figure 10-1). Because the chromosphere is roughly 1000 times fainter than the photosphere, we can see it with our unaided eyes only during a total solar eclipse, when the moon covers the brilliant photosphere. Then, the chromosphere flashes into view as a thin line of pink just above the photosphere. The word *chromosphere* comes from the Greek word *chroma,* meaning "color."

The chromosphere produces an emission spectrum, and Kirchhoff's laws tell us the chromosphere must be an excited, low-density gas. Atoms in the lower chromosphere are ionized, and atoms in the higher layers of the chromosphere are even more highly ionized. That is, they have lost more electrons. From this, astronomers can tell that the temperature in the chromosphere just above the photosphere is about 10,000 K and that it rises rapidly to over 500,000 K in the upper chromosphere (Figure 10-3a).

Solar astronomers can take advantage of some simple physics to study the chromosphere. The gases of the chromosphere are transparent to nearly all visible light, but atoms in the gas are very good at absorbing photos of specific wavelengths. This produces the dark absorption lines. A photon at one of those wavelengths is very unlikely to escape from deeper layers. A **filtergram** is a photograph made using light in one of the dark absorption

a

b

Granule

Sinking gas Rising gas

Figure 10-2
(a) A visible-light photo of the sun's surface shows granulation. Each granule is about the size of Texas and lasts 10 to 20 minutes before it is replaced by new granules. Doppler shifts show that the centers of granules are rising and the edges are sinking. *(NOAO)*
(b) This model of granulation explains granulation as the tops of convection currents just below the photosphere. Heat flows upward as rising currents of hot gas and downward as sinking currents of cool gas. The rising currents heat the solar surface in small regions that we see as granules.

Figure 10-3
The chromosphere. (a) If we could place thermometers in the sun's atmosphere, we would discover that the temperature increases from 6000 K at the photosphere to 10^6 K at the top of the chromosphere. (b) A filtergram shows that the chromosphere is not uniform. Spicules appear like black grass blades at the edges of supergranules over twice the diameter of Earth. *(© 1971 NOAO/NSO)*

a

b

lines. Those photons can only have escaped from higher in the atmosphere. In this way filtergrams reveal detail in the upper layers of the chromosphere.

Filtergrams such as the one shown in Figure 10-3b reveal **spicules,** flamelike structures 100 to 1000 km in diameter, extending up to 10,000 km above the photosphere and lasting from 5 to 15 minutes. These spicules appear to be cool regions (about 10,000 K) extending up into the much hotter corona (about 500,000 K). Seen at the edge of the solar disk, these spicules blend together and look like flames covering a burning prairie, but filtergrams of spicules located near the center of the solar disk show that they spring up around the edges of regions called **supergranules.** Over twice the diameter of Earth, supergranules appear to be caused by heat from larger convection cells deeper below the photosphere—further evidence of the flow of energy out of the sun's interior.

We will see evidence later in this chapter that the solar surface is dominated by a strong magnetic field, and much of that field emerges from below the photosphere at the edges of the supergranules. Thus, the spicules are clearly governed by magnetic fields that are rooted far below the photosphere.

The Solar Corona

The outermost part of the sun's atmosphere is called the corona, after the Greek word for "crown." During a total solar eclipse, the moon covers the brilliant photosphere, and the corona shines with a pearly glow not even as bright as the full moon. (See Figure 1-22b.) Although the corona is visible to the eye only during total solar eclipses, telescopes of special design can photograph the corona of the uneclipsed sun (Figure 10-4). The best images are those taken from high-flying aircraft or from space, and they can trace the corona out to 12 solar radii or more, well over 10 percent of the way to Earth's orbit.

The spectrum of the corona can tell us a great deal about the corona and simultaneously illustrate how astronomers can analyze a spectrum. Some of the light from the outer corona is just reflected sunlight, and that must mean the outer corona contains dust particles that reflect sunlight in all directions. Some of the coronal light consists of a continuous spectrum, and that ap-

Figure 10-4
Streamers in the solar corona. An image from the Solar Maximum Mission satellite (inset) has been computer-enhanced to produce a false-color image that reveals subtle variations in brightness. *(NASA/JPL)*

pears to form when sunlight from the photosphere is scattered off of free electrons in the ionized coronal gas. We might expect this light to reproduce the absorption spectrum of the photosphere, but it does not, because the coronal gas has a temperature over 1 million K, and the electrons travel very fast. When photons scatter off of these high-speed electrons, the photons suffer very large Doppler shifts that smear out any absorption lines to produce a continuous spectrum.

Superimposed on the corona's continuous spectrum are emission lines of highly ionized gases. In the lower corona, the atoms are not as highly ionized as they are at higher altitudes, and this tells us that the temperature of the corona rises with altitude. Just above the chromosphere, the temperature is about 500,000 K; but in the outer corona the temperature can be as high as 2 million K or more.

The spectrum of the corona tells us that it is exceedingly hot gas, but it is not very bright. Its density is very low, only 1 to 10 atoms/cm^3, better than the best vacuum on Earth. Because of this low density, the hot gas does not emit much radiation.

Astronomers have wondered for years how the corona and chromosphere can be so hot. Heat flows from hot regions to cool regions, so how can the heat from the photosphere, with a temperature of only 5800 K, flow out into the much hotter chromosphere and corona? The answer appears to be the magnetic field that emerges from the sun around supergranules. Stirred by turbulence below the photosphere, the magnetic field extending outward whips about, and that churns the chromosphere and corona and heats them. Thus energy flows outward in this instance as the motion of magnetic fields.

Gas from the solar atmosphere follows the magnetic fields pointing outward and flows away from the sun in a breeze called the solar wind. Like an extension of the corona, the solar wind blows past Earth at 300 to 800 km/s with gusts as high as 1000 km/s. Thus Earth is bathed in the corona's hot breath.

Do other stars have chromospheres and coronas like the sun's? Ultraviolet spectra taken by orbiting space telescopes such as the IUE (see Chapter 3) suggest that the answer is yes. The spectra of many stars contain emission lines in the far ultraviolet that could have been formed only in the low-density, high-temperature gases of the upper chromosphere and lower corona. Thus, the sun, for all its complexity, seems to be a normal star.

Helioseismology

It seems that the light we receive from the sun can tell us nothing about the layers below the photosphere, but solar astronomers have found a way to explore the sun's interior by **helioseismology,** the study of the modes of vibration of the sun. Just as geologists can study Earth's interior by observing how sound waves produced by earthquakes are reflected and transmitted by the layers of Earth's interior, so too can solar astronomers explore the sun's interior by studying how vibrations (sound waves) travel through the sun.

As these vibrations rise from the interior and reach the surface of the sun, we see the photosphere move up and down in a complicated pattern that depends on the frequency of the sound waves (Figure 10-5). The Doppler shifts caused by the moving surface, although very small, can tell solar astronomers which frequency vibrations are present. Although 10 million different frequencies are possible, some penetrate deeper than others, and conditions in the sun's interior layers can weaken or strengthen the waves. By observing which of the millions of frequencies actually occur, solar astronomers can determine the temperature, density, pressure, composition, and motion of the sun's internal layers.

Helioseismology sounds almost magical, but we can understand it better if we think of a duck pond. If we stood at the shore of a duck pond and looked down

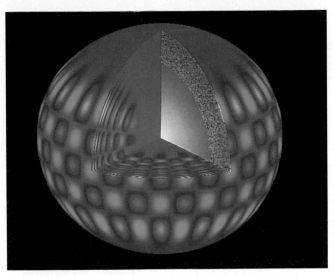

Figure 10-5
Helioseismology is the study of the modes of vibration of the sun. This computer image shows one of nearly 10 million possible modes of oscillation. Red regions are receding and blue approaching. Such studies can reveal details about the sun's interior. *(NOAO)*

at the water, we would see ripples arriving from all parts of the pond. Because every duck on the pond contributes to the ripples, we could, in principle, study the ripples near the shore and draw a map showing the position and velocity of every duck on the pond. Of course, it would be difficult to untangle all the different ripples, but all of the information would be there, lapping at the rocks at our feet.

Just as we can map the ducks on a pond, helioseismologists can study the vibrations in the surface of the sun and deduce the characteristics of the sun's interior. Because there are so many possible modes of vibration, they need large masses of data in order to separate the frequencies. The Global Oscillation Network Group (GONG) has set up small telescopes around the world to observe the sun nonstop for up to three years. Needless to say, supercomputers are needed to analyze the data and map the sun's interior. This new form of solar astronomy is producing exciting results.

R E V I E W *Critical Inquiry*

How deeply into the sun can we see?

This is a simple question, but it has a very interesting answer. To analyze this question we must avoid being misled by the words we use. When we look into the layers of the sun, our sight does not really penetrate *into* the sun. Rather, our eyes record photons that have escaped *from* the sun and traveled outward through the layers of the sun's atmosphere. If we observe at a wavelength at the center of a dark absorption line, then the photosphere and lower chromosphere are opaque, photons can't escape to our eyes, and the only photons we can see come from the upper chromosphere. What we see are the details of the upper chromosphere—a filtergram. On the other hand, if we observe at a wavelength that is not easily absorbed (a wavelength between spectral lines), the atmosphere is more transparent, and photons from deeper inside the photosphere can escape to our eyes.

By choosing the proper wavelength, solar astronomers can observe different depths in the sun's photosphere and chromosphere. But the corona is so thin and the gas below the photosphere so dense that this method doesn't work in these regions. How can we observe the corona and the deeper layers of the sun?

So far we have thought of the sun as a static, unchanging ball of gas with energy flowing outward from the interior and through the atmospheric layers. In fact, the sun is a highly variable body whose appearance is constantly changing. It is now time to think of the active sun.

10-2 Solar Activity

Solar activity refers to features on the sun that change over minutes, days, or years. We will use spectroscopic analysis to explore this aspect of solar astronomy, and we will discover that solar activity is shaped by magnetic fields and driven by the powerful flow of energy rising from the sun's interior.

Sunspots and Active Regions

Solar activity is often visible with even a small telescope, but we should exercise great caution in observing the sun. Not only is sunlight intense, but it also includes infrared radiation (heat), which can burn our eyes. It is not safe to look at the sun directly, and it is even more dangerous to look at the sun through a telescope, which concentrates the sunlight. Figure 10-6a

Figure 10-6
Looking through a telescope at the sun is dangerous, but you can always view the sun safely with a small telescope by projecting its image on a white screen (a). If you sketch the location and structure of sunspots on successive days (b), you will see the rotation of the sun and gradual changes in the size and structure of sunspots.

a   b

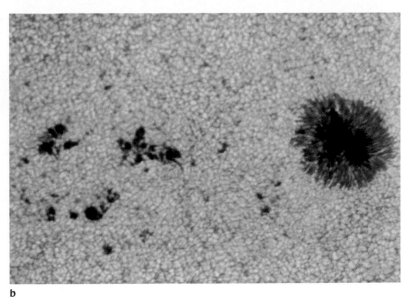

a b

Figure 10-7
(a) This visible-light photograph of a sunspot is shown digitally enhanced to reveal radial structure in the penumbra and in the umbra. These markings are caused by the magnetic field in the sunspot. *(Courtesy William Livingston)* (b) A dark sunspot slightly larger than Earth (right) is part of a group of spots. Notice the granulation of the solar surface throughout the photo. *(NOAO)*

illustrates a safe way to observe the sun with a small telescope. Such observations often reveal evidence of solar activity in the form of **sunspots,** dark regions on the sun's surface. If you observe sunspots day after day, you will notice the rotation of the sun and slow changes in the structure of the sunspots (Figure 10-6b).

If we look carefully at sunspots, we see that they consist of a dark center called the umbra and an outer border called the penumbra (Figure 10-7a). Sunspots come in a wide range of sizes, but we might observe that the average spot is about 40 seconds of arc in diameter in our telescope. The small-angle formula (By the Numbers 1-2) tells us that such a sunspot is about twice the diameter of Earth. Such spots last a week or so and tend to occur in pairs. A large group of sunspots is called an **active region,** may contain up to 100 spots, and could last as long as two months (Figure 10-7b).

Sunspots look dark because they are cooler than the photosphere. By noting which atoms are excited in the spectra of sunspots, astronomers can tell that the umbra has a temperature of about 4240 K compared with 5800 K for the photosphere. Because the amount of black body radiation emitted depends on the temperature to the fourth power (By the Numbers 4-1), this small difference in temperature makes a big difference in brightness, and the spot looks quite dark compared to the photosphere. In fact, a sunspot umbra emits quite a bit of light. If the sun were removed and only an average-size sunspot were left behind, it would glow a brilliant orange-red and would be brighter than the full moon.

The key to understanding sunspots lies in the **Zeeman effect,** the splitting of single spectral lines into multiple components through the influence of a magnetic field (Figure 10-8). When an atom is in a magnetic field, the electron orbits are altered, and the atom is able to absorb a number of different wavelength photons even though it was originally limited to a single wavelength. In the spectrum, we see single lines split into multiple components, with the separation between components proportional to the strength of the magnetic field.

Using the Zeeman effect, astronomers find that the field in a typical sunspot is about 1000 times stronger than the sun's average field (Figure 10-9a). Apparently

a b

Figure 10-8
(a) The slit of a spectrograph has been placed across a complex sunspot. Thus, light from the sunspot enters the spectrograph. (b) The resulting spectrum shows a typical absorption line split into three components (arrow) by the magnetic field in the sunspot. *(NOAO)*

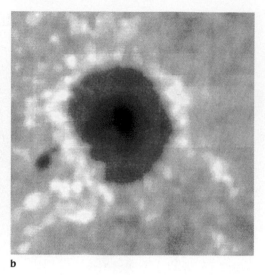

a   b

Figure 10-9

(a) A magnetogram, a magnetic map of the sun constructed using the Zeeman effect, confirms that the active regions around sunspots (white) contain powerful magnetic fields. In a visible-light photo of the sun taken on this day we would see sunspot groups in each of these active regions. *(NOAO)* (b) This infrared image of a sunspot shows a brightening of the photosphere around the sunspot apparently caused by the escape of energy that was unable to emerge through the strongly magnetic sunspot. *(Dan Gezari, NASA/Goddard)*

this powerful magnetic field inhibits gas motion below the photosphere, and rising gas cannot deliver its heat to the surface. Thus, the area cools and we see a sunspot. Infrared observations of sunspots suggest that the heat that did not emerge through the sunspot is deflected and causes a slight increase in the temperature of the photosphere around the sunspot (Figure 10-9b).

The total number of sunspots visible on the sun is not constant. In 1843, the German amateur astronomer Heinrich Schwabe noticed that the number of sunspots varies with a period of about 11 years. This is now known as the sunspot cycle (Figure 10-10a). At sunspot maximum there are often as many as 100 spots visible at any one time, but at sunspot minimum there are only a few small spots. A sunspot minimum occurred in the mid-1990s, with a maximum in the very first years of the 21st century.

At the beginning of each sunspot cycle, the spots begin to appear in the sun's middle latitudes about 35° above and below the sun's equator. As the cycle proceeds, the spots appear at lower latitudes, until, near the end of the cycle, they are appearing within 5° of the sun's equator. If we plot the latitude of the appearance of sunspots over time, the diagram takes on the appearance of butterfly wings (Figure 10-10b). Such diagrams are now known as **Maunder butterfly diagrams,** named after E. Walter Maunder of the Greenwich Observatory.

Maunder also noticed that records show very few sunspots from 1645 to 1715. Modern studies show that this **Maunder minimum** (Figure 10-10a) coincided with

Figure 10-10

(a) The number of sunspots varies over an 11-year period. Note that between about 1645 and 1715, the Maunder minimum, there were very few sunspots. (b) The Maunder butterfly diagram shows that the spots first occur at higher latitudes and later nearer the solar equator.

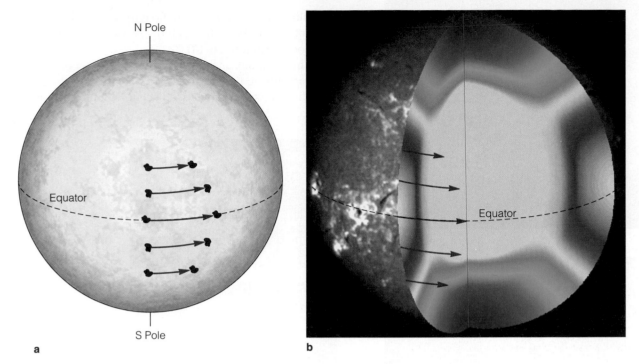

N Pole

Equator

S Pole

a

Equator

b

Figure 10-11
(a) At the photosphere, the sun rotates faster at the equator than at higher latitudes. If we started five sunspots in a row, they could not stay aligned as the sun rotates. (b) Helioseismology shows how the interior of the sun rotates relative to the surface. In this cutaway diagram, the photosphere at the equator rotates the fastest and the poles the slowest. *(Courtesy Kenneth Libbrecht, Caltech)*

a period of reduced solar activity and may be linked to the "little ice age," a period of unusually cool weather in Europe and America from about 1430 to 1850.

The Magnetic Cycle

The key to the sunspot cycle is the sun's rotation, which is complicated by the fact that the equatorial region rotates faster than do regions at the higher latitudes (Figure 10-11). At the equator the sun rotates once every 25 days, but at a latitude of 45° one rotation takes 27.8 days. The rotation of the sun also varies with depth, with the deep interior rotating more slowly than the equator. This **differential rotation** is clearly linked to the magnetic cycle.

The dynamo effect can generate a magnetic field in a rotating mass of electrically conducting material if the material is circulating. The gases of the sun are highly ionized, so they are very good conductors. Helioseismology shows that the sun's differential rotation only extends inward to the bottom of the convection zone. Inside that radius, the sun rotates as a solid body. The rapid circulation caused by the change in rotation rate near the bottom of the convection zone seems to be the source of the sun's magnetic field.

Zeeman effect measurements of sunspots give us evidence that the sun's magnetic field follows a cycle. Sunspots tend to occur in pairs, and the magnetic field around the pair resembles that around a bar magnet with one end magnetic north and the other end magnetic south (Figure 10-12). During alternate sunspot cycles, the polarity of these magnetic fields is reversed. Thinking of the rotation of the sun, we can refer to one spot as the leading spot and the other as the trailing spot. We might notice that the leading spot of each pair in the sun's northern hemisphere is a north pole and that each trailing spot is a south pole. South of the sun's equator we would find the polarities reversed. If we watched for an entire sunspot cycle, we would discover that the overall polarity reverses cycle to cycle.

This magnetic cycle is not fully understood, but the **Babcock model** (named for its inventor) explains the magnetic cycle as a progressive tangling of the solar magnetic field. Because the electrons in an ionized gas are free to move, the gas is a very good conductor of electricity, and any magnetic field in the gas is "frozen" into the gas. If the gas moves, the magnetic field must move with it. Thus the sun's magnetic field is frozen into its gases, and the differential rotation wraps this field around the sun like a long string caught on a hubcap. Rising and sinking gas currents twist the field into ropelike tubes, which tend to float upward. Where these magnetic tubes burst through the sun's surface, sunspot pairs occur (Figure 10-13).

The Babcock model explains the reversal of the sun's magnetic field from cycle to cycle. As the magnetic field becomes tangled, adjacent regions of the sun's

a

b

Figure 10-13

The Babcock model suggests that the differential rotation of the sun winds up the magnetic field. When the field becomes tangled and bursts through the surface, it forms sunspot pairs. For the sake of clarity, only one line in the sun's magnetic field is shown here.

Figure 10-12

(a) This far-ultraviolet image of the sun shows arched structures above active regions. *(SOHO/EIT, ESA, and NASA)* (b) Iron filings and a bar magnet reveal the arched shape of magnetic lines of force. *(Author)* (c) The polarity of sunspot groups in the sun's southern hemisphere is always the reverse of that in the northern hemisphere.

While many textbooks describe science as the process of testing hypotheses by observation and experiment, you should not think that every astronomer approaches the telescope expecting to make an observation that will disprove long-held beliefs and trigger a revolution in science. Then what is the daily grind of science really about?

First, many observations and experiments merely confirm already-tested hypotheses. The biologist knows that all worker bees in a hive are sisters, but a careful study of the DNA from different workers further confirms that hypothesis. By repeatedly confirming a hypothesis,

scientists build confidence in the hypothesis and may be able to extend it to a wider application. Of course, there is always the chance that a new observation or experiment will disprove the hypothesis, but that is usually very unlikely. Much of the daily grind of science is confirmation.

Another aspect of routine science is consolidation, the linking of a hypothesis to other well-studied phenomena. Chemists may understand certain kinds of carbon molecules shaped like rings, but by repeated study they find a carbon molecule shaped like a hollow sphere. To consolidate their findings, they must show that the chemical bonding in the two mol-

ecules follows the same rules and that the molecules have certain properties in common. No hypothesis is overthrown, but the chemists consolidate their knowledge and understand carbon molecules better.

The Babcock model of the solar magnetic cycle is an astronomical example of the scientific process. Solar astronomers know that the model explains some solar features but has shortcomings. Although most astronomers don't expect to discard the entire model, they work through confirmation and consolidation to better understand how the solar magnetic cycle works and how it is related to cycles in other stars.

surface are dominated by magnetic fields that point in different directions. After about 11 years of tangling, the field becomes so complex that adjacent regions of the solar surface begin changing their magnetic field to agree with neighboring regions. Quickly the entire field rearranges itself into a simpler pattern, and differential rotation begins winding it up to start a new cycle. But the newly organized field is reversed, and the next sunspot cycle begins with magnetic north replaced by magnetic south. Thus the complete magnetic cycle is 22 years long and the sunspot cycle is 11 years long.

Notice the power of a scientific model. The Babcock model may in fact be incorrect in some or all details, but it gives us a framework on which to organize

all of the complex solar activity. Even though our models of the sky (see Chapter 1) and the atom (see Chapter 4) were only partially correct, they served as organizing themes to guide our thinking. Similarly, although the precise details of the solar magnetic cycle are not yet understood, the Babcock model gives us a general picture of the behavior of the sun's magnetic field (Window on Science 10-1).

If the sun is truly a representative star, we might expect to find similar magnetic cycles on other stars, but stars other than the sun are too distant for spots to be directly visible. Some stars, however, vary in brightness over a period of days in a way that reveals they are marked with dark spots believed to resemble sunspots (Figure 10-14). Other stars have spectral features that vary over periods of years, suggesting that they are subject to magnetic cycles much like the sun's. In fact, some stars are subject to sudden flares that may resemble solar eruptions.

Solar Prominences and Flares

A **prominence** is composed of hot gas trapped in magnetic fields extending above active regions into the lower corona. We can tell that the gas in a prominence

Figure 10-14
This computer-generated map shows the distribution of dark spots over the surface of the star HR1099. The map was constructed by analyzing the Doppler shifts in the spectrum of the rotating star. Although more extensive than sunspots, these dark regions are believed to be related to magnetic activity similar to that responsible for sunspots. *(Courtesy Steven Vogt, Artie Hatzes, and Don Penrod)*

a   b

Figure 10-15
(a) A loop prominence near the edge of the sun shows by its shape the magnetic fields that control the gas. *(Sacramento Peak Observatory)* (b) This far-ultraviolet photograph of the sun shows a very large arched prominence extending out into the corona. Note the complexity of the magnetic field. *(SOHO/EIT, ESA, and NASA)*

is hot because it is ionized and produces an emission spectrum. At visible wavelengths, hydrogen emission lines give prominences a bright pink color, and they are visible during a total solar eclipse protruding beyond the edge of the moon. (See Figure 1-22b.) We can be sure that prominences are dominated by magnetic fields because of their arched shapes (Figure 10-15).

We can identify two kinds of prominences. Eruptive prominences burst out of active regions and in a few hours extend into the lower corona. Quiescent prominences sometimes hang in the lower corona for many days, sometimes with gas streaming downward into the active region along magnetic fields.

Although we think of the gas trapped in a prominence as hot and excited, it is not as hot as the gas in the lower corona. Thus prominences are complex objects that originate in the hot gas of an active region and extend up into the even hotter regions of the corona.

A solar **flare** is a violent outburst on the solar surface that grows to a maximum for a few minutes and decays in an hour or less. During that time it emits vast amounts of X-ray, ultraviolet, and visible radiation, plus streams of high-energy protons and electrons. Figure 10-16a shows an ultraviolet image of a flare, while the sequence in Figure 10-16b illustrates the violence of a flare. A large flare can release energy equivalent to 10 billion megatons of TNT. (Large hydrogen bombs on Earth typically release about 10 megatons of energy.)

Solar flares are linked to the sun's magnetic field. Flares almost always occur near active regions, where the magnetic field is strong and twisted into great arches. Small flares can recur over and over at the same place;

a large active region may experience 100 small flares a day. Flares are believed to occur when arches of magnetic force encounter each other and reconnect in such a way that they cancel each other out. This releases tremendous energy stored in the magnetic fields as X rays and as high-speed electrons that heat the gas nearby. During a major solar flare, the gas can be heated to over 5 million K, and X-ray observations show that the gas in a flare can become so hot that some nuclear reactions can occur. So-called white light flares, which are detectable in visible-light photographs, occur only about once a year.

Solar flares can have important effects on Earth. The X-ray and ultraviolet radiation reaches Earth in only 8 minutes and increases the ionization in Earth's upper atmosphere. This alters the reflection of radio signals and can interfere with communications. Flares can eject high-energy particles at a third the speed of light, but most particles ejected from flares are slower and reach Earth hours or days after the flare as gusts in the solar wind. Such gusts can interact with Earth's magnetic field to produce auroral displays in the upper atmosphere (Figure 10-17).

The Active Corona

Even the corona of the sun takes part in the magnetic cycle. At sunspot minimum, when there are few active regions and the magnetic field is not strongly tangled, eclipse observers see a small, slightly flattened corona. But at sunspot maximum, when the solar magnetic field produces many active regions, eclipse observers are treated to a blazing corona that is nearly circular.

Figure 10-16
(a) This ultraviolet image of the edge of the sun shows an outburst of hot gas extending up into the corona produced when two magnetic fields just above the photosphere re-combined, canceled each other out, and released their energy in a solar flare. *(NASA)*
(b) This image sequence recorded about 5 minutes apart shows the disturbance from a solar flare spreading outward at 50 km/s. It was equivalent to an earthquake 40,000 times stronger than the quake that destroyed San Francisco in 1906. *(SOHO/MDI, ESA, and NASA)*

b

We don't need to depend on total solar eclipses to study the corona. Specialized telescopes on Earth and ultraviolet and X-ray telescopes in space can image the corona every day. The results show that the corona is composed of streamers shaped by the solar magnetic field. Some parts of the corona are filled with hot gas trapped by magnetic fields that loop back down to the solar surface. In other regions the magnetic field does not loop back, and the hot gas streams away from the sun. These regions are called **coronal holes** because they appear as dark regions in ultraviolet and X-ray images of the sun (Figure 10-18). There are permanent coronal holes at the sun's north and south poles, but coronal holes in other regions come and go as they are shaped by the changing magnetic field.

The corona is strongly affected by matter ejected from active regions on the sun's surface. When arcs of magnetic field reconnect, tremendous energy can be released and great bursts of hot gas are expelled into the corona, as shown in Figure 10-19. If one of these eruptions strikes Earth, it can have dramatic effects on communications and electrical power networks.

It should not surprise us that Earth is affected by disturbances in the corona. The solar wind is derived from the coronal holes, so it is not an exaggeration to say that Earth orbits within the expanding corona.

Figure 10-17
Auroras, commonly seen at high latitudes, sometimes look like widespread glowing clouds or great curtains of light in the night sky. Both are visible in this photograph of auroras seen in southern Australia. *(David Miller)*

a

b

Figure 10-18

(a) This X-ray image of the sun was made in the 1970s by astronauts working on the space station *Skylab*. It shows intense X-ray emission from the corona above active regions and a dark coronal hole snaking down the center of the disk. *(NASA Skylab)* (b) A more recent X-ray image of the sun made during a rocket flight above the atmosphere shows how coronal X-ray emission follows regions of magnetic activity and often takes on looped shapes resembling magnetic fields. *(IBM Research and Smithsonian Astrophysical Observatory)*

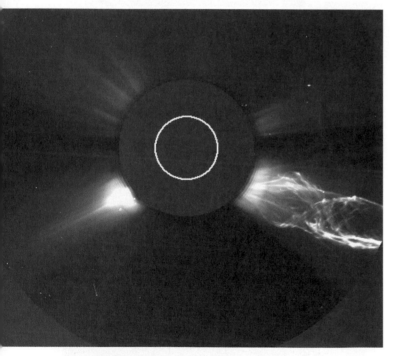

Figure 10-19

In this ultraviolet image of the corona, the disk of the sun (white circle) is hidden behind a circular screen. A violent eruption near the surface of the sun has ejected hot gases out into the corona toward the lower right. Such matter usually escapes from the sun completely. *(SOHO/LASCO, ESA, and NASA)*

REVIEW **Critical Inquiry**

What kind of activity would the sun have if it didn't rotate differentially?

This is a really difficult question, because we can see only one star close up and thus have no other examples. Nevertheless, we can make an educated guess by thinking about the Babcock model. If the sun didn't rotate differentially, with its equator traveling faster than higher latitudes, then the magnetic field would not get twisted and rise through the surface to form sunspot pairs. Of course, the magnetic field would not get progressively more wound up, so there would also be no magnetic cycle. Loops of magnetic field might not rise above the surface to trap ionized gases and form prominences, and twists and kinks might never form in the field; and thus there would be no flares. In fact, with no differential rotation to stir up the magnetic field, the chromosphere and corona might not be heated to such high temperatures.

This is very speculative, but sometimes in the critical analysis of ideas it helps to change a single important factor and try to understand what might happen. For example, what do you think the sun would be like if it had no convection inside?

Powerful activity sweeps across the surface of the sun like magnetic weather. Where does that energy come from? How does the sun make energy?

10-3 Nuclear Fusion in the Sun

Astronomers often use the wrong words to describe energy generation in the sun and stars. Astronomers will say, "The star ignites hydrogen burning." We use the word *ignite* to mean "catch on fire," and we use *burn* to mean "on fire." What goes on inside stars isn't really burning in the usual sense.

The sun is a star, and it is powered by nuclear reactions that occur near its center. The energy produced keeps the interior hot and the gas is totally ionized. That is, the electrons are not attached to the atomic nuclei, and the gas is an atomic soup of rapidly moving particles colliding with each other at high velocity. When we discuss nuclear reactions inside stars, we refer to atomic nuclei and not to atoms.

How exactly can the nucleus of an atom yield energy? The answer lies in the forces that hold the nuclei together.

Nuclear Binding Energy

The sun generates its energy by breaking and reconnecting the bonds between the particles *inside* atomic nuclei. This is quite different from the way we generate energy by burning wood in a fireplace. The process of burning wood extracts energy by breaking and reconnecting chemical bonds between atoms in the wood. Chemical bonds are formed by the electrons in atoms, and we saw in Chapter 4 that the electrons are bound to the atoms by the electromagnetic force. Thus chemical energy originated in the force that binds electrons.

There are only four known forces in nature: the force of gravity, the electromagnetic force, the **strong force,** and the **weak force.** The weak force is involved in the radioactive decay of certain kinds of nuclear particles, and the strong force binds together atomic nuclei. Thus nuclear energy comes from the strong force.

Nuclear power plants on Earth generate energy through **nuclear fission** reactions that split uranium nuclei into less massive fragments. A uranium nucleus contains a total of 235 protons and neutrons, and it splits into a range of fragments containing roughly half as many particles. Because the fragments produced are more tightly bound than the uranium nuclei, energy is released during uranium fission (Figure 10-20).

Stars make energy in **nuclear fusion** reactions that combine light nuclei into heavier nuclei. The most common reaction, including that in the sun, fuses hydrogen nuclei (single protons) into helium nuclei (two protons and two neutrons). Because the nuclei produced are more tightly bound than the original nuclei, energy is released. Notice in Figure 10-20 that both fusion and fission reactions move downward in the diagram toward

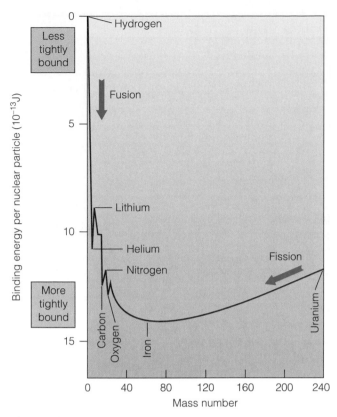

Figure 10-20
The red line in this graph shows the binding energy, the energy that holds an atomic nucleus together, for all the different atoms plotted by atomic mass, the number of protons and neutrons in their nucleus. Both fission and fusion nuclear reactions move downward in the diagram (arrows) toward more tightly bound nuclei. Iron has the most tightly bound nucleus, so no nuclear reactions can begin with iron and release energy.

more tightly bound nuclei. Thus both produce energy by releasing the binding energy of atomic nuclei.

Hydrogen Fusion

The sun fuses four hydrogen nuclei to make one helium nucleus. Because one helium nucleus contains 0.7 percent less mass than four hydrogen nuclei, it seems that some mass vanishes in the process. In fact, that mass is converted to energy, and we could figure out how much by using Einstein's famous equation, $E = mc^2$ (By the Numbers 10-1).

Creating one helium nucleus makes only a small amount of energy, hardly enough to raise a housefly 0.001 inch into the air. Only by concentrating many reactions in a small area can nature produce significant results. A single kilogram (2.2 lb) of hydrogen, for instance, converted entirely to energy, would produce enough power to raise an average-sized mountain 10 km (6 miles) into the air! The sun has a voracious energy appetite and needs 10^{38} reactions per second, transforming 5 million tons of mass into energy every second, just to balance its own gravity. This might sound as if the sun is losing mass at a furious rate, but in its entire 10-billion-year lifetime, the sun will convert less than 0.07 percent of its mass into energy.

These fusion reactions can occur only when the nuclei of two atoms get very close to each other. Because atomic nuclei carry positive charges, they repel each other with an electrostatic force called the Coulomb force. Physicists commonly refer to this repulsion between nuclei as the **Coulomb barrier.** To overcome this barrier, atomic nuclei must collide violently. Violent collisions are rare unless the gas is very hot, in which case the nuclei move at high speeds and collide violently. (Remember, an object's temperature is just a measure of the speed with which its particles move.)

Thus nuclear reactions in the sun take place only near the center, where the gas is hot and dense. A high temperature ensures that collisions between nuclei are violent enough to overcome the Coulomb barrier, and a high density ensures that there are enough collisions, and thus enough reactions, to meet the sun's energy needs.

We can symbolize this process with a simple nuclear reaction:

$$4\ {}^1\text{H} \rightarrow {}^4\text{He} + \text{energy}$$

In this equation, ${}^1\text{H}$ represents a proton, the nucleus of the hydrogen atom, and ${}^4\text{He}$ represents the nucleus of a helium atom. The superscripts indicate the approximate weight of the nu-

clei. The actual steps in the process are more complicated than this convenient summary suggests. Instead of waiting for four hydrogen nuclei to collide simultaneously, a highly unlikely event, the process can proceed step-by-step in a chain of reactions called the proton–proton chain.

The **proton–proton chain** is a series of three nuclear reactions that builds a helium nucleus by adding to protons one at a time. This process is efficient at temperatures above 10,000,000 K. The sun, for example, manufactures its energy in this way.

The three reactions in the proton–proton chain are:

$$\begin{aligned} {}^1\text{H} + {}^1\text{H} &\rightarrow {}^2\text{H} + e^+ + \nu \\ {}^2\text{H} + {}^1\text{H} &\rightarrow {}^3\text{He} + \gamma \\ {}^3\text{He} + {}^3\text{He} &\rightarrow {}^4\text{He} + {}^1\text{H} + {}^1\text{H} \end{aligned}$$

In the first reaction, two hydrogen nuclei (two protons) combine to form a heavy hydrogen nucleus, emitting a particle called a positron (a positively charged electron, symbolized e^+) and another called a neutrino (ν). In the second reaction, the heavy hydrogen nucleus absorbs another proton and, with the emission of a gamma ray (γ), becomes a lightweight helium nucleus (${}^3\text{He}$). Finally, two light helium nuclei combine to form a normal helium nucleus and two hydrogen nuclei. Because the last reaction needs two ${}^3\text{He}$ nuclei, the first and second reactions must each occur twice for each ${}^4\text{He}$ produced (Figure 10-21). The net result of this chain reac-

Figure 10-21

The proton–proton chain combines four protons (at left) to produce one helium nucleus (at right). Energy appears as gamma rays and as positrons, which combine with electrons to convert their mass into energy. Neutrinos escape, carrying away about 2 percent of the energy.

tion is the transformation of four hydrogen nuclei into one helium nucleus plus energy.

The energy appears in the form of gamma rays, positrons, and neutrinos. The gamma rays are photons that are absorbed by the surrounding gas before they can travel more than a few centimeters. This heats the gas. The positrons produced in the first reaction combine with free electrons, and both particles vanish, converting their mass into gamma rays. Thus, the positrons also help keep the center of the star hot. The neutrinos, however, are particles that resemble photons except that they almost never interact with other particles. The average neutrino could pass unhindered through a lead wall 1 ly thick. Thus, the neutrinos do not help heat the gas but race out of the star, carrying away roughly 2 percent of the energy produced.

It is time to ask the critical question that lies at the heart of science. What is the evidence to support our theoretical explanation of how the sun makes its energy? The search for that evidence introduces us to one of the great problems of modern astronomy.

The Solar Neutrino Problem

The center of a star seems forever hidden from us, but the sun is transparent to **neutrinos** because these subatomic particles almost never interact with normal matter. Nuclear reactions in the sun's core produce floods of neutrinos that rush out of the sun and off into space. If we could detect these neutrinos, we could probe the sun's interior.

Because neutrinos almost never interact with atoms, we never feel the flood of over 10^{12} solar neutrinos that flows through our bodies every second. Even at night, neutrinos from the sun rush through Earth as if it weren't there, up through our beds, through us, and onward into space. Obviously we are lucky to be transparent to neutrinos, but it means that neutrinos are extremely hard to detect. Certain nuclear reactions, however, can be triggered by a neutrino of the right energy; and in the late 1960s, chemist Raymond Davis, Jr., began using such a reaction to detect solar neutrinos.

Davis filled a 100,000-gallon tank with the cleaning fluid perchloroethylene (C_2Cl_4). Theory predicts that about once a day, a solar neutrino will convert a chlorine atom in the tank into radioactive argon, which can be detected later by its radioactive decay. To protect the detector from cosmic rays from space, the tank was buried nearly a mile deep in a South Dakota gold mine

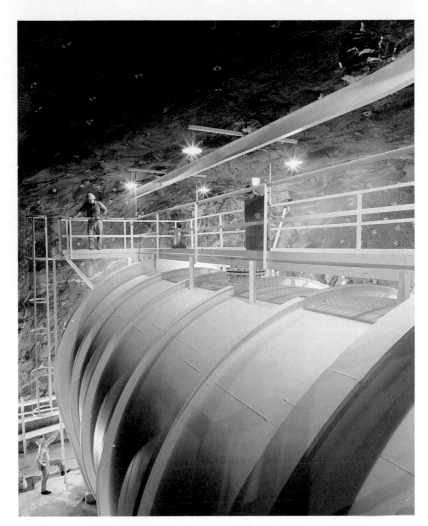

Figure 10-22
The Davis solar neutrino experiment counts neutrinos from the sun using 100,000 gallons of cleaning fluid held in a tank nearly a mile underground. A miniscule fraction of the solar neutrinos that pass through the cleaning fluid convert chlorine atoms into argon atoms that can be counted because they are radioactive. *(Brookhaven National Laboratory)*

(Figure 10-22). Of course, the mile of rock overhead has no effect on the neutrinos.

The result of the Davis experiment startled astronomers. The cleaning fluid detects too few neutrinos—not one neutrino per day as predicted by models of the sun, but about one every three days. The experiment has been refined, tested, and calibrated for three decades; but it has not found the missing neutrinos. In 1988, the Japanese detector, Kamiokande II, buried deep in a lead mine, confirmed the deficiency in solar neutrinos. The detectors count roughly one-third of the expected number of solar neutrinos.

The missing neutrinos left astronomers wondering if they misunderstood the nuclear reactions that power the sun. Over the years since the Davis experiment first

Scientists like to claim that every scientific belief is based on evidence, that every theory has been tested, and that the moment a theory fails a test, it is discarded or revised. The truth is much more complicated than that, and the solar neutrino problem is a good illustration. If the detection of solar neutrinos contradicts the theory of stellar structure, why hasn't the theory been abandoned?

While scientists do indeed have tremendous respect for evidence, they also have a faith in theories that have been tested successfully many times. If a theory has been tested and confirmed over and over, they may even begin to call it a natural law, and that means scientists have great faith in its truth. That is, they have confidence that the theory or law is a good description of how nature works.

Nevertheless, it is not unusual for an experiment or an observation to contradict well-established theories. In many cases, the experiments and observations are simple mistakes, or they have not been interpreted correctly. Scientists resist abandoning a well-tested theory even when an observation continues to contradict it. If confidence in the theory is stronger than the evidence, scientists begin testing the evidence. Can it be right? Do we understand it correctly? Of course, if the evidence cannot be impeached and it continues to contradict the theory, scientists must eventually abandon or modify the theory no matter how many times it has previously been tested and confirmed. This ultimate reliance on evidence is the distinguishing characteristic of science.

It is only human nature to hang onto the principles you have come to trust, and that confidence in well-tested scientific principles helps scientists avoid rushing to faulty judgments. For example, claims for perpetual-motion machines occasionally crop up in the news, but the world's scientists don't instantly abandon the laws of energy and motion pending an analysis of the latest claim. Of course, if such a claim did prove true, the entire structure of scientific knowledge would come crashing down, but because the known laws of energy and motion have been well tested and no perpetual-motion machine has ever been successful, scientists know which way to bet. Like the keel on a ship, confidence in well-tested theories and laws keeps the scientific boat from rocking before every little breeze.

Some forms of faith must be absolute and unshakable—religious faith, for example. But scientific faith may be better described as scientific confidence because it must be open to change. If, ultimately, a single experiment or observation conclusively contradicts our most cherished law of nature, we as scientists must abandon that law and find a new way to understand nature. We can see this scientific confidence at work in many controversies, from the origin of the human race to the meaning of IQ measurements; but one of the best examples is the solar neutrino problem. While we struggle to understand the origin of solar neutrinos, we continue to have confidence that we do indeed understand how stars work.

---

revealed a deficiency in solar neutrinos, other neutrino detectors have been built, and they confirm the basic observation. We count fewer solar neutrinos than theory predicts.

The solar neutrino problem has been one of the great scientific mysteries of the late twentieth century (Window on Science 10-2), and two kinds of solutions were proposed. Some scientists wondered if there are errors in the models of the sun, while others suggested that we don't yet understand neutrinos well enough to count them with certainty. After decades of research, the second suggestion seems more likely.

Physicists know that neutrinos come in three types, called "flavors," named electron, muon, and tau neutrinos. The Davis experiment can detect only electron neutrinos, so a group of scientists suggested a way that neutrinos might change back and forth—oscillate—among the three flavors while the neutrinos traveled through matter. If the electron neutrinos produced inside the sun could oscillate among the three flavors as they sped out of the sun, we would only detect a third as many as we expected. The other two-thirds of the electron neutrinos would have become tau or muon neutrinos, which the Davis experiment could not count.

To catch neutrinos in the act of oscillating, scientists used the Super Kamiokande neutrino detector, which contains 12.5 million gallons of hyper-pure water and is buried under a mountain in Japan. High-energy particles striking Earth's atmosphere from space should break up atoms and shower Earth with twice as many muon neutrinos as electron neutrinos. From under the mountain, the detector found the predicted numbers of muon and electron neutrinos coming from above, but equal numbers of muon and electron neutrinos coming up through Earth. This suggests that the neutrinos oscillated as they traveled through Earth's dense matter, just as the theory suggests solar neutrinos oscillate as they travel through the dense matter in the sun.

Neutrino oscillation would solve the solar neutrino mystery, but it is also exciting because it requires that neutrinos have mass. Neutrinos are so numerous in the universe that even a very small mass for each one could exert enough gravitational attraction to affect the evolution the universe as a whole. We will discuss this problem in Chapter 18.

The solar neutrino problem may now be solved by neutrino oscillation, and that gives astronomers added confidence that they understand how the sun makes its energy.

REVIEW Critical Inquiry

Why does nuclear fusion require that the gas be very hot?

Only under certain conditions do the nuclei of atoms fuse together to form a new nucleus. Inside a star, the gas is ionized, which means the electrons have been stripped off the atoms, and the nuclei are bare and have a positive charge. For hydrogen fusion, the nuclei are single protons. These atomic nuclei repel each other because of their positive charges, so they must collide with each other violently to overcome that repulsion and get close enough together to fuse. If the atoms in a gas are moving rapidly, we say it has a high temperature, and so nuclear fusion requires that the gas have a very high temperature. If the gas is cooler than about 10 million K, hydrogen can't fuse because the protons don't collide violently enough to overcome the repulsion of their positive charges.

The sun requires high temperature to make energy by nuclear fusion. Why does it need high density?

The sun is a star, and we now understand how it makes its energy and how that energy, flowing outward, drives solar activity. Now we turn to a larger problem. What are other stars like?

Summary

The atmosphere of the sun consists of three layers: the photosphere, chromosphere, and corona. The photosphere, or visible surface, is a thin layer of low-density gas that is the level in the sun from which visible photons most easily escape. It is marked by granulation, a pattern produced by gas currents rising from below the photosphere.

The chromosphere is most easily visible during total solar eclipses, when it flashes into view for a few seconds. It is a thin, hot layer of gas just above the photosphere, and its brilliant pink color is caused by the emission lines in its spectrum.

Filtergrams of the chromosphere reveal large jets called spicules extending up into the corona.

The corona is the sun's outermost atmospheric layer. It is composed of a very low density, very hot gas extending at least 30 solar radii from the sun. Its high temperature—up to 2×10^6 K—is believed to be maintained by the magnetic field and the rotation of the sun. The outer parts of the corona merge with the solar wind, a breeze of low-density ionized gas streaming away from the sun. Thus Earth, which is bathed in the solar wind, is orbiting within the solar atmosphere.

Sunspots are the most prominent example of solar activity. A sunspot seems dark because it is slightly cooler than the rest of the photosphere. The average sunspot is about twice the size of Earth and contains magnetic fields about 1000 times stronger than the sun's average field. Sunspots are thought to form because the magnetic field inhibits rising currents of hot gas and allows the surface to cool.

The average number of sunspots varies over a period of about 11 years and appears to be related to a 22-year magnetic cycle. The sunspot cycle does not repeat exactly each cycle, and the decades 1645 to 1715, known as the Maunder minimum, seem to have been a time when solar activity was very low.

Alternate sunspot cycles have reversed magnetic polarity, and this has been explained by the Babcock model of the magnetic cycle. In this model the differential rotation of the sun winds up the magnetic field. Tangles in the field rise to the surface and cause sunspot pairs. When the field becomes strongly tangled, it reorders itself into a simpler but reversed field, and the cycle starts over.

Prominences and flares are other examples of solar activity. Prominences occur in the chromosphere; their arch shapes show that they are formed of ionized gas trapped in the magnetic field. Flares, too, seem to be related to the magnetic field. They are sudden eruptions of X-ray, ultraviolet, and visible radiation, and high-energy particles that occur among the twisted magnetic fields around sunspot groups. Flares are important because they can have dramatic effects on Earth such as communications blackouts and auroras.

Activity in the corona is also guided by the magnetic field. The corona seems to be composed of streamers of thin, hot gas escaping from the magnetic field. In some regions of the corona, the magnetic field does not loop back to the sun, and the gas escapes unimpeded. These regions are called coronal holes and are believed to be the source of the solar wind.

The sun generates its energy near its center where the temperature and density are high enough for nuclear fusion reactions that combine hydrogen to make helium. Observations of too few neutrinos coming from the sun's core have been difficult to explain, but neutrino oscillation may explain the missing neutrinos.

New Terms

granulation	filtergram
convection	spicule

supergranule

helioseismology

sunspot

active region

Zeeman effect

Maunder butterfly diagram

Maunder minimum

differential rotation

Babcock model

prominence

flare

coronal hole

strong force

weak force

nuclear fission

nuclear fusion

Coulomb barrier

proton–proton chain

neutrinos

Review Questions

1. Why can't we see deeper than the photosphere?
2. What evidence do we have that granulation is caused by convection?
3. How are granules and supergranules related? How do they differ?
4. How can a filtergram reveal structure in the chromosphere?
5. What evidence do we have that the corona has a very high temperature?
6. What heats the chromosphere and corona to high temperature?
7. How are astronomers able to explore the layers of the sun below the photosphere?
8. What evidence do we have that sunspots are magnetic?
9. How does the Babcock model explain the sunspot cycle?
10. What does the spectrum of a prominence tell us? What does its shape tell us?
11. How can solar flares affect Earth?
12. Why does nuclear fusion require high temperatures?
13. Why does nuclear fusion in the sun occur only near the center?
14. How can astronomers detect neutrinos from the sun?
15. How can neutrino oscillation explain the solar neutrino problem?

Discussion Questions

1. What energy sources on Earth cannot be thought of as stored sunlight?
2. What would the spectrum of an auroral display look like? Why?

3. What observations would you make if you were ordered to set up a system that could warn astronauts in orbit of dangerous solar flares? Such a system exists.

Problems

1. The radius of the sun is 0.7 million km. What percentage of the radius is taken up by the chromosphere?
2. The smallest detail visible with ground-based solar telescopes is about 1 second of arc. How large a region does this represent on the sun? (*Hint:* Use the small-angle formula.)
3. What is the angular diameter of a star like the sun located 5 ly from the earth? Is the Hubble Space telescope able to detect detail on the surface of such a star?
4. If a sunspot has a temperature of 4200 K and the solar surface has a temperature of 5800 K, how many times brighter is the surface compared to the sunspot? (*Hint:* Use the Stefan-Boltzmann law, Chapter 4.)
5. A solar flare can release 10^{25} J. How many megatons of TNT would be equivalent? (*Hint:* A 1-megaton bomb produces about 4×10^{15} J.)
6. The United States consumes about 2.5×10^{19} J of energy in all forms in a year. How many years could we run the United States on the energy released by the solar flare in Problem 5?
7. Neglecting energy absorbed or reflected by our atmosphere, the solar energy hitting 1 square meter of Earth's surface is 1360 J/s (the solar constant). How long does it take a baseball diamond (90 ft on a side) to receive 1 megaton of solar energy? (*Hint:* See Problem 5.)
8. How much energy is produced when the sun converts 1 kg of mass into energy?
9. How much energy is produced when the sun converts 1 kg of hydrogen into helium? (*Hint:* How does this problem differ from Problem 8?)
10. A 1-megaton nuclear weapon produces about 4×10^{15} J of energy. How much mass must vanish when a 5-megaton weapon explodes?

Critical Inquiries for the Web

1. Do disturbances in one layer of the solar atmosphere produce effects in other layers? We have seen that filtergrams are useful in identifying the layers of the solar atmosphere and the structures within them. Visit a Web site that pro-

vides daily solar images, choose today's date (or one near it), and examine the sun in several wavelengths to explore the relation between disturbances in various layers.

2. Figure 10-17 shows the auroral displays caused by interaction of particles from the sun with our magnetosphere and ultimately our atmosphere. Explore the Web to find out how auroral activity is affected as solar activity rises and falls through the solar cycle. What changes in auroral visibility occur during this cycle? In what other ways can the increased activity associated with a solar maximum affect Earth?

 Go to the Brooks/Cole Astronomy Resource Center (www.brookscole.com/astronomy) for critical thinking exercises, articles, and additional readings from InfoTrac College Edition, Brooks/Cole's online student library.

CHAPTER ELEVEN

The Properties of Stars

If it can't be expressed in figures,

it is not science; it is opinion.

Robert Heinlein
The Notebooks of Lazarus Long

Guidepost

Science is based on measurement, but measurement in astronomy is very difficult. Even with the powerful modern telescopes described in Chapter 3, it is impossible to measure directly such simple parameters as the diameter of a star. This chapter shows how we can use the simple observations that are possible, combined with the basic laws of physics, to discover the properties of stars.

With this chapter we leave our sun behind and begin our study of the billions of stars that dot the sky. In a sense, the star is the basic building block of the universe. If we hope to understand what the universe is, what our sun is, what our Earth is, and what we are, we must understand stars.

Once we know how to measure these basic stellar data, we will be ready to trace the life stories of the stars from birth to death.

238

The stars are unimaginably remote. The nearest star is the sun, only 150 million km (93 million miles) away, so close that light takes only about eight minutes to reach Earth. The next nearest star is nearly 300,000 times farther away, a distance so great that the starlight takes over four years to reach Earth. These distances are so large we can express them in light-years. A light-year is the distance light travels in one year—about 9.5 trillion kilometers or about 5.9 trillion miles.

In spite of their great distances from us, the stars are the key to the secrets of the universe. The universe is filled with stars, and if we are to understand the universe, we must discover how stars are born, live, and die. We begin our study in this chapter by gathering data about the intrinsic properties of the stars—those properties inherent in the nature of the stars. In later chapters, we will use those data to deduce the life stories of different kinds of stars (Figure 11-1).

Figure 11-1
The central question of this chapter—what are stars like?—is manifest in this photo of the Trifid Nebula. Near the center of the cloud of gas and dust, new hot stars have been born. The hundreds of stars scattered in the background assure us that this is a common process. To understand how stars are born, evolve, and die, we must know the basic properties of the stars. *(NOAO and Nigel Sharp)*

Unfortunately, determining a star's intrinsic properties is quite difficult. When we look at a star through a telescope, we see only a point of light that tells us almost nothing about the star's energy production, temperature, diameter, or mass. Because we cannot visit stars, we can only observe from Earth and unravel their properties through the analysis of starlight. One of the reasons astronomy is interesting is that it contains so many such puzzles, each demanding a different method of solution.

To simplify our task in this chapter, we will concentrate on four intrinsic stellar properties. Our goals will be to find out how much energy stars emit, how hot their surfaces are, how large they are in diameter, and how much mass they contain. These four parameters will give us an overview of the nature of stars and provide us with the data we need to consider the birth, the lives, and the deaths of the stars.

Although we begin with four goals firmly in mind, we immediately meet a short detour. To find out how much energy a star emits, we must know how far away it is. If at night we see bright lights approaching on the highway, we cannot tell whether the lights are the intrinsically bright headlights of a distant truck or the intrinsically faint lights on a pair of nearby bicycles. Only

when we know the distance to the lights can we judge their intrinsic brightness. In the same way, to find the intrinsic brightness of a star, and thus the amount of energy it emits, we must know its distance from us. Our short detour will provide us with a method of measuring stellar distances.

Having reached our four goals, we will pause to consider the average densities of stars, an intrinsic property that is easily determined once we know a star's size and mass. The average densities of different kinds of stars will be helpful when we consider the internal structure of stars in the next two chapters.

We will conclude this chapter by considering the frequency of stellar types—that is, which kinds of stars are common and which are rare. This information, like all of the data in this chapter, is aimed at helping us understand what stars are.

11-1 Measuring the Distances to Stars

Distance is the most important and the most difficult measurement in astronomy, and astronomers have found many different ways to estimate the distance to stars. Yet each of those ways depends on a direct geometrical method that is much like the method surveyors would use to measure the distance across a river they cannot cross. We will begin by reviewing this method and then apply it to stars.

The Surveyor's Method

To measure the distance across a river, a team of surveyors begins by driving two stakes into the ground. The distance between the stakes is the baseline of the measurement. The surveyors then choose a landmark on the opposite side of the river, a tree perhaps, thus establishing a large triangle marked by the two stakes and the tree. Using their surveyor's instruments, they sight the tree from the two ends of the baseline and measure the two angles on their side of the river (Figure 11-2).

Knowing two angles of this large triangle and the length of the side between them, the surveyors can then find the distance across the river by simple trigonometry. Another way to find the distance is to construct a scale drawing. For example, if the baseline were 50 m and the angles were 66° and 71°, they could draw a line 50 mm long to represent the baseline. Using a protractor, they could construct angles of 66° and 71° at each end of the baseline, and then extend the two sides until they met at C, the location of the tree. Measuring the height of the triangle in the drawing, they would find it to be 64 mm high and would thus conclude that the distance across the river to the tree is 64 m.

The Astronomer's Method

To find the distance to a star, we must use a very long baseline, the diameter of Earth's orbit. If we took a photograph of a nearby star and then waited six months, Earth would have moved halfway around its orbit. We could then take another photograph of the star at a point in space 2 AU (astronomical units) from the point where the first photograph was taken. Thus our baseline would equal the diameter of Earth's orbit, or 2 AU.

We would then have two photographs of the same part of the sky taken from

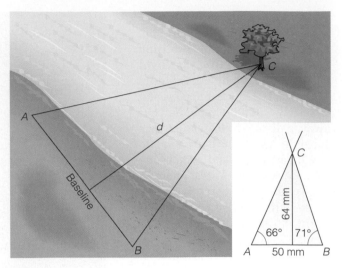

Figure 11-2
Surveyors can find the distance d across the river by measuring the baseline and the angles A and B and then constructing a scale drawing of the triangle.

slightly different locations in space. If we examined the photographs, we would discover that the star was not in exactly the same place in the two photographs. This apparent shift in the position of the star is called *parallax* (Figure 11-3).

Parallax is the apparent change in the position of an object due to a change in the location of the ob-

Figure 11-3
We can measure the parallax of a nearby star by photographing it from two points along Earth's orbit. For example, we might photograph it now, and again in six months. Half of the star's total change in position from one photograph to the other is its stellar parallax p.

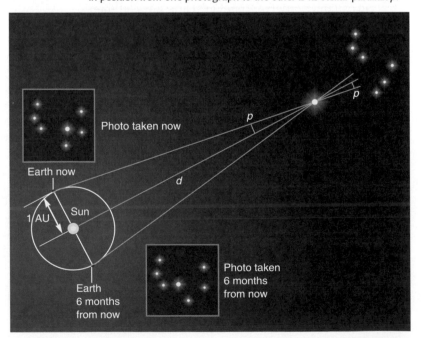

By the Numbers 11-1

Parallax and Distance

We wish to find the distance to a star from its measured parallax. To see how this is done, imagine that we observe Earth from the star. Figure 11-4 shows that the angular distance from the sun to Earth would equal the star's parallax p. To find the distance, we recall that the small-angle formula (see By the Numbers 1-2) relates an object's angular diameter, its linear diameter, and its distance. In this case, the angular diameter is p and the linear diameter is 1 AU. Then the small-angle formula, rearranged slightly, tells us that the distance to the star in AU is equal to 206,265 divided by the parallax in seconds of arc:

$$d = \frac{206,265}{p}$$

Because the parallaxes of even the nearest stars are less than 1 second of arc, the distances in AU are inconveniently large numbers. To keep the numbers manageable, astronomers have defined the parsec as their unit of distance in a way that simplifies the arithmetic. One parsec equals 206,265 AU, so the equation becomes:

$$d = \frac{1}{p}$$

Thus a parsec is the distance to an imaginary star whose parallax is 1 second of arc.

Example: The star Altair has a parallax of 0.20 second of arc. How far away is it? *Solution:* The distance in parsecs equals 1 divided by 0.2, or 5 pc:

$$d = \frac{1}{0.2} = 5 \text{ pc}$$

Because 1 pc equals about 3.26 ly, Altair is about 16.3 ly away.

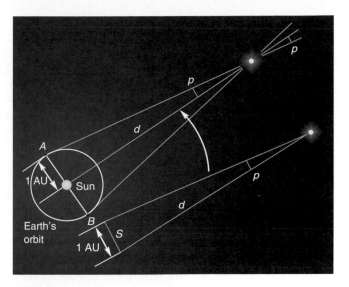

Figure 11-4

To measure the parallax of a star we must examine the long, thin triangle formed by Earth, the sun, and the star. The short side of the triangle, S, is the baseline of our measurement, 1 AU. If we were located at the star and looked back at Earth, the angular distance from Earth to the sun would equal the parallax, p. This means we can use the small-angle formula (see By the Numbers 1-2) to find the star's distance, d, using its parallax.

Astronomers measure the parallax, and surveyors measure the angles at the ends of the baseline, but both measurements tell us the same thing—the shape of the triangle and thus the distance to the object in question.

Measuring the small angle p is very difficult. The nearest star, α Centauri, has a parallax of only 0.76 second of arc, and the more distant stars have even smaller parallaxes. To see how small these angles are, hold a piece of paper edgewise at arm's length. The thickness of the paper covers an angle of about 30 seconds of arc.

We cannot use scale drawings to find the distances to stars because the distances are so large and the angles so small. Even for the nearest star, the triangle would have to be 300,000 times longer than it was wide. If the baseline in our drawing were 1 cm, the triangle would have to be about 3 km long. (By the Numbers 11-1 describes how we could find the distance from the parallax without drawing scale triangles.)

The distances to the stars are so large that it is not convenient to use astronomical units. As By the Numbers 11-1 explains, when we measure distance via parallax, it is convenient to use the unit of distance called a **parsec (pc).** One parsec is 206,265 AU, roughly 3.26 ly (light-years).*

---

*The parsec is used throughout astronomy because it simplifies the calculation of distance. However, there are instances in which the light-year is also convenient. Consequently, the chapters that follow use either parsecs or light-years as convenience and custom dictate.

server. We saw in Chapter 2 that parallax is an everyday experience. Our thumb, held at arm's length, appears to shift position against a distant background when we look with first one eye and then with the other (see Figure 2-4). In this case, the baseline is the distance between our eyes, and the parallax is the angle through which our thumb appears to move when we change eyes. The farther away we hold our thumb, the smaller the parallax. If we know the length of the baseline and measure the parallax, we can calculate the distance from our eyes to our thumb.

Because the stars are so distant, their parallaxes are very small angles, usually expressed in seconds of arc. The quantity that astronomers call **stellar parallax (p)** is half the total shift of the star, as shown in Figure 11-3.

Every measurement has some uncertainty, and the uncertainty in a parallax measurement made with ground-based telescopes is at best about 0.002 second of arc. If we measure a parallax of 0.002 second of arc, the uncertainty is as big as the quantity we are trying to measure. To say that another way, our uncertainty would be 100 percent. To limit our uncertainty to only 10 percent, we can't measure parallaxes smaller than about 0.02 second of arc, which corresponds to a distance of 50 pc. Thus ground-based parallax measurements can't go farther than about 50 pc, and only about 10,000 stars have measured parallaxes. Modern astronomy has been based on a small sample of parallax observations that astronomers struggled to measure through Earth's turbulent atmosphere.

A space telescope would provide a view from above Earth's atmosphere, so the European Space Agency launched the satellite Hipparcos in 1989 to measure stellar parallax. The satellite accumulated parallax data over the following four years, and three more years were needed to reduce the data. In June 1997, the data were published as catalogues, one containing 120,000 stars with parallaxes accurate to 0.001 second of arc and another containing over a million stars with parallaxes as accurate as ground-based observations. The Hipparcos data are giving astronomers a new insight into the nature of stars.

REVIEW Critical Inquiry

Why are parallax measurements made from space better than parallax measurements made from Earth?

At first we might suppose that a satellite in orbit can measure the parallax of the stars better because the satellite is closer to the stars, but when we recall the immense distances to the stars, we see that being in space doesn't really put the satellite significantly closer to the stars. Rather, the satellite is above Earth's atmosphere. When we try to measure parallax, the turbulence in Earth's atmosphere blurs the star images and smears them out into blobs roughly 1 second of arc in diameter. It isn't possible to measure the position of these fuzzy images more accurately than about 0.02 second of arc, so we can't measure parallax smaller than that. A parallax of 0.02 second of arc corresponds to a distance of 50 pc so we can't measure parallax accurately beyond that distance.

A satellite in orbit, however, is above Earth's atmosphere, so the only blurring in the star images is that produced by diffraction in the optics. In other words, the star images are very sharp, and a satellite in orbit can measure the positions of stars and thus their parallaxes to high accuracy.

If a satellite can measure parallaxes as small as 0.002 second of arc, then how far are the most distant stars it can measure? To put this distance into scale, express it as a percent of the diameter of our galaxy (Introduction).

Distance is the key to the secrets of the stars. Now that we know how to find parallaxes, we are ready to think about stellar brightness.

II-2 Intrinsic Brightness

Our eyes tell us that some stars look brighter than others, but unless we know how distant a star is, we cannot judge its intrinsic brightness. It might be a very distant star emitting tremendous amounts of light or a nearby star emitting much less light. To discover the true brightnesses of the stars, we must know how brightness depends on distance.

Brightness and Distance

When we look at a bright light, our eyes respond to the visual-wavelength energy falling on the eye's retina, which tells us how bright the object looks. Thus, brightness is related to the flux of energy entering our eye. **Flux** is the energy in joules (J) per second falling on one square meter. (We commonly refer to 1 J/s as 1 watt.)

If we placed a screen 1 meter square near a light-bulb, a certain amount of flux would fall on the screen (Figure 11-5). If we moved the screen twice as far from the bulb, the light that previously fell on the screen would be spread to cover an area four times larger, and a spot on the screen would receive only one-fourth as much light. If we tripled the distance to the screen, the spot would receive only one-ninth as much light. Thus the flux we receive from a light source depends inversely on the square of the distance to the source. This is known as the inverse square relation. (We first encountered the inverse square relation in Chapter 2, where it was applied to the strength of gravity.)

Absolute Visual Magnitude

If all the stars were the same distance away, we could compare one with another and decide which was emitting more light and which less. Of course, the stars are scattered at different distances, and we can't shove them around to line them up for comparison. If, however, we know the distance to a star, we can use the inverse square relation to calculate the brightness the star would have at some standard distance. Astronomers take 10 pc as the standard distance and refer to the in-

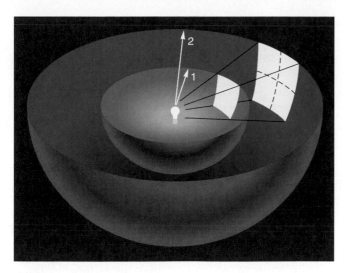

Figure 11-5
The inverse square relation. A light source is surrounded by spheres with radii of 1 unit and 2 units. The light falling on an area of 1 m² on the inner sphere spreads to illuminate an area of 4 m² on the outer sphere. Thus, the brightness of the light source is inversely proportional to the square of the distance.

trinsic brightness of the star as its **absolute visual magnitude (M_v),** which is the apparent visual magnitude the star would have if it were 10 pc away.

For example, suppose we know that a star is 100 pc away and has an apparent magnitude of 8.5. Because it is 10 times farther than the standard distance, it must

$m - M_v$	d (in pc)
0	10
1	16
2	25
3	40
4	63
5	100
6	160
7	250
8	400
9	630
10	1000
⋮	⋮
15	10,000
⋮	⋮
20	100,000
⋮	⋮

Table 11-1 Distance Moduli

look 100 times fainter than it would at the standard distance. We recall from our discussion of the magnitude system (Chapter 1) that a factor of 100 times in brightness is the same as a difference of 5 magnitudes, so the star must look 5 magnitudes fainter than its absolute magnitude. Thus the absolute magnitude of the star must be 8.5 minus 5, or 3.5. (By the Numbers 11-2 expresses this relation in a simple equation.)

The symbol for absolute visual magnitude is a capital M with a subscript v. The subscript tells us it is a

---

By the Numbers 11-2

Absolute Magnitude

Apparent magnitude tells us how bright a star looks (see By the Numbers 1-1), but absolute magnitude tells us how bright the star really is. The absolute visual magnitude M_v of a star is the apparent visual magnitude of the star if it were 10 pc away. If we know a star's apparent magnitude and its distance from us, we can calculate its absolute magnitude. The equation that allows this calculation relates apparent magnitude m, distance in parsecs d, and absolute magnitude M_v:

$$m - M_v = -5 + 5\log_{10}(d)$$

Sometimes it is convenient to rearrange the equation and write it in this form:

$$d = 10^{(m-M_v+5)/5}$$

It is the same equation, so we can use whichever form is most convenient in a given problem. This equation shows that the difference between apparent and absolute magnitude depends only on the distance to the star.

The quantity $m - M_v$ is called the **distance modulus,** a measure of how far away the star is. If the star is very far away, the distance modulus is large; if the star is close, the distance modulus is small. We could use the preceding equation to make a table of distance moduli (Table 11-1).

If we know the distance to a star, we can find its distance modulus from the table. If we subtract the distance modulus from the apparent magnitude, we get the star's absolute magnitude.

Example: Deneb is 490 pc away and has an apparent magnitude of 1.26. What is its absolute magnitude? *Solution:* From the table, we find that Deneb's distance modulus is about 8.5. Thus, its absolute magnitude is about 1.26 minus 8.5, or about −7.2. Deneb is intrinsically a very bright star. If it were only 10 pc away, it would dominate the night sky, shining 5.8 magnitudes or over 200 times more brightly than Sirius.

visual magnitude based only on the wavelengths of light we can see. Other magnitude systems are based on other parts of the electromagnetic spectrum, such as the infrared and ultraviolet.

The intrinsically brightest stars known have absolute magnitudes of about −8, and the faintest about +15 or fainter. The nearest star to the sun, α Centauri, is only 1.4 pc away, and its apparent magnitude is 0.0, indicating that it looks bright in the sky. Its absolute magnitude, however, is 4.39, telling us it is not intrinsically very bright. Because we know the distance to the sun and can measure its apparent magnitude, we can find its absolute magnitude—about 4.78. If the sun were only 10 pc away from us, it would look no brighter than the faintest star in the handle of the Little Dipper.

Our detour to find the distance to stars has led us to absolute magnitude, a measure of the true brightness of the stars. We are now ready to reach the first of our four goals in this chapter.

Luminosity

The first of our four goals for this chapter was to find out how much energy the stars emit. With the absolute magnitudes of the stars in hand, we can now compare other stars with our sun and find their **luminosities (L)**, the total amount of energy they emit per second.

We know the sun has an absolute magnitude of 4.78, and we know how much energy the sun emits because we can measure it as it falls on Earth's surface. The luminosity of the sun is 4×10^{26} J/s. If a star has an absolute magnitude that is 5 magnitudes brighter than the sun, then that star must emit 100 times more energy per second.

However, one correction is necessary. The absolute visual magnitude of a star refers only to visible light, but luminosity includes energy of all wavelengths. Hot stars emit a great deal of ultraviolet radiation we cannot see, and cool stars emit infrared. Thus astronomers must make a correction that depends on the star's temperature. With that correction, we can use the absolute magnitudes of the stars to calculate their luminosities.

We can express luminosity in two ways. For example, we can say that the star Capella, which is 100 times more luminous than the sun, has a luminosity of 100 solar luminosities, or we can multiply by the luminosity of the sun and write the luminosity of Capella as 4×10^{28} J/s.

When we look at the stars in the night sky, they look much the same, yet some are astonishingly more luminous than others. The most luminous stars are roughly a million times more luminous than the sun, and the least luminous are 10,000 times less luminous than the sun. Clearly, the stars are much more than just twinkling points in the sky.

Although we had to make a detour to find the distances to the stars, we have now reached our first goal. We have found the luminosities of the stars. When we reach the next goal, we will know even more about the stars.

11-3 Spectra of Stars

Our quest is to understand stars, and the spectrum of a star is a treasure of information. Most important, stellar spectra can tell us the precise temperatures of stars. That will help us find their diameters later in this chapter.

The Star Thermometer

When we discussed spectra in Chapter 4, we discovered that the spectrum of a star is an absorption spectrum formed by the gas at the surface of a star absorbing photons to produce dark spectral lines. Which lines we see depends on the temperature of the gases at the surface of the star. Stellar spectra are precision star thermometers.

To understand the importance of temperature, we can begin with the spectrum of hydrogen. The spectral lines of hydrogen that fall in the visible and near-ultraviolet part of the spectrum are called the **Balmer lines,** and three are visible to our eyes. A deep red line is called H-alpha, a blue line is H-beta, and a violet line is H-gamma. (See Figure 4-7.) Other Balmer lines in the ultraviolet are easily photographed but are not visible to our eyes. These three visible lines explain the purple-pink color of excited hydrogen gas that we see in the solar chromosphere (Figure 1-22) and in clouds of gas in space (Figure 11-1). We are seeing the blended light of the three visible Balmer lines.

Balmer lines are produced by hydrogen at a medium temperature of roughly 10,000 K. In a hot star, the hydrogen atoms are too excited, and some may be ionized; that is, they have lost their electron. Thus hot stars have weak Balmer lines. If the gas is cool, the hydrogen atoms are not excited enough to absorb the Balmer wavelength photons. Consequently, cool stars have weak Balmer lines. Figure 11-6a shows how the strength of the Balmer lines depends on temperature.

Other atoms behave the same way that hydrogen does, producing lines of maximum strength at some intermediate temperature, but that temperature is different for each atom. For example, ionized calcium (calcium atoms that have lost one electron) produces two lines in the near-ultraviolet that are strongest if the temperature is around 5000 K. We can add a curve for the strength of ionized calcium lines to our graph as in Figure 11-6b.

If we add more curves to our graph to represent the temperature dependence of other atoms, we get a powerful tool to study the stars. Our graph (Figure 11-6c) is a precision star thermometer. We can find a star's temperature by comparing its spectrum to our graph. For example, if we see in a star's spectrum, very weak Balmer lines and strong lines of ionized helium, we know the star is very hot. If we see medium-strength Balmer lines and weak lines of ionized calcium, we know the star is about 7000 K, a bit hotter than the sun. If a stellar spectrum contains dark bands produced by the titanium oxide molecule (TiO) we know immediately that the star is very cool. Only in the coolest stars can these molecules avoid being broken up by collisions with other atoms.

Our star thermometer is a precision instrument. In Chapter 4 we discovered that we could estimate a star's temperature from its color. Red stars are cool, and blue stars are hot. Stellar spectra can give us accurate measurements of the temperatures of stellar surfaces. The hottest stars have surface temperatures of 40,000 K and the coolest 2000 K. Compare these with the surface temperature of the sun, about 5800 K.

Spectral Classification

We have seen that the strengths of spectral lines depend on the surface temperature of the star. From this we can predict that all stars of a given temperature

Figure 11-6

(a) The strength of the Balmer lines in a stellar spectrum depends on the temperature of the star. (b) The curves for lines of hydrogen and ionized calcium reach maximum strength at different temperatures. (c) The strengths of the spectral lines produced by each atom, ion, and molecule depend on the temperature of the star. Astronomers can compare the strengths of lines in a stellar spectrum with a diagram such as this to find the temperature of the star.

should have similar spectra. If we learn to recognize the pattern of spectral lines produced by a 6000-K star, for instance, we need not use Figure 11-6c every time we see that kind of spectrum. We can save time by classifying stellar spectra rather than analyzing each one individually.

Table 11-2 Spectral Classes

Spectral Class	Approximate Temperature (K)	Hydrogen Balmer Lines	Other Spectral Features
O	40,000	Weak	Ionized helium
B	20,000	Medium	Neutral helium
A	10,000	Strong	Ionized calcium weak
F	7500	Medium	Ionized calcium weak
G	5500	Weak	Ionized calcium medium
K	4500	Very weak	Ionized calcium strong
M	3000	Very weak	Titanium oxide strong

The first widely used classification system was devised by astronomers at Harvard during the 1890s and 1900s. One of them, Annie J. Cannon, personally inspected and classified the spectra of over 250,000 stars. The spectra were first classified in groups labeled A through Q, but some groups were later dropped, merged with others, or reordered. The final classification includes the seven **spectral classes,** or **types,** still used today: O, B, A, F, G, K, M.*

This sequence of spectral types, called the **spectral sequence,** is important because it is a temperature sequence. The O stars are the hottest, and the temperature continues to decrease down to the M stars, the coolest. For maximum precision, astronomers divide each spectral class into 10 subclasses. For example, spectral class A consists of the subclasses A0, A1, A2, . . . A8, A9. Next comes F0, F1, F2, and so on. This finer division gives us a star's temperature to an accuracy within about 5 percent. The sun, for example, is not just a G star, but a G2 star, with a temperature of 5800 K.

We classify a star by the lines and bands in its spectrum as shown in Table 11-2. For example, if it has weak Balmer lines and lines of ionized helium, it must be an O star. This table is based on the same information we used in Figure 11-6c.

Figure 11-7 illustrates the spectra of stars of different spectral types. Notice that the spectra extend into the near ultraviolet to reveal the two important calcium lines called H and K. The human eye cannot see wavelengths shorter than about 420 nm.

With the rise of computers, astronomers began displaying spectra as graphs of intensity versus wavelength. Such graphs show more detail than photographic plates, and they can be measured directly. Figure 11-8 shows graphical spectra for stars of different spectral types. In these spectra, absorption lines are sharp dips in the curve. Compare Figures 11-7 and 11-8 to find similar lines and bands in the spectra, and notice how detailed the graphical spectra are.

Not only can a star's spectrum tell us the surface temperature of the star, but it can give us other information. In Chapter 4 we discussed the Doppler effect that allows us to measure radial motion, and, in Chapter 10, we discussed the Zeeman effect that we used to measure magnetic fields. Spectra can also give us information about the chemical composition of stars, but that information is confused by temperature, so the analysis must be done carefully.

Chemical Composition

Identifying the elements in a star by identifying the lines in the star's spectrum is a relatively straightforward procedure. For example, two dark absorption lines appear in the yellow region of the solar spectrum at the wavelengths 589 nm and 589.6 nm. The only atom that can produce this pair of lines is sodium, so we must conclude that the sun contains sodium. Over 90 elements in the sun have been identified this way.

However, just because the spectral lines characteristic of an element are missing, we cannot conclude that the element itself is absent. For example, the hydrogen Balmer lines are weak in the sun's spectrum, yet 90 percent of the atoms in the sun are hydrogen. The reason for this apparent paradox is that the sun is too cool to produce strong Balmer lines. Similarly, an element's spectral lines may be absent from a star's spectrum because the star is too hot or too cool to excite those atoms to the energy levels that produce visible spectral lines.

To derive accurate chemical abundances, we must use the physics that describes the interaction of light and matter to analyze a star's spectrum, take into account the star's temperature, and calculate the amounts

*Generations of astronomy students have remembered the spectral sequence using the mnemonic "Oh, Be A Fine Girl (Guy), Kiss Me."

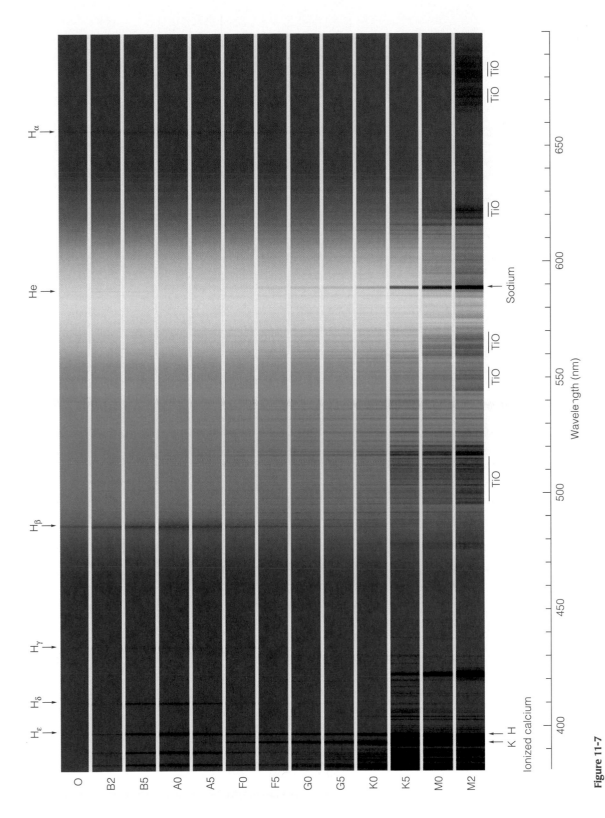

Figure 11-7
These stellar spectra were produced by computer simulation of real spectra. The spectra run from the hot O star at the top to the cool M2 star at the bottom. In the spectra of the hot stars we see weak Balmer lines and barely visible helium lines. The Balmer lines are strongest about A0 but are very weak in cool stars. The two ultraviolet lines of ionized calcium are strong in cooler stars, while sodium lines and titanium oxide (TiO) bands are strong in the spectra of the coolest stars. These spectra have been extended past the short-wavelength limit for the human eye, about 420 nm, to show the H and K lines of calcium. *(Courtesy Roger Bell and Michael Briley)*

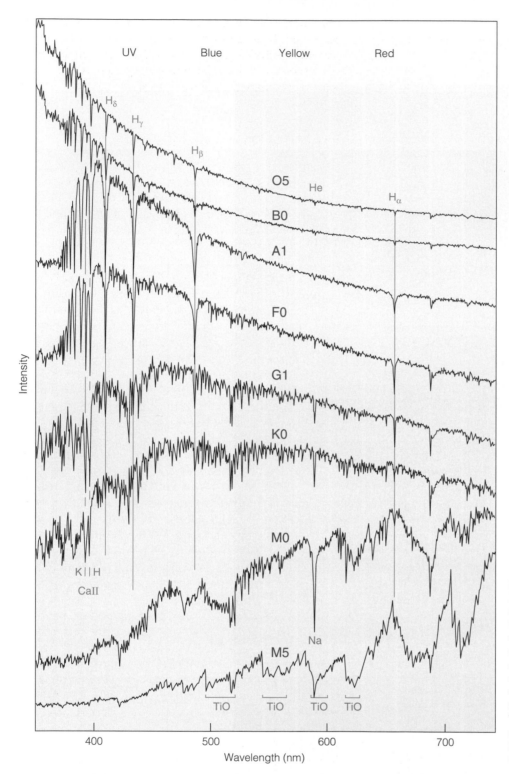

UV   Blue   Yellow   Red

H_δ

H_γ

H_β

O5

He

B0

A1

H_α

F0

G1

K0

M0

K||H

CaII

M5

Na

TiO   TiO   TiO   TiO

Intensity

400   500   600   700

Wavelength (nm)

Figure 11-8
Modern digital spectra show how stellar spectra depend on spectral class. Here spectra are represented by graphs of intensity versus wavelength, and dark absorption lines appear as sharp dips in intensity. Hydrogen Balmer lines are strongest at about A0, while lines caused by ionized calcium (CaII) are strong in K stars. Bands produced by TiO molecules are strong in the coolest stars. Compare these spectra with Figures 11-6c and 11-7. *(Courtesy NOAO, G. Jacoby, D. Hunter, and C. Christian)*

of the elements present in the star. Such results show that nearly all stars have compositions similar to the sun's—about 91 percent of the atoms are hydrogen, and 8.9 percent are helium, with small traces of heavier elements (Table 11-3). We will use these results in later chapters when we analyze the life stories of the stars, the history of our galaxy, and the origin of the universe.

R E V I E W   Critical Inquiry

Why are helium lines weak and calcium lines strong in the visible spectrum of the sun?

To analyze this problem, we must recall that the ability of an atom or ion to absorb light depends on tempera-

Table 11-3 The Most Abundant Elements in the Sun

Element	Percentage by Number of Atoms	Percentage by Mass
Hydrogen	91.0	70.9
Helium	8.9	27.4
Carbon	0.03	0.3
Nitrogen	0.008	0.1
Oxygen	0.07	0.8
Neon	0.01	0.2
Magnesium	0.003	0.06
Silicon	0.003	0.07
Sulfur	0.002	0.04
Iron	0.003	0.1

ture. Helium is quite abundant in the sun, but the surface of the sun is too cool to excite helium atoms and enable them to easily absorb visible-wavelength photons. On the other hand, calcium is easily ionized, and calcium atoms that have lost an electron are very good absorbers of photons at temperatures like those at the sun's surface. Thus, calcium lines are strong in the solar spectrum even though calcium ions are rare, and helium lines are weak even though helium atoms are common.

If we are to analyze a star's spectrum and consider the strength of spectral lines, we must take into account the temperature of the star. Why don't we need to take temperature into account when we use the Doppler effect to measure a star's radial velocity?

We began this chapter by finding the distances to the stars, and that allowed us to reach our first goal, to discover the luminosities of stars. Now we have found a way to measure stellar temperatures, and that will allow us to achieve our third goal, to find the diameters of the stars.

11-4 The Diameters of Stars

Our third goal in this chapter is to find the diameter of the stars, because diameter is a fundamental property. We know little about stars until we know their diameters. Are they all the same size as our sun, or are some larger and some smaller? We certainly can't see their diameters through a telescope; the stars are much too small for us to resolve their disks and measure their diameters. But there is a way to find out how big stars really are. If we know their temperature and luminosity, we can find their diameters. That relationship will introduce us to the most important diagram in astronomy, where we will discover the family relations among the stars.

Luminosity, Radius, and Temperature

The luminosity and temperature of a star can tell us its diameter if we understand the two factors that affect a star's luminosity: surface area and temperature. For example, you can eat dinner by candlelight because the candle flame has a small surface area, and although it is very hot, it cannot radiate much heat; it has a low luminosity. However, if the candle flame were 12 ft tall, it would have a very large surface area from which to radiate, and, although it would be no hotter than a normal candle flame, its luminosity would drive you from the table.

In a similar way, a star's luminosity is proportional to its surface area. A hot star may not be very luminous if it has a small surface area, although it could be highly luminous if it were larger. Even a cool star could be luminous if it had a large surface area. (See By the Numbers 11-3.) Because of this dependence on both temperature and surface area, we can use stellar luminosities to determine the diameters of stars if we can separate the effects of temperature and surface area.

The **Hertzsprung-Russell (H-R) diagram,** named after its originators, Ejnar Hertzsprung and Henry Norris Russell, is a graph that separates the effects of temperature and surface area on stellar luminosities and enables us to sort the stars according to their diameters. Before we discuss the details of the H-R diagram (as it is often called), let us look at a similar diagram we might use to sort automobiles.

We can plot a diagram such as Figure 11-9 to show horsepower versus weight for various makes of cars. And in so doing, we find that in general the more a car weighs, the more horsepower it has. Most cars fall somewhere along the sequence of cars running from heavy, high-powered cars to light, low-powered models. We might call this the main sequence of cars. But some cars have much more horsepower than normal for their weight—the sport or racing models—and the economy models have less power than normal for cars of the same weight. Just as this diagram helps us understand the different kinds of autos, the H-R diagram helps us understand the kinds of stars.

The H-R diagram relates the intrinsic brightness of stars to their surface temperatures (Figure 11-10). We may plot either absolute magnitude or luminosity on the vertical axis of the graph, because both refer to intrinsic brightness. Spectral type is related to temperature, so we may plot either spectral type or temperature on

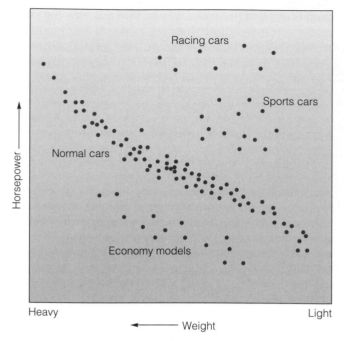

Figure 11-9
We could analyze automobiles by plotting their horsepower versus their weight and thus reveal relationships between various models. Most would lie somewhere along the main sequence of "normal" cars.

the horizontal axis. Technically, only graphs of absolute magnitude versus spectral type are H-R diagrams. However, we will refer to plots of luminosity versus either spectral type or surface temperature by the generic term *H-R diagram*.

A point on an H-R diagram shows a star's luminosity and surface temperature. Points near the top of the diagram represent very luminous stars, and points near the bottom represent very faint stars. Points on the left represent hot stars, and points on the right represent cool stars. Notice that the location of a star in the H-R diagram has nothing to do with its location in space. Also, as we will see in later chapters, changes in a star's luminosity and surface temperature as it ages mean that its position on the H-R diagram changes, but this has nothing to do with the star's actual motion through space.

Dwarfs, Giants, and Supergiants

The **main sequence** is the region of the H-R diagram running from upper left to lower right, which includes roughly 90 percent of all stars. These are the "ordinary" stars. As we might expect, the hot main-sequence stars are brighter than the cool main-sequence stars. The sun is a medium-temperature main-sequence star.

Just as sports cars do not fit in with the normal cars in Figure 11-9, some stars do not fit in with the main-

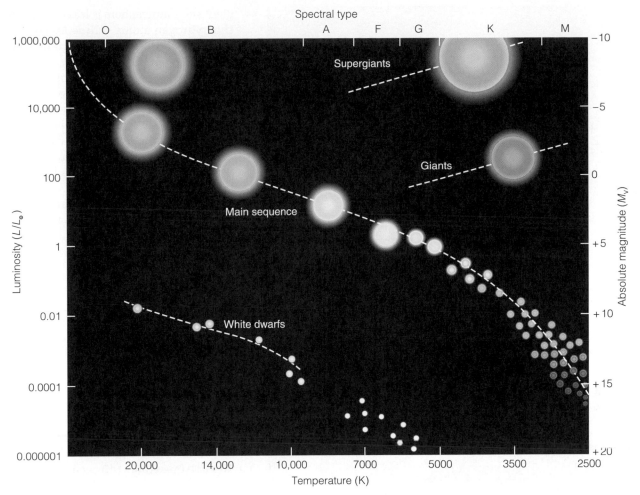

Figure 11-10
An H-R diagram is a graph of luminosity versus temperature on which stars are represented as points. Spectral types have been added at the top to show how they are related to temperature. Absolute magnitude has been added at the right to show its relationship to luminosity. Star diameters are not to scale in this schematic diagram.

sequence stars in Figure 11-10. The **giant stars** lie at the upper right in the H-R diagram. These stars are cool, radiating little energy per square meter. Nevertheless, they are highly luminous because they have enormous surface areas, hence the name *giant stars.* In fact, we can estimate the size of these giants with a simple calculation. Notice that giants are about 100 times more luminous than the sun even though they have about the same surface temperature. Thus, they must have about 100 times more surface area than the sun. This means they must be about 10 times larger in diameter than the sun (Figure 11-11).

Near the top of the H-R diagram we find the **supergiants.** These exceptionally luminous stars are 10 to 1000 times the diameter of the sun. Betelgeuse in Orion is a supergiant (Figure 11-11). If it magically replaced the sun at the center of our solar system, it would swallow up Mercury, Venus, Earth, and Mars, and would just

reach Jupiter. μ Cephei is believed to be over three times larger than Betelgeuse. It would reach nearly to the orbit of Uranus.

In the lower left of the H-R diagram are the economy models, stars that are very faint even though they are hot. Clearly such stars must be small. These are the **white dwarf stars,** about the size of Earth (Figure 11-11).

Luminosity Classification

We can tell from a star's spectrum what kind of star it is. Main-sequence stars are rather small and have dense atmospheres, but giant and supergiant stars are larger and have lower-density atmospheres.

Some spectral lines are particularly sensitive to density. If the atoms collide often in a dense gas, their energy levels become distorted and the spectral lines

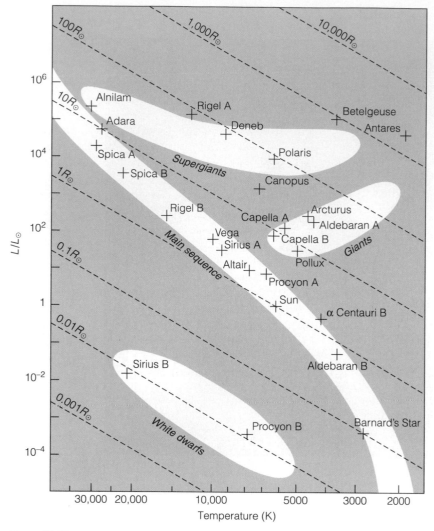

Figure 11-11
An H-R diagram showing the luminosity and temperature of many well-known stars. The dashed lines are lines of constant radius. (Individual stars that orbit each other are designated A and B, as in Spica A and Spica B.)

giant star's atmosphere is less dense and the hydrogen atoms collide less often. In the spectrum of a supergiant star, the hydrogen lines are very narrow.

Thus, we can look at a star's spectrum and classify its luminosity. Although these are **luminosity classes,** the names refer to the sizes of stars because size is the dominating factor determining luminosity. Supergiants, for example, are very luminous because they are very large.

The luminosity classes are represented by the roman numerals I through V, with supergiants further subdivided into types Ia and Ib, as follows:

Luminosity Classes

Ia   Bright supergiant

Ib   Supergiant

II   Bright giant

III   Giant

IV   Subgiant

V   Main-sequence star

Using letters for subclasses, we can distinguish between the bright supergiants (Ia) such as Rigel (β Orionis) and the regular supergiants (Ib) such as Polaris, the North Star. The star Adhara (ε Canis Majoris) is a bright giant (II), Capella (α Aurigae) is a giant (III), and Altair (α Aquilae) is a subgiant (IV). The sun is a main-sequence star (V). The luminosity class appears after the spectral type, as in G2 V for the sun. White dwarfs don't enter into this classification because their spectra are peculiar.

We can plot the positions of the luminosity classes on the H-R diagram (Figure 11-13). Remember that these are rather broad classifications. A star of luminosity class III may lie slightly above or below the line labeled III. The lines are only approximate.

are broadened. Hydrogen lines are an example. In the spectrum of a main-sequence star, the hydrogen lines are broad because the star's atmosphere is dense and the hydrogen atoms collide often. In the spectrum of a giant star, the lines are narrower (Figure 11-12), because the

Figure 11-12
Differences in widths and strengths of spectral lines distinguish the spectra of supergiants, giants, and main-sequence stars. *(Adapted from H. A. Abt, A. B. Meinel, W. W. Morgan, and J. W. Tapscott,* An Atlas of Low-Dispersion Grating Stellar Spectra, *Kitt Peak National Observatory, 1968)*

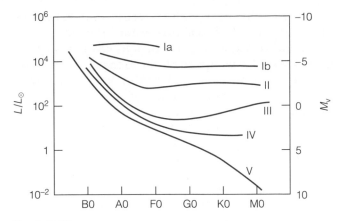

Figure 11-13
The approximate location of the luminosity classes on the H-R diagram.

Spectroscopic Parallax

The luminosity classification is subtle and not too accurate, but it is an important tool in modern astronomy. As we will see in the next section, it gives us a way to find the distances to stars.

We can measure the parallax of nearby stars, but most stars are too distant to have measurable parallaxes. We can find the distances to these stars if we can photograph their spectra and determine their luminosity classes in a process called **spectroscopic parallax**—the estimation of the distance to a star from its spectral type, luminosity class, and apparent magnitude. Spectroscopic parallax does not actually involve measuring parallax, but it does tell us the distance to the star.

The method of spectroscopic parallax depends on the H-R diagram. If we photograph the spectrum of a star, we can determine its spectral class, and that tells us its horizontal location in the H-R diagram. We can also determine its luminosity class by looking at the widths of its spectral lines, and that tells us the star's vertical location in the diagram. Once we plot the point that represents the star in the H-R diagram, we can read off its absolute magnitude. As we have seen earlier in this chapter, we can find the distance to a star by comparing its apparent and absolute magnitudes.

For example, Spica is classified B1 V, and its apparent magnitude is +1. We can plot this star in an H-R diagram such as Figure 11-13, where we would find it should have an absolute magnitude of about −3. Therefore, its distance modulus is 1 minus −3, which equals 4, and the distance (taken from Table 11-1) is about 63 pc.

REVIEW **Critical Inquiry**

What evidence do we have that giant stars really are bigger than the sun?

We can find stars that have the same spectral type as the sun, but they are clearly more luminous than the sun. Capella, for example, is a G star with an absolute magnitude of 0. If it is a G star, then it must have about the same temperature as the sun, but its absolute magnitude is 5 magnitudes brighter than the sun's. A difference of 5 magnitudes corresponds to an intensity ratio of 100, so Capella must be about 100 times more luminous than the sun. If it has the same surface temperature as the sun but it is 100 times more luminous, then it must have a surface area that is 100 times bigger than the sun's. Because the surface area of a sphere is proportional to the square of the radius, Capella must be 10 times larger in radius. That is clear observational evidence that Capella is a giant star.

In Figure 11-11, we see that Procyon B is a white dwarf only slightly warmer than the sun but about 10,000 times less luminous than the sun. Use the same logical procedure followed above to prove that the white dwarfs must be small stars.

The first three goals of this chapter have been achieved; we know how to find the luminosities, temperatures, and diameters of stars. The next step is to find the masses of the stars.

II-5 The Masses of Stars

Our fourth goal is to find out how much matter stars contain. That is, we want to know their mass. Do they all contain about the same amount of mass as our sun, or are some more massive and others less? Unfortunately, it's difficult to determine the mass of a star. Looking through a telescope at a star, we see only a point of light that tells us nothing about the mass of the star. Before we can measure the masses of stars, we must discuss the motions of **binary stars,** pairs of stars that orbit each other.

Binary Stars in General

The key to finding the mass of a binary star is understanding orbital motion, which provides clues to mass.

In Chapter 2, we illustrated orbital motion by imagining a cannon firing a cannonball from a high mountain (see Figure 2-24). If Earth's gravity didn't act on the cannonball, it would follow a straight-line path and leave Earth forever. Because Earth's gravity pulls it away from its straight-line path, the cannonball follows a curved path around Earth—an orbit. In the case of two stars orbiting each other, it is the mutual gravitation of the stars pulling on each other that pulls them away from straight-line paths and makes them follow closed orbits around a point between the stars.

Each star in a binary system moves in its own orbit around the system's **center of mass,** the balance point of the system. If the stars were connected by a massless rod and placed in a uniform gravitational field such as that near Earth's surface, the system would balance at its center of mass like a child's seesaw (Figure 11-14a). If one star were more massive than its companion, then the massive star would be closer to the center of mass and would travel in a smaller orbit, while the lower-mass star would whip around in a larger orbit (Figure 11-14b). The ratio of the masses of the stars M_A/M_B always equals the inverse of the ratio of the sizes of the orbits r_B/r_A. If one star has an orbit twice as large as the other star's orbit, then it must be half as massive. Getting the ratio of the masses is easy, but that doesn't tell us the individual masses of the stars, which is what we really want to know. That takes further analysis.

To find the mass of a binary star system, we must know the size of the orbits and the orbital period—the

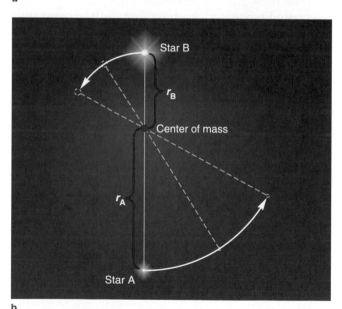

Figure 11-14
(a) The center of mass between two stars in a binary system is the balance point. If we could connect the stars with a massless rod, they would balance like a seesaw at the center of mass. (b) As stars in a binary star system revolve around each other, the line connecting them always passes through the center of mass. The more massive star is always closer to the center of mass, and the ratio of the masses of the stars M_A/M_B equals the inverse of the ratio of the sizes of the orbits r_B/r_A.

By the Numbers 11-4

The Masses of Binary Stars
According to Newton's laws of motion and gravity, the total mass of two stars orbiting each other is related to the average distance a between them and their orbital period P. If the masses are M_A and M_B, then:

$$M_A - M_B = \frac{a^3}{P^2}$$

In this formula, we will measure a in AU, P in years, and the mass in solar masses.

Notice that this formula is related to Kepler's third law of planetary motion. Almost all of the mass of the solar system is in the sun. If we apply this formula to any planet in our solar system, the total mass is 1 solar mass. Then the formula becomes $P^2 = a^3$, which is Kepler's third law.

In other star systems, the total mass is not necessarily 1 solar mass, and this gives us a way to find the masses of binary stars. If we can find the average distance between the two stars in AU and their orbital period in years, the sum of the masses of the two stars is just a^3/P^2.

Example A: If we observe a visual binary with a period of 32 years and an average separation of 16 AU, what is the total mass?

Solution: The total mass equals $16^3/32^2$, or 4 solar masses.

Example B: In the previous example, suppose star A is 12 AU away from the center of mass and star B is 4 AU away. What are the individual masses?

Solution: The ratio of the masses must be 12:4. What two numbers add up to 4 and have the ratio 12:4? Star B must be 3 solar masses, and star A must be 1 solar mass.

Scientists can rarely observe the things they really want to know, so they must construct chains of inference. We can't observe the mass of stars directly, so we must find a way to use what we can observe, orbital period and angular separation, and step-by-step figure out the parameters we need to reach our goal. In this case, the goal is the masses of the two stars. We follow the chain of inference from the observable parameters to the unobservable quantities we want to know.

Chains of inference can be mathematical, as when geologists use earthquakes to calculate the temperature and density of Earth's interior. There is no way to drill a hole to Earth's center and lower a thermometer or recover a sample, so the internal temperature and density are not directly observable. Nevertheless, the speed of vibrations from distant earthquakes depends on the temperature and density of the rock they pass through. Geologists can't measure the speed of the vibrations deep inside Earth; but they can measure the delay in the arrival time at different locations on the surface, and that allows them to work their way back to the speed and, finally, the temperature and density.

Chains of inference can also be non-mathematical. Biologists studying the migration of whales can't follow individual whales for years at a time, but they can observe feeding and mating in different locations; take into consideration food sources, ocean currents, and water temperatures; and construct a chain of inference that leads back to the seasonal migration pattern for whales.

This chapter contains a number of chains of inference because it describes how we know what stars are like. Almost all sciences use chains of inference. If you can link the observable parameters step-by-step to the final conclusions, you have gained a strong insight into the nature of science.

---

length of time the stars take to complete one orbit. The smaller the orbits are and the shorter the orbital period is, the stronger the stars' gravity must be to hold each other in orbit. For example, if two stars whirl rapidly around each other in small orbits, then their gravity must be very strong to prevent their flying apart. Such stars would have to be very massive. By knowing the size of the orbits and the orbital period, we could figure out how much mass the stars contain. Calculations such as those shown in By the Numbers 11-4 will yield the total mass, which, combined with the ratio of the masses found from the relative sizes of the orbits, can tell us the individual masses of the stars.

Actually, figuring out the mass of a binary star system is not as easy as it might seem from the examples in By the Numbers 11-4. The orbits of the stars are often elliptical and tipped at an unknown angle to our line of sight, and astronomers must find ways to correct for that distortion of the orbits. In addition, astronomers analyzing binary stars must find the distances to the stars so they can estimate the true size of the orbits in astronomical units. Notice that finding the masses of binary stars requires a number of steps to get from what we can observe to what we really want to know, the masses. Constructing such sequences of steps is an important part of science (Window on Science 11-1).

Although there are many different kinds of binary stars, three types are especially important for determining stellar masses. We will discuss these separately in the next sections.

Visual Binaries

In a **visual binary,** such as Sirius and its companion, the two stars are separately visible in the telescope (Figure 11-15). Only a pair with large orbits can be separated visually, for if the orbits are small, the star images blend together in the telescope and we see only a single point of light. Because visual binaries have such large orbits, they also have long orbital periods. Some take hundreds or even thousands of years to complete a single revolution.

Astronomers study visual binaries by measuring the position of the two stars directly at the telescope or on photographic plates. In either case, they need measurements over many years to map the orbits. Figure 11-15 shows the orbits of the visual binary Sirius and its white-dwarf companion. The frames in the figure show the observed location of the stars since 1960. The orbital period is about 50 years.

More than half of all stars are members of multiple systems, so binaries are common. Few, however, can be analyzed completely. Many pairs of binary stars are widely separated and can be measured easily. But these binary systems have large orbits, and that means very long orbital periods. Earth's astronomers have been measuring visual binaries for only a bit more than a century, so it isn't possible to map the orbits of these long-period visual binary stars, determine the shapes of their orbits, and find their masses. Of course, some

Figure 11-15
Sirius and its white dwarf companion (arrow) are shown in the top frame in a 1960 photograph. Spikes are caused by diffraction in the telescope. *(Lick Observatory photograph)* Sequential measurements of the positions of the two stars reveal the shapes and sizes of their orbits. The intersection of the lines connecting the stars at different times defines the center of mass. The more massive star has the smaller orbit.

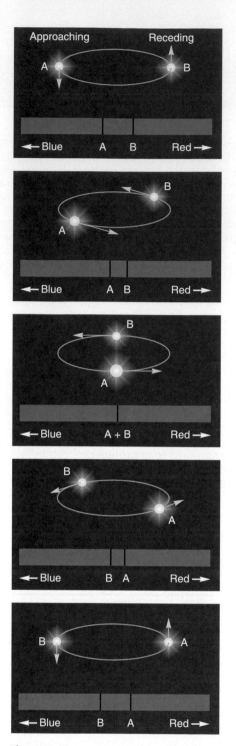

Figure 11-16
As the stars of a spectroscopic binary revolve around their common center of mass, they alternately approach and recede from Earth. The Doppler shifts cause their spectral lines to move back and forth across each other.

visual binaries have short periods—less than the lifetime of an astronomer. These short-period visual binaries, however, are pairs of stars that orbit close to each other, and that makes them exceedingly difficult to measure because the images of the two stars are difficult to resolve in the telescope. Thus astronomers search for other ways to analyze binary star systems.

Spectroscopic Binaries

If the stars of a binary system are close together, the telescope, limited by diffraction and seeing, shows us a single point of light. Only by taking a spectrum, which is formed by light from both stars and contains spectral lines from both, can we tell that there are two stars present and not one. Such a system is called a **spectroscopic binary.**

Because the stars in a spectroscopic binary orbit each other, they alternately approach and recede from us, as in Figure 11-16. As one star comes toward us, its spectral lines are Doppler-shifted toward the blue. The other star is moving away from us, and its spectral lines are shifted toward the red. Half an orbit later, the star that was approaching is receding. As we watch the spectrum of the binary system, we see the spectral lines split into two parts that move apart and then back together as the stars follow their orbits (Figures 11-16 and 11-17).

We can find the orbital period of a spectroscopic binary by observing its changing radial velocities. We can also determine the velocity of each star. If we know the velocity and the length of time it takes for the star to complete one orbit, we can multiply to find the circumference of the orbit. From that we can find the radius of the orbit. Thus, we would know most of what we need to find the masses of the stars—the size and period of the orbits. One important detail is still missing, however. We don't know how the orbits are inclined to our line of sight.

We can find the inclination of a visual binary because we can see the two stars moving along their orbits. In a spectroscopic binary, however, we cannot see the individual stars, find the inclination, or untip the orbits. The velocities we observe are not the true orbital velocities but only the part of that velocity directed radially toward or away from Earth. Because we cannot find the inclination, we cannot correct these radial velocities to their true orbital velocities. Therefore, we cannot find the true masses. All we can find from a spectroscopic binary is a lower limit to the masses.

More than half of all stars are binary, and most of those are spectroscopic binaries. Thus, it is not unusual to find the lines in a star's spectrum splitting apart and moving back together in the telltale behavior of a spectroscopic binary (Figure 11-18).

a

b

Figure 11-17
(a) At the bend of the handle of the Big Dipper lies a pair of stars, Mizar and Alcor. Through a telescope we discover that Mizar has a fainter companion and so is a member of a visual binary. (b) Spectra of Mizar recorded at different times show that it is itself a spectroscopic binary. (Compare with Figure 11-16.) In fact, both the faint companion to Mizar and the nearby star Alcor are also spectroscopic binaries. *(The Observatories of the Carnegie Institution of Washington)*

So far we have discussed an ideal spectroscopic binary in whose spectrum lines of both stars appear. In many spectroscopic binaries, however, one of the stars is significantly fainter than its companion, and the lines of the fainter star are invisible in the spectrum. These systems are even more difficult to analyze and provide even less information about stellar masses.

Eclipsing Binaries

Rare among binary stars are those with orbits tipped so the stars cross in front of each other as seen from Earth. Imagine a model of a binary star system in which a cardboard disk represents the orbital plane, as in Figure 11-19. If the orbits are seen edge-on from Earth, then the two stars cross in front of each other. The small star crosses in front of the large star and then, half an orbit later, the larger star crosses in front of the small star. We say the stars are eclipsing each other, and we call the system an **eclipsing binary.**

Figure 11-18
Fourteen spectra of the spectroscopic binary HD80715 are shown as graphs of the intensity of light across a short range of wavelengths. Dips in the graph show the location of dark absorption lines. In this series of spectra spanning 3.677 days, we see how a single line of iron (arrow in first spectrum) is split into two components (arrows in third spectrum) by the orbital motion of the stars. *(Adapted from data courtesy of Samuel C. Barden and Harold L. Nations)*

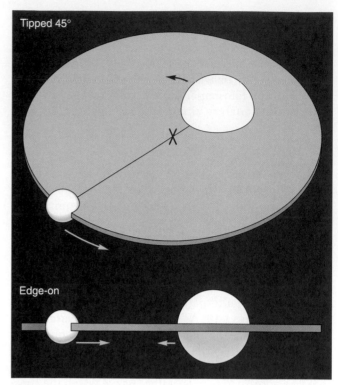

Figure 11-19
Imagine a model of a binary system with balls for stars and a disk of cardboard for the plane of the orbits. Only if we view the system edge-on do we see the stars cross in front of each other. These eclipsing binary systems are rare.

Seen from Earth, the two stars are not resolvable—that is, they are not visible separately. The system looks like a single point of light. But when one star moves in front of the other star, part of the light is blocked, and the total brightness of the system decreases. Figure 11-20 shows a smaller star moving in an orbit around a larger star, first eclipsing the larger star as it crosses in front and then being eclipsed as it moves behind. The resulting variation in brightness is recorded in a graph called the **light curve** of the system (Figure 11-21). Such light curves contain tremendous amounts of information about the masses, sizes, and shapes of the stars and can even reveal the presence of spots on the surfaces of the stars.

Once the light curve of an eclipsing binary has been observed, we can construct a chain of inference that leads to the masses. We can find the orbital period easily, and if we could get spectra showing the Doppler shifts of the two stars, we could find the orbital velocity. Then we could find the size of the orbits and the masses of the stars. The inclination of the orbits poses no problem because we know that we must be observing the orbits nearly edge on. Otherwise, we would not see the stars eclipse each other.

Eclipsing binaries are especially important because they enable us to measure the diameters of the stars. From the light curve we can tell how long it took for the small star to cross the large star. Multiplying this time interval by the orbital velocity of the small

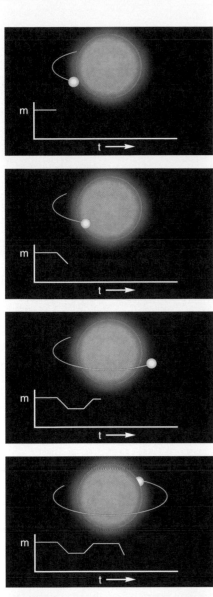

Figure 11-20
As the stars in an eclipsing binary cross in front of each other, the total brightness of the system changes. In this example, a small hot star orbits a giant cool star.

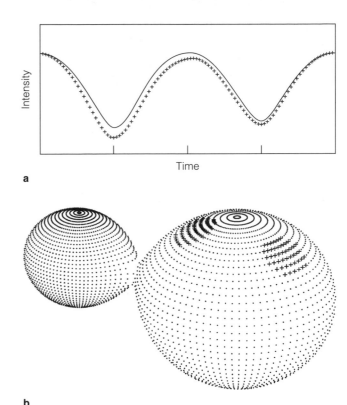

a

b

Figure 11-21
The light curve of an eclipsing binary star can contain large amounts of information. (a) The light curve of VW Cephei (lower curve) shows that the two stars are so close together their gravity has distorted their shapes and they actually touch each other. Slight distortions in the light curve are caused by dark spots, much like sunspots, near the pole of the larger star. The upper curve shows what the light curve would look like if there were no spots. (b) Through the detailed analysis of such light curves, astronomers can construct a model showing the size and shape of the stars and the location of the dark spots. *(Graphics created with Binary Maker 2.0)*

star gives us the diameter of the larger star. We can also determine the diameter of the small star by noting how long it took to disappear behind the edge of the large star. For example, if it took 300 seconds for the small star to disappear while traveling 500 km/s relative to the large star, then it must be 150,000 km in diameter.

Of course, there are complications due to the inclination and eccentricity of orbits, but often these effects can be taken into account, and the system can tell us not only the masses of its stars but also their diameters.

Algol (β Persei) is one of the best-known eclipsing binaries because its eclipses are visible to the naked eye. Normally, its magnitude is about 2.15, but its brightness drops to 3.4 in eclipses that occur every 68.8 hours. Although the nature of the star was not recognized until 1783, its periodic dimming was probably known to the ancients. *Algol* comes from the Arabic for "the demon's head," and it is associated in constellation mythology with the severed head of Medusa, the sight of whose serpentine locks turned mortals to stone (Figure 11-22). Indeed, in some accounts, Algol is the winking eye of the demon.

a

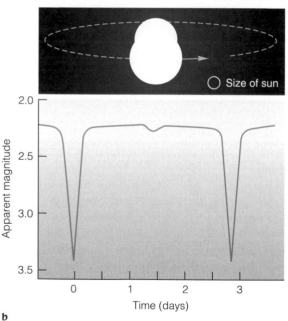

b

Figure 11-22

(a) The eclipsing binary Algol is the star on the demon's forehead in this 1837 engraving of Perseus and the head of the gorgon Medusa. *Algol* comes from the Arabic for "the demon's head." *(From Duncan Bradford,* Wonders of the Heavens, *Boston: John B. Russell, 1837)* (b) The light curve of Algol reveals it is a B8 V star orbiting a smaller star only slightly hotter than the sun. The partial eclipses, which are visible to the naked eye, are deepest when part of the hot star is hidden and are shallow when part of the cooler star is hidden (as shown).

Binary stars give us a powerful tool with which to discover the masses of the stars. With that information, we are now ready to assemble our data and begin trying to understand the secrets of the stars.

11-6 The Family of Stars

We have achieved the four goals we set for ourselves at the start of this chapter. We know how to find the luminosities, temperatures, diameters, and masses of stars. Now we can put that data (Window on Science 11-2) together to paint a family portrait of the stars; and like most family portraits, both similarities and differences are important clues to the history of the family. As we begin trying to understand how stars are born and how they die, we ask a simple question: What is the average star like? Answering that question is both challenging and illuminating.

Mass, Luminosity, and Density

The H-R diagram is filled with patterns that give us clues as to how stars are born, how they age, and how they die. When we add our data, we see traces of those patterns.

In a simple sense, science is the process by which we look at data and search for relationships that tell us how nature works. But as a result, science sometimes requires large amounts of data. For example, astronomers need to know the masses and luminosities of many stars before they can begin looking for relationships.

Compiling basic data is one of the common forms of scientific study. This work may not seem very exciting to an outsider, but scientists often love their work not so much because they want to know nature's secrets but because they love the process of studying nature. Using a microscope or a telescope is fun. Gathering plants in the rain forest or geological samples from a cliff face can be

tremendously exciting and satisfying. It is sometimes the process of science that is most rewarding, and enjoyment of the process can lead scientists to gather significant amounts of information.

Solving a single binary star to find the masses of the stars does not tell an astronomer a great deal about nature, but solving a binary is like solving a puzzle. It is fun and it is satisfying. Over the years, many astronomers have added their results to the growing data file on stellar masses. We can now analyze that data to search for relationships between the masses of stars and other parameters such as diameter and luminosity.

The history of science is filled with stories of hard-working scientists who com-

piled large amounts of data that later scientists used to make important discoveries. For example, Tycho Brahe spent 20 years recording the positions of the stars and planets (Chapter 2). He must have loved his work to spend 20 years on his windy island, but he did not live to use his data. It was his successor, Johannes Kepler, who used Brahe's data to discover the laws of planetary motion.

Whatever science you study, you will encounter accumulations of measurements and observations that have been compiled over the years, including everything from the hardness of rocks to the attention span of infants. Determining these basic scientific data is as much a part of science as is testing a hypothesis.

If we add the masses of selected stars to an H-R diagram such as Figure 11-23, we discover that the most massive stars are about 55 to 100 times the mass of the sun, and the least massive are slightly less than 0.1 solar mass. Among main-sequence stars, the most massive are the O and B stars, the so-called upper-main-sequence stars. As we run our eye down the main sequence, we find lower-mass stars. The lowest-mass are the K and M stars, the lower-main-sequence stars. Thus a star's location along the main sequence depends on its mass (Figure 11-23).

Stars that do not lie on the main sequence do not appear to be arranged in any particular pattern according to mass. Some giants and supergiants are quite massive, while others are no more massive than our sun. White dwarfs have about the same mass, somewhere in the narrow range of 0.5 to about 1 solar mass.

Because of the systematic ordering of mass along the main sequence, these stars obey a **mass–luminosity relation**—the more massive a star is, the more luminous it is (Figure 11-24). In fact, the mass–luminosity relation can be expressed as a simple formula. (See By the Numbers 11-5.) Giants and supergiants do not follow the mass–luminosity relation very closely, and white dwarfs not at all. In the next chapters, the mass–luminosity relation will help us understand how stars generate their energy.

Though mass alone does not reveal any pattern among giants, supergiants, and white dwarfs, density

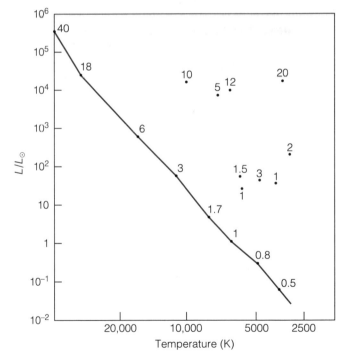

Figure 11-23
Stellar masses plotted in the H-R diagram. Giants and supergiants consist of stars of various masses, but masses along the main sequence are ordered. The most massive stars are the most luminous.

Figure 11-24
The mass–luminosity relation shows that the more massive a main-sequence star is, the more luminous it is. The open circles represent white dwarfs, which do not obey the relation. The red line represents the equation in By the Numbers 11-5.

does. Once we know a star's mass and diameter, we can calculate its average density by dividing its mass by its volume. Stars are not uniform in density but are most dense at their center and least dense near their surface. The center of the sun, for instance, is about 100 times as dense as water; its density near the visible surface is about 3400 times less dense than Earth's atmosphere at sea level. A star's average density is intermediate between its central and surface densities. The sun's average density is approximately 1 gm/cm^3—about the density of water.

Main-sequence stars have average densities similar to the sun's, but giant stars, being large, have low average densities, ranging from 0.1 to 0.01 gm/cm^3. The enormous supergiants have still lower densities, ranging from 0.001 to 0.000001 gm/cm^3. These densities are thinner than the air we breathe, and if we could insulate ourselves from the heat, we could fly an airplane through these stars. Only near the center would we be in any danger, for there the material is very dense—about 3,000,000 gm/cm^3.

The white dwarfs have masses about equal to the sun's, but are very small, only about the size of Earth. Thus, the matter is compressed to densities of 3,000,000 gm/cm^3 or more. On Earth, a teaspoonful of this material would weigh about 15 tons.

Density divides stars into three groups. Most stars are main-sequence stars with densities like the sun's. Giants and supergiants are very low density stars, and white dwarfs are high-density objects. We will see in later chapters that these densities reflect different stages in the evolution of stars.

Surveying the Stars

What is the most common kind of star? To answer that question we must go beyond the H-R diagram and actually take a survey of large numbers of stars. If we assume the sun is located in a typical place in the disk of our galaxy, then we can survey stars in the solar neighborhood and assume the results apply to stars in general. Unfortunately, there are three problems that make our survey difficult.

All we want to know is the number of stars of different kinds in every 1,000,000 cubic parsecs, but finding that information means we must observe the spectrum of each star to find its spectral type and measure its distance to find its absolute magnitude. A sphere containing 1,000,000 pc^3 is 62 pc in radius. Until recently, astronomers couldn't measure such small parallaxes accurately, but data from the Hipparcos satellite is now available. Before that astronomers had to estimate such distances using spectroscopic parallax.

Of course, one way to avoid uncertainties in the distances is to observe stars only near the sun, but that is where the second problem comes in. Some stars, such as the lower-main-sequence M stars and white dwarfs, have very low luminosities. Even if they are only a few dozen parsecs away, they are very hard to find. We could miss large numbers of these faint stars in our survey.

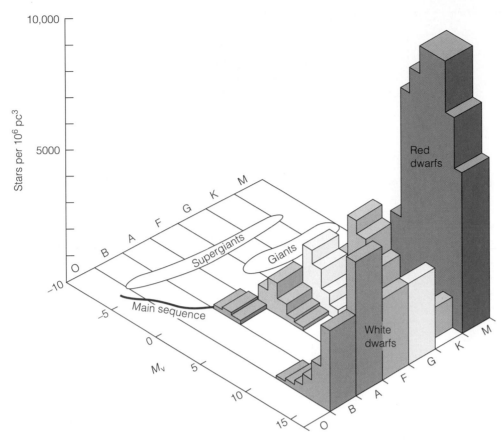

Figure 11-25

In this histogram, bars rise from an H-R diagram to represent the number of stars of given spectral type and absolute magnitude found in 1 million cubic parsecs. The main-sequence M stars are the most common stars, followed closely by the white dwarfs. The upper-main-sequence stars and the giants and supergiants are so rare their bars are not visible in this diagram.

The third problem is that the most luminous stars are very rare. There are no O stars at all within 62 pc of the sun, and we must extend our survey to a very great distance before we find many of these stars. At such distances, parallaxes are too small to be measured, and we have to find distance in other, less accurate ways, such as through spectroscopic parallax.

The Frequency of Stellar Types

In spite of the difficulties in surveying stars, astronomers have been able to use statistical methods to study the abundance of stars in the solar neighborhood. These studies help us understand what the average star is like by telling us the numbers of different kinds of stars in space. Shown in Figure 11-25, these data show that some kinds of stars are extremely rare, and others are very common. Compare Figure 11-25 with Figure 11-10 to locate the main-sequence stars, giants, white dwarfs, and supergiants.

The lowest-mass main-sequence stars, being faint and cool, are called **red dwarfs.** Notice how common

these stars are; there are about 65,000 in every million cubic parsecs. White dwarfs are also very common. Fortunately for us, these stars are very faint, for if they were as luminous as supergiants, the sky would be filled with their glare, and we would see hardly anything else. As it is, the most common stars are so faint they are hard to find even with large telescopes.

Notice that the more luminous stars are rare. The stars become less common as we go up the main sequence; A stars are rare, and type O and B stars are so rare we cannot plot them on this diagram. There are only about 0.036 of the type O stars per million cubic parsecs. That is, we would have to search through about 30,000,000 pc^3 to find an O star. Put yet another way, only 1 star in 4 million is an O star. Giants and supergiants are also very rare. Every million cubic parsecs contains only about 100 giants and only 0.07 supergiant. Luckily, these very rare kinds of stars are also very luminous—we can see them from great distances. If they were as faint as the main-sequence M stars, we might never know they existed.

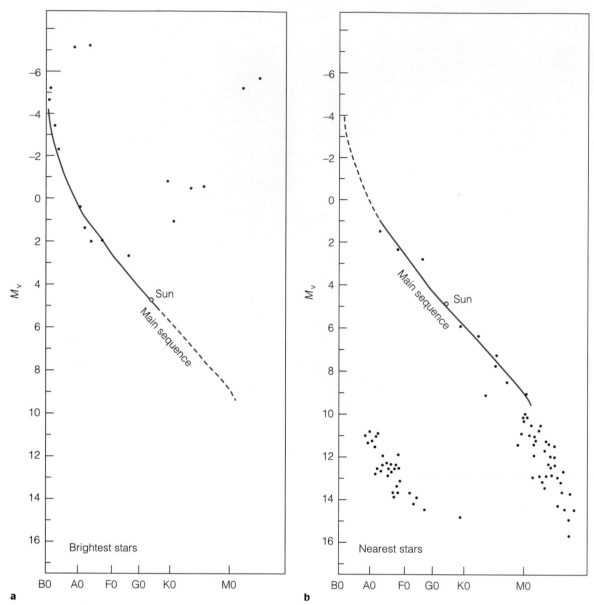

Figure 11-26

Representative H-R diagrams of (a) the brightest stars in the sky and (b) the nearest stars in space show that the most common kind of star is a lower-main-sequence M star or a white dwarf. Of the bright stars we see in the sky, many are giants and supergiants, and none are lower-main-sequence M stars or white dwarfs.

We can see the consequences of the luminosities of the stars on any clear night. The stars we see in the sky with our unaided eyes are in fact very rare, highly luminous stars (Figure 11-26a). They are far away, and we can see them at all only because they are so very luminous. What we don't see in our sky are the common, low-luminosity stars. In a sample of the nearest stars, nearly all are red dwarfs or white dwarfs (Figure 11-26b), but we cannot see those stars with our unaided eyes because they are so faint. Thus the family relations among the stars determine what we see when we look at the night sky.

REVIEW Critical Inquiry

Why is it difficult to be sure how common O and B stars are?

In our survey of the stars, the number of O and B stars per million cubic parsecs is difficult to determine because these stars are quite rare. We know that O and B stars are not common because there are none near us in space. If we take our survey by counting stars within 63 pc of the sun, a distance that would include a million cubic parsecs, we find no O or B stars that close. To find these stars we must look to great distances, and at such distances we can't measure their distance by parallax. Rather we must use the spectroscopic parallax method, which is uncertain. Thus, the chain of inference we use to find the number of sunlike stars in space does not work well for these more luminous stars.

Now consider the chain of inference for the most common stars. Why is it difficult to be sure just how common these stars are?

We have reached our four goals in this chapter. We know the luminosities, temperatures, diameters, and masses of the stars. But we have discovered a puzzle. The most massive stars are luminous and rare, while the least massive stars are faint and common. Why does this happen? We search for the answer in the next chapter.

Summary

Our goal in this chapter was to characterize the stars by finding their luminosities, temperatures, diameters, and masses. Before we could begin, we needed to find the distances to stars.

We can measure the distance to the nearer stars by observing their parallaxes. The more distant stars are so far away that their parallaxes are unmeasurably small. To find the distances to these stars, we must use spectroscopic parallax. Stellar distances are commonly expressed in parsecs. One parsec is 206,265 AU—the distance to an imaginary star whose parallax is 1 second of arc.

Once we know the distance to a star, we can find its intrinsic brightness, expressed as its absolute magnitude or its luminosity. A star's absolute magnitude is the apparent magnitude we would see if the star were only 10 pc away. The luminosity of the star is the total energy radiated in 1 second, usually expressed in terms of the luminosity of the sun.

By classifying the stars according to spectral type, we found their temperatures, and that allowed us to plot them according to intrinsic brightness and surface temperature in an H-R diagram. Roughly 90 percent of stars fall on the main sequence, with the most massive at the top and the least massive at the bottom. Giants and supergiants are large stars and lie above the main sequence. White dwarfs are small and hot and lie below the main sequence.

The only direct way we can find the mass of a star is by studying binary stars. When two stars orbit a common center of mass, we can find their masses by observing the period and sizes of their orbits.

Given the mass and diameter of a star, we can find its average density. On the main sequence, the stars are about as dense as the sun, but the giants and supergiants are very low density stars. Some are much thinner than air. The white dwarfs, lying below the main sequence, are tremendously dense.

The mass–luminosity relation says that the more massive a star is, the more luminous it is. Main-sequence stars follow this rule closely, the most massive being the upper-main-sequence stars and the least massive the lower-main-sequence stars. Giants and supergiants do not follow the relation precisely, and white dwarfs not at all.

A survey in the neighborhood of the sun shows us that lower-main-sequence stars are the most common type. The hot stars of the upper-main-sequence are very rare. Giants and supergiants are also rare, but white dwarfs are quite common, although they are faint and hard to find.

New Terms

stellar parallax (p)	supergiant star
parsec (pc)	white dwarf star
flux	luminosity class
absolute visual magnitude (M_v)	spectroscopic parallax
distance modulus ($m - M_v$)	binary stars
luminosity (L)	center of mass
Balmer lines	visual binary
spectral class or type	spectroscopic binary
spectral sequence	eclipsing binary
Hertzsprung-Russell (H-R) diagram	light curve
main sequence	mass–luminosity relation
giant star	red dwarf

Review Questions

1. Why are parallax measurements limited to the nearest stars?

2. Why is the Hipparcos satellite able to make more accurate parallax measurements than are ground-based telescopes?

3. What do the words *absolute* and *visual* mean in the definition of absolute visual magnitude?

4. What does luminosity measure that is different from what absolute visual magnitude measures?

5. Why does the luminosity of a star depend on both its radius and its temperature?

6. How can we be sure that giant stars really are larger than main-sequence stars?

7. Why do we conclude that white dwarfs must be very small?

8. What observations would we make to classify a star according to its luminosity? Why does that method work?

9. Why does the orbital period of a binary star depend on its mass?

10. What observations would you make to study an eclipsing binary star?

11. Why don't we know the inclination of a spectroscopic binary? How do we know the inclination of an eclipsing binary?

12. How do the masses of stars along the main sequence illustrate the mass–luminosity relation?

13. Why is it difficult to find out how common the most luminous stars are? The least luminous stars?

14. What is the most common kind of star?

15. If you look only at the brightest stars in the night sky, what kind of star are you likely to be observing? Why?

Discussion Questions

1. If someone asked you to compile a list of the nearest stars to the sun based on your own observations, what measurements would you make, and how would you analyze them to detect nearby stars?

2. The sun is sometimes described as an average star. Is that true? What is the average star really like?

Problems

1. If a star has a parallax of 0.050 second of arc, what is its distance in parsecs? in light-years? in AU?

2. If a star has a parallax of 0.016 second of arc and an apparent magnitude of 6, how far away is it, and what is its absolute magnitude?

3. Complete the following table.

m	M_v	d (pc)	p (sec of arc)
—	7	10	—
11	—	1000	—
—	−2	—	0.025
4	—	—	0.040

4. If a main-sequence star has a luminosity of 400 $L_\odot$, what is its spectral type? (*Hint:* See Figure 11-13).

5. If a star has an apparent magnitude equal to its absolute magnitude, how far away is it in parsecs? in light-years?

6. If a star is 10 times the radius of the sun and half as hot, what will its luminosity be? (*Hint:* See By the Numbers 11-3.)

7. An O8 V star has an apparent magnitude of +1. Use the method of spectroscopic parallax to find the distance to the star. Why might this distance be inaccurate?

8. In the following table, which star is brightest in apparent magnitude? most luminous in absolute magnitude? largest? farthest away?

Star	Spectral Type	m
a	G2 V	5
b	B1 V	8
c	G2 Ib	10
d	M5 III	19
e	White dwarf	15

9. What is the total mass of a visual binary system if its average separation is 8 AU and its period is 20 years?

10. What is the mass ratio of the stars in Figure 11-14b? in Figure 11-15?

11. Assume that the stars in Figure 11-14b have a separation of 18 AU and a period of 30 years. What is the total mass? What is the mass ratio? What are their individual masses?

12. If the orbital velocity of the eclipsing binary in Figure 11-20 is 153 km/s and the smaller star becomes completely eclipsed in $2\frac{1}{2}$ hours, what is its diameter?

13. What is the luminosity of a 4-solar-mass star? of a 9-solar-mass star? of a 7-solar-mass star?

Critical Inquiries for the Web

1. Hertsprung-Russell diagrams allow astronomers to represent the distribution of stellar properties. Examine H-R diagrams at various Web sites and determine which types of stars are most (and least) numerous in our galaxy.

2. What if Algol were oriented such that its stars did not eclipse each other as seen from Earth? Use the Internet to find information about Algol and binary stars in general, and discuss how astronomers could still determine that it is a multiple star system.

Exploring *The Sky*

1. *The Sky* can tell us many interesting things about stars.

   a. Determine the apparent magnitude, the parallax, distance in parsecs and light years, proper motion (in mas/yr, or milli-arc-seconds per year), and spectral classification for the following stars: Sirius, Aldebaran, Vega, Deneb, Betelguese, Antares, Spica, Altair. Which is the brightest? The furthest? Which has the largest proper motion? Determine the luminosity classification of each.

   How to proceed: use **Find** to obtain the **Object Information** window.

   b. Determine the visual magnitude (**M_v**) of each of the stars listed in the previous exercise. Use the formula given in By the Numbers 11-2 (p. 243).

Go to the Brooks/Cole Astronomy Resource Center (www.brookscole.com/astronomy) for critical thinking exercises, articles, and additional readings from InfoTrac College Edition, Brooks/Cole's online student library.

CHAPTER TWELVE

The Formation and Structure of Stars

Jim he allowed [the stars] was made, but I allowed they happened. Jim said the moon could'a laid them; well, that looked kind of reasonable, so I didn't say nothing against it, because I've seen a frog lay most as many, so of course it could be done.

Mark Twain
The Adventures of Huckleberry Finn

Guidepost

Over the last two chapters we have painted a general picture of what the sun and other stars are like. In this chapter we begin to see how stars really work.

The formation and structure of stars is critically important if we want to understand how nature works, but something else happens in this chapter. Here we see how science works. The insides of stars seem, at first, beyond human understanding, but this chapter shows how astronomers use the properties of light and the basic rules of physics to build theories about the way stars work. Even more important, in this chapter we see how scientists use evidence to test and refine theories.

Testing theories against evidence is the basic skill required of the scientist, and we will use it over and over in the seven chapters that follow.

Figure 12-1
Where we find thick clouds of gas and dust, we usually find young stars, a correlation that suggests that stars are born from these clouds. Here, the great dust cloud known as the Horsehead Nebula (0.55 pc from nose to ear) is silhouetted against glowing gas excited by ultraviolet radiation from hot stars imbedded in the dense clouds to the left. (NOAO and Nigel Sharp)

Stars exist because of gravity. They form because gravity makes clouds of gas contract, and they spend long, stable lives generating nuclear energy at their centers and balancing their own gravity with the pressure of the hot gas in their interiors. In the end, stars die because they exhaust their fuel supply and can no longer withstand the force of their own gravity.

A star can remain stable only by maintaining great gas pressure in its interior. Gravity tries to make the stellar gas contract, but if the internal temperature is high enough, the pressure of the gas pushes outward just enough to balance gravity. Thus, a star is a battlefield where gas pressure and gravity struggle for dominance. If gas pressure wins, the star must expand, but if gravity wins, the star must contract.

Only by generating tremendous amounts of nuclear energy can a star keep its interior hot enough to maintain the gravity–pressure balance. The sun, for example, generates enough energy in 1 second to power Earth's civilization at its current level for 2 million years.

To understand how stars form, we must understand how they generate their energy and how that energy eventually escapes from the star. Thus, we must explore the nuclear reactions that generate energy in the cores of stars, and we must explore the internal structure of

stars, which resists the inward pull of the star's gravity. These explorations will allow us to imagine the slow evolution of stable stars over billions of years.

How can we know what stars are like when we can't see inside them and we don't live long enough to see them evolve? The answer lies in the nature of science. By constructing theories that describe how nature works and then testing those theories against evidence from observations, we can unravel some of nature's greatest secrets. We can even understand the origin and structure of the stars.

12-1 The Birth of Stars

The key to understanding star formation is the correlation between young stars and clouds of gas and dust. Where we find the youngest groups of stars, we also find large clouds of gas illuminated by the hottest and brightest of the new stars (Figure 12-1). This leads us to suspect that stars form from such clouds, just as raindrops condense from the water vapor in a thundercloud. To study the formation of stars, we must examine these cold clouds of gas that float between the stars and understand how they contract, heat up, and become stars.

The Interstellar Medium

Astronomers can describe the gas and dust between the stars, the **interstellar medium,** in some detail. About 75 percent of the mass of the gas is hydrogen, and 25 percent is helium; there are also traces of carbon, nitrogen, oxygen, calcium, sodium, and heavier atoms. Roughly 1 percent of the mass is made up of microscopic dust about the size of the particles in cigarette smoke. The dust seems to be made mostly of carbon and silicates (rocklike minerals) mixed with or coated with frozen water. The average distance between dust grains is about 150 m.

This interstellar material is not uniformly distributed through space; it consists of a complex tangle of cool, dense clouds pushed and twisted by currents of hot, low-density gas. Although the cool clouds contain only 10 to 1000 atoms/cm^3 (fewer than any vacuum on Earth), we refer to them as dense clouds in contrast with the hot, low-density gas that fills the spaces between clouds. That thin gas contains only about 0.1 atom/cm^3.

The preceding two paragraphs describe the interstellar medium, but we should not accept these facts blindly. Science is based on evidence, so we should demand to know what observational evidence supports these facts. How do we know there is an interstellar medium, and how do we know its properties?

In some cases, the interstellar medium is easily visible as clouds of gas. Astronomers refer to any cloud of gas and dust as a **nebula** (from the Latin word for "cloud"). Such nebulae (plural) give us clear evidence of an interstellar medium. If a star is hot enough, it can ionize the gas of the nebula and produce an **emission nebula**—a nebula whose spectrum is an emission spectrum. The Balmer lines of hydrogen are strong in the spectra of these nebulae, showing that hydrogen is abundant. In fact, the three visible lines of hydrogen blend together to produce the pink color that is typical of emission nebulae. In contrast, some nebulae look blue, and their spectrum is the reflected spectrum of nearby stars. This light is scattered by dust in the nebula, and thus the nebulae are called **reflection nebulae.** Because tiny dust grains scatter blue light better than red light, reflection nebulae are typically blue (Figure 12-2). This is clear evidence that the dust grains are

a

b

Figure 12-2

(a) The Pleiades star cluster, just visible to the naked eye, is surrounded by a faint reflection nebula visible only in long-exposure photographs. The nebula is produced by dust reflecting the starlight. *(Palomar Obs/California Institute of Technology)* (b) This infrared mosaic, made up of images from the Infrared Astronomy Satellite, shows the curving wake left by the Pleiades as it passes through the interstellar medium. The + marks show the locations of the brightest stars, which heat the nebula and make it expand to produce a low-density cavity tracing the path of the cluster. *(Courtesy Richard E. White; image rendered by Duncan Chesley of American Image, Inc.)*

tiny—with diameters roughly equal to the wavelength of visible light.

Some nebulae make their presence known by obscuring our view of more distant stars. If a cloud is unusually dense and dusty, it totally obscures our view of the stars beyond, and we see it as a dark cloud (Figure 12-3a). Such clouds are typically not spherical, but torn and twisted like wisps of smoke distorted by a breeze.

a

b

Figure 12-3
Some interstellar clouds of gas and dust are so dense we cannot see through them and they appear as regions nearly free of stars. (a) In the case of Barnard's S Nebula, only a few stars are visible through the dark cloud. *(Courtesy Rudolph E. Schild)* (b) Stars that are visible through interstellar dust will look redder than normal because blue photons, having shorter wavelengths, are more likely to be scattered by the dust grains. Thus, more red photons reach our telescope, and such stars look redder. Note the reddened stars visible through the dusty nebula in (a).

a

b

Figure 12-4
Interstellar absorption lines can be recognized in two ways. (a) The B0 supergiant ε Orionis is much too hot to show spectral lines of once-ionized calcium (CaII), yet this short segment of its spectrum reveals narrow, multiple lines of CaII (tick marks) that must have been produced in the interstellar medium. *(The Observatories of the Carnegie Institution of Washington)* (b) Spectral lines produced in the atmospheres of stars are much broader than the spectral lines produced in the interstellar medium. *(Adapted from a diagram by Binnendijk)* In both (a) and (b) the multiple interstellar lines are produced by separate interstellar clouds with slightly different radial velocities.

If a cloud is less dense, starlight may be able to penetrate it, and we can see stars through the cloud; but the stars look dimmer because the cloud scatters some of the light. The same scattering of blue light by tiny dust grains that makes reflection nebulae appear blue makes stars seen through interstellar dust appear red (as explained in Figure 12-3b). This is called **interstellar reddening,** and it can give us important clues about the dust grains and their sizes.

The interstellar medium also reveals its presence by forming interstellar absorption lines in the spectra of distant stars (Figure 12-4). As starlight travels through the interstellar medium, gas atoms of elements such as calcium and sodium absorb photons of certain wavelengths, producing narrow interstellar absorption lines. We can be sure these lines originate in the interstellar medium because they appear in the spectra of O and B stars—stars that are too hot to form calcium and sodium absorption lines in their own atmospheres. Also, the narrowness of the interstellar lines tells us they could not have been formed in the hot atmospheres of the stars. In a hot gas, the atoms move rapidly, and the Doppler

shifts of the different atoms smear out the spectral line and make it broad. The interstellar medium is very cold, 10 to 50 K, so the atoms are not moving very rapidly at all, and they form extremely sharp spectral lines. In fact, the narrow widths of the spectral lines are clear evidence of the temperature of the interstellar gas. Often interstellar lines appear as multiple lines because the light from the star passed through a number of clouds on its way to Earth (see Figure 12-4b).

Observations at nonvisible wavelengths give us valuable evidence about the interstellar medium. About 1 percent of the interstellar medium is dust, and although it is very cold, the dust can radiate in the far infrared. This would be a negligible amount of radiation except that the interstellar medium contains such vast quantities of dust. Infrared telescopes can map the interstellar medium by its infrared radiation. For example, the Infrared Astronomy Satellite discovered a faint, wispy network of dusty clouds covering the entire sky (Figure 12-5a). This **infrared cirrus** is quite cold, and its wispy nature shows how the interstellar medium is in constant motion. In contrast, X-ray observations can detect regions of very hot gas apparently produced by exploding stars (Figure 12-5b), and it appears that the sun lies just inside one of these cavities filled with hot gas. Ultraviolet observations provide information about the distribution of the gas in the interstellar medium, its chemical composition, and its temperature. The inter-

a

b

Figure 12-5

Nonvisible observations of the interstellar medium. (a) The Infrared Astronomy Satellite observed the sky in the far infrared and discovered the glow of the infrared cirrus, a wispy distribution of dust slightly warmed by starlight. At temperatures only slightly above absolute zero, the dust radiates black body radiation in the far infrared. The infrared cirrus is spread all across the sky. *(NASA/IPAC, Courtesy Deborah Levine)* (b) This ROSAT X-ray map of the constellation Cygnus reveals a yellow arch of gas clouds outlining a bubble of high-temperature gas (dashed line) spanning 450 ly. This hot gas has been produced by the explosive deaths of many massive stars. *(Courtesy Steve Snowden and Max-Planck Institute for Extraterrestrial Physics, Germany)*

stellar medium is even detectable in the radio part of the spectrum. Radio astronomers have found many different radio signals at the precise frequencies emitted by certain molecules. These are the radio equivalent of emission lines in the spectra of visible light. Thus, the radio observations tell us that some of the atoms in space have linked together to form molecules.

We have no shortage of evidence concerning the interstellar medium. Clearly, space is not empty, and from the correlation between young stars and clouds of gas and dust we can suppose that stars form from these clouds. But how can the icy cold interstellar clouds create hot stars? The answer is gravity.

The Formation of Stars from the Interstellar Medium

To study the formation of stars, we must continue comparing theory with evidence. The theory of gravity predicts that the combined gravitational attraction of the atoms in a cloud of gas will squeeze the cloud, pulling every atom toward the center. Thus, we might expect that every cloud would eventually collapse and be-

come a star; however, the thermal energy in the cloud resists collapse. Temperature is a measure of the motion of the atoms or molecules in a material—in a hot gas, the atoms move more rapidly than do those in a cool gas. The interstellar clouds are very cold, but even at a temperature of only 10 K, the average hydrogen atom moves about 0.5 km/s (1100 mph). This thermal motion would make the cloud drift apart if gravity were too weak to hold it together.

The densest interstellar clouds contain about 1000 atoms/cm^3, include from a few hundred to a billion solar masses, and have temperatures as low as 10 K. In such clouds hydrogen can exist as molecules (H_2) rather than as atoms. Although hydrogen molecules cannot be detected by radio telescopes, the clouds can be mapped by the emission of carbon monoxide molecules (CO) present in small amounts in the gas. Consequently, these clouds are known as **molecular clouds,** and the largest ones are called giant molecular clouds. Stars form in these clouds when the densest parts of the clouds become unstable and contract under the influence of their own gravity.

Most clouds do not appear to be gravitationally unstable, but such a cloud colliding with a **shock wave** (the astronomical equivalent of a sonic boom) can be compressed and disrupted into fragments (Figure 12-6). Theoretical calculations show that some of these

fragments can become dense enough to collapse and form stars.

Thus, most interstellar clouds may not collapse and form stars until they are triggered by the compression of a shock wave. Happily for this hypothesis, space is filled with shock waves. Supernova explosions (exploding stars described in the next chapter) produce shock waves that compress the interstellar medium, and recent observations show young stars forming at the edges of such shock waves. Another source of shock waves may be the ignition of very hot stars. If one part of a cloud forms a massive star, the sudden flare of radiation as the star "turns on" could push against other parts of the cloud, compress the gas, and trigger more star formation. Even the collision of two clouds can produce a shock wave and trigger star formation.

Although these are important sources of shock waves, the dominant trigger of star formation in our galaxy may be the spiral pattern itself. In Figure I-11, we noted that our galaxy contains spiral arms, and many astronomers believe the arms are marked by shock waves traveling through the interstellar medium. If they are, then the collision of an interstellar cloud with a spiral arm could trigger the formation of new stars.

Astronomers have found a number of giant molecular clouds where stars are forming in a repeating cycle. Both high-mass and low-mass stars form in such a cloud, but when the massive stars form, their intense radiation or eventual supernovae explosions push back the surrounding gas and compress it. This compression in turn can trigger the formation of more stars, some of which will be massive. Thus a few massive stars can drive a continuing cycle of star formation in a giant molecular cloud.

While low-mass stars do form in such clouds along with massive stars, low-mass stars also form in smaller clouds of gas and dust. Because lower-mass stars have lower luminosities and do not develop quickly into supernovae explosions, low-mass stars alone cannot drive a continuing cycle of star formation.

Once part of a cloud is triggered into collapse, its gravity draws each atom toward the center. The infalling atoms pick up speed as they fall until, by the time the gas becomes dense enough for the atoms to collide often, they are traveling very fast. Because temperature is a measure of the speed of the atoms, the temperature of the gas increases. Thus, the contraction of an interstellar cloud heats the gas by converting gravitational energy into thermal energy.

Such a collapsing cloud of gas does not form a single object; because of instabilities, it fragments, producing an association of ten to a few hundred stars. An **association** is a group of stars not gravitationally bound to one another. Thus the association drifts apart within a few million years. The sun probably formed in such a cluster about 5 billion years ago.

Figure 12-6
An interstellar gas cloud has such a low density that it is unlikely to form stars without an outside stimulus. Computer models show that a passing shock wave (red line) can compress and fragment a cloud, driving some regions to densities high enough to trigger star formation (bottom). From first frame to last, this figure spans about 6 million years.

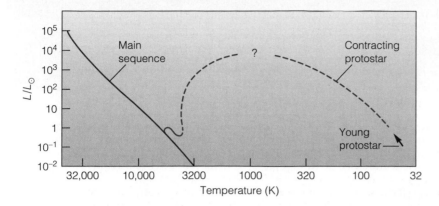

Figure 12-7
A contracting protostar, hidden deep inside a dust cloud, begins very cool and very faint (arrow). The thermal insulation of the dust makes the exact behavior of the protostar uncertain.

The Evolution of Protostars

To follow star formation further, we must concentrate on a fragment of a collapsing cloud as it forms a star. At first, unopposed by internal pressure, the gas atoms fall inward, picking up speed and increasing density and temperature. This free-fall contraction changes the cold gas into a warm **protostar,** an object that will become a star.

When a star or protostar changes, we say it evolves, and we can follow that evolution in the H-R diagram. As a star's temperature and luminosity change, its location in the H-R diagram shifts as well. We can draw an arrow or line called an **evolutionary track** to show how the star moves in the diagram. Of course, the motion of the point in the diagram that represents the star is not related in any way to the star's motion through space.

Stars begin contracting from the interstellar gas, which is very cold. Thus, to trace the contraction of a protostar, we must extend the H-R diagram to the right to include very low temperatures. Figure 12-7, which includes temperatures as low as 32 K, shows the initial collapse of a 1-solar-mass cloud fragment as its temperature rises to more than 1000 K. The exact process is poorly understood because of the complex effect of the dust in the cloud. The dusty cloud also hides the protostar from sight during its contraction. If we could see it, it would be a luminous red object a few thousand times larger than the sun, and we would plot it in the red-giant region of the H-R diagram.

The protostar's original free-fall contraction is slowed by the increasing density and internal pressure, but because the protostar is not yet hot enough to begin nuclear reactions, it must contract further as it radiates energy from its surface. This contraction is much slower than the free-fall contraction.

Throughout its contraction, the protostar converts its gravitational energy into thermal energy. Half of this energy radiates into space, but the remaining half raises the internal temperature. As the internal temperature

climbs, the gas becomes ionized, a mixture of positively charged atomic nuclei and free electrons. When the center gets hot enough, nuclear reactions begin generating energy, the protostar halts its contraction, and, having blown away its cocoon of gas and dust, it becomes a stable, main-sequence star.

The time a protostar takes to contract from a cool interstellar gas cloud to a main-sequence star depends on its mass. The more massive the star, the stronger its gravity and the faster it contracts (Figure 12-8). The sun took about 30 million years to reach the main sequence, but a 15-solar-mass star can contract in only 160,000

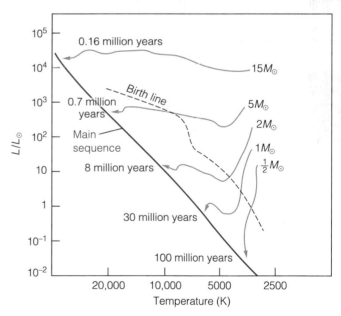

Figure 12-8
The more massive a protostar is, the faster it contracts. A 1-$M_\odot$ star requires 30 million years to reach the main sequence. (Recall that $M_\odot$ means "solar mass.") The dashed line is the birth line, where contracting protostars first become visible as they dissipate their surrounding clouds of gas and dust. Compare with Figure 12-7, which shows the formation of a star of about 1-$M_\odot$ as a dashed line up to the birth line and as a solid line from the birth line to the main sequence. *(Illustration design by author)*

The fundamental work of science is the testing of theories by comparison with facts. As we think about science, we need to clearly distinguish between facts and theories. The facts are the evidence against which we test the theories.

Scientific facts are those observations or experimental results of which we are confident. An astronomer makes observations of stars, a botanist collects samples of related plants, and a chemist performs an experiment to measure the rate of a chemical reaction. A fact could be a precise measurement, such as the mass of a star expressed as a specific number, or a fact could be a simple observation, such as that a certain butterfly no longer visits a certain mountain valley. In each case, the scientist is gathering facts.

A theory, however, is a conjecture as to how nature works. If we are uncertain of the theory, we might call it a hypothesis. In any case, these conjectures are not facts; they are attempts to explain how nature works. In a sense, a theory or a hypothesis is a story that scientists have made up to explain how nature works in some specific case. These stories can be wonderfully detailed and ingenious, but without evidence they are nothing more than hunches.

When one of these stories is tested against the facts and confirmed, we have more confidence that the story is probably right. The more a story is tested successfully, the more confidence we have in it. The facts represent reality, and every theory or hypothesis must be repeatedly tested against reality.

We can't test one theory against another theory. Theories are not evidence; they are conjectures. If we were allowed to test one theory against another theory, we might fall into the trap of circular reasoning. "Elves make the flowers bloom. I know that elves exist because the flowers bloom." Here two theories are used to confirm each other, and it leads us to nonsense. Only facts can be evidence.

Nor can we use a theory to deduce facts. We can use a theory to make predictions that we can test against facts, but the predictions themselves can never be certainties, so they can't be facts. The only way to get facts is to consult nature and make direct measurements or observations.

As we study different problems in astronomy, we must carefully distinguish between facts and theories. Facts are the basic building blocks of all science, and they come only from the careful study of nature. To use the words of Galileo, we must "read the book of nature."

---

years. Conversely, a star of 0.2 solar mass takes 1 billion years to reach the main sequence.

The story we have told, the story of the formation of a star from the interstellar medium, is based on theory. By understanding what the interstellar medium is like and by understanding how the laws of physics work, astronomers have been able to tell the story of how stars must form. But we can't accept a scientific theory without testing it, and that means we must compare the theory with the evidence. That constant checking of theories against evidence is the distinguishing characteristic of science, so it is time to carefully separate theory from evidence (Window on Science 12-1) and ask how much of this star formation story can be observed.

Observations of Star Formation

In astronomy, evidence means observations. Thus, we must ask what observations confirm our theories of star formation. Unfortunately, a protostar is not easy to observe. The protostar stage is less than 0.1 percent of a star's total lifetime, so although that is a long time in human terms, we cannot expect to find many stars in the protostar stage. Furthermore, protostars form deep inside clouds of dusty gas that absorb any light the protostar might emit. Only when the protostar is hot enough to drive away its enveloping cloud of gas and dust do we see it. The **birth line** in the H-R diagram in Figure 12-8 is the line where contracting protostars first become visible, and protostars cross the birth line shortly before they reach the main sequence. Thus the early evolution of a protostar is hidden from our sight.

Although we cannot see protostars at visible wavelengths, we can detect them in the infrared. The dust absorbs light from the protostar and grows warm, and warm dust radiates copious infrared. Infrared observations such as those made by the Infrared Astronomy Satellite reveal many bright sources that are probably protostars buried in dust clouds.

T Tauri stars, named after the first example found, variable star T in Taurus, appear to be protostars just clearing away the surrounding gas and dust. T Tauri stars vary in brightness, and spectroscopic observations show gas flowing away in strong stellar winds comparable to the sun's solar wind. In some cases, gas appears to be flowing into the star. All T Tauri stars are sources of infrared radiation, which shows that the stars are still accompanied by warm dust. In some cases, evidence suggests the dust and gas surround the star in a rotating disk.

a

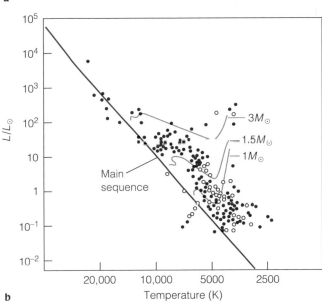

b

Figure 12-9
(a) One of the youngest star clusters known, NGC2264 is only a few million years old. *(Anglo-Australian Telescope Board)* (b) The H-R diagram of the cluster shows that many of its lower-mass stars have not yet reached the main sequence. T Tauri stars (open circles) are common and fall between the birth line and the main sequence (see Figure 12-8). Blue evolutionary tracks are derived from theoretical models of contracting protostars. *(Adapted from data by M. Walker)*

terstellar matter. The excited clouds glow, and as the jets of gas fluctuate, the clouds change in brightness and shape. Herbig-Haro objects HH34S and HH34N are clearly related to a young star located halfway between them (Figure 12-10), and the hottest part of the jet is visible pointing toward HH34S.

When a young star blasts gas away in two jets shooting in opposite directions, astronomers refer to it as a **bipolar flow.** According to recent theoretical work, such beams of gas appear to be caused by the formation of a spinning disk of gas around a forming star. As additional gas falls into the disk, it drags in a magnetic field from the interstellar medium, and the resulting interaction between the magnetic field and the rotating disk blows beams of gas outward along the axis of rotation (Figure 12-11). We will discover in later chapters that both disks and bipolar flows are common—they occur in dying stars and exploding galaxies as well as in young stars.

We know that many infrared objects, Herbig-Haro objects, and T Tauri stars are young because they are located in regions where hot, blue stars are found. Because these stars have such short lives, they too must be young, a fact that implies that the entire region is a site of active star formation.

Some very young star clusters such as NGC2264 contain large numbers of T Tauri stars. This cluster formed so recently that the less massive stars, contracting more slowly than the massive stars, have not yet reached the main sequence (Figure 12-9). Scattered throughout the cluster are a number of T Tauri stars, whose temperature and luminosity place them to the right of the main sequence near the birth line, just where we would expect to find contracting protostars.

T Tauri stars are often accompanied by **Herbig-Haro objects** (named after astronomers George Herbig and Guillermo Haro)—small nebulae that can change their brightness and shape over only a few years. Apparently, Herbig-Haro objects occur where jets of gas flowing away from young stars smash into clouds of in-

Figure 12-10

At the center of this image, a newborn star is emitting a powerful jet toward the left and, presumably, a less powerful jet to the right. Where these jets run into the interstellar medium, they excite the two Herbig-Haro objects HH34S and HH34N. *(Reinhard Mundt, Calar Alto 3.5-m telescope)* (Inset) This Hubble Space Telescope image shows that the jet emitted by the protostar is irregular and turbulent. *(J. Hester, Arizona State University, and NASA)*

Figure 12-11

(a) Star formation begins in a gas cloud when dense cores begin to form as gravitation draws gas and dust into condensations. (b) The infalling matter forms a dense, hot, central object destined to become a star at the center of a larger opaque cocoon of gas and dust. Rotation forces some of the matter to fall into a disk rotating around the central body. (c) The central object, now a protostar, and the rotating disk eventually begin to expel gas along the axis of rotation while matter continues to fall into the disk. The bipolar flows, guided and perhaps driven by magnetic fields (red), push out of the surrounding cloud of gas and dust to produce oppositely directed jets.

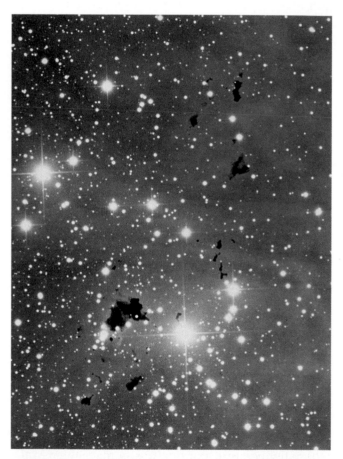

Figure 12-12

Bok globules are small, dense, dusty clouds that are either contracting to become stars or are the remnants of clouds that have formed stars. These dark globules are in the nebula IC2944. *(Anglo-Australian Telescope Board)*

Where we find the youngest stars, we often also find objects called **Bok globules** (named after the astronomer Bart Bok)—small clouds only about 1 ly in diameter that contain 10 to 1000 solar masses of gas and dust. Because these clouds produce no light of their own, they are only seen silhouetted against bright nebulae (Figure 12-12). Some astronomers believe the globules are collapsing clouds that will become stars. Infrared observations, including those made by the Infrared Astronomy Satellite, show that some globules do contain infrared sources that appear to be contracting protostars.

Because most of the stages of star formation occur deep inside dense clouds of gas and dust, we can't see what is happening. It would be nice if we could excavate such a cloud the way an archaeologist would excavate an ancient temple by stripping away the layers of dirt. A Hubble Space Telescope image of part of the Eagle Nebula reveals nature doing exactly that (Figure 12-13). Hot stars in and near the nebula are evaporating the dust and driving away the gas of the densest parts of the nebula, revealing small globules of denser gas and dust presumably drawn together by their own gravity. At least some of these seem to be forming stars. In part because these objects were found in the Eagle Nebula, astronomers have enjoyed calling them EGGs—evaporating gaseous globules. Whatever we call them, they are evidently stars exposed in the act of forming.

All of these observations confirm our theoretical model of contracting protostars. Although star formation still holds many mysteries, the general process seems clear. In at least some cases, interstellar gas clouds are compressed by passing shock waves, and the clouds' gravity, acting unopposed, draws the matter inward to form protostars.

a

b

Figure 12-13
(a) The Eagle Nebula contains dark clouds of gas and dust which are evaporating as hot, nearby stars vaporize the dust and drive away the gas to expose EGGs, evaporating gaseous globules. Some of the EGGs appear to be contracting to form protostars, and at least one seems to contain a newborn star (arrow). Compare with Figure 12-6. *(Jeff Hester and Paul Scowen, Arizona State University and NASA)* (b) A larger view of the Eagle Nebula reveals the hot stars at upper right and the evaporating dark clouds at lower left. *(NOAO)*

R E V I E W   Critical Inquiry

If the interstellar medium is so cold, how can we detect it at infrared wavelengths?

The interstellar medium is certainly very cold. The interstellar clouds have temperatures only 10 to 50 Kelvin above absolute zero. Our discussion of black body radiation (Chapter 4) told us that the energy emitted per second by a square meter on the surface of a black body depends only on T^4. A dust grain at such a low temperature radiates very little energy per second per square meter; and, of course, a dust grain has a tiny surface area.

So it does seem strange at first that we can detect interstellar clouds of gas and dust through the infrared radiation emitted by the dust grains. But recall that the mass of such a cloud might be a million solar masses, and that dust makes up about 1 percent of the cloud, about 10,000 solar masses of dust. An interstellar dust grain has a diameter roughly equal to the wavelength of light, so it has a tiny surface area compared with the sun. But 10,000 solar masses of dust have a surface area roughly 10^{20} times that of the sun. So even though a single dust grain emits very little infrared energy, the vast number of dust grains in the interstellar medium are very luminous at infrared wavelengths.

The interstellar medium is so cold that we don't really have a word for it. "Cold," "icy," or "frigid" can't express the awful cold of a few degrees above absolute zero. How can such cold material get hot enough to ignite nuclear fusion in the hearts of protostars?

Gravity converts the cold gas of the interstellar medium into hot protostars. A protostar becomes a star when it becomes hot enough to make its own energy through nuclear fusion.

12-2 Fusion in Stars

The sun generates energy through hydrogen fusion in the proton–proton chain (Chapter 10). Other main-sequence stars fuse hydrogen, but some use a different process, and aging stars fuse other fuels.

The CNO Cycle

Main-sequence stars more massive than the sun fuse hydrogen into helium using the **CNO cycle,** a hydrogen fusion process that uses carbon, nitrogen, and oxygen as stepping-stones. The CNO cycle begins with a carbon nucleus, transforms it first into a nitrogen nucleus, then into an oxygen nucleus, and then back to a carbon nucleus. The carbon is unchanged in the end, but along the way four hydrogen nuclei are fused to make a helium nucleus plus energy, just as in the proton–proton chain.

The CNO cycle begins with a carbon nucleus combining with a hydrogen nucleus. Because a carbon nucleus has a charge six times higher than hydrogen, the

Coulomb barrier is high, and temperatures higher than 16,000,000 K are required to make the CNO cycle work. The center of the sun is not quite hot enough. In stars more massive than about 1.1 solar masses, the cores are hotter, and those stars use the CNO cycle and not the less efficient proton–proton chain.

Heavy-Element Fusion

At later stages in the life of a star, when it has exhausted its hydrogen fuel, it may fuse other nuclear fuels. As we learned previously, the ignition of these fuels requires high temperatures, because the nuclei have large positive charges. The particles must travel at high velocities to overcome the electrostatic repulsion and force the particles close enough together to react. Helium fusion requires a temperature of at least 100 million K, and carbon fusion requires 600 million K.

We can summarize the helium-fusion process in two steps:

$$^{4}\text{He} + {}^{4}\text{He} \rightarrow {}^{8}\text{Be} + \gamma$$
$$^{8}\text{Be} + {}^{4}\text{He} \rightarrow {}^{12}\text{C} + \gamma$$

Because a helium nucleus is called an alpha particle, these reactions are commonly known as the **triple-alpha process.** Helium fusion is complicated by the fact that beryllium-8, produced in the first reaction of the process, is very unstable and may break up into two

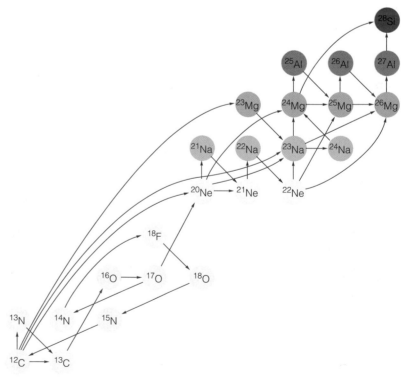

Figure 12-14
Carbon fusion involves many possible reactions that build numerous heavy atoms.

helium nuclei before it can absorb another helium nucleus. Three helium nuclei can also form carbon directly, but such a triple collision is unlikely.

At temperatures above 600,000,000 K, carbon fuses rapidly in a complex network of reactions illustrated in Figure 12-14, where each arrow represents a different nuclear reaction. The process is complicated because nuclei can react by adding a proton, a neutron, or a helium nucleus, or by combining directly with other nuclei. Unstable nuclei can decay by ejecting an electron, a positron, or a helium nucleus, or by splitting into fragments. The complexity of this process makes it difficult to determine exactly how much energy will be generated and how many heavy atoms will be produced.

Reactions at still higher temperatures can convert magnesium, aluminum, and silicon into yet heavier atoms. These reactions involving heavy elements will be important in the study of the deaths of massive stars in the next chapter.

The Pressure–Temperature Thermostat

Nuclear reactions in stars manufacture energy and heavy atoms under the supervision of a built-in thermostat that keeps the reactions from erupting out of control. That thermostat is the relation between gas pressure and temperature.

In a star, the nuclear reactions generate just enough energy to balance the inward pull of gravity. Consider what would happen if the reactions began to produce too much energy. The extra energy would raise the internal temperature of the star; and because the pressure of the gas depends on its temperature, the pressure would also rise, causing the star to expand. Expansion of the gas would cool the star slightly, slowing the nuclear reactions. Thus the star has a built-in regulator that keeps the nuclear reactions from occurring too rapidly.

The same thermostat also keeps the reactions from dying down. Suppose the nuclear reactions began to produce too little energy. Then the inner temperature would decrease, lowering the pressure and allowing gravity to compress the star slightly. As the gas was compressed, it would heat up, increasing the nuclear energy generation until the star regained stability.

The stability of a star depends on this relation between gas pressure and temperature. If an increase or decrease in temperature produces a corresponding change in pressure, then the thermostat functions correctly and the star is stable. We will see in this chapter how the thermostat accounts for the mass–luminosity relation. In the next chapter, we will see what happens to a star when the thermostat breaks down completely and the nuclear fires rage unregulated.

REVIEW Critical Inquiry

Why does the fusion of heavier atoms require higher temperatures than the fusion of hydrogen?

Nuclear fusion happens when atomic nuclei collide and fuse to form a new nucleus. Inside a star, the gas is ionized, so the nuclei have no orbiting electrons, and thus the nuclei have positive charges. Two objects with positive charges will repel each other in what we have called the Coulomb barrier. To start fusion, the nuclei must collide at high enough velocity to overcome that barrier. Heavier atomic nuclei such as helium and carbon have higher positive charges, so their Coulomb barriers are higher. That requires that the gas be hotter so the collisions between nuclei will be violent enough to overcome the barrier.

Nuclear fusion in stars follows rules that are amazingly simple. For example, what rule sees that a star will produce exactly the energy needed?

The law of gravity and the rules of nuclear fusion determine how stars work and allow us to understand what the inside of a star is like. In other words, we can describe the internal structure of a star.

12-3 Stellar Structure

The nuclear fusion at the center of stars heats their interior, creates high gas pressure, and thus balances the inward force of gravity. If there is a single idea in modern astronomy that can be called critical, it is that balance. Stars are simple, elegant power sources held together by their own gravity and supported by their nuclear fusion.

Having discussed the birth of stars and their nuclear power sources, we can now consider the structure of a star—that is, the variation in temperature, density, pressure, and so on from the surface of the star to its center. A star's structure depends on how it generates its energy, on four simple laws of structure, and on what it is made of.

It will be easier to think about stellar structure if we imagine that the star is divided into concentric shells like those in an onion. We can then discuss the temperature, density, pressure, and so on in each shell. Of course, these shells do not really exist; stars have no separable layers.

The Laws of Mass and Energy

The first two laws of stellar structure have something in common—they are both conservation laws. They tell us that certain things cannot be created out of nothing or vanish into nothing. Such conservation laws are powerful tools to help us understand how nature works.

The **conservation of mass** law is a basic law of nature, which we can apply to the structure of stars. It says that the total mass of the star must equal the sum of the masses of its shells. This is like saying the weight of a cake must equal the sum of the weight of its layers.

The **conservation of energy** law is another basic law of nature. It says that the amount of energy flowing out of the top of a layer in the star must be equal to the amount of energy coming in at the bottom plus whatever energy is generated within the layer. That means that the energy leaving the surface of the star, its luminosity, must equal the sum of the energies generated in all of the layers inside the star.

These two laws may seem so familiar and so obvious we hardly need to state them, but they are important clues to the structure of stars. The third law of stellar structure is familiar because we have been using it under a different name in the preceding sections.

Hydrostatic Equilibrium

The law of **hydrostatic equilibrium** says that, in a stable star, the weight of the material pressing downward on a layer must be balanced by the pressure of the gas in that layer. Hydrostatic equilibrium is just a new name for the gravity–pressure balance discussed earlier. *Hydro* (from the Greek word for "water") implies a fluid, the gases of a star, and *static* implies that the fluid is stable, neither expanding nor contracting.

The law of hydrostatic equilibrium can prove to us that the temperature and density must increase as we descend into a star. Near the surface, there is little weight pressing down on the gas, so the pressure must be low, implying a low temperature and density. But as we go deeper into the star, the weight becomes greater, so the pressure, and therefore the temperature and density, must also increase (Figure 12-15).

Of course, how hydrostatic equilibrium actually works depends on what an object is made of. Nearly all stars are made of hydrogen and helium gas, with some trace of heavier elements. To understand how a star works, we must describe how that gas responds to changes in temperature and pressure. Hydrostatic equilibrium also applies to planets, including Earth, but Earth is made of rock and metal, so again, understanding how hydrostatic equilibrium supports Earth requires that Earth scientists know how rock and metal respond to changes in temperature and pressure. We will see in later chapters how the structure of a star such as a white dwarf changes dramatically when its gas behaves differently from that in most normal stars.

Although the law of hydrostatic equilibrium can tell us some things about the inner structure of stars,

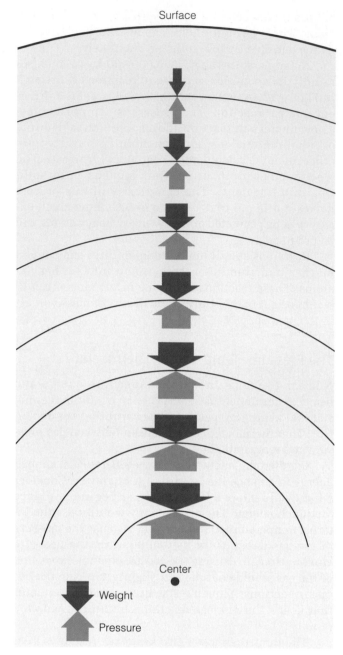

Figure 12-15
Hydrostatic eqlibrium says the pressure in each layer must balance the weight on that layer. As a result, pressure and temperature must increase from the surface of a star to its center.

we need one more law to completely describe a star. We need a law that describes the flow of energy from the center to the surface.

Energy Transport

The surface of a star radiates light and heat into space and would quickly cool if that energy were not replaced. Because the inside of the star is hotter than the surface, energy must flow outward from the core,

Figure 12-16
The three modes by which energy may be transported from the flame of a candle, as shown here, are the three modes of energy transport within a star.

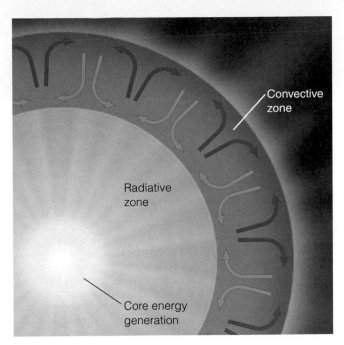

Figure 12-17
A cross section of the sun. Near the center, nuclear fusion reactions generate high temperatures. Energy flows outward through the radiation zone as photons. In the cooler, more opaque outer layers, the energy is carried outward by rising convection currents of hot gas (red) and sinking currents of cooler gas (blue).

where it is generated, to the surface, where it radiates away. This flow of energy through the shells determines their temperature, which, as we saw previously, determines how much weight each shell can balance. To understand the structure of a star, we must understand how energy moves from the center through the shells to the surface.

The law of **energy transport** says that energy must flow from hot regions to cooler regions by conduction, convection, or radiation.

Conduction is the most familiar form of heat flow. If you hold the bowl of a spoon in a candle flame, the handle of the spoon grows warmer. Heat, in the form of motion among the molecules of the spoon, is conducted from molecule to molecule up the handle, until the molecules of metal under your fingers begin to move faster and you sense heat (Figure 12-16). Thus, conduction requires close contact between the molecules. In most stars matter is gaseous, so conduction is unimportant. Conduction is significant in white dwarfs, which have tremendous internal densities.

The transport of energy by radiation is another familiar experience. Put your hand beside a candle flame, and you can feel the heat. What you actually feel are infrared photons radiated by the flame (Figure 12-16). Because photons are packets of energy, your hand grows warm as it absorbs them.

Radiation is the principal means of energy transport in the sun's interior (Figure 12-17). Photons are absorbed and re-emitted in random directions over and over as they work their way outward. This process breaks the high-energy photons common near the sun's center into large numbers of low-energy photons in the cooler outer layers. Thus, the energy of a single high-energy photon at the center of the sun takes about 1 million years to reach the sun's surface, where it emerges as roughly 1600 photons of visible light.

The flow of energy by radiation depends on how difficult it is for the photons to move through the gas. If the gas is cool and dense, the photons are more likely to be absorbed or scattered, and so the radiation does not get through easily. We would call such a gas opaque. In a hot, thin gas, the photons can get through more easily; such a gas is less opaque. The **opacity** of the gas, its resistance to the flow of radiation, depends strongly on its temperature.

If the opacity is high, radiation cannot flow through the gas easily, and it backs up like water behind a dam. When enough heat builds up, the gas begins to churn as hot gas rises upward and cool gas sinks downward. This is convection, the third way energy can move in a star. Convection is a common experience; the wisp of smoke rising above a candle flame travels upward in a small convection current (Figure 12-16). If you hold your hand above the flame, you can feel the rising current of hot gas. In stars, energy may be carried upward by rising currents of hot gas hundreds or thousands of miles in diameter.

Convection is important in stars because it carries energy and because it mixes the gas. Convection currents flowing through the layers of a star tend to homogenize the gas, giving it a uniform composition throughout the convective zone. As you might expect,

One of the most powerful tools in science is the mathematical model, a group of equations carefully designed to mimic the behavior of the object that scientists want to study. Astronomers build mathematical models of stars using only four equations, but other systems are much more complicated and may require many more equations.

Many sciences use mathematical models. Medical scientists have built mathematical models of the nerves that control the human heart, and physicists have built mathematical models of the inside of an atomic nucleus. Economists have built mathematical models of certain aspects of economic systems, such as the municipal bond market, and Earth scientists have built mathematical models of Earth's atmosphere. In each case, the mathematical model allows the scientists to study something that is difficult to study in reality. The model can reveal regions we cannot observe, speed up a slow process, slow down a fast process, or allow scientists to perform experiments that would be impossible in reality. Astronomers, for example, can change the abundance of different chemical elements in a model star to see how its structure depends on its composition.

Many of these mathematical models require very large and fast computers—in some cases, supercomputers. A modern desktop computer takes only seconds to compute a stellar model; but a model of the motions of the planets of our solar system millions of years into the future would require the largest and fastest computers in the world. Most astronomers are apt computer programmers, and some are experts. A few astronomers have built their own highly specialized computers to compute specific kinds of models. Thus, the mathematical model is one of the most important tools in astronomy.

As is true for any scientific model, a mathematical model is only as reliable as the assumptions that go into its creation. In Window on Science 1-1, we saw that the celestial sphere is an adequate model of the sky for some purposes but breaks down if we extend it too far. So too with mathematical models. We can think of a mathematical model as a numerical expression of one or more theories. While such models can be very helpful, they are always based on theories and assumptions and so must be compared with the real world at every opportunity. As always in science, there is no substitute for a careful comparison with reality.

---

this mixing affects the fuel supply of the nuclear reactions, just as the stirring of a campfire makes it burn more efficiently.

The four laws of stellar structure are summarized in Table 12-1. These laws, properly understood, can tell us how stars are born, how they live, and how they die.

Stellar Models

The laws of stellar structure, described in general terms in the previous section, can be written as mathematical equations. By solving those equations in a special way, astronomers can build a mathematical model of the inside of a star.

If we wanted to build a model of a star, we would have to divide the star into about 100 concentric shells, and then write down the four equations of stellar structure for each shell. We would then have 400 equations that would have 400 unknowns, namely, the temperature, density, mass, and energy flow in each shell. Solving 400 equations simultaneously is not easy, and the first such solutions, done by hand before the invention of the electronic computer, took months of work. Now a properly programmed computer can solve the equations in a few seconds and print a table of numbers that represent the conditions in each shell of the star. Such a table is a **stellar model** (Window on Science 12-2).

The table shown in Figure 12-18 is a model of the sun. The bottom line, for radius equal to 0.00, represents the center of the sun, and the top line, for radius equal to 1.00, represents the surface. The other lines tell us the temperature and density in each shell, the mass inside each shell, and the fraction of the sun's luminosity flowing outward through the shell. We can use the table to understand the sun. The bottom line tells us the temperature at the center of the sun is about 15 million Kelvin. At such a high temperature the gas

Table 12-1 The Four Laws of Stellar Structure

I.	Conservation of mass	Total mass equals sum of shell masses.
II.	Conservation of energy	Total luminosity equals sum of energies generated in each shell.
III.	Hydrostatic equilibrium	The weight on each layer is balanced by the pressure in that layer.
IV.	Energy transport	Energy moves from hot to cool by conduction, radiation, or convection.

$R/R_\odot$	T $(10^6\ \text{K})$	Density (g/cm^3)	$M/M_\odot$	$L/L_\odot$
1.00	0.006	0.00	1.00	1.00
0.90	0.60	0.009	0.999	1.00
0.80	1.27	0.035	0.996	1.00
0.70	1.80	0.12	0.990	1.00
0.60	2.42	0.40	0.97	1.00
0.50	3.42	1.3	0.92	1.00
0.40	4.74	4.1	0.82	1.00
0.30	6.65	13.	0.63	0.99
0.20	9.35	36.	0.34	0.91
0.10	12.65	85.	0.073	0.40
0.00	14.62	134.	0.000	0.00

$$\frac{dM}{dr} = 4\pi r^2 \rho$$

$$\frac{dL}{dr} = 4\pi r^2 \rho e$$

$$\frac{dP}{dr} = -\frac{GM}{r^2}\rho$$

$$\frac{dT}{dr} = \frac{-3}{16\pi ac}\frac{\bar{\kappa}\rho}{T^3}\frac{L}{r^2}$$

Figure 12-18

A stellar model is a table of numbers that represents conditions inside a star. Such tables can be computed using the four laws of stellar structure, shown here in mathematical form. The table in this figure describes the interior of the sun. *(Illustration design by author)*

is highly transparent, and energy flows as radiation. Nearer the surface, the temperature is lower, the gas is more opaque, and the energy is carried by convection. This is confirmed by the evidence of convection visible on the sun's surface (Chapter 10).

Stellar models also let us look into the star's past and future. In fact, we can use models as time machines to follow the evolution of stars over billions of years. To look into a star's future, for instance, we use a stellar model to determine how fast the star uses its fuel in each shell. As the fuel is consumed, the chemical composition of the gas changes and the amount of energy generated declines. By calculating the rate of these changes, we can predict what the star will look like at any point in the future.

Although this sounds simple, it is actually a highly challenging problem involving nuclear and atomic physics, thermodynamics, and sophisticated computational methods. Only since the 1950s have electronic computers made the rapid calculation of stellar models possible, and the advance of astronomy since then has been heavily influenced by the use of such models to study the structure and evolution of stars. Our summary of star formation in this chapter is based on thousands of stellar models. We will continue to rely on theoretical models as we study main-sequence stars in the next section and the deaths of stars in the next chapter.

R E V I E W Critical Inquiry

What would happen if the sun stopped generating energy?

Stars are supported by the outward flow of energy generated by nuclear fusion in their interiors. That energy keeps each layer of the star just hot enough so that the gas pressure can support the weight of the layers above. Each layer in the star must be in hydrostatic equilibrium; that is, the inward weight is balanced by outward pressure. If the sun stopped making energy in its interior, nothing would happen at first, but over many thousands of years the loss of energy from its surface would reduce the sun's ability to withstand its own gravity, and it would begin to contract. We wouldn't notice much for 100,000 years or so, but eventually the sun would lose its battle with gravity.

Stars are elegant in their simplicity: just a cloud of gas held together by gravity and warmed by nuclear fusion. We can learn a lot about the inside of stars by understanding only a few principles. For example, how does hydrostatic equilibrium tell us the inside of stars must be dense and hot?

We have traced the birth of stars from the first instability in the interstellar medium to the final equilibrium of nuclear fusion and gravity. Now we must look carefully at these stable stars and ask two questions. First, what evidence do we have that stars really work this way? And second, how long can these stars remain stable? The answers to these two questions will give us a new insight into the lives of the stars.

12-4 Main-Sequence Stars

When a contracting protostar begins to fuse hydrogen, it stops contracting and becomes stable. The location of these stable stars in the H-R diagram is marked by the main sequence. Because 90 percent of all stars lie on the main sequence, it is important that we understand their structure.

The Mass–Luminosity Relation

Main-sequence stars obey a simple rule—the more massive a star is, the more luminous it is. That rule, the mass–luminosity relation discussed in the previous chapter, is the key to understanding the stability of stars. In addition, the mass–luminosity relation can be explained by the theory of stellar structure, giving us direct observational confirmation of the theory.

To understand the mass–luminosity relation, we must consider the law of hydrostatic equilibrium, which says that pressure must balance weight, and the pressure–temperature thermostat, which regulates energy production. We have seen that a star's internal pressure stays high because generation of thermonuclear energy keeps its interior hot. Because more massive stars have more weight pressing down on the inner layers, their interiors must have high pressures and thus must be hot. For example, the temperature at the center of a 15-solar-mass star is about 34,000,000 K, more than twice the central temperature of the sun.

Because massive stars have hotter cores, their nuclear reactions rage more fiercely. That is, their pressure–temperature thermostat is set higher. One gram of material at the center of a 15-solar-mass star fuses over 3000 times more rapidly than 1 g of material at the center of the sun. The rapid reactions in massive stars make them more luminous than the lower-mass stars. Thus the mass–luminosity relation results from the requirement that a star support its weight by generating nuclear energy.

This also tells us why the main sequence must have a lower end. Masses below about 0.08 solar mass cannot raise their central temperatures high enough to ignite hydrogen fusion. A mass of 0.08 solar mass corresponds to about 80 Jupiter masses, and astronomers have speculated that nature must make such failed stars. They would be small, not much larger than Jupiter, and warm from their contraction. Such objects should emit copious infrared radiation and consequently have been called **brown dwarfs.** Careful searches for such stars are turning up scores of examples. The star GL229 is orbited by an object about 20 times the mass of Jupiter. The low mass of the object and the presence of methane in its spectrum confirm that it cannot be a star, and so it is presumably a brown dwarf. Such objects are difficult to find and study because of their very low temperature, but the theories of stellar formation do seem to be confirmed. Nature does indeed make brown dwarfs, apparently when a forming star has a mass too low to ignite hydrogen fusion.

Now that we know how stars form, and how they maintain their stability through the mass–luminosity relation, we can predict the evolution of main-sequence stars.

The Life of a Main-Sequence Star

While a star is on the main sequence, it is stable, so we might think its life would be totally uneventful. But a main-sequence star balances its gravity by fusing hydrogen, and as the star gradually uses up its fuel, that balance must change. Thus, even the stable main-sequence stars are changing as they consume their hydrogen fuel.

Recall that hydrogen fusion combines four nuclei into one. Thus, as a main-sequence star fuses its hydrogen, the total number of particles in its interior decreases. Each newly made helium nucleus can exert the same pressure as a hydrogen nucleus, but because the gas has fewer nuclei, its total pressure is less. This unbalances the gravity–pressure stability, and gravity squeezes the core of the star more tightly. As the core contracts, its temperature increases and the nuclear reactions run faster, releasing more energy. This additional energy flowing outward through the envelope forces the outer layers to expand. As the star becomes larger, it becomes more luminous, and eventually the expansion begins to cool the surface.

As a result of these gradual changes in main-sequence stars, the main sequence is not a sharp line across the H-R diagram but rather a band (shaded in Figure 12-19). Stars begin their stable lives fusing hydrogen on the lower edge of this band, the **zero-age main sequence (ZAMS),** but gradual changes in luminosity and surface temperature move the stars upward and to the right. By the time they reach the upper edge of the main sequence (the dashed line in Figure 12-19), they have exhausted nearly all the hydrogen in their centers. Thus we find main-sequence stars scattered throughout this band at various stages of their main-sequence lives.

The sun is a typical main-sequence star, and as it undergoes these gradual changes, Earth will suffer. When the sun began its main-sequence life about 5 billion years ago, it was only about 70 percent as lumi-

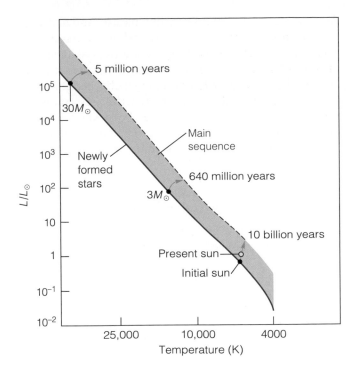

Figure 12-19
The main sequence is not a line but a band (shaded). Newly formed stars begin fusing hydrogen on the lower edge (red). As hydrogen fusion changes their composition, the stars slowly move across the band. Notice that a star of 3 solar masses crosses the band in less than a tenth the time the sun will take, and that a star of 30 solar masses will cross the band in less than a thousandth the time the sun will take.

nous as it is now, and by the time it leaves the main sequence in another 5 billion years, the sun will have twice its present luminosity. This will raise the average temperature on Earth by at least 19°C (34°F). As this happens over the next few billion years, the polar caps will melt, and Earth's climate will change dramatically. Life on Earth may not survive this change in the sun.

The average star spends 90 percent of its life on the main sequence. This explains why 90 percent of all stars are main-sequence stars—we are most likely to see a star during that long, stable period while it is on the main sequence. To illustrate: Imagine that you photograph, at a single instant, a crowd of 20,000 people. We all sneeze now and then, but the act of sneezing is very short compared to a human lifetime, so we would expect to find very few people in our photograph in the act of sneezing. By contrast, because the main-sequence phase is very long compared to a stellar lifetime, we can expect to see many stars on the main sequence.

The number of years a star spends on the main sequence depends on its mass (Table 12-2). Massive stars consume fuel rapidly and live short lives, but low-mass stars conserve their fuel and shine for billions of years. For example, a 25-solar-mass star will exhaust its hydrogen and die in only about 7 million years. Very-low-

Table 12-2 Main-Sequence Stars

Spectral Type	Mass (Sun = 1)	Luminosity (Sun = 1)	Years on Main Sequence
O5	40	405,000	1×10^6
B0	15	13,000	11×10^6
A0	3.5	80	440×10^6
F0	1.7	6.4	3×10^9
G0	1.1	1.4	8×10^9
K0	0.8	0.46	17×10^9
M0	0.5	0.08	56×10^9

mass stars, the red dwarfs, use up their fuel so slowly that they last for 200 to 300 billion years. Because the universe seems to be only 10 to 15 billion years old, red dwarfs must still be in their infancy. (By the Numbers 12-1 explains how we can quickly estimate the life expectancies of stars from their masses.)

Nature makes more low-mass stars than high-mass stars, but this fact is not sufficient to explain the vast numbers of low-mass stars that fill the sky. An additional factor is the stellar lifetime. Because low-mass stars live long lives, there are more of them in the sky than massive stars. Look at Figure 11-25, and notice how much more common the lower-main-sequence stars are than the massive O and B stars. The main-sequence K and M stars are so faint they are difficult to locate, but they are very common. The O and B stars are luminous and easy to locate, but because of their fleeting lives, there are never more than a few on the main sequence at any one time.

REVIEW Critical Inquiry

Why are main-sequence O stars so rare compared to the sun?

In Section 11-6, we discussed the family of stars and noticed that space contains only about 0.036 main-sequence O stars per million cubic parsecs, but that there are about 2000 main-sequence G stars per million cubic parsecs (see Figure 11-25). Dividing, we discover that main-sequence G stars are over 50,000 times more common than O stars. Our discussion of the lives of main-sequence stars gives us a way to analyze these numbers. O stars live very short lifetimes. Our simple formula predicts a 20-solar-mass O star has a life expectancy of about 6 million years. The sun, however, will live a total of about 10 billion years or about 1600 times longer than an O star. Assuming nature makes equal numbers of different kinds of stars, we would expect to see about

1600 times more main-sequence G stars than O stars. But we see over 50,000 times more, so there must be another factor involved. Nature must be making vastly more lower-mass stars than massive stars. By comparing theory and observation, our analysis has revealed something new about how nature makes stars.

Now we can understand why O stars are so rare. What factors must you take into consideration to explain why main-sequence M stars are even more common than G stars?

Science is a way of understanding nature, and although we use theories, we must always go back to reality and compare our theories with observations. We have constructed an elaborate theory to explain how stars are born and reach stability, and we have checked it a number of times with observations; but there is one fascinating and beautiful nest of newborn stars that we should study before we proceed. The Great Nebula in Orion is our final reality check on the formation of stars.

12-5 The Orion Nebula

On a clear winter night, you can see with your naked eye the Great Nebula of Orion as a fuzzy wisp in Orion's sword. With binoculars or a small telescope it is striking, and through a large telescope it is breathtaking. At the center lie four brilliant blue-white stars known as the Trapezium, and surrounding them are the glowing filaments of a nebula more than 8 pc across (Figure 12-20). Like a great thundercloud illuminated from within, the churning currents of gas and dust suggest immense power. A deeper significance lies hidden, figuratively and literally, behind the visible nebula, for

Figure 12-20
The Great Nebula in Orion is a glowing cloud of gas and dust over 8 pc in diameter. The hydrogen glows red because it is ionized by the hottest of the stars of the Trapezium (arrow), a cluster of young, hot stars just visible at the very center of the nebula. Compare with Figure 12-23. *(Daniel Good)*

Figure 12-21
Disks of gas and dust around young stars near the hot stars of the Trapezium in the Orion Nebula. The first four images show disks that are being evaporated by the heat from the nearby hot stars with the resulting gas being blown away to produce elongated tails. The last two images show dust disks seen silhouetted against the bright nebula. Bars in images are labeled in astronomical units. The diameter of our solar system taken as the diameter of Pluto's orbit is about 80 AU. *(John Bally, Dave Devine, and Ralph Sutherland and NASA)*

radio and infrared astronomers have discovered a vast dark cloud just beyond the visible nebula—a cloud in which stars are now being created.

Evidence of Young Stars

We should not be surprised to find star formation in Orion. Many of the stars in the constellation are hot, blue, upper-main-sequence stars that have short lifetimes. They must have formed recently. The region is also rich in T Tauri stars, which are known pre-main-sequence stars. The Hubble Space Telescope has found that the stars are young and surrounded by disks (Figure 12-21). Furthermore, an infrared image of the region shows streamers of gas rushing away from a star that has recently become luminous. All of these observations are evidence of star formation.

In fact, the glowing nebula itself is evidence of recent star formation. The nebula is an ionized hydrogen region excited by young stars, the Trapezium stars, which reached the main sequence no more than a million years ago. The most massive, a star of about 30 solar masses, must consume its fuel at a tremendous rate to support its large mass. Thus, it is hot and luminous—over 40,000 K and over 300,000 solar luminosities. The other stars are too cool to affect the surrounding gas very much, but the massive star is so hot it radiates large amounts of ultraviolet radiation (Figure 12-22). These

Figure 12-22
Photons with wavelengths shorter than 91.2 nm have enough energy to ionize hydrogen. The O6 star is the only star in the Trapezium hot enough to produce appreciable ionization.

Figure 12-23

The central 2.5 ly of the Great Nebula in Orion shown in this Hubble Space Telescope image contains the four stars of the Trapezium (arrow) and a complex tangle of gas clouds. *(John Bally, Dave Devine, Ralph Sutherland and NASA)* The inset at right shows the boxed area of the nebula at the infrared wavelength of 12.4 microns. Undetectable at visible wavelengths, the infrared sources are evidence of star formation going on in the molecular cloud behind the visible nebula. *(Dan Gezari, Dana Backman, and Mike Werner)*

photons have enough energy to ionize hydrogen near the Trapezium. Thus, the glowing gas of the nebula tells us that at least one massive star lies inside, and such massive stars are very short-lived. It must be very young.

The Orion Nebula looks impressive, but it is nearly a vacuum, containing a mere 6000 atoms/cm^3. For comparison, the interstellar medium contains about 1 atom/cm^3 and air at sea level about 10^{19} atoms/cm^3. Infrared observations show that the nebula also contains sparsely scattered dust warmed by the stars to about 70 K.

The Molecular Cloud

The significance of the Orion Nebula was established when observations at infrared and radio wavelengths revealed dense clouds of gas and dust just beyond the Great Nebula. At these wavelengths, hot stars and ionized gas are invisible, but cool, dense gas and dust can be detected. In space, ultraviolet photons break up most molecules, but hidden deep in the dark clouds, where ultraviolet photons cannot penetrate, molecules such

as carbon monoxide have formed. These molecules emit radio photons of characteristic wavelengths that allow radio astronomers to map these molecular clouds. Dust heated by stars radiates strongly at infrared wavelengths, and thus infrared telescopes can map the location of the dust.

These clouds are significantly different from the Great Nebula. The molecular clouds contain at least 10^6 atoms/cm^3 and large amounts of dust. Because the clouds seem warmest near their centers, astronomers conclude they are contracting.

The Infrared Objects

Infrared observations reveal strong infrared sources hidden within the dusty molecular cloud behind the visible nebula. Invisible at optical wavelengths, these objects are evidently stars in pre-main-sequence stages, wrapped deep in cocoons of dust. Three especially interesting objects lie almost directly behind the visible nebula (Figure 12-23).

The Becklin-Neugebauer object (BN object), named after its discoverers, was first thought to be a protostar, but infrared and radio observations show that the gas near its center is ionized and flowing outward. Thus, the object is probably a hot B star that has just reached the main sequence.

The Kleinmann-Low nebula is much cooler than the BN object. The KL nebula is thought to be a contracting cloud of gas in an early stage of star formation.

Near the B-N object and the K-L nebula lies a strong infrared source called IRc2. Radio observations show that the object is surrounded by a thick, expanding disk of gas and is ejecting a powerful bipolar flow. Theory predicts that the object cannot sustain such loss of mass for more than about 1000 years, so it must be a very young protostar.

Telling the Story of Star Formation in Orion

The goal of science is to use theory supported by evidence to describe how nature works. Modern astronomers can now tell the story of the Orion Nebula with some confidence. The history of star formation in Orion is complex, but we can find evidence that the entire region has been a site of star formation for at least three generations of massive stars. The ages of the stars tell the story. The stars at Orion's west shoulder are about 12 million years old, while the stars of Orion's belt are about 8 million years old. The stars of the Trapezium are no older than 2 million years. Apparently, star formation began near the west shoulder and the massive stars that formed there triggered star formation near the belt. That star formation triggered the formation of the stars we see in the Great Nebula. Like a grass fire, star formation has swept across Orion from northwest to southeast.

We can read the story of the Great Nebula in the Trapezium stars and the molecular cloud that lies behind it. Just a few million years ago, the Trapezium stars were contracting protostars buried within a molecular cloud. As they approached the main sequence, their temperatures increased, and their radiation drove away their cocoons. But not until the 30-solar-mass star turned on was there sufficient ultraviolet radiation to ionize the gas. The ionization transformed the cloud from a cold, dusty cocoon into a hot, transparent bubble of gas expanding away from the Trapezium. We can see evidence of this expansion in the Doppler shifts of spectral lines and in the twisted filaments of gas within the nebula.

Once the hot star ionized the gas, star formation in the Trapezium region stopped. In more distant parts of the cloud, the contraction of the gas into protostars continues. Indeed, the ionized gases of the visible nebula pushing against the remains of the cloud may have triggered the collapse of more protostars. We now see the

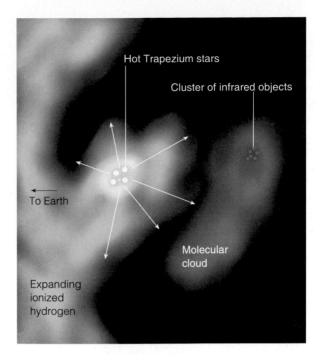

Figure 12-24
A side view of the Great Nebula would show the newly formed Trapezium stars heating and driving away the surrounding gas as new stars form in the molecular cloud behind the visible nebula.

front of the cloud torn by the expanding ionized gases around the Trapezium cluster, while deeper within the cloud more protostars are forming (Figure 12-24).

In the next few thousand years, the familiar outline of the Great Nebula will change, and a new nebula may form as the protostars in the molecular clouds reach the main sequence and the most massive become hot enough to ionize the surrounding gas. Thus, the Great Nebula in Orion and its parent molecular cloud show how the formation of stars continues even today.

REVIEW **Critical Inquiry**

What did Orion look like to the ancient Egyptians, to the first humans, and to the dinosaurs?

The Egyptian civilization had its beginning only a few thousand years ago, and that is not very long in terms of the history of Orion. The stars we see in the constellation are hot and young, but they are a few million years old, so the Egyptians saw the same constellation we see. (They called it Osiris.) Even the Orion Nebula hasn't changed very much in a few thousand years, and Egyptians may have admired it in the dark skies along the Nile.

Our oldest human ancestors lived about 3 million years ago, and that was about the time that the youngest stars in Orion were forming. Thus, they may have looked up and seen some of the stars we see, but some stars have formed since that time. Also, the Great Nebula is excited by the Trapezium stars, and they are not more than a few million years old, so our early ancestors probably didn't see the Great Nebula.

The dinosaurs saw something quite different. The last of the dinosaurs died about 65 million years ago, long before the birth of the brightest stars in the constellation we see. The dinosaurs, had they had the brains to appreciate the view, might have seen bright stars along the Milky Way, but when we think back millions of years, we must remember that the stars are moving through space and the sun is orbiting the center of our galaxy. The night sky above the dinosaurs contained totally different star patterns.

Of course, a giant molecular cloud can't continue to spawn new stars forever. What processes limit star formation in a molecular cloud and will eventually end star formation in the Orion region?

If stars are born, then how do they die? We will begin that story in the next chapter.

Summary

We know there is an interstellar medium because of observational evidence. We can see emission, reflection, and dark nebulae, and we can detect interstellar absorption lines in the spectra of distant stars. The dust in the interstellar medium makes distant stars look fainter and redder than we would expect.

The interstellar medium contains large, dense, cool clouds of gas in which hydrogen exists as molecules rather than as atoms. These molecular clouds are sites of star formation. Such clouds can be triggered into collapse by collision with a shock wave, which compresses and fragments the gas cloud, producing a cluster of protostars.

A contracting protostar is large and cool and would appear in the red-giant region of the H-R diagram were it not surrounded by a cocoon of dust and gas. The dust in the cocoon absorbs the protostar's light and reradiates it as infrared radiation. Many infrared sources are probably protostars. T Tauri stars may be cocoon stars in later stages as the cocoon is absorbed by the protostar or driven into space. Very young star clusters, like NGC2264, contain large numbers of T Tauri stars. Herbig-Haro objects may be fragments of cocoon nebula. Some protostars have been found emitting bipolar flows of gas as they expel material from their surroundings.

As a protostar's center grows hot enough to fuse hydrogen, it settles onto the main sequence to begin its long hydrogen-fusion phase. If it is a low-mass star, it cannot become very hot and must generate energy by the proton–proton chain. But if it is a massive star, it will become hot enough to use the CNO cycle. In either case, the nuclear reactions are regulated by the pressure–temperature thermostat in the star's core.

Almost everything we know about the internal structure of stars comes from mathematical stellar models. The models are based on four simple laws of stellar structure. The first two laws say that mass and energy must be conserved and spread smoothly through the star. The third, hydrostatic equilibrium, says the star must balance the weight of its layers by its internal pressure. The fourth says energy can only flow outward by conduction, convection, or radiation.

The mass–luminosity relation is explained by the requirement that a star support the weight of its layers by its internal pressure. The more massive a star is, the more weight it must support and the higher its internal pressure must be. To keep its pressure high, it must be hot and generate large amounts of energy. Thus the mass of a star determines its luminosity. The massive stars are very luminous and lie along the upper main sequence. The less massive stars are fainter and lie lower on the main sequence.

How long a star can stay on the main sequence depends on its mass. The more massive a star is, the faster it uses up its hydrogen fuel. A 25-solar-mass star will exhaust its hydrogen and die in only about 7 million years, but the sun can last for 10 billion years.

The Great Nebula in Orion is an active region of star formation. The bright stars we see in the center of the nebula formed within the last few million years, and infrared telescopes detect protostars buried inside the molecular cloud that lies behind the visible nebula.

New Terms

interstellar medium	bipolar flow
nebula	Bok globule
emission nebula	CNO cycle
reflection nebula	triple-alpha process
interstellar reddening	conservation of mass
infrared cirrus	conservation of energy
molecular cloud	hydrostatic equilibrium
shock wave	energy transport
association	opacity
protostar	stellar model
evolutionary track	brown dwarf
birth line	zero-age main sequence
T Tauri star	(ZAMS)
Herbig-Haro object	

Review Questions

1. What evidence do we have that the spaces between the stars are not empty?

2. What evidence do we have that the interstellar medium contains both gas and dust?

3. Why would an emission nebula near a hot star look red, but a reflection nebula near its star looks blue?

4. Why do astronomers rely heavily on infrared observations to study star formation?

5. How are Herbig-Haro objects related to star formation?

6. How do the proton–proton chain and the CNO cycle resemble each other? How do they differ?

7. Why does the CNO cycle require a higher temperature than the proton–proton chain?

8. How does the pressure–temperature thermostat control the nuclear reactions inside stars?

9. Step-by-step, explain how energy flows from the sun's core to Earth.

10. Why is there a mass–luminosity relation?

11. Why is there a lower limit to the mass of a main-sequence star?

12. Why does a star's life expectancy depend on its mass?

13. What evidence do we have that star formation is happening right now in the Orion Nebula?

Discussion Questions

1. When we see distant streetlights through smog, they look dimmer and redder than they do normally. But when we see the same streetlights through fog or falling snow, they look dimmer but not redder. Use your knowledge of the interstellar medium to discuss the relative sizes of the particles in smog, fog, and snow compared to the wavelength of light.

2. If planets form as a natural by-product of star formation, which do you think are more common—stars or planets?

Problems

1. The interstellar medium dims starlight by about 1.9 magnitudes/1000 pc. What fraction of photons survive a 1000-pc trip? (*Hint:* See By the Numbers 1-1.)

2. A small Bok globule has a diameter of 20 seconds of arc and is 1000 pc from Earth. What is its diameter?

3. If a giant molecular cloud is 30 pc in diameter and drifts relative to neighboring clouds at 20 km/s, how long will it take to travel its own diameter?

4. A protostar's dust cocoon radiates most strongly at a wavelength of 30 microns. What is the temperature of the dust? (*Hint:* See By the Numbers 4-1.)

5. Gas in a bipolar flow travels as fast as 300 km/s. How long would it take to travel 1 light-year?

6. In the model shown in Figure 12-18, how much of the sun's mass is hotter than 12,000,000 K?

7. At what wavelength does gas at the sun's center radiate most strongly? (*Hint:* See By the Numbers 4-1.)

8. What is the life expectancy of a 16-solar-mass star?

9. How massive could a star be and still survive for at least 1 billion years?

10. At 40,000 K the hottest star in the Orion Nebula radiates most strongly at what wavelength? (*Hint:* See By the Numbers 4-1.)

Critical Inquiries for the Web

1. Astronomers have been searching for brown dwarfs for years, but few candidates have been identified. Search the Web for information on efforts to locate these substellar objects. Why are they so difficult to detect? List any likely brown dwarfs found so far.

Exploring *The Sky*

1. The following objects are gas and dust clouds (star-forming regions) that can easily be viewed through small telescopes: M42, M20, M8, and M17. The "M" stands for Messier, a person who cataloged objects that look fuzzy in a telescope. M42 is the 42nd object in the Messier catalog, which contains 103 objects.

   Determine the following information about these objects: their common name, the constellation in which they reside, their magnitude, and their angular size. Which of these objects can you see with the unaided eye?

   How to proceed: Within the **Find** window, locate the **Common Names** window, and select **Messier Objects.**

Go to the Brooks/Cole Astronomy Resource Center (www. brookscole.com/astronomy) for critical thinking exercises, articles, and additional readings from InfoTrac College Edition, Brooks/Cole's online student library.

CHAPTER THIRTEEN

The Deaths of Stars

Natural laws have no pity.

Robert Heinlein
The Notebooks of Lazarus Long

Guidepost

As you read this chapter, take a moment to be astonished and proud that we humans know how stars die. Finding the mass of stars is difficult, and understanding the invisible matter between the stars takes ingenuity; but we human beings have used what we know about stars and what we know about the physics of energy and matter to figure out how stars die.

The death of stars is important to us because our own sun will die, but it is also important because the explosive deaths of massive stars contaminate the universe with atoms heavier than helium. The atoms we are made of were created inside stars. If stars didn't die, we would not exist.

In the chapters that follow, we will see that some matter from dying stars becomes trapped in dead ends—neutron stars and black holes—but some escapes into the interstellar medium and is incorporated into new stars and the planets that circle them. Thus, the deaths of stars are part of a cycle of stellar birth and death that includes our sun, our planet, and us.

The stars are dying. Occasionally astronomers see a new star, a **nova,** appear in the sky, grow brighter, fade after a few weeks, and vanish. The brighter and longer lasting **supernovae** (such as Tycho Brahe's supernova discussed in Chapter 2) are so rare that centuries pass between supernovae bright enough to see with the unaided eye. In this chapter we will discover that these "new" stars mark the explosive deaths of old stars.

The mass of a star is critical in determining its fate. Massive stars use up their nuclear fuel at a furious rate and die after only a few million years, while the lowest-mass stars use their fuels sparingly and may be able to live hundreds of billions of years. Massive stars die in violent explosions, but low-mass stars die quiet deaths. To follow the stars to their graves, we will sort them into three categories according to their masses: low-mass red dwarfs, medium-mass sunlike stars, and massive upper-main-sequence stars. These stars lead dramatically different lives and die in different ways (Figure 13-1).

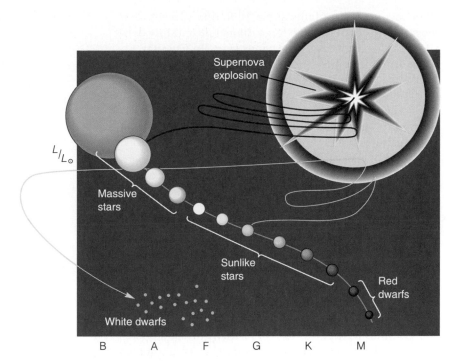

Figure 13-1
Three different modes of stellar death are summarized in this schematic H-R diagram. The lowest-mass stars, the red dwarfs, could live much longer than the present age of the universe. Medium-mass stars like the sun become giant stars and then collapse to become white dwarfs. The most massive stars evolve rapidly to become supergiants and explode as supernovae, leaving behind either a neutron star or a black hole.

The three end-states of stellar evolution—white dwarfs, neutron stars, and black holes—represent the final victory of gravity. Understanding how stars die means we must first understand how stars delay their inevitable collapse. We begin at the beginning of the end, the evolution of giant stars.

13-1 Giant Stars

Main-sequence stars generate energy by fusing hydrogen into helium. This hydrogen fusion lasts a long time, and the star remains on the main sequence for 90 percent of its total existence as a star. When the hydrogen is exhausted, however, the star begins to evolve rapidly. It can delay its final collapse by fusing elements heavier than hydrogen and expanding to become a giant star, but that stage is short-lived. It can remain a giant star for only about 10 percent of its total lifetime; then it must die. Thus, the giant-star stage is the first step in the death of a star.

Expansion into a Giant

The nuclear reactions in a main-sequence star's core fuse hydrogen to produce helium. Because the core is cooler than 100,000,000 K, the helium cannot fuse in nuclear reactions, so it accumulates at the star's center like ashes in a fireplace. Initially, this helium ash has little effect on the star, but as hydrogen is exhausted and the stellar core becomes almost pure helium, the star loses the ability to generate nuclear energy. Because it is the energy generated at the center that opposes gravity and supports the star, the core begins to contract as soon as the energy generation starts to die down.

Although the contracting helium core cannot generate nuclear energy, it does grow hotter because it converts gravitational energy into thermal energy (see Chapter 12). The rising temperature heats the unprocessed hydrogen just outside the core, hydrogen that was never before hot enough to fuse. When the temperature of the surrounding gas becomes high enough, hydrogen begins to fuse in a spherical layer or shell around the exhausted core of the star. Like a grass fire burning outward from an exhausted campfire, the hydrogen-fusion shell creeps outward consuming fresh hydrogen, leaving helium ash behind and increasing the mass of the helium core.

The flood of energy from the hydrogen fusion shell pushes toward the surface, heating the outer layers of the star, and making them expand. When this happens to stars like the sun, they become giant stars 10 to 100 times the diameter of the sun, and the most massive stars become supergiants some 1000 times larger than the sun. This explains the large diameters and low densities of the giant and supergiant stars. In Chapter 11, we found that giant and supergiant stars were 10 to 1000 times larger in diameter than the sun and from 10 to 10^6 times less dense. Now we understand that these stars were once normal main-sequence stars that expanded to large size and low density when hydrogen shell fusion began.

The expansion of the envelope dramatically changes the star's location in the H-R diagram. As the outer layers of the star expand outward, the surface cools, and the point that represents the star in the H-R diagram moves quickly to the right (in less than a million years for a star of 5 solar masses). A medium-mass star like the sun becomes a red giant. As the radius of the star continues to increase, the enlarging surface area makes the star more luminous, moving its point upward in the H-R diagram (Figure 13-2). Aldebaran, the glowing red eye of Taurus the Bull, is such a red giant, with a diameter 25 times that of the sun but with a surface temperature only half that of the sun. A more massive star also expands, and its surface cools, moving the point that represents it in the H-R diagram to the right across the top. Such a massive star becomes a supergiant. The supergiant Rigel looks very blue to our eyes. With a surface temperature of 12,000 K, Rigel is not among the hottest stars, but because of the limited wavelength sensitivity of the human eye, all such stars look quite blue (see Figure 4-6). The red supergiant Betelgeuse is much cooler than Rigel.

Degenerate Matter

Although the hydrogen-fusion shell can force the envelope of the star to expand, it cannot stop the contraction of the helium core. Because the core is not hot enough to fuse helium, gravity squeezes it tighter, and it becomes very small. If we represent the helium core of a 5-solar-mass star with a tennis ball, the outer envelope of the star would be about the size of a baseball stadium. Yet the core would contain about 12 percent of the star's mass compressed to very high density. When gas is compressed to such extreme densities, it begins to behave in astonishing ways that can alter the evolution of the star. Thus, to follow the story of stellar evolution, we must consider the behavior of gas at extremely high densities.

Normally, the pressure in a gas depends on its temperature. The hotter the gas is, the faster its particles move and the more pressure it exerts. But the gas inside a star is ionized, so there are two kinds of particles: atomic nuclei and free electrons. If the gas is compressed to very high densities, for example, in the core of a giant star, the difference between these two kinds of particles becomes important.

Figure 13-2
The evolution of a massive star moves the point that represents it in the H-R diagram to the right of the main sequence into the region of the supergiants such as Rigel and Betelgeuse. The evolution of a medium-mass star moves its point in the H-R diagram into the region of the giants such as those shown here.

If the density is very high, the particles of the gas are forced close together, and two laws of quantum mechanics become significant. First, quantum mechanics says that the moving electrons confined in the star's core can have only certain amounts of energy, just as the electron in an atom can occupy only certain energy levels (see Chapter 4). We can think of these permitted energies as the rungs of a ladder. An electron can occupy any rung, but not the spaces between.

The second quantum mechanical law (called the Pauli Exclusion Principle) says that two identical electrons cannot occupy the same energy level. Because electrons spin in one direction or the other, two electrons can occupy an energy level if they spin in opposite directions. That level is then completely filled, and a third electron cannot enter because, whichever way it spins, it will be identical to one or the other of the two electrons already in the level. Thus, no more than two electrons can occupy the same energy level.

A low-density gas has few electrons per cubic centimeter, so there are plenty of energy levels (Figure 13-3). If a gas becomes very dense, however, nearly all of the lower energy levels may be occupied, and the gas is termed *degenerate*. In such matter, a moving electron cannot slow down, because slowing down would decrease its energy, and there are no open energy levels

One of the most interesting lessons of science is that the behavior of very small things often determines the structure and behavior of very large things. In degenerate matter, the quantum-mechanical behavior of electrons helps determine the evolution of giant stars. Such links between the very small and the very large are common in astronomy, and we will see in later chapters how the nature of certain subatomic particles may determine the fate of the entire universe.

This is a second reason why astronomers study atoms. Recall that the first reason is that atoms interact with light, and astronomers must understand that interaction in order to analyze the light. The second reason is the link between the very small and the very large. It would be impossible to understand the evolution of the largest stars if we failed to understand how degenerate electrons behave. Similarly, astronomers studying galaxies must understand how microscopic dust in space reddens starlight, and astronomers studying planets must know how atoms form crystals.

As you study any science, look for the way very small things affect very large things. Sociologists and psychologists know that the mass behavior of very large groups of people can depend on the behavior of a few key individuals. Biologists study the visible consequences of atomic bonds in molecules we call genes, and meteorologists know that tiny changes in temperature in one part of the world can affect worldwide climate weeks or months later.

Nature is the sum of many tiny parts, and science is the study of nature. Thus, one way to do science is to search out the tiny causes that determine the way nature behaves on the grandest scales.

---

for it to drop down to. It can speed up only if it can absorb enough energy to leap to the top of the energy ladder, where there are empty energy levels.

Such **degenerate matter** has two properties that can affect the core of a star. First, the degenerate gas resists compression. To compress the gas, we must push against the moving electrons, and changing their motion means changing their energy. That requires tremendous effort, because we must boost them to the top of the energy ladder. Thus, degenerate matter, though still a gas, takes on the consistency of hardened steel.

Second, the pressure of degenerate gas does not depend on temperature but rather on the speed of the electrons, which cannot be changed without tremendous effort. The temperature depends on the motion of all the particles in the gas, both electrons and nuclei. If we add heat to the gas, most of that energy goes to speed up the motions of the nuclei, and only a few electrons can absorb enough energy to reach the empty energy levels at the top of the energy ladder. Thus changing the temperature of the gas has almost no effect on the pressure.

These two properties of degenerate matter become important when stars end their main-sequence lives and approach their final collapse (Window on Science 13-1). Eventually, many stars collapse into white dwarfs, and we will discover that these tiny stars are made of degenerate matter. But long before that, the cores of many giant stars become so dense that they are degenerate, a situation that can produce a cosmic bomb.

Helium Fusion

Hydrogen fusion in main-sequence stars leaves behind helium ash, which cannot begin fusing. Helium nuclei have a positive charge twice that of a proton, and so they must collide at higher velocity to overcome the repulsion between nuclei. The temperature for hydrogen fusion is too cool to fuse helium. As the star becomes a giant star fusing hydrogen in a shell, the inner core of

Low-density gas (nondegenerate)   High-density gas (degenerate)

Figure 13-3
Electron energy levels are arranged like rungs on a ladder. In a low-density gas many levels are open, but in a degenerate gas all lower energy levels are filled.

helium contracts and grows hotter. It may even become degenerate, but when it finally reaches a temperature of 100,000,000 K, it begins to fuse helium nuclei to make carbon (see Chapter 12).

How a star begins helium fusion depends on its mass. Stars more massive than about 3 solar masses contract rapidly, their helium-rich cores heat up, and helium fusion begins gradually. But less massive stars evolve more slowly, and their cores contract so much the gas becomes degenerate. On Earth, a teaspoon of the gas would weigh as much as an automobile. In this degenerate matter, the pressure does not depend on temperature, and that means the pressure–temperature thermostat does not regulate energy production. When the temperature becomes hot enough, helium fusion begins to make energy, and the temperature rises, but pressure does not increase because the gas is degenerate. The higher temperature increases the helium fusion even further, and the result is a runaway explosion called the **helium flash** in which, for a few minutes, the core of the star can generate more energy per second than does an entire galaxy.

Although the helium flash is sudden and powerful, it does not destroy the star. In fact, if you were observing a giant star as it experienced the helium flash, you would probably see no outward evidence of the erup-

tion. The helium core is quite small (Figure 13-4), and all of the energy of the explosion is absorbed by the distended envelope. In addition, the helium flash is a very short-lived event in the life of a star. In a matter of minutes to hours, the core of the star becomes so hot it is no longer degenerate, the pressure–temperature thermostat brings the helium fusion under control, and the star proceeds to fuse helium steadily in its core (Figure 13-5).

The sun will experience helium fusion, but stars less massive than about 0.4 solar mass never get hot enough to ignite helium. Stars more massive than 3 solar masses ignite helium before their contracting cores become degenerate.

Whether a star experiences a helium flash or not, the ignition of helium in the core changes the structure of the star. The star now makes energy in its helium-fusion core and in its hydrogen-fusion shell. The energy flowing outward from the core can halt the contraction of the core, and the distended envelope of the star contracts and grows hotter. Thus, the point that represents the star in the H-R diagram moves downward and to the left toward the hot side of the H-R diagram.

Helium fusion produces carbon and oxygen atoms in the core, atoms that require much higher temperatures to fuse. Thus, as the helium fuel is used up, carbon and oxygen atoms accumulate in an inert core. Once again, the core contracts, heats up, and a helium-fusion shell ignites below the hydrogen-fusion shell. The star now makes energy in two fusion shells, it quickly expands, and its surface cools once again. The point that represents the star in the H-R diagram moves back to the right, completing a loop (Figure 13-5).

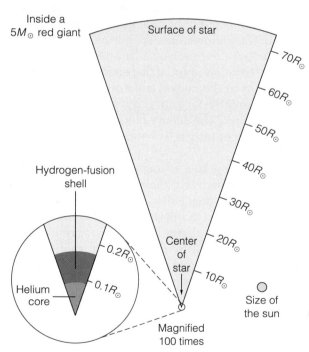

Figure 13-4
When a star runs out of hydrogen at its center, the core contracts to a small size, becomes very hot, and begins nuclear fusion in a shell (red). The outer layers of the star expand and cool. The red giant star shown here has an average density much lower than the air at Earth's surface. Here $M_\odot$ stands for the mass of the sun and $R_\odot$ stands for the radius of the sun. *(Illustration design by author)*

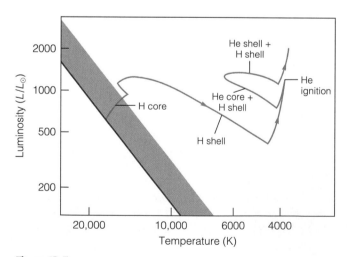

Figure 13-5
An expanded region of the H-R diagram shows the evolution of a star through helium fusion. After exhausting hydrogen in its core, the star ignites a hydrogen-fusion shell and leaves the main sequence as it expands to become a giant star. It eventually ignites helium in its core (in a helium flash for some stars) and then settles down to fuse helium in its core and hydrogen in a shell. Later it will fuse helium in a shell inside the hydrogen-fusion shell (top).

The evolution of a star after helium exhaustion is uncertain, but the general course is clear. The inert carbon–oxygen core contracts and becomes hotter. Stars more massive than about 3 solar masses can reach temperatures of 600,000,000 K and ignite carbon fusion. Subsequent contraction may be able to fuse oxygen and silicon as well as other heavy elements.

Eventually, the star must exhaust all of its fuels and die in a final collapse. Exactly how the star collapses depends on which fuels it has consumed and on the strength of its gravity. Both of these factors depend on the star's mass, so we must divide the stars into two groups, low-mass stars and high-mass stars.

Now we can understand why giant stars are so rare (see Figure 11-25). A star spends about 90 percent of its lifetime on the main sequence and only 10 percent as a giant star. During any particular human lifetime, only a fraction of the visible stars will be passing through the giant stage.

REVIEW Critical Inquiry

How are giant stars different from main-sequence stars?

We have seen that main-sequence stars begin the death process by expanding and cooling to become giant or supergiant stars, and in our discussion we discovered that these bloated stars are fundamentally different from main-sequence stars. Stars like the sun fuse hydrogen in their cores, and the energy flows outward to support the weight of the star. The pressure–temperature thermostat regulates the rate of energy production to produce a stable main-sequence star. But a giant star has used up the hydrogen in its core, and it can no longer maintain that main-sequence stability. Hydrogen fusion continues in a shell, but the core of helium ash contracts while the envelope of the star expands and cools. Later, helium fusion will begin in the core of the star, soon to be replaced by an inert carbon–oxygen core and a helium-fusion shell below the hydrogen-fusion shell. Thus, the internal structure of a giant star is dramatically different from the simple structure inside a main-sequence star. One of the pitfalls of critical analysis is to assume that similar consequences must arise from similar causes; in this case, main-sequence stars and giants look very similar, but they are quite different inside.

Consider a 5-solar-mass star and the sun. They look quite similar, and both will become giant stars, yet their internal evolution includes an important difference, the helium flash. What makes the difference?

We have now explored the structure of giant stars, so we are ready to divide the stars into three categories and tell the story of each kind of star. We begin with the low- and medium-mass stars.

13-2 The Deaths of Low- and Medium-Mass Stars

Because contracting stars heat up by converting gravitational energy into thermal energy, low-mass stars cannot get very hot. This limits the fuels they can ignite. In Chapter 12, we saw that protostars less massive than 0.08 solar mass cannot get hot enough to ignite hydrogen. Consequently, this section will concentrate on stars more massive than 0.08 solar mass but no more than a few times the mass of the sun.

Structural differences divide the low-mass stars into two subgroups—very-low-mass red dwarfs and medium-mass stars such as the sun. The critical difference between the two groups is the extent of interior convection. If the star is convective, fuel is constantly mixed, and the resulting evolution is drastically altered.

Red Dwarfs

Stars less massive than about 0.4 solar mass have two advantages. First, they have very small masses, and thus they have very little weight to support. Their pressure–temperature thermostats are set low, and they consume their hydrogen fuel very slowly. Our discussion of the life expectancies of stars in Chapter 12 predicted that the red dwarfs could live very long lives.

The red dwarfs have a second advantage because they are totally convective. That is, they are stirred by circulating currents of hot gas rising from the interior and cool gas sinking inward. This means the stars are mixed like a pot of soup that is constantly stirred as it cooks. Hydrogen is consumed and helium accumulates uniformly throughout the star, and that means the star is not limited to the fuel in its core. It can use all of its hydrogen to prolong its life on the main sequence.

Because a red dwarf is mixed by convection, it cannot develop an inert helium core surrounded by unprocessed hydrogen. Then it can never ignite a hydrogen shell and thus cannot become a giant star. Rather, nuclear fusion converts hydrogen into helium, which the small star cannot fuse because it cannot get hot enough. Thus, what we know about stellar evolution tells us that these red dwarfs should use up nearly all of their hydrogen and live very long lives on the lower main sequence. They could survive for a hundred billion years—longer than the present age of the universe. When they finally run out of hydrogen, they should gradually contract, grow hotter, and become white dwarfs.

This discussion seems almost pointless. Who cares how the red dwarfs die if the universe is so young that not a single red dwarf has evolved to become a white

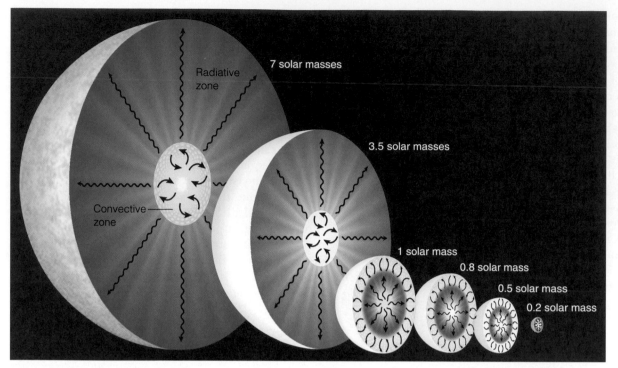

Figure 13-6

Inside stars. The more massive stars have small convective interiors and radiative envelopes. Stars like the sun have radiative interiors and convective envelopes. The lowest-mass stars are convective throughout. The "cores" of the stars where nuclear fusion occurs (not shown) are smaller than the interiors. *(Illustration design by author)*

dwarf? In fact, space is full of white dwarfs. We can see only those within a few hundred parsecs, but we see so many we must conclude that our galaxy contains 10 billion or so. If none of those white dwarfs are the remains of low-mass red dwarfs, then there must be some other, faster way for nature to make white dwarfs.

Medium-Mass Stars

Stars between about 4 and 0.4 solar mass,* including the sun, will die as white dwarfs. These medium-mass stars can ignite hydrogen shells, become giants, and fuse helium, but they can't get hot enough to fuse carbon, the next fuel after helium. When they reach that impasse, they collapse and become white dwarfs. There are two keys to the evolution of these sunlike stars: the lack of complete mixing and mass loss.

The interiors of medium-mass stars are not completely mixed (Figure 13-6). Stars with a mass of 1.1 solar masses or less have no convection near their centers, so they are not mixed at all. Stars with a mass greater than 1.1 solar masses have small zones of con-

vection at their centers, but this mixes no more than about 12 percent of the star's mass. Thus medium-mass stars, whether they have convective cores or not, are not mixed, and the helium ash accumulates in an inert helium core surrounded by unprocessed hydrogen. When this core contracts, the unprocessed hydrogen ignites in a shell and swells the star into a giant.

In the giant stage, the core of the star contracts and the envelope expands. The star fuses helium in its core and then in a shell surrounding a core of carbon and oxygen. This core contracts and grows hotter; but because the star has too low a mass, the core cannot get hot enough to ignite carbon fusion. Thus the carbon–oxygen core is a dead end for these medium-mass stars.

All of this discussion is based on theoretical models of stars and our general understanding of how stars evolve. Does it really happen? We need observational evidence to confirm our theoretical discussion, and the gas that is lost from these giant stars gives us visible testimony that sunlike stars do indeed die in this way.

Planetary Nebulae

When a medium-mass star like the sun becomes an aging giant, it can expel its atmosphere to form one of the most interesting objects in astronomy, a **planetary nebula,**

---

*This mass limit is uncertain, as are many of the masses quoted here. The evolution of stars is highly complex, and such parameters are difficult to specify.

IC 3568

Ring Nebula

a

b

Figure 13-7
The geometry of planetary nebulae. (a) The planetary nebula IC3568 appears to be a spherical shell of gas expelled from a dying star. *(H. Bond, B. Ballick, and NASA)* (b) The Ring Nebula looks spherical, but detailed studies show it is a cylinder of expelled gas that happens to be pointed at Earth. *(Hubble Heritage Team, AURA/STScI/NASA)*

so-called because, through a small telescope, it looks like the greenish-blue disk of a planet such as Uranus (Figure 13-7). In fact, a planetary nebula has nothing to do with planets. It is gas expelled by a dying star.

Basic observations tell us a great deal about planetary nebulae. In photographs, some are roughly spherical shells of gas centered on hot stars. From estimates of their distances and their angular diameters, we can find their typical radii—from 0.2 ly to 3 ly. The spectra of the nebulae contain emission lines, which tells us the gas is low-density, excited gas. Doppler shifts show that the nebulae are expanding away from their central stars at 10 to 20 km/s. If we divide the radius of a nebula by its velocity of expansion, we get its age, and planetary nebulae tend to be 1000 to 10,000 years old. Evidently, older nebulae become mixed into the interstellar medium and disappear.

To understand how planetary nebulae form, we begin by thinking of the spherical shape. Just as the sun expels its solar wind, a giant star expels a strong stellar wind, which astronomers can detect as blue-shifted lines in the star's spectrum. During the final centuries of its life, a giant star expels even more of its surface layers, but this gas is not easily visible in photographs because it is not ionized. Soon, however, the star expels the last surface gases and exposes deeper, hotter layers. The intense radiation drives gas near the star violently outward to overtake and sweep up the gas lost earlier, much as a snowplow sweeps up snow. Short-wavelength radiation from the hot surface ionizes the expanding shell of gas, and, as if someone flipped a switch, it lights up like a great neon sign, a planetary nebula.

The best images reveal that many planetary nebulae are not spherical (Figure 13-7b and 13-8). A number of things could cause these asymmetries. Some of the first gas expelled may form a thick disk around the star's equator. Later, when the high speed gas is expelled, the disk could confine the expanding gases into two oppositely directed flows. A second star orbiting the aging giant star or even planets could have a similar effect. Some of the planetary nebulae are very complex, which suggests there are undiscovered processes at work.

Planetary nebulae don't survive long, but we see roughly 1500 of them, and that must mean their formation is a common event. Most medium-mass stars end their lives by producing planetary nebulae. The sun will die in this way within 5 billion years. Stars up to a few times the mass of the sun can produce planetary nebulae if they have strong stellar winds and expel much of their mass before they end their lives.

Once an aging giant star blows its surface into space, the remaining interior collapses into a small, intensely hot star. It consists of the carbon–oxygen core surrounded by the helium- and hydrogen-fusion shells and topped by a shallow atmosphere of hydrogen. Such a central star in a planetary nebula is very hot at first—in some cases, over 100,000 K—and we find it plotted far to the upper left in the H-R diagram (Figure 13-9). As the central star cools, the point that represents it moves downward and to the right in the diagram and eventually reaches the region of the white dwarfs. Thus it seems that white dwarfs are the remains of sunlike stars that ejected their surfaces to produce planetary nebulae.

This suggests a second way we can search for evidence of the deaths of medium-mass stars: We can study white dwarfs.

White Dwarfs

In Chapter 11, we surveyed the stars and discovered that white dwarfs are the second most common kind of star (Figure 11-25). Only red dwarfs are more abundant. Now we can recognize these white dwarfs as the remains of medium-mass stars that fused hydrogen and

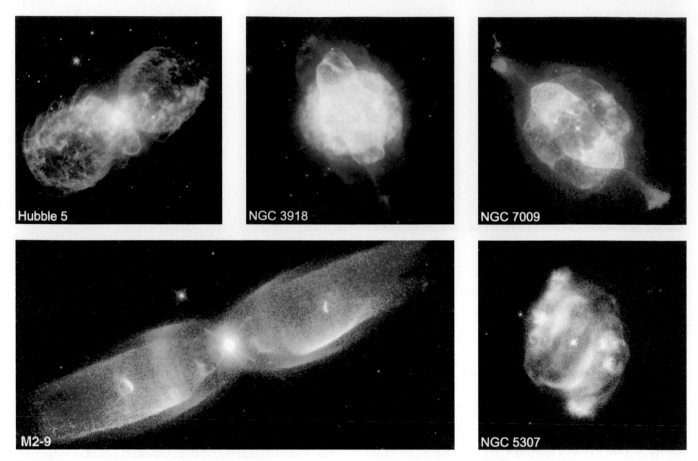

Figure 13-8
When the outflowing gas creating a planetary nebula is focused in opposite directions it can produce the complex structures seen here. *(H. Bond, B. Ballick, and NASA)*

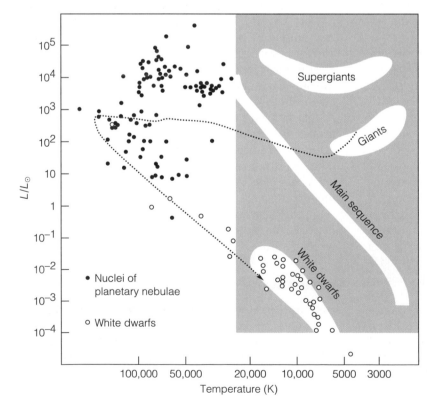

Figure 13-9
Our customary H-R diagram (shaded) must be expanded to show the evolution of a star after it ejects a planetary nebula. Central stars of planetary nebulae (filled circles) are hotter and more luminous than most white dwarfs (open circles). The dotted line shows the evolution of a 0.8 $M_\odot$ star as it collapses toward the white-dwarf region. *(Adapted from diagrams by C. R. O'Dell and S. C. Vila)*

helium, failed to ignite carbon, drove away their envelopes to form planetary nebulae, and collapsed and cooled to form white dwarfs. The billions of white dwarfs in our galaxy must be the remains of medium-mass stars.

The first white dwarf discovered was the faint companion to Sirius. In that visual binary system, the bright star is Sirius A. The white dwarf, Sirius B, is 10,000 times fainter than Sirius A. The orbital motions of the stars (shown in Figure 11-15) tell us that the white dwarf's mass is about 1 solar mass, and its blue-white color tells us that its surface is hot, about 25,000 K. Although it is very hot, it has a very low luminosity, so it must have a small surface area (see By the Numbers 11-3)—in fact, its diameter is about 92 percent of Earth's. Dividing its mass by its volume reveals that it is very dense—over 3×10^6 g/cm^3. On Earth, a teaspoonful of Sirius B material would weigh more than 15 tons. Thus, basic observations and simple physics lead us to the conclusion that white dwarfs are astonishingly dense.

A normal star is supported by energy flowing outward from its core, but a white dwarf has no internal energy source. It has exhausted its hydrogen and helium and cannot get hot enough to ignite carbon. Thus, there is no energy flow to balance gravity. The star contracts until it becomes degenerate. A white dwarf is supported not by energy flowing outward but by the refusal of its electrons to pack themselves into a smaller volume.

The interior of a white dwarf is mostly carbon and oxygen ions immersed in a whirling storm of degenerate electrons. Theory predicts that as the star cools, these ions will lock together to form a crystal lattice, so there may be some truth in thinking of aging white dwarfs as great crystals of carbon and oxygen. Near the surface, where the pressure is lower, a layer of ionized gases makes up a hot atmosphere, which, because of the strong gravity, is pulled down into a shallow layer. If Earth's atmosphere were equally shallow, people on the top floors of skyscrapers would have to wear space suits.

Clearly, a white dwarf is not a normal star. It generates no nuclear energy, is almost totally degenerate, and, except for a thin layer at its surface, contains no gas. Instead of calling a white dwarf a "star," we can call it a "compact object." In the next chapter, we will discuss two other kinds of compact objects—neutron stars and black holes.

A white dwarf's future is bleak. As it radiates energy into space, its temperature gradually falls, but it cannot shrink any smaller because its degenerate electrons cannot get closer together. This degenerate matter is a very good thermal conductor, so heat flows to the surface and escapes into space, and the white dwarf gets fainter and cooler, moving downward and to the right in the H-R diagram. Because the white dwarf contains a tremendous amount of heat, it needs billions of

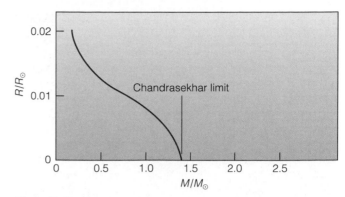

Figure 13-10
The more massive a white dwarf is, the smaller its radius. Stars more massive than the Chandrasekhar limit of 1.4 $M_\odot$ cannot be white dwarfs.

years to radiate that heat through its small surface area. Eventually, such objects may become cold and dark, so-called **black dwarfs.** Our galaxy is not old enough to contain black dwarfs. The coolest white dwarfs in our galaxy are about the temperature of the sun.

Perhaps the most interesting thing we have learned about white dwarfs has come from mathematical models. The equations predict that if we added mass to a white dwarf, its radius would *shrink* because added mass would increase its gravity and squeeze it tighter. If we added enough to raise its total mass to about 1.4 solar masses, its radius would shrink to zero (Figure 13-10). This is called the **Chandrasekhar limit** after Subrahmanyan Chandrasekhar, the astronomer who discovered it. It seems to imply that a star more massive than 1.4 solar masses could not become a white dwarf unless it shed mass in some way.

Stars do lose mass. Young stars have strong stellar winds, and aging giants also lose mass (Figure 13-11). This suggests that stars more massive than the Chandrasekhar limit can eventually die as white dwarfs if they reduce their mass enough. Theoretical models suggest that even stars as massive as 8 solar masses may be able to shed enough mass to get under the Chandrasekhar limit before they collapse. Thus a wide range of medium-mass stars eventually die as white dwarfs.

REVIEW Critical Inquiry

How will the sun die?

The sun's death process will begin in a few billion years, when it exhausts the hydrogen in its core. It will then expand to become a giant star fusing helium. Although the surface of the sun will become cooler, it will expand until it has such a large surface that it will be about 100

a

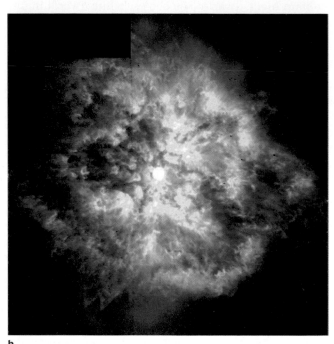

b

Figure 13-11

Many stars lose mass as they age. (a) This far-infrared map of the constellation Orion shows clouds of gas and dust heated and pushed away by hot young stars. The Orion nebula is the brightest area at lower right, and the large ring at upper right is expanding away from the O8 giant star λ Orionis. *(NASA/IPAC, courtesy Deborah Levine)* (b) The massive star WR124 is ejecting mass in a violent stellar wind forming chaotic arcs and blobs. *(Grosdidier and Moffat, University of Montreal; Joncas, Universite Laval; Acker, Observatorie de Strasbourg; and NASA)*

times more luminous than it is now. (Incidentally, this will evaporate Earth's atmosphere and oceans.) As a giant star, the sun will have a strong solar wind carrying gas into space and pushing back the interstellar medium. Eventually, the loss of mass will expose deeper layers in the sun. These extremely hot layers will ionize the gas around the sun and propel it outward to scoop up previously expelled gases and produce a planetary nebula. Soon the last remains of the sun will collapse inward and slowly cool to form a small, dense white dwarf. The surface will be very hot but so small that the sun will be about 100 times fainter than it is now (and any cinder that remains of Earth will fall into a deadly deep freeze).

This story of stellar evolution tells us that our sun will someday create a planetary nebula and a white dwarf. At that point, what will have become of all the hydrogen the sun once contained in its core?

Medium-mass stars die by producing beautiful planetary nebulae and tiny white dwarfs, but we have discovered a limit—1.4 solar masses. What happens to massive stars that can't shed mass fast enough? We will address that question in the next section.

13-3 The Deaths of Massive Stars

We have seen that low- and medium-mass stars die relatively quietly as they exhaust their hydrogen and helium and then eject their surface layers to form planetary nebulae. In contrast, massive stars live spectacular lives (Figure 13-12a) and die in violent explosions that destroy the star.

Nuclear Fusion in Massive Stars

Stars on the upper main sequence have too much mass to die as white dwarfs, but their evolution begins much like that of their lower-mass cousins. They consume the hydrogen in their cores and ignite hydrogen shells; as a result, they expand and become giants or, for the most massive stars, supergiants. Their cores contract and fuse helium first in the core and then in a shell, producing a carbon–oxygen core.

Unlike medium-mass stars, the massive stars are able to ignite carbon fusion at a temperature of about 1 billion Kelvin. For some stars, the core becomes degenerate before the carbon ignites, and thus the carbon ignites explosively in a **carbon detonation,** an event similar to but more violent than the helium flash. Carbon fusion produces more oxygen and neon, but as soon as the carbon is exhausted in the core, the core contracts and carbon ignites in a shell. This pattern of core ignition and shell ignition continues with fuel after fuel, and the star develops a layered structure (Figure 13-12b), with a hydrogen-fusion shell above a helium-fusion shell above a carbon-fusion shell above . . . After carbon fuses, oxygen, neon, and magnesium fuse to make silicon and sulfur, and then the silicon fuses to make iron.

The fusion of these nuclear fuels goes faster and faster as the massive star evolves rapidly. Recall that massive stars must consume their fuels rapidly to support their great weight, but other factors also cause the heavier fuels like carbon, oxygen, and silicon to fuse at increasing speed. For one thing, the amount of energy released per fusion reaction decreases as the mass of the fusing atom increases. To support the weight, the star must fuse oxygen much faster than it fused hydrogen. Also, there are fewer atoms in the core of the star by the time heavy atoms begin to fuse. Four hydrogens made a helium atom, and three heliums made a carbon, so there are 12 times fewer atoms of carbon available for fusion than there were hydrogen. This means heavy-element fusion goes very fast in massive stars (Table 13-1). Hydrogen fusion can last 7 million years in a 25-solar-mass star, but that same star will fuse its oxygen in 6 months and its silicon in a day.

Table 13-1 Heavy-Element Fusion in a 25 $M_\odot$ Star

Fuel	Time	Percentage of Lifetime
H	7,000,000 years	93.3
He	500,000 years	6.7
C	600 years	0.008
O	0.5 years	0.000007
Si	1 day	0.00000004

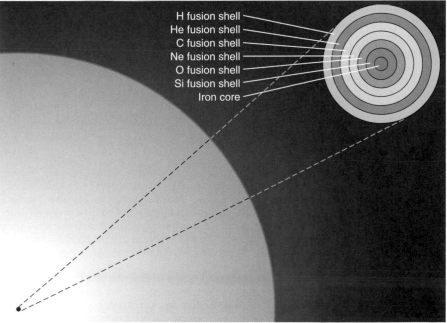

a

b

H fusion shell
He fusion shell
C fusion shell
Ne fusion shell
O fusion shell
Si fusion shell
Iron core

Figure 13-12
Massive stars evolve rapidly. (a) The star η Carinae is believed to be about 100 times more massive than the sun. It was a bright star 150 years ago, but it has faded since; and this Hubble Space Telescope image shows that it is ejecting mass rapidly in two expanding clouds. Poorly understood, it may someday explode as a supernova. *(Jon Morse, University of Colorado, NASA)* (b) On the verge of a supernova explosion, this red supergiant is almost as large as the orbit of Jupiter. The core (magnified 100,000 times here) is about the size of Earth and contains concentric layers of fusing gases with an inert iron core at the center.

Supernova Explosions

Theoretical models of evolving stars combined with nuclear physics allow astronomers to describe what happens inside a massive star when the last nuclear fuels are exhausted. It begins with iron atoms and ends in cosmic violence.

Silicon fusion produces iron, the most tightly bound of all atomic nuclei (see Figure 10-20). Nuclear fusion is able to release energy by combining less tightly bound nuclei into a more tightly bound nucleus, but iron is the limit. Once the gas in the core of the star has been converted to iron, there are no nuclear reactions that can combine iron atoms and release energy. Thus, the iron core is a dead end in the evolution of a massive star.

As a star develops an iron core, energy production begins to decline, and the core contracts. For atoms less massive than iron, this contraction would heat the gas and ignite new fusion fuels, but nuclear reactions involving iron tend to remove energy from the core in two ways. First, the iron atoms begin capturing electrons and breaking into smaller nuclei. The degenerate electrons helped support the core, so the loss of the electrons allows the core to contract even faster. Second, temperatures are so high that the average photon is a high-energy gamma ray, and these gamma rays are absorbed by atomic nuclei, which break into even smaller particles. The removal of the gamma rays further cools the core and allows it to contract more. With the loss of the electrons and the gamma rays, the core of the star collapses inward in less than a tenth of a second.

This collapse happens so rapidly that our most powerful computers are unable to predict the details. Thus models of supernova explosions contain many approximations. Nevertheless, it seems clear that the collapsing core of the star must quickly become a neutron star or a black hole, the subjects of the next chapter, while the envelope of the star is blasted apart in a supernova explosion.

To understand how the inward collapse of the core can produce an outward explosion, we must think about a traffic jam. The collapse of the innermost part of the degenerate core allows the rest of the core to fall inward, and this creates a tremendous traffic jam as all of the nuclei fall toward the center. It is as if every car owner in Indiana suddenly tried to drive into downtown Indianapolis. There would be a traffic jam not only downtown but also in the suburbs; and as more cars arrived, the traffic jam would spread outward. Similarly, as the inner core falls inward, a shock wave (a "traffic jam") develops and begins to move outward.

The shock wave moves outward through the star aided by two additional sources of energy. First, when the iron atoms in the core are disrupted, they produce a flood of neutrinos. In fact, for a short time the core produces more energy per second than all of the stars in all of the visible galaxies in the universe; and 99 percent of that energy is in the form of neutrinos. This flood of neutrinos rushes out of the core and helps heat the gas and accelerate the outward-bound shock wave. Second, the tremendous heat in the core of the star triggers giant convection currents, and intensely hot gas rushes outward from the interior (Figure 13-13). Again, this convection carries energy out into the envelope and helps drive the shock wave outward. Within a few hours, the shock wave bursts through the surface of the star and blasts it apart.

The supernova we see from Earth is the brightening of the star as its distended envelope is blasted outward by the shock wave. As months pass, the cloud of gas expands, thins, and fades, but the rate at which some supernovae fade tells astronomers that the gas is enriched with short-lived radioactive nuclei such as nickel-56. The gradual decay of these nuclei can keep the gas hot and prevent the supernova from fading rap-

Figure 13-13
Supernova explosions do not occur symmetrically. This color graphic, made from a mathematical model (Window on Science 12-2) of a supernova explosion, shows a cross section of the core of an exploding star only 0.102 second after the in-falling material bounces off of the core. The square is only 150 km on a side, and black arrows show the motion of the gas. The model reveals that the bouncing material becomes violently turbulent (blue) as it rushes outward to meet the rest of the star, which is falling inward (red at upper right). The innermost core of the star is forming a neutron star (red at lower left). To include the surface of the star at this scale, this diagram would need to be about 140 km (90 mi) in diameter. *(Courtesy Adam Burrows, John Hayes, and Bruce Fryxell)*

idly. The presence of these radioactive elements tells us that the density and temperature inside the exploding star were so high that nuclear reactions took place and built large numbers of heavy atoms.

Astronomers working with the largest and fastest computers are using modern theory to try to understand how supernovae explode. The companion to theory is observation, so we should ask what observational evidence supports this story of the deaths of massive stars.

Observations of Supernovae

In A.D. 1054, Chinese astronomers saw a "guest star" appear in the constellation we know as Taurus the Bull. The star quickly became so bright it was visible in the daytime, and then, after a month, it slowly faded, taking almost two years to vanish from sight. When modern astronomers turned their telescopes to the location of the "guest star," they found a cloud of gas about 1.35 pc in radius expanding at 1400 km/s. Projecting the expansion back in time, they concluded the expansion must have begun about 900 years ago, just when the "guest star" made its visit. Thus, the nebula, now called the Crab Nebula because of its shape (Figure 13-14a), marks the site of the A.D. 1054 supernova.

Supernovae are rare. Only a few have been seen with the naked eye in recorded history. Arab astronomers saw one in A.D. 1006, and the Chinese saw one in A.D. 1054. European astronomers observed two—one in A.D. 1572 (Tycho's supernova) and one in A.D. 1604 (Kepler's supernova). In addition, the guest stars of A.D. 185, 386, 393, and 1181 may have been supernovae.

In the centuries following the invention of the astronomical telescope in 1609, no supernova was bright enough to be visible to the naked eye. Supernovae are sometimes seen in distant galaxies, but they are faint and hard to study. Then, in the early morning hours of February 24, 1987, astronomers around the world were startled by the discovery of a naked-eye supernova still growing brighter in the southern sky (Figure 13-14b). The supernova, known officially as SN1987A, is only 53,000 pc away in the Large Magellanic Cloud, a small satellite galaxy to our own Milky Way. This first naked-eye supernova in 383 years is giving astronomers a ringside seat to the most spectacular event in stellar evolution.

One observation of SN1987A is critical in that it confirms the theory of core collapse. At 2:35 A.M. EST on February 23, 1987, nearly four hours before the supernova was first seen, a storm of neutrinos swept through Earth.* Instruments buried in a salt mine under Lake

*Within a few seconds of that moment, roughly 20 trillion neutrinos from the supernova passed harmlessly through your body.

a

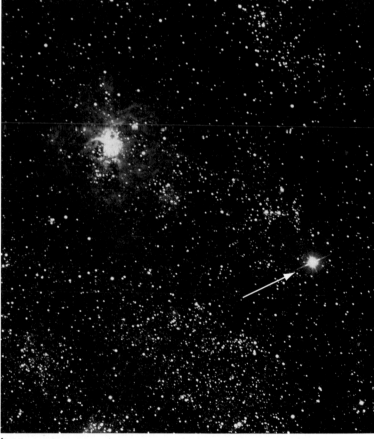

b

Figure 13-14
(a) The Crab Nebula is the remains of the supernova of A.D. 1054. At visual wavelengths, the glow of the nebula is woven through a shell of expanding filaments. *(California Institute of Technology)*
(b) Supernova 1987A (arrow) in the Large Magellanic Cloud was the first supernova visible to the naked eye since 1604. The Tarantula Nebula is at upper left, and the supernova is the bright image at the right. *(NOAO)*

Figure 13-15
A supernova (arrow) is visible in this Hubble Space Telescope photograph of the Whirlpool Galaxy. (Dark marks across the image are caused by missing data.) *(Robert Kirshner, Harvard, and NASA)*

Figure 13-16
Type I supernovae decline rapidly at first and then more slowly, but Type II supernovae pause for about 100 days before beginning a steep decline. Supernova 1987A was odd in that it did not rise directly to maximum brightness. These light curves have been adjusted to the same maximum brightness. Generally, Type II supernovae are about 2 magnitudes fainter than Type I.

Erie and in a lead mine in Japan, though designed for another purpose, recorded 19 neutrinos in less than 15 seconds. Neutrinos are so difficult to detect, the 19 neutrinos actually detected mean that some 10^{17} neutrinos must have passed through the detectors in those 15 seconds. Furthermore, the neutrinos were arriving from the direction of the supernova. Thus, astronomers conclude that the burst of neutrinos was released when the iron core collapsed, and the supernova was seen hours later when the shock wave blasted its surface into space.

Most supernovae are seen in distant galaxies (Figure 13-15), and careful observations show that there are fundamentally two kinds. Type I supernovae are more luminous, reaching a maximum luminosity about 4 billion times that of the sun. They decline rapidly at first and then more slowly. Type II supernovae become only about 0.6 billion times more luminous than the sun and decline in a more irregular way (Figure 13-16). Spectra of Type II supernovae show hydrogen lines, but spectra of Type I do not. These observational clues make sense when we consider an alternative way a star can collapse.

Type II supernovae are apparently produced by the collapse of the iron core in a massive star. SN1987A was a Type II supernova, although its light curve is not typical (Figure 13-16). It was produced by the explosion of a hot, blue supergiant rather than the usual cool, red supergiant (see Figure 13-1). Evidently, the star was a red supergiant a few thousand years ago but had contracted and heated up slightly, becoming smaller, hot-

ter, and bluer before it exploded. Scientists believe that most Type II supernovae are caused by the collapse of red supergiants. Spectra of Type II supernovae contain hydrogen lines because the envelopes of giant and supergiant stars are rich with unfused hydrogen.

Type I supernovae are a related phenomenon in that they too are caused by the collapse of a star, but in this case the star is a white dwarf. A typical white dwarf has a mass of 0.6 solar mass and is supported by the degenerate matter in its interior. A white dwarf in a binary star system can gain mass from its companion star (a process we will discuss in detail later in this chapter). It is also possible for a pair of white dwarfs orbiting each other to merge. In either case, a white dwarf could find itself with more mass than the Chandrasekhar limit—more mass than the degenerate matter can support—and then it must collapse. The sudden compression and heating of the degenerate matter triggers violent nuclear fusion, a supernova explosion that destroys the star. Spectra of Type I supernovae lack hydrogen lines—evidently because white dwarfs contain very little hydrogen.

Although the supernova explosion fades to obscurity in a year or two, an expanding shell of gas marks the site of the explosion. The gas, originally expelled at 10,000 to 20,000 km/s, may carry away a fifth of the mass of the star. The collision of that expanding gas with the surrounding interstellar medium can sweep up even more gas and excite it to produce a **supernova remnant,** the nebulous remains of a supernova explosion (Figure 13-17a).

Supernova remnants look quite delicate and do not survive very long—a few tens of thousands of years—before they gradually mix with the interstellar medium and vanish. The Crab Nebula is a young remnant, only

a

b

c

Figure 13-17

(a) The Cygnus Loop is a supernova remnant created when a massive star exploded 20,000 years ago. *(Caltech)* (b) This radio map of the 300-year-old remnant Cas A was made with the VLA radio telescope. Cas A is also detectable in X rays. *(NRAO)* (c) Some of the energy emitted by supernova remnants is synchrotron radiation, produced when an electron spirals around a magnetic field and radiates its energy away as photons. Usually synchrotron radiation is detected at radio wavelengths, but the Crab Nebula also produces synchrotron radiation at visible wavelengths.

about 900 years old and about 2.7 pc in diameter. The Cygnus Loop (see Figure 13-17a) is an estimated 20,000 years old, with a diameter of 40 pc. Some supernova remnants are visible only at radio and X-ray wavelengths (Figure 13-17b and Window on Science 13-2). They have become too tenuous to emit detectable light, but the collision of the expanding gas shell with the interstellar medium can generate radio and X-ray radiation. We saw in Chapter 12 that the compression of the interstellar medium by supernova remnants can also trigger star formation.

The radio emission coming from the Crab Nebula is **synchrotron radiation** (Figure 13-17c). This form of radiation is produced by rapidly moving electrons spiraling through magnetic fields; and in the case of the

Crab Nebula, the electrons are so energetic that they also emit visible light. This leaves us with a puzzle. The Crab Nebula is 900 years old, so the electrons should have radiated away their energy long ago. The Crab Nebula must contain a powerful energy source to maintain the synchrotron radiation. We will search for that energy source in the next chapter.

Evolution and Star Clusters

Before we rush ahead with our story of stellar evolution, we should pause to test our theories against reality. We can now arrange the stars by mass and describe the different stages in their lives (Figure 13-18), but do stars really evolve this way? We live such short lives compared to the stars that it is hard to test our theories.

Astronomers use computers to add false color to images and reveal things our eyes cannot see. Radio, infrared, ultraviolet, and X-ray radiation are not visible to our eyes, so data collected at these wavelengths can be displayed in images that are given colors visible to our eyes. The choice of colors is usually arbitrary, but the variation in color across an image does have meaning. A radio astronomer might choose shades of red through blue to display a radio map of a supernova remnant, and although the choice of colors was arbitrary, we would be able to identify the regions of strongest radio emission as the reddest regions on the map.

Astronomers also use false color to enhance visible-wavelength photographs. By converting the image to digital form and assigning colors to different levels of intensity, the astronomer can bring out subtle details that might not otherwise be visible. This is a common technique in other fields as well. Physicians often apply false color to medical X rays and CAT scans. Biologists use false color to analyze microscope photographs, and geologists use false color to study images of Earth recorded by Earth satellites.

Even a simple photograph of an astronomical object such as the Crab Nebula is a kind of false-color image. The nebula is too faint to see with the human eye, and even if we flew in a spaceship to within a short distance of the nebula, it would still be too faint to see without special equipment. We could not see its shape or its colors. The photograph we see is a time exposure made using special filters, and thus it reveals a part of nature that is invisible to our eyes. Even a black-and-white photograph of a nebula shows us an aspect of the universe we can never see with our unaided eyes. In that sense, all astronomical photographs are false-color images. Their virtue is not that they are colorful, but that they are meaningful.

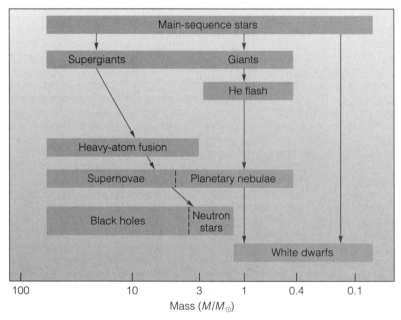

Figure 13-18

How a star evolves depends on its mass. The lowest-mass stars become white dwarfs, while stars like the sun become giants and then white dwarfs. Mass loss can move a star to the right in this diagram, as shown for the massive star. *(Illustration design by author)*

Nevertheless, all of science is based on evidence, so we should demand evidence to support these theories. Star clusters give us dramatic confirmation that stars do indeed evolve as our theories predict.

The stars in a star cluster all form at about the same time, but they include a range of masses. Like entrants in a footrace, the stars quickly sort themselves out. The most massive stars evolve and die quickly, while lower-mass stars evolve more slowly. When we look at the H-R diagram of a star cluster, we can see stellar evolution made visible.

Suppose we follow the evolution of a star cluster by making H-R diagrams like frames in a film (Figure 13-19). Our first frame shows the cluster only 10^6 years after it began forming, and already the most massive stars have reached the main sequence, consumed the hydrogen in their cores, and moved off to become supergiants. The medium- to low-mass stars have not yet reached the main sequence.

Because evolution is such a slow process, we cannot make the time step between frames equal, or we would fill over 1000 pages with nearly identical diagrams. Instead, we increase the time step by a factor of 10 with each frame. Thus, the second frame shows the cluster after 10^7 years, and the third after 10^8 years.

By the third frame, all massive stars have died, and stars slightly more massive than the sun are beginning to leave the main sequence. Notice that the lowest-mass stars have finally begun to fuse hydrogen. Only after 10^{10} years will the sun begin to swell into a giant.

These five frames were made from theoretical models of stellar evolution, but they compare very well with H-R diagrams of real star clusters. That is, our theory

Figure 13-19

A series of theoretical H-R diagrams, like frames in a film, illustrates the evolution of a cluster of stars. Massive stars approach the main sequence faster, live shorter lives, and die sooner than lower-mass stars. Compare with Figure 13-20.

compares well with reality. For example, NGC2264 (Figure 13-20a) is only a few million years old and still has many of its lower-mass stars contracting toward the main sequence. The Pleiades (Figure 13-20b), a cluster visible to the naked eye in Taurus, is older than NGC2264, dating back about 10^8 years. Compare these

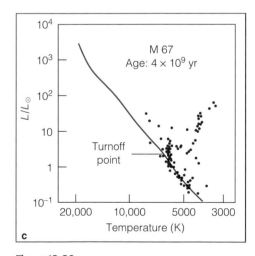

Figure 13-20

H-R diagrams of real star clusters compare well with H-R diagrams computed from theoretical models as shown in Figure 13-19.

younger clusters with M67 (Figure 13-20c), a faint cluster of stars about 4×10^9 years old.

We can estimate the age of a star cluster by noting the point at which its stars turn off the main sequence and move toward the red-giant region. The masses of the stars at this **turnoff point** tell us the age of the cluster because those stars are on the verge of exhausting their hydrogen-fusion cores. Thus the life expectancy of the stars at the turnoff point equals the age of the cluster (see By the Numbers 12-1).

Star clusters confirm by their H-R diagrams that astronomers do understand how stars are born, evolve, and die. The theory of stellar evolution is one of the great accomplishments of 20th-century astronomy. It gives us insights into how nature converts the cold, dark interstellar medium into heat and light.

REVIEW Critical Inquiry

Why does a Type II supernova explode?

A Type II supernova occurs when a massive star reaches the end of its usable fuel and develops an iron core. The iron is the final ash produced by nuclear fusion, and it cannot fuse to produce energy because iron is the most stable element. When energy generation begins to fall, the star contracts; but because iron can't ignite, there is no new energy source to stop the contraction. In a fraction of a second, the core of the star falls inward and a shock wave moves outward. Aided by a flood of neutrinos and sudden convection, the shock wave blasts the star apart, and we see it brighten as its surface gases expand into space.

Type II supernovae are easy to recognize because their spectra contain hydrogen lines. Use what you know about Type I supernovae to explain why the spectra of these supernovae do not contain visible hydrogen lines.

Supernovae mark the death of massive stars, and white dwarfs mark the death of less massive stars. But so far we have discussed only single stars. At least half of all stars are binary. How do stars that are members of binary systems die? That is the subject of the next section.

13-4 The Evolution of Binary Systems

Stars in binary systems can evolve independently of each other if they orbit at a large distance from each other. In this situation, one of the stars can swell into a giant and collapse without disturbing its companion. But some binary stars are as close to each other as 0.1 AU, and when one of those stars begins to swell into a giant, its companion can suffer in peculiar ways.

These interacting binary stars are interesting in their own right. The stars share a complicated history and can evolve to experience strange and violent phenomena. But such systems are also important because they can help us understand the ultimate fate of stars and certain observed phenomena, such as the temporary appearance of new stars in the sky. In the next chapter, we will use interacting binary stars as tools to help us find black holes.

Mass Transfer

Binary stars can sometimes interact by transferring mass from one star to the other. Of course, the gravitational field of each star holds its mass together, but the gravitational fields of the two stars, combined with the rotation of the binary system, define a teardrop-shaped volume around each star called the **Roche lobe** (Figure 13-21). Matter inside a star's Roche lobe is gravitationally bound to the star. The size of the Roche lobes depends on the mass of the stars and on the distance between the stars. If the stars are far apart, the lobes are very large, and the stars easily control their own mass.

Like the lobes of a giant dumbbell, the Roche lobes in a binary star meet at the **Lagrangian point** somewhere between the stars. The Lagrangian point connects the two Roche lobes, and if matter can leave a star

Figure 13-21
The Roche lobes around a pair of binary stars represent the two volumes of space that the stars can control with their gravity. The Roche lobes meet at the Lagrangian point. If matter can leave a star and reach the Lagrangian point, it can overflow into the other Roche lobe and eventually fall into the other star.

and reach the Lagrangian point, it can flow into the other star. Thus, the Lagrangian point is the connection through which the stars can transfer matter.

In general, there are only two ways matter can escape from a star and reach the Lagrangian point. First, if a star has a strong stellar wind, some of the gas blowing away from the star can pass through the Lagrangian point and be captured by the other star. Second, if an evolving star expands so far that it fills its Roche lobe, which can occur if the stars are close together and the lobes are small, then matter can overflow through the Lagrangian point onto the other star. Mass transfer driven by a stellar wind tends to be slow, but mass can be transferred rapidly by an expanding star.

Evolution with Mass Transfer

Mass transfer between stars can affect the evolution of the stars in surprising ways. In fact, this is the explanation of a problem that puzzled astronomers for many years.

In some binary systems, the less massive star has become a giant, while the more massive star is still on the main sequence. If higher-mass stars evolve faster than lower-mass stars, how do the lower-mass stars in such binaries manage to leave the main sequence first? This is called the Algol paradox, after the binary system Algol.

Mass transfer explains how this could happen. Imagine a binary system that contains a 5-solar-mass star and a 1-solar-mass companion (Figure 13-22). The two stars formed at the same time, so the higher-mass star will evolve faster and leave the main sequence first. When it expands into a giant, it can fill its Roche lobe and transfer matter to the low-mass companion. Thus the higher-mass star could evolve into a lower-mass star, and the companion could gain mass and become a high-mass star still on the main sequence. Thus, we might find a system such as Algol containing a 5-solar-mass main-sequence star and a 1-solar-mass giant.

The first four frames of Figure 13-22 show mass transfer producing a system like Algol. The last frame shows an additional stage in which the giant star collapses to form a white dwarf, and the massive companion evolves to become a giant transferring matter back onto the white dwarf. Such systems can become the site of tremendous explosions. To see how this can happen, we must consider in detail how mass falls into a star.

Accretion Disks

Matter flowing from one star to another cannot fall directly into the star. Rather, because of conservation of angular momentum, it must flow into a whirling disk around the star.

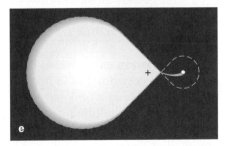

Figure 13-22
The evolution of a close binary system. As the more massive star (at the right) evolves (a), it fills its Roche lobe (b) and begins to transfer mass through the Lagrangian point to its companion. The companion grows more massive (c), and the star losing mass shrinks (d). Eventually the star on the right in this illustration can collapse into a white dwarf, while the star on the left evolves into a giant and begins transferring mass back to the white dwarf (e).

Figure 13-23
Skaters demonstrate conservation of angular momentum when they spin faster by drawing their arms and legs closer to their axes of rotation.

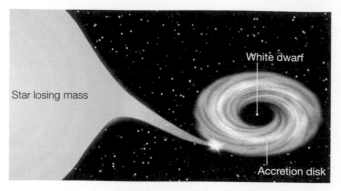

Figure 13-24
Matter falling into a compact object forms a whirling accretion disk. Friction and tidal forces can make the disk very hot.

Angular momentum refers to the tendency of a rotating object to continue rotating. All rotating objects possess some angular momentum, and in the absence of external forces, an object maintains (conserves) its total angular momentum. Ice skaters take advantage of conservation of angular momentum by starting spins slowly with arms extended and then drawing their arms closer to their axis of rotation. As their mass is concentrated closer to the axis of rotation, they spin faster (Figure 13-23). The same effect causes the slowly circulating water in a bathtub to spin in a whirlpool as it approaches the drain.

Mass transferred through the Lagrangian point in a binary system toward a white dwarf must conserve its angular momentum. Thus it must flow into a rapidly rotating whirlpool called an **accretion disk** around the white dwarf (Figure 13-24).

Two important things happen in an accretion disk. First, the gas in the disk grows very hot due to friction and tidal forces. The disk also acts as a brake, ridding the gas of its angular momentum and allowing it to fall into the white dwarf. The temperature of the gas in the inner parts of an accretion disk can exceed 1,000,000 K and emit intense X rays. In addition, the matter falling inward from the accretion disk can cause a violent explosion if it accumulates on a white dwarf.

Novae

At the beginning of this chapter we saw that the word "nova" refers to a new star that appears in the sky for a while and then fades away. The evidence tells us that it is not a new star at all. After a nova fades, astronomers can photograph the spectrum of the object. Invariably, they find a closely spaced binary star containing a normal star and a white dwarf. A nova is evidently an explosion involving a white dwarf in a binary star system.

Observations can tell us how novae occur. As the explosion begins, spectra show blue-shifted absorption lines, which tells us the gas is dense and coming toward us at a few thousand kilometers per second. After a few days, the spectral lines change to emission lines, which tells us the gas has thinned, but the blue shifts remain, showing that a cloud of debris has been ejected into space.

We can understand nova explosions as the result of mass transfer from a normal star through the Lagrangian point into an accretion disk around the white dwarf. As the matter loses its angular momentum in the accretion disk, it settles inward onto the surface of the white dwarf and forms a layer of unused nuclear fuel—mostly hydrogen. As the layer deepens, it becomes denser and hotter until the hydrogen fuses in a sudden explosion that blows the surface off of the white dwarf. Although the expanding cloud of debris contains less than 0.0001 solar mass, it is hot and its expanding surface area makes it very luminous. Nova explosions can become 100,000 times more luminous than the sun. As the debris cloud expands, cools, and thins over a period of months and years, the nova fades from view.

The explosion of its surface hardly disturbs the white dwarf and its companion star. Mass transfer quickly resumes, and a new layer of fuel begins to accumulate. How fast the fuel builds up depends on the rate of mass transfer. Accordingly, we can expect novae to repeat each time an explosive layer accumulates. Many novae take thousands of years to build an explosive layer, but some take only decades (Figure 13-25).

Figure 13-25
Nova T Pyxidis erupts about every two decades, expelling shells of gas into space. The shells of gas are visible from ground-based telescopes, but the Hubble Space Telescope reveals much more detail. The shell consists of knots of excited gas that presumably form when a new shell collides with a previous shell. *(M. Shara and R. Williams, STScI; R. Gilmozzi, ESO; and NASA)*

Ground

HST

REVIEW **Critical Inquiry**

How does modern astronomy explain the Algol paradox?

When we encounter a paradox in nature, it is usually a warning that we don't understand things as well as we thought we did. The Algol paradox is a good example. The binary star Algol contains a lower-mass giant star and a more massive main-sequence star. Because the two stars must have formed together, they are the same age; and the more massive star should have evolved first and left the main sequence. But in binary systems such as Algol, the lower-mass star has left the main sequence, or so it seems.

We can understand this paradox if we add mass transfer to our theory of stellar evolution. The first star to leave the main sequence must be the more massive star; but if the stars are close together, the star that is initially more massive can fill its Roche lobe and transfer mass back to its companion. The companion can grow more massive and the giant can grow less massive. Thus, the lower-mass giant star we see in the system may have originally been more massive and may have evolved away from the main sequence, leaving its companion behind to increase in mass as it decreased.

Mass transfer can explain the Algol paradox, and it can also explain the violent explosions called novae. How does mass transfer explain why novae can explode over and over in the same binary system?

The study of the deaths of stars has led us to discover astonishing objects of unbelievable density, temperature, and violence—all consequences of the victory of gravity over matter. But we have not considered the strangest circumstance of all. What happens when degenerate matter can't support the weight of a dying star—that is, when the mass of the compact object exceeds the Chandrasekhar limit? The answer is in the title of the next chapter.

Summary

When a star's central hydrogen-fusion reactions cease, its core contracts and heats up, igniting a hydrogen-fusion shell and swelling the star into a cool giant. The contraction of the star's core ignites helium first in the core and later in a shell. If the star is massive enough, it can eventually fuse carbon and other elements.

If a star's mass lies between about 0.4 and 3 solar masses, its helium core becomes degenerate before the helium ignites. In degenerate gas, pressure does not depend on temperature, so there is no pressure–temperature thermostat to control the reactions. As a result, the core explodes in a helium flash. All of the energy produced is absorbed by the star. A similar thing happens when carbon ignites in more massive stars. This carbon detonation is much more powerful than the helium flash.

How a star evolves depends on its mass. Stars less massive than about 0.4 solar mass are completely mixed and will have very little hydrogen left when they die. They cannot ignite a hydrogen shell or a helium core, so they will eventually become white dwarfs. Medium-mass stars between about 0.4 and 4 solar masses become giants and fuse helium but cannot fuse carbon. They produce planetary nebulae and become white dwarfs.

Stars as massive as 8 solar masses may lose enough mass to eject planetary nebulae and die as white dwarfs, but more massive stars suffer a different fate. The most massive stars fuse nuclear fuels up to iron but cannot generate further nuclear energy because iron is the most tightly bound of all atomic nuclei. When an iron core forms in a massive star, the core collapses and triggers a supernova explosion that expels the outer layers of the star to form an expanding supernova remnant. The first supernova visible to the naked eye since 1604 was discovered in February 1987.

We can see evidence of stellar evolution in the H-R diagrams of clusters of stars. Beginning their evolution at about the same time, the stars evolve in different ways, depending on their masses. The most massive evolve off of the main sequence first and are followed later by progressively less massive stars. Thus, we can estimate the age of a star cluster from the turnoff point in its H-R diagram.

Close binary stars evolve in complex ways because they can transfer mass from one star to the other as they evolve. This can explain why some binary systems contain a main-sequence star more massive than its giant companion—the Algol paradox. Also, mass transfer into an accretion disk around a white dwarf can produce X rays from the hot disk and can trigger nova explosions.

New Terms

nova	supernova remnant
supernova	synchrotron radiation
degenerate matter	turnoff point
helium flash	Roche lobe
planetary nebula	Lagrangian point
black dwarf	angular momentum
Chandrasekhar limit	accretion disk
carbon detonation	

Review Questions

1. Why does helium fusion require a higher temperature than hydrogen fusion?

2. How can the contraction of an inert helium core trigger the ignition of a hydrogen-fusion shell?

3. Why does the expansion of a star's envelope make it cooler and more luminous?

4. Why is degenerate matter so difficult to compress?

5. How does the presence of degenerate matter in a star trigger the helium flash?

6. Why don't red dwarfs become giant stars?

7. What causes an aging giant star to produce a planetary nebula?

8. Why can't a white dwarf contract as it cools? What is its fate?

9. Why can't a white dwarf have a mass greater than 1.4 solar masses?

10. How can a star of as much as 8 solar masses form a white dwarf when it dies?

11. How can the inward collapse of the core of a massive star produce an outward explosion?

12. What is the difference between a Type I and a Type II supernova?

13. How can star clusters confirm our theories of stellar evolution?

14. How can we understand the Algol paradox?

15. What is the difference between a supernova explosion and a nova explosion?

Discussion Questions

1. How do we know the helium flash occurs if it cannot be observed? Can we accept something as real if we can never observe it?

2. False-color radio images and time-exposure photographs of astronomical images show us aspects of nature we can never see with our unaided eyes. Can you think of common images in newspapers or on television that reveal phenomena we cannot see?

Problems

1. If the sun becomes a giant star and expands to a radius 100 times its present radius, what will its density be? (*Hint:* The volume of a sphere is $\frac{4}{3}\pi r^3$.)

2. If a giant star 100 times the diameter of the sun were 1 pc from us, what would its angular diameter be? (*Hint:* Use the small-angle formula.)

3. The Ring Nebula in Lyra is a planetary nebula with an angular diameter of 72 seconds of arc and a distance of 5000 ly. What is its linear diameter? (*Hint:* Use the small-angle formula.)

4. If the Ring Nebula is expanding at a velocity of 15 km/s, typical of planetary nebulae, how old is it? (*Hint:* 1 ly = 9×10^{12} km and 1 y = 3.15×10^7 s.)

5. A planetary nebula has increased in radius by 0.6 second of arc in 20 years, and Doppler shifts show it is expanding at 20 km/s. How far away is it? (*Hint:* First figure out how many parsecs it has increased in radius in 20 years. Then use the small-angle formula.)

6. The Crab Nebula is now 1.35 pc in radius and is expanding at 1400 km/s. About when did the supernova occur?

7. The supernova remnant Cassiopeia A is expanding in radius at about 0.5 second of arc per year. Doppler shifts show it is expanding at about 5700 km/s. How far away is the nebula?

8. Cassiopeia A has a radius of about 2.5 minutes of arc. If it is expanding at 0.5 second of arc per year, when did the supernova occur? (There is no record of a supernova being seen at that time.)

9. If a star cluster contains 500 stars and is 25 pc in diameter, what is the average distance between stars? (*Hints:* What share of the volume of the cluster surrounds each star? The volume of a sphere is $\frac{4}{3}\pi r^3$.)

10. If the stars at the turnoff point in a star cluster have masses of about 4 solar masses, how old is the cluster?

Critical Inquiries for the Web

1. As seen in Figure 13-8, there is an incredible diversity of appearance for planetary nebulae. Browse the Web for images and information on these dying stars, and discuss why there is a range of shapes of planetary nebulae that we see.

2. Naked-eye supernovae in our galaxy are rare, but astronomers have noted supernovae in other galaxies for years. Look for summaries of observations of recent supernovae. Are similar numbers of Type I and Type II being seen? Compare the number of supernovae seen during the last few years with that of two decades ago. Why are we finding so many more supernovae in recent years than in the past?

Exploring *The Sky*

1. The following objects are a sample of planetary nebulae that can be seen with small telescopes: M57, M27, M97. The "M" stands for Messier, who cataloged objects that look fuzzy in a telescope. M57 is the 57th object in the Messier catalog, which contains 103 objects. Determine the following information about these objects: their common name, the constellation in which they reside, their magnitude, and their angular size. Which of these objects can you see with the unaided eye?

 Go to the Brooks/Cole Astronomy Resource Center (www. brookscole.com/astronomy) for critical thinking exercises, articles, and additional readings from InfoTrac College Edition, Brooks/Cole's online student library.

Neutron Stars and Black Holes

Almost anything is easier to get into

than out of.

Agnes Allen

Guidepost

The preceding two chapters have traced the story of stars from their birth as clouds of gas in the interstellar medium to their final collapse. This chapter finishes that story by discussing the two most extreme of the three end states of stellar evolution: neutron stars and black holes.

How strange and wonderful that we humans can talk about places in the universe where gravity is so strong it bends space, slows time, and curves light back on itself! To carry on these discussions, astronomers have learned to use the language of relativity. Throughout this chapter, remember that only the most sophisticated application of general relativity allows us to understand these strange bodies.

This chapter ends the story of individual stars, but it does not end the story of stars. The next chapter extends that story to the giant communities that stars live in—the galaxies.

Gravity always wins. However a star lives, it must eventually die in one of three final states—white dwarfs, neutron stars, or black holes. These objects, often called compact objects, are small monuments to the power of gravity. Almost all of the energy available has been squeezed out of compact objects, and we find them in their final, high-density states.

In this chapter, we must compare evidence and hypothesis with great care. Theory predicts the existence of these objects; but by their nature, they are difficult to detect. Is theory right? Do these objects really exist? To confirm the theories, astronomers have searched for real objects that could be identified as having the properties predicted by theory. That is, they have tried to find real neutron stars and real black holes. No medieval knight ever rode off on a more difficult quest, but modern astronomers have found objects that do seem to be neutron stars and black holes. We can judge their success by critically analyzing the evidence to see if it does confirm the theory.

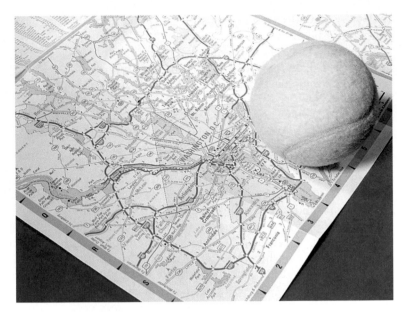

Figure 14-1
A tennis ball and a road map illustrate the relative size of a neutron star. Such an object, containing slightly more than the mass of the sun, would fit with room to spare inside the beltway around Washington, D.C. *(Photo by author)*

14-1 Neutron Stars

A **neutron star** is a star of a few solar masses compressed to a radius of about 10 km. Its density is so high that the matter is stable only as a fluid of neutrons. Theory predicts that such an object would spin a number of times a second, be nearly as hot at its surface as the inside of the sun, and have a magnetic field a billion times stronger than Earth's. Two questions should occur to us immediately. First, how could any theory predict such a wondrously unbelievable star? And second, do these neutron stars really exist?

Theoretical Prediction of Neutron Stars

The neutron was discovered in the laboratory in 1932, and its properties suggested something fantastic. Neutrons spin in much the way that electrons do, which means that neutrons must obey the Pauli Exclusion Principle. In that case, if neutrons are packed tightly enough together they can become degenerate just as electrons do. White dwarfs are supported by degenerate electrons, and in 1932, the Russian physicist Lev Landau predicted that neutron stars might exist supported by degenerate neutrons. Of course, the inside of a neutron star would have to be much denser than the inside of a white dwarf.

Only two years later, in 1934, Walter Baade and Fritz Zwicky provided a pedigree for neutron stars. They suggested that some of the stellar explosions seen in history were caused by the collapse of a massive star in an explosion they called a supernova. What was left of the core of the star, they suggested, was a small, high-density neutron star.

Atomic physics gives us an explanation of how the collapsing core of a massive star could form a neutron star. As the core of the star begins to collapse, the density quickly reaches that of a white dwarf, but the weight is too great to be supported by degenerate electrons. The collapse of the core continues, and the atomic nuclei are broken apart by gamma rays. Almost instantly, the increasing density forces the freed protons to absorb electrons and become neutrons. In a fraction of a second, the collapsing core becomes a contracting ball of neutrons. As we saw in the previous chapter, the envelope of the star is blasted away in a supernova explosion. The remaining core of the star is left behind as a neutron star.

Theoretical calculations predict that a neutron star will be only 10 or so kilometers in radius (Figure 14-1) and will have a density of about 10^{14} g/cm^3. On Earth, a sugar-cube-sized lump of this material would weigh 100 million tons. This is roughly the density of the atomic nucleus, and we can think of a neutron star as matter with all of the empty space squeezed out of it.

How massive can a neutron star be? That is a critical question, and a difficult one to answer because we don't know the strength of pure neutron material. Scientists can't make such matter in the laboratory, so its properties must be predicted theoretically. The most widely accepted calculations suggest that a neutron star cannot be more massive than 2 to 3 solar masses. If a neutron star were more massive than that, the degenerate neutrons would not be able to support the weight, and the object would collapse (presumably into a black hole).

Simple physics, the physics we have used in previous chapters to discuss normal stars, predicts that neutron stars should be hot, spin rapidly, and have strong magnetic fields. We have seen that contraction heats the gas in a star. As the gas particles fall inward, they pick up speed, and when they collide, their high speeds become thermal energy. The sudden collapse of the core of a massive star to a radius of 10 km should heat it to millions of degrees. Furthermore, neutron stars should cool slowly because the heat can escape only from the surface, and neutron stars are so small they have little surface from which to radiate. Thus, basic theory predicts that neutron stars should be very hot.

The conservation of angular momentum predicts that neutron stars should spin rapidly. All stars rotate to some extent because they form from swirling clouds of interstellar matter. As such a star collapses, it must rotate faster because it conserves angular momentum. Recall that we see this happen when ice skaters spin slowly with their arms extended and then speed up as they pull their arms closer to their bodies (see Figure 13-23). In the same way, a collapsing star must spin faster as it pulls its matter closer to its axis of rotation. If the sun collapsed to a radius of 10 km, its period of rotation would decrease from 25 days to about 0.001 second. We might expect the collapsed core of a massive star to rotate 10 or 100 times a second.

Basic theory also predicts that a neutron star should have a powerful magnetic field. Whatever magnetic field a star has is frozen into the star. The gas of the star is ionized, and that means the magnetic field cannot move easily through the gas. When the star collapses, the magnetic field is carried along and squeezed into a smaller area, which could make the field a billion times stronger. Because some stars have magnetic fields over 1000 times stronger than the sun's, we might expect a neutron star to have a magnetic field as much as a trillion times stronger than the sun's. For comparison, that is about 10 million times stronger than any magnetic field ever produced in the laboratory.

Thus, basic physics predicted the properties of neutron stars, but it also predicted that such objects should be difficult to observe. Neutron stars are very hot, so black body radiation predicts they will radiate most of their energy in the X-ray part of the spectrum, radiation that could not be observed in the 1940s and 1950s because astronomers could not put their telescopes above Earth's atmosphere. Also, the small surface areas of neutron stars mean that they will be faint objects. Thus astronomers of the mid-20th century were not

Figure 14-2
The first pulsar, CP1919, was discovered November 28, 1967, when its regularly spaced pulses were noticed in the output from a radio telescope. The period is 1.33730119 seconds. Pulses are marked by ticks.

surprised that none of the newly predicted neutron stars were found. Neutron stars were, at that point, entirely theoretical objects. Progress came not from theory but from observation.

The Discovery of Pulsars

In November 1967, Jocelyn Bell, a graduate student at Cambridge University, England, found a peculiar pattern on the paper chart from a radio telescope. Unlike other radio signals from celestial bodies, this was a series of regular pulses (Figure 14-2). At first she and the leader of the project, Anthony Hewish, thought the signal was interference, but they found it day after day in the same place in the sky. Clearly, it was celestial in origin.

Another possibility, that it came from a distant civilization, led them to consider naming it LGM for Little Green Men. But within a few weeks, the team found three more objects in other parts of the sky pulsing with different periods. The objects were clearly natural, and the team dropped the name LGM in favor of **pulsar**—a contraction of *pulsing star*.

As more pulsars were found, astronomers argued over their nature. Periods ranged from 0.033 to 3.75 seconds and were nearly as exact as an atomic clock. Months of observation showed that many of the periods were slowly growing longer by a few billionths of a second per day. Thus, whatever produced the regular pulses had to be highly precise, nearly as exact as an atomic clock, but it had to gradually slow down.

It was easy to eliminate possibilities. Pulsars could not be stars. A normal star, even a small white dwarf, is too big to pulse that fast. Nor could a star with a hot spot on its surface spin fast enough to produce the pulses. Even a small white dwarf would fly apart if it spun 30 times a second.

The pulses themselves gave the astronomers a clue. The pulses lasted only about 0.001 second. This places an upper limit on the size of the object producing the pulse. If a white dwarf blinked on and then off in that interval, we would not see a 0.001-second pulse. The near side of the white dwarf would be about 6000 km closer to us, and light from the near side would arrive 0.022 second before the light from the bulk of the white dwarf. Thus its short blink would be smeared out into a longer pulse. This is an important principle in astronomy—an object cannot change its brightness appreciably in an interval shorter than the time light takes to cross its diameter. If pulses from pulsars are no longer than 0.001 second, then the object cannot be larger than 300 km (190 miles) in diameter.

Of all the objects astronomers could think of, only neutron stars were small enough and energetic enough to be pulsars. In fact, neutron stars are so small they can't vibrate slowly enough, but they can spin as fast as 1000 times a second without flying apart. Thus astronomers were pleased but not quite surprised when, in late 1968, a pulsar was found at the center of the Crab Nebula (Figure 14-3). The Crab Nebula is a known supernova remnant, and theory predicts a supernova could leave behind a neutron star. The Crab Nebula pulsar was the missing link between pulsars and neutron stars.

Over 1000 pulsars are now known, and astronomers have strong evidence that they are spinning neutron stars. We have identified pulsars as neutron stars, but how does such an object generate pulses? To answer that question, we must build a model of a pulsar.

A Model Pulsar

Although the complex physics of spinning neutron stars is not yet fully understood, astronomers can build a model to unify and test hypotheses, observations, and theories. In Chapter 1, we discussed the celestial sphere as a model that helps us think about the sky, and we concluded that a scientific model does not have to be true to be useful. In contrast, the modern model of a pulsar has been refined so carefully, it may now be close to a true description of a pulsar.

Pulsars do not really pulse. Rather, the spinning neutron star emits beams of electromagnetic radiation that sweep around the sky. When a beam sweeps over us, we detect a pulse, just as sailors see a pulse of light when the beam from a lighthouse sweeps over their ship. This explanation of the pulses is the **lighthouse theory.**

Astronomers do not yet understand how a rotating neutron star produces beams, but the tremendous mag-

Figure 14-3

The Crab Nebula is located at the spot in the sky where Chinese astronomers saw a supernova in A.D. 1054, so the nebula is well understood as a supernova remnant. The pulsar at its center (arrow) is so energetic its pulses are detectable at visual wavelengths. The inset image was taken with the Hubble Space Telescope and shows energy from the pulsar exciting the gas of the inner nebula to produce changing wisps and waves. *(Caltech; inset, Jeff Hester and Paul Scowen, Arizona State University, NASA)*

netic field must be involved. The rapidly rotating field should act as a generator and create a powerful electric field capable of ripping charged particles, mostly electrons, out of the neutron star's surface. Thus the spinning neutron star must be surrounded by intense magnetic fields twisted into complicated shapes and filled with powerful currents of flowing electrons. This evidently results in high-energy beams of photons emitted like searchlights from the magnetic poles. If the magnetic axis is inclined with respect to the axis of rotation, as is the case with Earth, then the neutron star will sweep the beams around the sky (Figure 14-4).

Although neutron stars are small, they are powerful. The spectrum of the blue glow in the Crab Nebula reveals that it is synchrotron radiation. Recall that synchrotron radiation is emitted by high-speed electrons spiraling through a magnetic field. In the 900 years since the Crab supernova, such high-speed electrons should have radiated their energy away and slowed. The blue glow in the nebula should have faded soon after the supernova explosion. But the Crab Nebula continues to glow. Evidently, the synchrotron radiation in the nebula is powered by its neutron star.

Observations made by the Hubble Space Telescope provide dramatic evidence of energy flowing from the neutron star into the nebula. The inset in Figure 14-3 shows circular wisps of gas flowing away from the neutron star in its equatorial plane at about half the speed of light. In fact, the wisps change their shape and brightness from day to day. Above and below the image of the neutron star, the gas of the nebula glows as it is excited by beams of high-energy particles coming from the neutron star (Figure 14.5a).

A handful of supernova remnants are known that contain neutron stars, and many more neutron stars are known that do not lie in a supernova remnant (Figure 14-5b). Evidently, neutron stars are common.

The Evolution of Pulsars

Astronomers are beginning to understand how a pulsar ages. When it first forms, it is spinning very fast, perhaps nearly 100 times a second, and it contains a strong magnetic field. As it converts its energy of rotation into radiation, it gradually slows, and its magnetic field grows weaker. The average pulsar is apparently only a few million years old, and the oldest is about 10 million years old. By the time a pulsar gets older than that, it is rotating too slowly to generate detectable radio beams.

If a pulsar is young, we would expect it to spin fast, have a strong magnetic field, and consequently emit very strong beams of radiation. The Crab Nebula gives us an example of such a system. Only about 900 years old, the Crab pulsar is so powerful it emits photons of radio, infrared, visible, X-ray, and gamma-ray wavelengths (Figure 14-6). Careful measurements of its brightness with high-speed instruments show that it blinks twice for every rotation. One beam sweeps almost directly over us, and we detect a strong pulse. Half a rotation later, the edge of the other beam sweeps past us, and we detect a weaker pulse.

We would expect only the most energetic pulsars to produce short-wavelength photons and thus pulse at visible wavelengths. The Crab Nebula pulsar is young and powerful, and it produces visible pulses, and so does a pulsar called the Vela pulsar (located in the Southern Hemisphere constellation Vela). The Vela pulsar is fast, pulsing about 11 times a second, and like the Crab Nebula pulsar, the Vela pulsar is located inside a supernova remnant. Its age is estimated at about 11,000 years, young in terms of the average pulsar. Thus we suspect that pulsars are capable of producing optical pulses only when they are young.

We might expect to find the pulsars inside supernova remnants, but the statistics must be examined with care. Not every supernova remnant contains a pulsar,

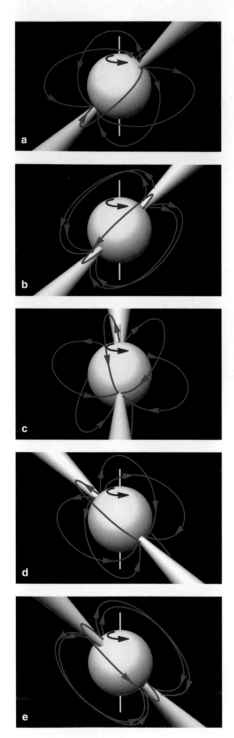

Figure 14-4
Schematic diagram of a neutron star (yellow green) with its powerful magnetic field (red). Beams of electromagnetic radiation blast out of the magnetic poles, and the rotation of the neutron star sweeps them around the sky like beams of light from a lighthouse.

a   b

Figure 14-5
(a) The neutron star at the center of the Crab Nebula is visible in this X-ray image. It has flung rings of highly
ionized gas up to a light-year away and is ejecting jets of high-energy particles perpendicular to the rings. On
Earth, we detect pulses from this spinning neutron star. *(NASA/CSC/SAO)* (b) In this visible wavelength photo, a
solitary neutron star (arrow) has outlasted its supernova remnant. With a temperature of 700,000 K, it is not
known to pulse. *(Fred Walter, State University of New York at Stony Brook, NASA)*

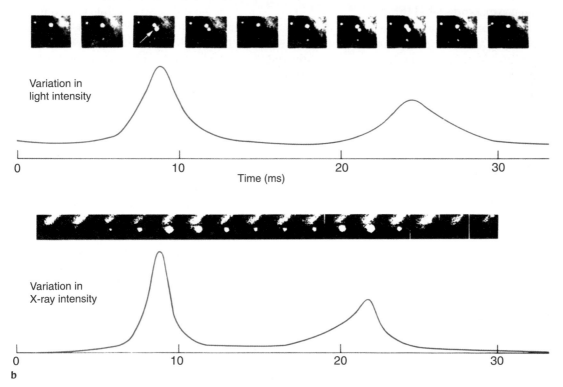

b

Figure 14-6
High-speed images of the Crab Nebula pulsar show it pulsing at visual wavelengths (a) and at X-ray wavelengths
(© AURA, Inc., NOAO, KPNO) (b) The period of pulsation is 33 milliseconds (ms), and each cycle includes two pulses
as its two beams of unequal intensity sweep over the earth. *(Courtesy F. R. Harnden, Jr., Harvard-Smithsonian Center
for Astrophysics, from* The Astrophysical Journal, *published by the University of Chicago Press; © 1984 The American
Astronomical Society)*

and not every pulsar is located inside a supernova remnant. Many supernova remnants probably contain pulsars whose beams never sweep over Earth, and it will be difficult to detect such pulsars. Also, some pulsars move through space at high velocity, which suggests that supernova explosions can occur asymmetrically, perhaps because of the violent convection in the exploding core. Such a violent, off-center explosion could give the neutron star a high velocity through space. Also, some supernovae probably disrupt binary systems and fling the stars apart at high velocity. In any case, pulsars are known to have such high velocities that many probably escape the disk of our galaxy. We should not be surprised that many neutron stars quickly leave their supernova remnant behind. And finally, we must remember that a pulsar can remain detectable for 10 million years or so, but a supernova remnant cannot survive more than about 50,000 years before it is mixed into the interstellar medium. Thus, we should not be surprised that most pulsars are not in supernova remnants, and most supernova remnants do not contain pulsars.

The explosion of supernova 1987A in February 1987 apparently formed a neutron star. The burst of neutrinos detected on Earth is commonly accepted as evidence of a neutron star. We would not have expected to see the neutron star at first because it would have been hidden at the center of the expanding cloud of gas ejected into space; but as the gas expanded and thinned, we might have been able to see the neutron star. Even if its beams didn't sweep over Earth, we might have been able to detect it from its X-ray and gamma-ray emission. Although, years after the explosion, no neutron star has been detected, astronomers continue to watch the site, hoping to see the youngest pulsar known.

Our theory predicts that newborn pulsars should blink rapidly and old pulsars should blink slowly, but the handful that blink the fastest may be quite old. One of the fastest known pulsars is catalogued as PSR 1937 + 21 in the constellation Vulpecula. It pulses 642 times a second and is slowing down only slightly. The energy stored in the rotation of a neutron star at this rate is equal to the total energy of a supernova explosion, so it seemed difficult at first to explain this pulsar. It now appears that PSR 1937 + 21 is an old neutron star that has gained mass and rotational energy from a companion in a binary system. Like water hitting a mill wheel, the matter falling on the neutron star has spun it up to 642 rotations per second. With its old, weak magnetic field, it slows down very slowly and will continue to spin for a very long time.

A number of other very fast pulsars have been found, and they are known generally as **millisecond pulsars** because their pulse periods are almost as short as a millisecond (0.001 s). This produces some fascinating physics, because the pulse period of a pulsar equals the rotation period of the neutron star. If a neutron star 10 km in radius spins 642 times a second, as in the case of PSR 1937 + 21, then the period is 0.0016 seconds, and the equator of the neutron star must be traveling about 40,000 km/s. That is fast enough to flatten the neutron star into an ellipsoidal shape and is nearly fast enough to break it up.

One reason pulsars are so fascinating is the extreme conditions we find in spinning neutron stars. To see natural processes of even greater violence, we have only to look at pulsars in binary systems.

Binary Pulsars

Of the hundreds of pulsars now known, a few are located in binary systems. These pulsars are of special interest because we can learn more about the neutron star by studying the orbital motions of the binary. Also, in some cases, mass can flow from the companion star onto the neutron star, and that produces high temperatures and X rays.

The first binary pulsar was discovered in 1974 when astronomers noticed that the pulse period of the pulsar PSR 1913 + 16 was changing (Figure 14-7). The period first grew longer and then grew shorter in a cycle that took 7.75 hours. Thinking of the Doppler shifts seen in spectroscopic binaries, the radio astronomers realized that the pulsar had to be in a binary system with a period of 7.75 hours. When the orbital motion of the pulsar carries it away from Earth, we see the pulse period slightly lengthened, just as the wavelength of light emitted by a receding source is lengthened. Then, when the pulsar rounds its orbit and approaches Earth, we see the pulse period slightly shortened. A few dozen of these binary pulsars have been found; and by analyzing the Doppler shifts in their pulse periods and making reasonable assumptions about the mass of the companion stars, astronomers can estimate the mass of the neutron stars. Such masses lie between 1.5 and 2 solar masses, in good agreement with our theory of neutron stars.

Binary pulsars are relatively quiet systems now, but they were once violently active. The companion star was probably once a giant star losing mass to the neutron star, and that matter falling into the neutron star would have liberated tremendous energy. A single marshmallow dropped onto the surface of a neutron star from a distance of 1 AU would hit with the impact of a 3-megaton nuclear bomb. Even a small amount of matter flowing into a neutron star can generate high temperatures and release X rays and gamma rays.

As an example of such an active system, we can examine Hercules X-1. It emits pulses of X rays with a period of about 1.2 seconds, but every 1.7 days the pulses vanish for a few hours (Figure 14-8). Astronomers can understand this system by comparing it to an eclipsing binary star. Hercules X-1 seems to contain a 2-solar-

a

a

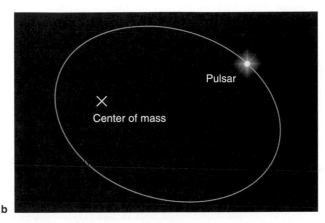

b

Figure 14-7
(a) The period of pulsation of pulsar PSR 1913 + 16 varies over an interval of 7.75 hours. If we treat this interval as the result of the Doppler effect, the resulting radial velocity curve resembles that of a spectroscopic binary. (b) Fitting an orbital solution (solid line) to the data points reveals that the pulsar revolves in an elliptical orbit around the center of mass of a binary star system.

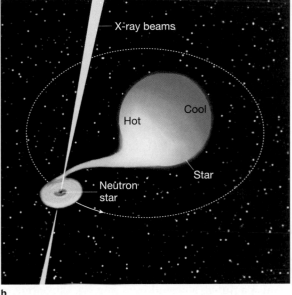

b

Figure 14-8
(a) X-ray observations of Hercules X-1 show that the X-ray pulses disappear periodically (insets), and a graph of X-ray intensity versus time looks like the light curve of an eclipsing binary. The pulsar in the system is eclipsed by its companion every 1.7 days. *(Insets: J. Trümper, Max-Planck Institute)* (b) Detailed analysis allows this artist's conception of the Hercules X-1 system. Matter flows from a stellar companion into an accretion disk around a neutron star producing X rays, which heat the near side of the star to 20,000 K compared to only 7000 K on the far side.

mass star and a neutron star that orbit each other with a period of 1.7 days. Matter flowing from the normal star into the neutron star becomes trapped in the neutron star's magnetic field and generates X rays that the spinning neutron star beams into space. We receive a pulse of X rays every time a beam points our way. The X rays shut off completely every 1.7 days when the neutron star is eclipsed behind the normal star. Hercules X-1 is a complex system with many different high-energy processes going on simultaneously, and our quick analysis only serves to illustrate how complex and powerful such binary systems are during mass transfer.

X-ray source 4U18-20-30 (Figure 14-9) illustrates a number of processes. It consists of a neutron star orbiting a white dwarf with a period of only 11 minutes and a separation of only about a third the Earth–moon distance. Astronomers suggest that a neutron star collided with a giant star and went into orbit inside the star. Recall that the outer layers of a giant star have a very low density. The neutron star gradually ate away the giant star's outer layers, and the star collapsed to form a white dwarf. Matter still flows from the white dwarf through an accretion disk down to the surface of the neutron star, where it accumulates until it ignites to produce periodic bursts of X rays. Notice the similarity with the mechanism that produces nova explosions.

A pulsar called the Black Widow shows how a pulsar can get rid of its companion and become a solitary, millisecond pulsar. The pulsar produces 622 pulses per

a   b

Figure 14-9
(a) At visible wavelengths, the center of star cluster NGC6624 is crowded with stars. *(Ivan King and NASA/ESA)* (b) In the ultraviolet, one star stands out, an X-ray source consisting of a neutron star orbiting a white dwarf with a period of only 11 minutes. *(Ivan King and NASA/ESA)*

second as it orbits a low-mass companion. Spectra reveal that the blast of radiation and high-energy particles from the Black Widow is destroying the companion star as its surface boils away. Eventually, the Black Widow pulsar will be left as a solitary, millisecond pulsar.

Not all exotic neutron stars are violent, high-energy objects. One seems almost peaceful; it has planets.

Pulsar Planets

Finding planets orbiting stars other than the sun is very difficult, and only a few are known. Oddly, the first such planets were found orbiting a neutron star.

Because a pulsar's period is so precise, astronomers can detect tiny variations by comparison with atomic clocks. When astronomers checked pulsar PSR 1257 + 12, they found variations in the period of pulsation much like those caused by the orbital motion of the binary pulsar (Figure 14-10a). However, in the case of PSR 1257 + 12, the variations were much smaller, and when they were interpreted as Doppler shifts, it became evident that the pulsar was being orbited by at least two objects with planetlike masses of 3.4 and 2.8 Earth masses. The gravitational tugs of the planets make the pulsar wobble about the center of mass of the system by no more than 800 km, and that produces the tiny changes in period (Figure 14-10b).

Astronomers greeted this discovery with both enthusiasm and skepticism. As usual, they looked for ways to test the hypothesis. Simple gravitational theory predicts that the planets should interact and slightly modify each other's orbit. When the data were analyzed, that interaction was found, further confirming the hypothesis that the variations in the period of the pulsar are caused by planets (Window on Science 14-1). In fact, further data suggest the presence of two more planets orbiting the pulsar.

a

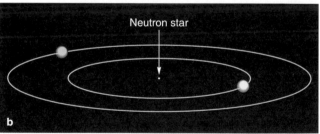

b

Figure 14-10
(a) The dots in this graph are observations showing that the period of pulsar PSR 1257 + 12 varies from its average value by a fraction of a billionth of a second. The blue line shows the variation that would be produced by two planets orbiting the pulsar. (b) As the planets orbit the pulsar, they cause it to wobble by less than 800 km, a distance that is invisibly small in this diagram.

Astronomers wonder how a neutron star can have planets. The innermost planet orbiting PSR 1257 + 12 is only 0.36 AU from the pulsar. Any planets that orbit a star would be lost or vaporized when the star exploded. Furthermore, a star about to explode as a supernova would be a large giant or a supergiant, and

No scientific theory or hypothesis can be proved correct. We can test a theory over and over by performing experiments or making observations, but we can never prove that the theory is absolutely true. It is always possible that we have misunderstood the theory or the evidence, and the next observation we make might disprove the theory. In that sense, we never learn anything new until we disprove a theory or hypothesis. Only at that point do we discover something we did not know before.

For example, we might propose the theory that the sun is mostly iron. We might test the theory by looking at the iron lines in the solar spectrum, and the strength of the lines would suggest our theory is right. Although our observation has confirmed our theory, we know nothing more than we did before. We might continue to confirm the theory over and over, but we still know nothing new and there is always the danger that we have misunderstood the evidence. Finally, we might study the formation of atomic spectra and discover that iron atoms are very good absorbers of photons, so the strong iron lines in the solar spectrum do not mean that the sun is made mostly of iron. In fact, the hydrogen lines mean that the sun is mostly hydrogen atoms. Now that we have disproved our hypothesis, we have learned something new.

The nature of scientific hypothesis testing can lead to two kinds of mistakes. Sometimes nonscientists will say, "You scientists just want to tear everything down—you don't believe anything." It is the nature of science to test every hypothesis, not because the scientist wants to tear things down, but because the scientist wants to know what is the most dependable description of nature. Others will say, "You scientists are never sure of anything." Again, the scientist knows that no theory can ever be proved correct. That the sun will rise tomorrow is very likely, but in the end it is still a theory.

People will say of an idea they dislike, "That is only a theory," as if a theory were simply a random guess. In fact, a theory can be a well-tested truth in which all scientists have great confidence. Yet we can never prove that any theory is absolutely true.

---

planets only a few AU distant would be inside such a large star and could not survive. It seems more likely that these planets are the remains of a stellar companion that was devoured by the neutron star. In fact, the pulsar is very fast (162 pulses per second), suggesting that it was spun up in a binary system.

We can imagine what these worlds might be like. Formed from the remains of dying stars, they might have chemical compositions richer in heavy elements than Earth. We can imagine visiting these worlds, landing on their surfaces, and hiking across their valleys and mountains. Above us, the neutron star would glitter in the sky, a tiny point of light.

Gamma Ray Bursts from Neutron Stars

In 1968, satellites that could detect gamma rays were put in orbit around Earth, and they immediately began picking up bursts of gamma rays coming from space. Typically one burst a day was seen lasting for seconds or minutes. These gamma-ray bursts have been a mystery for decades, but new observations suggest at least some of the bursts may be coming from neutron stars.

A 1992 theory proposes that some young neutron stars can have magnetic fields 1000 times stronger than the average neutron star. Dubbed **magnetars,** these objects would be tremendously hot, because the magnetic field would heat the surface of the neutron star to 10 million Kelvin. Shifts in the magnetic field would break the rigid crust of the neutron star in a violent starquake—equivalent to an earthquake on Earth. Such a starquake would release a great burst of high-energy photons.

Observations of two repeating gamma-ray bursters confirm that they are very hot and have powerful magnetic fields. This has been taken as confirmation that they are magnetars. One of the objects was the source of a burst of gamma rays that reached Earth on August 27, 1998, and temporarily ionized Earth's upper atmosphere, disrupting radio communication worldwide.

Because a gamma-ray burst is over so quickly and most do not appear to repeat, it has been hard to trace them to their source, but coordination between orbiting space telescopes and ground-based observatories has revealed that at least some gamma-ray bursters are located in very distant galaxies. To be detectable at such great distances, they must produce tremendous amounts of energy. One theory proposes that the gamma-ray bursts occur when two neutron stars orbiting each other in a binary system spiral together and collide. Such an event would release about the right amount of energy to be observed from Earth as a gamma-ray burst. Another theory suggests that massive stars collapsing directly into a black hole may emit powerful gamma-ray bursts. It seems likely that gamma-ray bursts are produed by more than one kind of event.

Gamma-ray bursts occur daily, but magnetars, colliding neutron stars, and massive supernovae seem unusual. In fact, these are common. Every galaxy contains vast numbers of neutron stars, and although massive stars make up a small fraction of stars, galaxies contain so many stars that massive stars are numerous. The gamma-ray bursts are so powerful they can be observed at great distances, so gamma-ray telescopes can "hear" one of these events almost anywhere in the universe. Thus it is reasonable to link gamma-ray bursts to the deaths of stars.

REVIEW Critical Inquiry

How can a neutron star be found at X-ray wavelengths?

First, we should remember that a neutron star will be very hot because of the heat released when it contracts to a radius of 10 km. It could easily have a surface temperature of 1,000,000 K, and Wien's law (By the Numbers 4-1) tells us that such an object will radiate most intensely at a very short wavelength typical of X rays. However, we know that the total luminosity of a star depends on its surface temperature and its surface area, and a neutron star is so small it can't radiate much energy and would be hard to find even with an X-ray telescope.

There is, however, a second way a neutron star can radiate X rays. If a normal star in a binary system loses mass to a neutron star companion, the inflowing matter will be heated to very high temperatures. It may form a very hot accretion disk that can radiate intense X rays easily detectable by X-ray telescopes orbiting above Earth's atmosphere.

If you discovered a pulsar, what observations would you make to determine whether it was young or old, single or a member of a binary system, alone or accompanied by planets?

Perhaps the strangest planets in the universe are those orbiting pulsars. But however strange a pulsar planet may be, we can imagine going to one. Our next topic, in contrast, seems beyond the reach of even our imaginations.

14-2 Black Holes

Although the physics of black holes is difficult to discuss without sophisticated mathematics, simple logic is sufficient to predict that they should exist. Our problem is to consider their predicted properties and try to confirm that they exist. What objects observed in the heavens could be real black holes? More difficult than the search for neutron stars, the quest for black holes is nevertheless meeting with success.

To begin our discussion of black holes, we must consider a simple question. How fast must an object travel to escape from the surface of a celestial body? The answer will lead us to black holes.

Escape Velocity

Suppose we threw a baseball straight up. How fast must we throw it if it is not to come down? Of course, gravity would pull back on the ball, slowing it, but if the ball were traveling fast enough to start with, it would never come to a stop and fall back. Such a ball would escape from Earth. The escape velocity is the initial velocity an object needs to escape from a celestial body (Figure 14-11).

Figure 14-11
Escape velocity, the velocity needed to escape from a celestial body, depends on mass. The escape velocity at the surface of a very small body would be so low we could jump into space. Earth's escape velocity is much larger, about 11 km/s (25,000 mph).

Whether we are discussing a baseball leaving Earth or a photon leaving a collapsing star, the escape velocity depends on two things: the mass of the celestial body and the distance from the center of mass to the escaping object. If the celestial body had a large mass, its gravity would be strong, and we would need a high velocity to escape; but if we began our journey farther from the center of mass, the velocity needed would be less. For example, to escape from Earth, a spaceship would have to leave its surface at 11 km/s (25,000 mph), but if we could launch spaceships from the top of a tower 1000 miles high, the escape velocity would be only 8.8 km/s (20,000 mph). If we could make an object massive enough and/or small enough, its escape velocity could be greater than the speed of light. Such an object could never be seen because light could never leave it.

Rev. John Mitchell, a British amateur astronomer, was the first person to realize that Newton's laws of gravity and motion contained this implication. In 1783, he pointed out that an object 500 times the radius of the sun but of the same density would have an escape velocity greater than the speed of light. Then, "all light emitted from such a body would be made to return towards it." Mitchell had discovered the black hole.

Schwarzschild Black Holes

If the core of a star collapses and contains more than about 3 solar masses, no known force can stop it (Figure 14-12). The object cannot stop when it reaches the size of a white dwarf, because degenerate electrons cannot support the weight. It cannot stop when it reaches the size of a neutron star, because degenerate neutrons cannot support the weight. No known force remains to stop the object from shrinking to zero radius.

As an object shrinks, its density and the strength of its surface gravity increase; and if an object shrinks to zero radius, its density and gravity become infinite. Mathematicians call such a point a **singularity;** but in physical terms it is difficult to imagine an object of zero radius. Some theorists believe that a singularity is impossible and that when we better understand the laws of physics, we will discover that the collapse halts before diameter zero. Astronomically, it seems to make little difference.

If the contracting core of a star becomes small enough, the escape velocity nearby is so large that no light can escape. We can receive no information about the object or about the volume of space near it, and we refer to this volume as a **black hole.** The boundary of this region is called the **event horizon,** because any event that takes place inside the surface is invisible to an outside observer (Figure 14-13). To see how such a region can exist, we must consider general relativity.

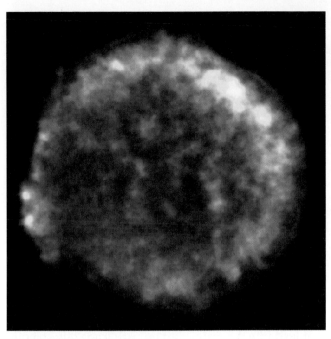

Figure 14-12
The collapse of a massive star causes a supernova explosion that ejects the outer layers of the star in an expanding shell while the core of the star may become so small it forms a black hole. Tycho's supernova occurred in 1572, and the expanding shell is shown in this X-ray image. No pulsar has been found, so the collapsing star may have formed a black hole. *(John P. Hughes, Rutgers Univ.)*

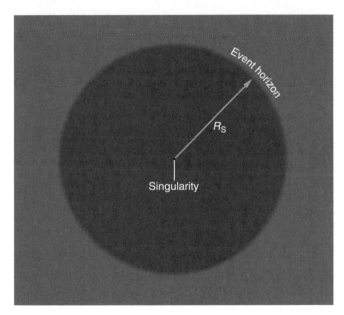

Figure 14-13
A black hole forms when an object collapses to a small size (perhaps to a singularity) and the escape velocity in its neighborhood is so great light cannot escape. The boundary of this region is called the event horizon because any event that occurs inside is invisible to outside observers. The radius of the region R_s is the Schwarzschild radius.

In 1916, Einstein published a mathematical theory of space and time that became known as the general theory of relativity. Einstein's equations showed that gravity could be described as a curvature of space–time, and almost immediately the astronomer Karl Schwarzschild found a way to solve the equations to describe the gravitational field around a single, nonrotating, electrically neutral lump of matter. That solution contained the first general relativistic description of a black hole; nonrotating, electrically neutral black holes are now known as Schwarzschild black holes. In recent decades, theorists such as Stephen W. Hawking have found a way to apply the mathematical equations of the general theory of relativity to charged, rotating black holes. For our discussion the differences are minor, and we may proceed as if all black holes were Schwarzschild black holes.

Schwarzschild's solution showed that if matter is packed into a small enough volume, then space–time is curved back on itself. Objects can still follow paths that lead into the region, but no path leads out, so nothing can escape, not even light. Thus, the inside of the black hole is totally beyond the view of an outside observer. The event horizon is the boundary between the isolated volume of space–time and the rest of the universe, and the radius of the event horizon is called the **Schwarzschild radius, R_s**—the radius within which an object must shrink to become a black hole (Figure 14-13).

Although Schwarzschild's work was highly mathematical, his conclusion is quite simple. The Schwarzschild radius (in meters) depends only on the mass of the object (in kilograms):

$$R_s = \frac{2GM}{c^2}$$

In this simple formula, G is the gravitational constant, M is the mass, and c is the speed of light. A bit of arithmetic shows that a 1-solar-mass black hole will have a Schwarzschild radius of 3 km, a 10-solar-mass black hole will have a Schwarzschild radius of 30 km, and so on (Table 14-1). Even a very massive black hole would not be very large.

Every object with mass has a Schwarzschild radius, but not every object is a black hole. For example, Earth has a Schwarzschild radius of about 1 cm, but it could become a black hole only if we squeezed it inside that radius. Fortunately, Earth will not collapse spontaneously into a black hole because its mass is less than the critical mass of about 3 solar masses. Only exhausted stellar cores more massive than this can form black holes under the sole influence of their own gravity. In this chapter, we are interested in black holes that might originate from the deaths of massive stars. These would have masses a bit larger than 3 solar masses. In later chapters, we will encounter black holes whose masses might exceed a million solar masses.

Do not think of black holes as giant vacuum cleaners that will pull in everything in the universe. A black hole is just a gravitational field, and at a reasonably large distance its force is quite small. If the sun were replaced by a 1-solar-mass black hole, the orbits of the planets would not change at all. The gravity of a black hole becomes extreme only when we approach close to it. Earth's gravitational field is not dangerous, because we can never come closer than 6400 km (4000 miles) to its center. But were Earth compressed into a black hole only a centimeter in radius, we could come much closer to the center of the gravitational field, and then we would feel extreme gravitation. Thus, we can imagine that the universe contains numerous black holes. So long as we and other objects stay a safe distance from the black holes, they will have no catastrophic effects.

A Leap into a Black Hole

Before we can search for real black holes, we must understand what theory predicts about the appearance of a black hole. To explore that idea, we can imagine that we leap, feet first, into a Schwarzschild black hole.

If we were to leap into a black hole of a few solar masses from a distance of an astronomical unit, the gravitational pull would not be very large, and we would fall slowly at first. Of course, the longer we fell and the closer we came to the center, the faster we would travel. Our wristwatches would tell us that we fell for about 65 days before we reached the event horizon.

Our friends who stayed behind would see something different. They would see us falling more slowly as we came closer to the event horizon because, as explained by general relativity, clocks slow down in curved space–time. This is known as **time dilation.** In fact, our friends would never actually see us cross the event horizon. To them we would fall more and more slowly until we seemed hardly to move. Generations later, our descendants could focus their telescopes on us and see us still inching closer to the event horizon. We, how-

Table 14-1 The Schwarzschild Radius		
	Mass ($M_\odot$)	R_s
Star	10	30 km
Star	3	9 km
Star	2	6 km
Sun	1	3 km
Earth	0.000003	0.9 cm

ever, would have sensed no slowdown and would conclude that we had crossed the event horizon after only about 65 days.

Another relativistic effect would make it difficult to see us with normal telescopes. As light travels out of a gravitational field, it loses energy, and its wavelength grows longer. This is known as the **gravitational red shift.** Although we would notice no effect as we fell toward the black hole, our friends would need to observe at longer and longer wavelengths in order to detect us.

While these relativistic effects seem merely peculiar, other effects would be quite unpleasant. Imagine again that we are falling feet first toward the event horizon of a black hole. We would feel our feet, which would be closer to the black hole, being pulled in more strongly than our heads. This is a tidal force, and at first it would be minor. But as we fell closer, the tidal force would become very large (Figure 14-14). Another tidal force would compress us as our left side and our right side both fell toward the center of the black hole. For any black hole with a mass like that of a star, the tidal forces would crush us laterally and stretch us longitudinally long before we reached the event horizon. The friction from such severe distortions of our bodies would heat us to millions of degrees, and we would emit X rays and gamma rays.

Some years ago a popular book suggested that we could travel through the universe by jumping into a black hole in one place and popping out of another somewhere far across space. That might make for good science fiction, but tidal forces would make it an unpopular form of transportation even if it worked.

Our imaginary leap into a black hole is not entirely frivolous. We now know how to find a black hole. Look for a strong source of X rays. It may be a black hole into which matter is falling.

The Search for Black Holes

Do black holes really exist? Beginning in the 1970s, astronomers searched for evidence that their theories were correct by trying to find one or more objects that were obviously black holes. That very difficult search is a good illustration of how the unwritten rules of science help us understand nature (Window on Science 14-2).

A black hole alone is totally invisible, but if gas is rapidly flowing into the black hole it will be heated and can emit X rays before the gas enters the event horizon. A solitary black hole in space would not gain much mass, but a black hole in a close binary system could drain mass from a companion star and form a hot accretion disk that would emit X rays as the gas spirals inward.

X-ray binaries, such as Hercules X-1, contain a neutron star and not a black hole. We know because neutron stars contain less than 3 solar masses. Thus we can

Figure 14-14
Leaping feet first into a black hole. A person of normal proportions (left) would be distorted by tidal forces (right) long before reaching the event horizon around a typical black hole of stellar mass. Tidal forces would stretch the body lengthwise while compressing it laterally. Friction from this distortion would heat the body to high temperatures.

search for black holes by looking for X-ray binaries with a compact object of more than 3 solar masses.

Cygnus X-1, the strongest X-ray source in Cygnus, was the first black hole candidate, but it isn't conclusive. It consists of a supergiant O star with a strong stellar wind carrying gas into a hot accretion disk around a compact object (Figure 14-15). Doppler shifts in the spectrum of the O star suggest the mass of the compact object is 10 to 15 solar masses, well above the neutron star limit. But it isn't conclusive evidence. The O star might be odd, so we can't estimate its mass accurately, and there may be a third star in the system confusing the analysis. Thus Cygnus X-1 isn't a conclusive example of a black hole.

X-ray telescopes have found a few dozen candidates. A few of these objects, such as the first two in

Fraud is actually quite rare in science. The nature of science makes fraud difficult, and the way scientists publish their research makes it almost impossible. In fact, we can think of science as a set of unwritten rules of behavior that have evolved to prevent scientists from lying to one another or to themselves, even by accident.

Suppose for a moment that we wanted to commit scientific fraud. We would have to invent data supposedly obtained from experiment or observation. We might invent X-ray data supposedly obtained by observing an X-ray binary star. Or if we were interested in theory, we might invent a fraudulent mathematical calculation of the physics going on in an X-ray binary. We might get away with it for a short time, but one of the most important rules in science is that good results must be reproducible. Other people must be able to repeat our observations, experiments, and calculations. In fact, most scientists routinely repeat other scientists' work as a way of getting started on a research topic. As soon as someone tries to repeat our fraudulent research, we will be caught. The more important a scientific result is, the sooner other scientists will repeat it, so we don't have much of a chance of getting away with scientific fraud. In this way, science is self-correcting.

Even if we could invent some convincing scientific research, we would probably have difficulty publishing it. When a scientist submits an article to a scientific journal, it is subject to peer review. That is, the editor of the journal sends the article to one or two other experts in the field for comment and suggestions. These reviewers often make helpful suggestions, but they may also point out errors that have to be fixed before the journal can publish the article. In some cases, an article may be so flawed the editor will refuse to publish it at all. If we submitted our fraudulent research on X-ray binaries, the reviewers would almost certainly notice things wrong with it, and it would never get into print.

Scientists know the rules, and they use them. If someone makes a big discovery and is interviewed by the press, scientists will begin asking, "Has this work been published in a peer-reviewed journal yet?" That is, they want to know if other experts have checked the work. Until research is published, it isn't official, and most scientists would treat the results with care.

Fraud isn't impossible in science. Some cases have even been in the national news. Big grant money is a terrible temptation. But in science, "the truth will out." Because of the way scientists reproduce research and because of the way research is published, scientific fraud is rare.

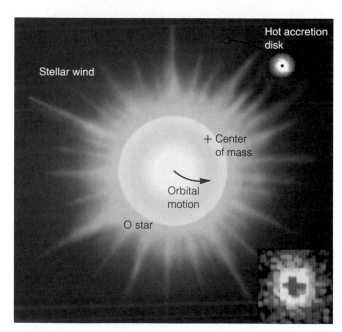

Figure 14-15
The X-ray source Cygnus X-1 is a supergiant O star and a compact object orbiting each other. Gas from the O star's stellar wind flows into the hot accretion disk, and the X rays we detect come from the disk. An X-ray image is shown in the inset. The masses can be estimated from the Doppler shifts produced in the spectrum of the O star as it circles the center of mass. *(J. Trümper, Max-Planck Institute)*

Table 14-2, contain massive stellar companions, giants, supergiants, or massive main-sequence stars. Such systems are difficult to analyze because the massive stars dominate the system. For example, LMC X-3 (LMC refers to the Large Magellanic Cloud, a small galaxy near our own) contains a massive B star and a compact object of 4 to 11 solar masses. One way we know the compact object is massive is that it distorts the B main-sequence star into an egg shape. As the system rotates, the light from the B star varies as we see the side of the egg and then the end. The light variation reveals the shape of the B star and thus the mass of the compact object.

Many black hole candidates are binaries containing a lower-main-sequence star. Such systems don't remain steady X-ray sources but suffer X-ray nova outbursts as matter flows rapidly into the accretion disk. A year or so after an outburst, the flow has stopped, the accretion disk has dimmed, and telescopes can detect the spectrum of the main-sequence star. The spectral type and Doppler motions reveal the mass of the compact object with little uncertainty. The low-mass star makes these systems easier to analyze than systems with massive stars.

Table 14-2 lists a few examples of these X-ray novae. V616 Mon, an old nova, erupted again in 1975. It con-

Table 14-2 Six Black Hole Candidates

Object	Companion Star	Orbital Period	Mass of Compact Object
Cygnus X-1	O supergiant	5.6 days	10–15 $M_\odot$
LMC X-3	B3 main-sequence	1.7 days	4–11 $M_\odot$
V616 Mon	K main-sequence	7.75 hours	3.3–4.2 $M_\odot$
V404 Cygni	K main-sequence	6.47 days	8–15 $M_\odot$
J1655-40	F-G main-sequence	2.61 days	4–5.2 $M_\odot$
QZ Vul	K main-sequence	8 hours	5–14 $M_\odot$

tains a main-sequence K star and a compact object that orbit with a period of 7.75 hours. The orbital motion and the distortion of the K star tell us the compact object must have a mass of 3.3 to 4.2 solar masses. V404 Cygni is a similar system containing a compact object of 8 to 15 solar masses. One of the best examples is J1655-40 (Nova Scorpii 1994). It contains a compact object of 4 to 5.2 solar masses. QZ Vul contains a compact object with a mass of at least 5 solar masses.

These candidates are more conclusive than Cygnus X-1, because their masses are more accurately known. There seems to be no way they could be neutron stars.

Another kind of evidence gives astronomers further confidence that black holes really exist. In a binary system, gas flowing toward a neutron star emits X rays as it spirals inward through the accretion disk and again as it hits the neutron star's surface. Observations show that when the compact object has a mass greater than 3 solar masses, we don't detect the X-rays coming from the impact on the surface. Why? Because the gas has crossed the event horizon and entered a black hole.

Astronomers continue to test the theory of black holes, but the evidence has grown very strong. Black holes really do exist. The problem now is to understand how these objects interact with matter flowing into them through accretion disks to produce X rays, gamma rays, and jets of matter.

Jets of Energy from Compact Objects

Our impression of a black hole might suggest that it is impossible to get any energy out of such an object. In later chapters, we will meet galaxies that extract vast amounts of energy from massive black holes, so we should pause here to see how a compact object can produce high-energy jets.

Whether a compact object is a black hole or a neutron star, it has a strong gravitational field. Any matter flowing into that field is accelerated inward, and because it must conserve angular momentum, it flows into an accretion disk made so hot by friction that the inner regions can emit X rays and gamma rays. Ther-

mal and magnetic processes can cause the inner parts of the accretion disk to eject high-speed jets of matter in opposite directions along the axis of rotation (Figure 14-16). This process recalls the bipolar flows ejected by

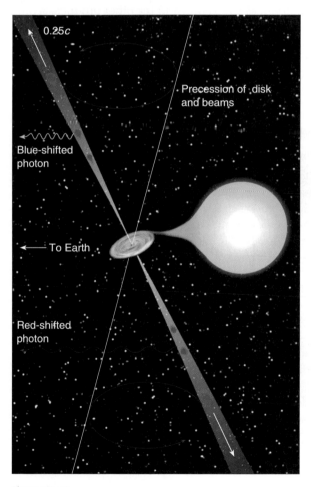

Figure 14-16
The generally accepted model of SS433 includes a compact object with a very hot accretion disk producing beams of radiation with embedded blobs of gas traveling at 25 percent the velocity of light. The precession of the disk swings the beams around a conical path every 164 days.

protostars, but it is much more powerful. As examples, the Vela pulsar is known to have a jet visible at X-ray wavelengths, and the black hole candidate J1655-40 is observed at radio wavelengths to be ejecting oppositely directed jets at 92 percent the speed of light.

One of the most powerful examples of this process is an X-ray binary called SS433. Its optical spectrum shows sets of spectral lines that are Doppler-shifted by about one-fourth the speed of light, with one set shifted to the red and one set shifted to the blue. The object is both receding and approaching at a fantastic speed. Furthermore, the two sets of lines shift back and forth across each other with a period of 164 days.

Apparently, SS433 is a binary system in which a compact object (a neutron star) pulls matter from its companion star and forms an extremely hot accretion disk. Jets of high-temperature gas blast away in beams aimed in opposite directions. As the disk precesses, it sweeps these beams around the sky once every 164 days, and we see light from gas trapped in both beams. One beam produces a red shift, and the other produces a blue shift. SS433 is a prototype that illustrates how the gravitational field around a compact object can produce powerful beams of radiation and matter. We will meet this phenomenon again when we study peculiar galaxies.

REVIEW Critical Inquiry

How can a black hole emit X rays?

Once a bit of matter falling into a black hole crosses the event horizon, no light or other electromagnetic radiation it emits can escape from the black hole. The matter becomes lost to our view. But if it emitted radiation before it crossed the event horizon, that radiation could escape, and we could detect it. Furthermore, the powerful gravitational field near a black hole stretches and distorts in-falling matter, and internal friction heats the matter to millions of degrees. Wien's law (By the Numbers 4-1) tells us that matter at such a high temperature should emit X rays. Any X rays emitted before the matter crosses the event horizon will escape, and thus we can look for black holes by looking for X-ray sources.

The search for black holes has succeeded in finding a few strong candidates, but the problem is being sure the binary system contains a black hole and not a neutron star. What observations would you make of an X-ray binary system to distinguish between a black hole and a neutron star?

Many stars collapse to form black holes or neutron stars in binary systems, but these are objects of only a few solar masses. We will see similar phenomena many

times more powerful when we explore the galaxies in Chapters 15, 16, and 17.

Summary

If the remains of a star collapse with a mass greater than the Chandrasekhar limit of 1.4 solar masses, then the object cannot reach stability as a white dwarf. It must collapse to the neutron star stage, with a radius of about 10 km and a density equal to that of an atomic nucleus. Such a neutron star can be supported by the pressure of its degenerate neutrons. But if the mass is greater than 2 to 3 solar masses, then the degenerate neutrons cannot stop the collapse, and the object must become a black hole.

Theory predicts that a neutron star should rotate very fast, be very hot, and have a strong magnetic field. Such objects have been identified as pulsars, sources of pulsed radio energy. Pulsars are evidently spinning neutron stars, which emit beams of radiation from their magnetic poles. As they spin, they sweep the beams around the sky; and if the beams sweep over Earth, we detect pulses. This is known as the lighthouse theory. The spinning neutron star slows as it radiates energy into space.

A few pulsars have been found in binary systems, which allows astronomers to estimate the masses of the pulsars. Such masses are consistent with the predicted masses of neutron stars. In some binary systems, such as Hercules X-1, mass flows into a hot accretion disk around the neutron star and causes the emission of X rays.

If a star collapses with a mass greater than 2 or 3 solar masses, then it must contract to a very small size—perhaps to a singularity, an object of zero radius. Near such an object, gravity is so strong that not even light can escape, and we call the region a black hole. The boundary of this region is called the event horizon. The Schwarzschild radius is the radius of this event horizon, amounting to only a few kilometers for black holes of stellar mass.

If we were to leap into a black hole, our friends who stayed behind would see two relativistic effects. They would see our clock slow relative to their own clock because of time dilation. Also, they would see our light red-shifted to longer wavelengths. We would not notice these effects, but we would feel powerful tidal forces that would deform and heat our mass until we grew hot enough to emit X rays. Any X rays we emitted before crossing the event horizon could escape.

To search for black holes, we must look for binary star systems in which mass flows into a compact object and emits X rays. If the mass of the compact object is greater than about 3 solar masses, then the object is presumably a black hole. A number of such objects have been located.

SS433 is an X-ray binary that illustrates how rapid mass transfer can generate tremendous energy. Apparently, mass flows rapidly into an accretion disk around a compact object, and the hot accretion disk ejects beams of radiation and gas in opposite directions at about 25 percent the speed of light.

New Terms

neutron star   black hole

pulsar   event horizon

lighthouse theory   Schwarzschild radius (R_s)

millisecond pulsar   time dilation

magnetar   gravitational red shift

singularity

Review Questions

1. How are neutron stars and white dwarfs similar? How do they differ?

2. Why is there an upper limit to the mass of neutron stars?

3. Why do we expect neutron stars to spin rapidly?

4. If neutron stars are hot, why aren't they very luminous?

5. Why do we expect neutron stars to have a powerful magnetic field?

6. Why did astronomers conclude that pulsars could not be pulsating stars?

7. What does the short length of pulsar pulses tell us?

8. How does the lighthouse theory explain pulsars?

9. What evidence do we have that pulsars are neutron stars?

10. Why would astronomers at first assume that the first discovered millisecond pulsar was young?

11. How can a neutron star in a binary system generate X rays?

12. If the sun has a Schwarzschild radius, why isn't it a black hole?

13. How can a black hole emit X rays?

14. What evidence do we have that black holes really exist?

15. How can mass transfer into a compact object produce jets of high-speed gas?

Discussion Questions

1. In your opinion, has the existence of neutron stars been sufficiently tested to be called a theory, or should it be called a hypothesis? What about black holes?

2. Why wouldn't an accretion disk orbiting a giant star get as hot as an accretion disk orbiting a compact object?

Problems

1. If a neutron star has a radius of 10 km and rotates 642 times a second, what is the speed of the surface at the neutron star's equator in terms of the speed of light? (*Hint:* The circumference of a circle is $2\pi r$.)

2. A neutron star and a white dwarf have been found orbiting each other with a period of 11 minutes. If their masses are typical, what is the average distance between them? (*Hint:* See By the Numbers 11-4.)

3. If Earth's moon were replaced by a typical neutron star, what would the angular diameter of the neutron star be, as seen from Earth? (*Hint:* See By the Numbers 1-2.)

4. What is the Schwarzschild radius of Jupiter (mass = 2×10^{27} kg)? of a human adult (mass = 75 kg)? (*Hint:* See Appendix A for the values of G and c.)

5. If the inner accretion disk around a black hole has a temperature of 10^6 K, at what wavelength will it radiate the most energy? (*Hint:* See By the Numbers 4-1.)

6. What is the orbital period of a bit of matter in an accretion disk 2×10^5 km from a 10-solar-mass black hole? (*Hint:* See By the Numbers 11-4.)

Critical Inquiries for the Web

1. Imagine that you are on a mission to explore one of the pulsar planets noted in the chapter. What would you find there? Look for information about pulsars and the known pulsar planets on the Web, and describe what you might encounter on such a mission.

2. What would you experience if you were to pilot a spacecraft near a black hole? Visit black hole-related Internet sites to determine what the gravitational effects and general environment would be. Also use the Internet to find the limits of human tolerance to strong gravitational forces. (*Hint:* Look for information about astronaut training, and find out how many Gs a human can withstand.) Use these sources to give a brief account about what your voyage would be like.

Exploring *The Sky*

1. The pulsar in the Crab Nebula was first discovered in 1967. Find and center this supernova remnant, also designated M1 (Messier 1). Determine the constellation in which the Crab Nebula resides, its magnitude, and its angular size. Is it possible to see this object with the unaided eye? View the Crab Nebula with high enough magnification to see the shape of the gas cloud, and compare with Figure 14-3. Note that *The Sky* does not show the pulsar itself.

 Go to the Brooks/Cole Astronomy Resource Center (www.brookscole.com/astronomy) for critical thinking exercises, articles, and additional readings from InfoTrac College Edition, Brooks/Cole's online student library.

CHAPTER FIFTEEN

The Milky Way Galaxy

A hypothesis or theory is clear, decisive, and positive, but it is believed by no one but the man who created it. Experimental findings, on the other hand, are messy, inexact things, which are believed by everyone except the man who did that work.

Harlow Shapley
Through Rugged Ways to the Stars

Guidepost

This chapter plays three parts in our cosmic drama. First, it introduces the concept of a galaxy; second, it discusses our home, the Milky Way Galaxy, naturally an object of our curiosity; third, it elaborates our story of stars by introducing us to galaxies, the communities in which stars exist.

Science is based on the interaction of theory and evidence, and this chapter will show a number of examples of astronomers using evidence to test theories. If the theories seem incomplete and the evidence contradictory, we should not be disappointed. Rather, we must conclude that the adventure of discovery is not yet over.

We struggle to understand our own galaxy as an example. We will extend the concept of the galaxy in the next two chapters, and then we will apply our understanding of galaxies to the study of the universe as a whole.

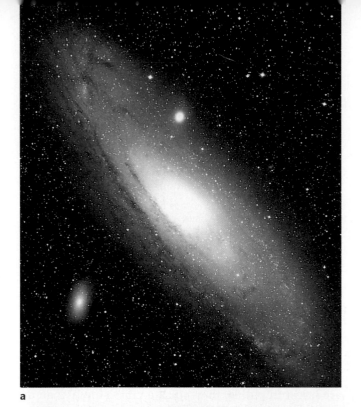

a

We are all fantastically wealthy. We live inside one of the largest star systems in the universe. Our Milky Way Galaxy is over 75,000 ly in diameter and contains over 100 billion stars. If Earth is the only inhabited planet in the galaxy, then the galaxy belongs to us; and, sharing equally, each person on Earth owns 15 stars plus assorted planets, moons, comets, and so on. Even if we must share with a few other inhabited planets, we are all rich beyond any dream. Of course, we lack transportation to visit our dominions, but we can still admire our galaxy. It is, after all, ours.

Almost every celestial object visible to our naked eyes is part of our Milky Way Galaxy. The only exception visible from Earth's northern hemisphere is the Andromeda Galaxy, just visible to our unaided eyes as a faint patch of light in the constellation Andromeda.* Our galaxy probably looks much like the Andromeda Galaxy (Figure 15-1a).

*Consult the star charts at the end of this book to locate the Milky Way and the Andromeda Galaxy in your night sky.

b

Figure 15-1
(a) This photograph of the Great Galaxy in Andromeda, a spiral galaxy about 2.3 million ly from us, shows us what our Milky Way Galaxy would look like if we could view it from such a great distance. *(Bill Schoening and Vanessa Harvey, NSF REU, NOAO, © AURA, Inc.)* (b) From inside, we see our galaxy as scattered nearby stars and distant clouds of stars in a faintly luminous path that circles the sky. Invisible from the brightly lit skies of cities, the band of the Milky Way is easily visible from a dark site. This artwork shows the location of a portion of the Milky Way near a few bright winter constellations. (See Figure 15-4 and the star charts at the end of this book to further locate the Milky Way in the sky.)

It isn't obvious that we live in a galaxy. We are inside, and what we see is a band of faint light circling the sky (Figure 15-1b). The ancient Greeks named that band *galaxies kuklos,* the "milky circle." The Romans changed the name to *via lactea,* "milky road" or "milky way." It was not until early in this century that astronomers understood that we live inside a great wheel of stars and that the universe is filled with other such star systems. Drawing on the Greek word for milk, we call them galaxies.

Our goal is to understand our galaxy—how it formed and evolved and why it takes the form it does. As always, we begin by gathering evidence for comparison with theories. In this case, our evidence includes the size, mass, and shape of our galaxy and the distribution of stars within it.

15-1 The Discovery of the Galaxy

It seems odd to say astronomers discovered something that is all around us; but until the 20th century, although everyone could see the Milky Way, no one knew what it was. Two hundred years ago, astronomers generally assumed that stars were uniformly scattered through space. One hundred years ago, astronomers understood that the stars we see lie in a great wheel-shaped star system, but they thought the sun was near the center and that the star system was not very large. Then, about one lifetime ago, a woman studying variable stars and a man studying star clusters unlocked one of nature's biggest secrets—that we live in a galaxy.

We follow the story of their discovery for two reasons. First, it is an important historical moment in human history—as dramatic as the Copernican realization that we live on a planet. Second, the story of the discovery of our galaxy illustrates how scientists build on the work of their predecessors and step-by-step refine their ideas about the natural world we live in.

Variable Stars

Although we think of stars as unchanging, astronomers have known for centuries that certain stars change their brightness, growing brighter, fainter, and brighter with periods that range from less than a day to hundreds of days (Figure 15-2a). These **variable stars** were not understood until the 20th century.

In Chapter 12 we saw that the layers of a star are supported against gravity by the flow of energy out of their interiors. Most stars reach a stable balance between gravity and energy, but in a certain region of the H-R diagram called the **instability strip** (Figure 15-2b), the temperature and size of the star are just right to make it unstable. A layer just below the surface of the star absorbs and releases energy like a spring, and the star pulsates like a beating heart—expanding and contracting and thus changing its brightness. Modern astronomers understand that the evolution of some giant stars can carry them through the instability strip, and thus they become variable stars while they cross the region of instability.

There are many kinds of variable stars, but two kinds are important to our story of the discovery of the Milky Way. The **RR Lyrae variable stars,** named after variable star RR in the constellation Lyrae, have periods of one-half to one day and are a bit less than 100 times as luminous as the sun. A related kind of variable star is the **Cepheid,** named after δ Cephei, the first star of this type to be found. Cepheids have periods ranging from 1 to 60 days, but they do not all have the same luminosity. More massive stars cross the instability strip higher in the diagram and are thus more luminous. Because they are larger in radius, they pulsate more slowly than do the lower-mass stars. Thus, there is a relation between the period and luminosity of Cepheid variable stars: The longer the period, the greater the luminosity.

In the early part of the 20th century, astronomers knew none of the stellar evolution we have summarized in the preceding paragraphs. Nevertheless, in 1912 the Harvard astronomer Henrietta Leavitt was photographing the sky and discovering large numbers of Cepheid and RR Lyrae variables when she discovered that the longer-period Cepheid variables were brighter than the short-period Cepheids. This **period–luminosity relation** has become a powerful tool for finding distance (Figure 15-2c). It was the key to the Milky Way Galaxy.

The Size of the Galaxy

By 1900, astronomers studying the distribution of the stars in the sky concluded that the band of the Milky Way was evidence that we lived near the center of a disk-shaped cloud of stars that was not very large. Many believed that the star system, as it was called, was isolated in an otherwise empty universe.

A young astronomer named Harlow Shapley (1885–1972) began the discovery of the true Milky Way when he noticed that different kinds of star clusters have different distributions in the sky. **Open star clusters** (Figure 15-3a) contain 100 to 1000 stars in a region 10 to 60 ly in diameter and are concentrated along the band of the Milky Way. The Pleiades (see Figure 12-2a) is an open cluster. **Globular star clusters** (Figure 15-3b) contain 100,000 to 1,000,000 stars crowded into a region

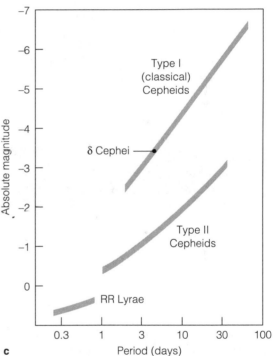

Figure 15-2

(a) Some stars are variable and pulsate in brightness. The period of pulsation is the time it takes the star to complete one cycle. (b) When the evolution of a giant star carries it through the instability strip (shaded) in the H-R diagram, the star becomes unstable and pulsates as a variable star. More massive stars are more luminous, larger, and have longer periods. (c) A modern diagram shows the period–luminosity relation for RR Lyrae stars and two kinds of Cepheid variable stars.

about 75 ly in diameter. Shapley noticed that globular clusters are highly concentrated in one part of the sky. Almost half of all known globular clusters are in or near the constellation Sagittarius (Figure 15-4).

Shapley assumed that the globular clusters were controlled by the combined gravitational field of the entire star system, in which case he could study the size and extent of the star system by studying the globular clusters. To do that, he needed to measure the distances to as many globular clusters as possible.

Globular clusters are much too far away to have measurable parallaxes, but they do contain variable stars. Shapley recalled Henrietta Leavitt's work on these stars, but she knew only their apparent magnitudes. Cepheids and RR Lyrae stars are relatively rare, and there are none near enough to have measurable parallaxes. Thus, Leavitt could not find the absolute magnitude of any of the variable stars she discovered. She knew only their apparent magnitudes. Shapley realized that he could find the distance to the globular clusters if he could find the absolute magnitude of the variable stars in the clusters, so he turned his attention to variable stars.

All stars are moving through space, and over periods of a few years these motions, called **proper motions,** can be detected as small shifts in the positions of the stars in the sky. The more distant stars have undetectably small proper motions, but the nearer stars have larger proper motions. Clearly, proper motions contain clues to distance. Although no Cepheids are close enough to have measurable parallaxes, Shapley found 11 with measured proper motions. Through a statistical process, he was able to find their average distance and thus their average absolute magnitude. That meant he could replace Leavitt's apparent magnitudes with absolute magnitudes on the period–luminosity diagram (see Figure 15-2c). That is, he knew how bright the variable stars really were.

a

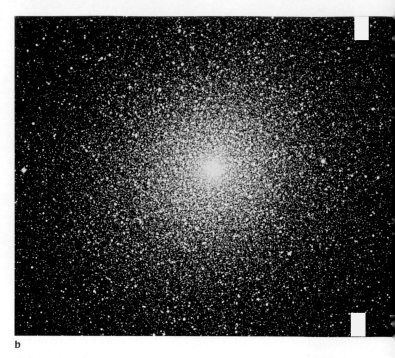

b

Figure 15-3
(a) The "Jewel Box" is an open star cluster containing only a few hundred stars in a region about 8 pc in diameter. *(NOAO)* (b) The object known as 47 Tucanae is a large globular cluster containing over 1 million stars in a region about 50 pc in diameter—slightly larger than the typical globular cluster. *(NOAO)*

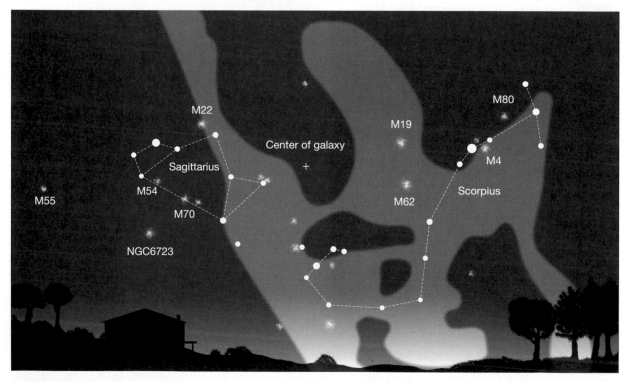

Figure 15-4
The globular clusters are concentrated toward the center of our galaxy. A few of the brighter globular clusters are shown here. The brightest are visible through binoculars on a dark night. The light-blue region shows the extent of the visible Milky Way. Constellations are shown as they appear above the southern horizon on a summer night as seen from latitude 40° N, typical for most of the United States.

Finally, he could find the distance to the globular clusters. He could identify the variable stars he found in the clusters (RR Lyrae stars) and he could find their apparent magnitude from his photographs. The comparison of apparent and absolute magnitude gave him the distance to the star cluster using the process we identified in Chapter 11 as spectroscopic parallax. Notice how Shapley proceeded step by step in his research; astronomers say he **calibrated** the variable stars for distance determination (Window on Science 15-1).

Shapley later wrote that it was late at night when he plotted the direction and distance to the globular clusters and found that, just as he had supposed, they formed a great swarm. But it was not centered on the sun. The center of the swarm lay many thousands of light-years in the direction of Sagittarius. Evidently the center of the star system was in Sagittarius (Figure 15-5). He called the only other person in the building, a cleaning lady, and the two stood looking at his graph as he explained that they were the only two people on Earth who understood that humanity lives, not at the center of a small star system, but in the suburbs of a vast wheel of stars, a galaxy.

Why did astronomers before Shapley think we lived near the center of a small star system? Space is filled with gas and dust that dims our view of distant stars. When we look toward the band of the Milky Way, we can see only the neighborhood near the sun. Most of the star system is invisible and, like travelers in a fog, we seem to be at the center of a small region. Shapley was able to see the globular clusters at greater distances because they lie outside the plane of the star system and are not dimmed very much by the gas and dust.

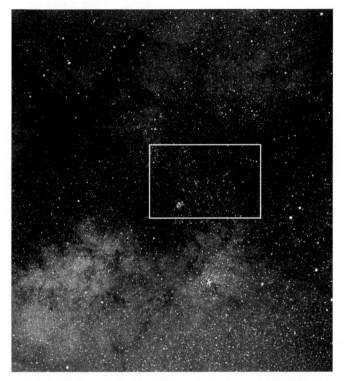

Figure 15-5
A photograph of the Milky Way near Sagittarius shows no evidence that this is the direction toward the center of the galaxy. Gas and dust block our view, and we see only relatively nearby stars and star clusters. Detailed studies show the center of the galaxy lies in the boxed area. *(Daniel Good)*

Building on Shapley's work, other astronomers began to suspect that some of the faint patches of light visible through telescopes were other galaxies like our own. Shapley and astronomer Heber Curtis met in 1920 in a famous debate known as the **Shapley-Curtis debate.**

Curtis claimed the faint objects were other galaxies, and Shapley argued they were not other galaxies but were nebulae in our star system. It isn't clear who won the debate, but in 1923 Edwin Hubble photographed individual stars in the Andromeda Galaxy, and in 1924 he identified Cepheids there. Clearly the faint patches were other star systems like our own.

How big is our galaxy? Modern studies tell us the disk is at least 75,000 ly in diameter, but there is more to our galaxy than just the disk. To get a true impression, we must conduct a careful analysis.

An Analysis of the Galaxy

Our galaxy, like many others, contains two components—a disk and a sphere (Figure 15-6). The differences between these components will give us clues to the formation of our galaxy.

The **disk component** consists of all matter confined to the plane of rotation—that is, everything in the disk itself. This includes stars, open star clusters, and nearly all of the galaxy's gas and dust. Because the disk contains lots of gas and dust, it is the site of most of the star formation in the galaxy, and so the disk is illuminated by the brilliant, blue, massive stars. Thus, the disk of the galaxy tends to be blue.

The size of the disk is uncertain. The thickness is uncertain because the disk does not have sharp boundaries. Stars become less crowded as we move away from the plane. Also, the thickness depends on what kind of object we study—O stars lie within a narrow disk only about 200 ly thick, but sunlike stars are more widely spread. The diameter of the disk is also difficult to determine. Gas and dust block our view in the plane of the galaxy so we cannot see to the center or to the edge. The best studies place the sun about 8.5 kpc from the center, where 1 kpc is a **kiloparsec,** or 1000 pc. We seem to be about two-thirds of the way from the center to the edge, so the diameter of our galaxy must be about 25 kpc. This is often given as 75,000 ly, which is a bit smaller than 25 kpc, but this illustrates the uncertainty in this number. We don't know it to better than 10 percent. This is the diameter of the luminous part of our galaxy, the part we would see from a distance. We will see later strong evidence that our galaxy is much larger than this but that the outer parts are not luminous.

One way to explore our galaxy is to use radio telescopes to map the distribution of un-ionized hydrogen. Fortunately for astronomers, neutral hydrogen in space is capable of radiating photons with a wavelength of 21 cm. This happens because the spinning proton and electron act like tiny magnets (Figure 15-7). When the spinning electron in the hydrogen atom reverses its spin, the magnetic fields reverse, and the atom is left

Figure 15-6
An artist's conception of our Milky Way Galaxy seen face-on and edge-on shows the shape and location of the disk, halo, and nuclear bulge. Note the position of the sun and the distribution of globular clusters.

with a tiny amount of excess energy, which it radiates as a photon with a wavelength of 21 cm. Radio astronomers can map our entire galaxy at 21 cm because these long-wavelength photons are unaffected by the microscopic dust grains that scatter photons of light. Thus, radio telescopes can "see" our entire galaxy while optical telescopes cannot.

Observations made at other wavelengths can also let us see through the dust clouds. Infrared photons have wavelengths long enough to be unaffected by the dust. Thus, a map of the sky at long infrared wavelengths reveals the disk of our galaxy (Figure 15-8).

The most striking features of the disk component are the **spiral arms**—long spiral patterns of bright stars, star clusters, gas, and dust. Such spiral arms are easily visible in other galaxies, and we will see later that our own galaxy contains a spiral pattern.

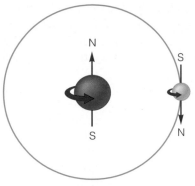

Same spins
Magnetic fields reversed

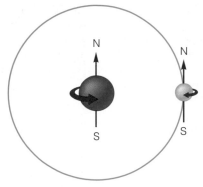

Opposite spins
Magnetic fields the same

Figure 15-7
Neutral hydrogen in space can emit radio energy with a wavelength of 21 cm. Both the proton and electron in a hydrogen atom spin and consequently have small magnetic fields. When they spin in the same direction, the magnetic fields are reversed; and when they spin in the opposite direction, the magnetic fields are aligned. If the spin of the electron flips from the higher to the lower energy state, the atom can radiate the excess energy as a photon of radio energy. Radio astronomers can detect this 21-cm radiation and map the galaxy.

Figure 15-8
Just as the inset map projects Earth's entire surface onto a flat oval, this infrared image from the IRAS satellite projects the entire sky onto a flat map. The plane of our galaxy runs horizontally from left to right with the center of the galaxy at the center of the oval. The red glow in this false-color image is produced by dust in the disk of the galaxy warmed by starlight. The thinness of the distribution of dust is evident in this image. (Warmer dust in our solar system is located along the ecliptic and produces the faint blue S-shaped curve.) *(NASA)*

The second component of our galaxy is the **spherical component,** which includes all matter in our galaxy scattered in a roughly spherical distribution around the center. This includes a large halo and the nuclear bulge.

The **halo** is a spherical cloud of thinly scattered stars and globular clusters. It contains only about 2 percent as many stars as the disk of the galaxy and contains very little gas and dust. Thus, no new stars are forming in the halo. In fact, the vast majority of the halo stars are old, cool, lower-main-sequence stars and giants much like the stars in globular clusters. Although we can map the halo of our galaxy by studying these stars, the halos of other galaxies are too faint to detect.

The **nuclear bulge** is the dense cloud of stars that surrounds the center of our galaxy. It has a radius of about 3 kpc and is slightly flattened. We cannot observe it easily because thick dust in the disk blocks our view, but observations at other wavelengths can penetrate the dust. The bulge seems to contain little gas and dust, and thus little star formation. Most of the stars are old, cool stars like those in the halo.

The vast numbers of stars in the disk, halo, and nuclear bulge lead us to raise a basic question. How massive is the galaxy?

The Mass of the Galaxy

When we needed to find the masses of stars, we studied the orbital motions of the pairs of stars in binary systems. To find the mass of the galaxy, we must look at the orbital motions of the stars within the galaxy. Every star in our galaxy follows an orbit around the center of mass of the galaxy, and thus we say the galaxy rotates. That rotation can reveal the total mass of the galaxy. Consequently, our discussion of the mass of the galaxy is also a discussion of the rotation of the galaxy and the orbits of the stars within the galaxy.

We can find the orbits of stars by finding how they move. Of course, we could use the Doppler effect to find a star's radial velocity (see By the Numbers 4-2). In addition, if we could measure the distance to and proper motion of a star, we could find the velocity of the star perpendicular to the radial direction. Combining all of this information, we could then find the shape of the star's orbit.

The orbital motions of the stars are strikingly different in the halo and in the disk (Figure 15-9). In the halo, each star and globular cluster follows its own highly inclined elliptical orbit. These orbits carry the stars and clusters far out into the spherical halo, where they move slowly, but when they fall back into the inner part of the galaxy, their velocities increase. Thus, motions in the halo do not resemble a general rotation but are more like the motions of a swarm of bees. In contrast, the stars in the disk of the galaxy move in the same direction in nearly circular orbits that lie in the plane of the galaxy. The sun is a disk star and follows a nearly circular orbit around the galaxy that never carries it out of the disk.

We can use the orbital motion of the sun to find the mass of the galaxy. By observing the radial velocity of other galaxies in various directions around the sky, astronomers can tell that the sun moves about 220 km/s in the direction of Cygnus, carrying Earth and the other planets of our solar system along with it. Because its orbit is a circle with a radius of 8.5 kpc, we can divide the circumference of the orbit by the velocity and find that the sun completes a single orbit in about 240 million years. Obviously, we notice no change in a time as short as a human lifetime.

If we think of the sun and the center of mass of our galaxy as two objects orbiting each other, we can find the mass of the galaxy (see By the Numbers 11-4). We need to divide the separation in AU cubed by the period in years squared to find the mass in solar masses. The radius of the sun's orbit is about 8.5 kpc, and the orbital period is 240 million years. Converting kiloparsecs to AU and doing the arithmetic tells us the galaxy must have a mass of about 100 billion solar masses.

This estimate is uncertain for a number of reasons. First, we don't know the radius of the sun's orbit with great certainty. We estimate the radius as 8.5 kpc, but we could be wrong by 10 percent or more, and this radius gets cubed in the calculation, so it makes a big difference. Second, this estimate of the mass includes only the mass inside the sun's orbit. Mass spread uniformly outside the sun's orbit will not affect its orbital motion. Thus 100 billion solar masses is a lower limit for the mass of the galaxy, but we don't know how much to increase our estimate to include the rest of the galaxy.

This estimate is also uncertain because it depends on our knowledge of the rotation of the galaxy, and that rotation is very complex. The motion of the stars near the sun shows that the disk does not rotate as a solid body. Each star follows its own orbit, and stars in some regions have shorter or longer orbital periods than the sun. This is called **differential rotation** (Figure 15-10) and is different from the rotation of a solid body—a record on a turntable, for instance. Whereas three spots lined up on a record would stay together as the record turned, three stars lined up in the galaxy will draw apart because some have shorter orbital periods.

A graph of the orbital velocity of stars at various orbital radii in the galaxy is called a **rotation curve** (Figure 15-11). If all of the mass in the galaxy were concentrated at its center, then orbital velocity would be high near the center and would decline as we move outward. This kind of motion has been called *Keplerian motion* because it follows Kepler's third law, as in the

a

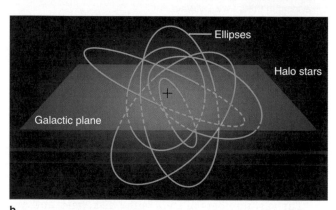

b

Figure 15-9

(a) Stars in the galactic disk have nearly circular orbits that lie in the plane of the galaxy. (b) Stars in the halo have randomly oriented, elliptical orbits.

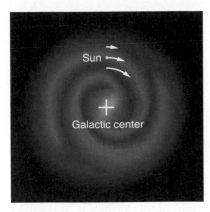

Figure 15-10
The differential rotation of the galaxy means that stars at different distances from the center have different orbital periods. In this example, the star just inside the sun's orbit has a shorter period and gains on the sun, while the star outside falls behind.

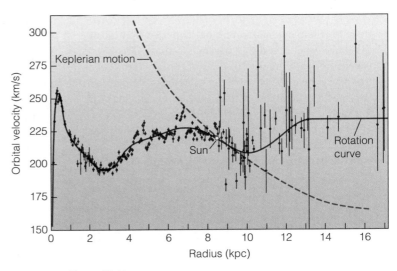

Figure 15-11
Radio observations of orbital velocities are plotted here with vertical error bars showing the uncertainty in each measurement. Measurements outside the sun's orbit are more difficult and thus more uncertain. The fitted rotation curve (solid line) does not decline outside the sun's orbit as we would expect for Keplerian motion (dashed line). Rather the curve is flat at great distance and this suggests the galaxy contains significant mass outside the orbit of the sun. *(Adapted from a diagram by Françoise Combes)*

case of our solar system, where nearly all of the mass is in the sun. Of course, the galaxy's mass is not all concentrated at its center, but if most of the mass is inside the orbit of the sun, then we would expect to see orbital velocities decline at greater distances. Many observations confirm, however, that velocities actually increase at greater distance; and this fact tells us that these larger orbits are enclosing more mass. In other words, significant amounts of matter in our galaxy lie in its outer parts. Although it is difficult to determine a precise edge to the visible galaxy, it seems clear that large amounts of matter are located beyond the traditional edge of the galaxy—the edge we would see if we journeyed into space and looked back at our galaxy. Thus, this matter must not be producing light, and astronomers refer to it as **dark matter.**

Some of this dark matter is evidently located in the disk of our galaxy, but the shape of the rotation curve and theoretical models of the rotation of the disk suggest that much of the dark matter is located in an extended **dark halo,** also called the **galactic corona.** Estimates suggest this dark halo extends as far as seven times the visible radius of our galaxy and contains 10 to 100 times the mass of the visible part of the Milky Way Galaxy.

Astronomers debate the nature of this dark matter. It cannot be normal stars, because they would produce light and we would see them. We can also eliminate neutron stars and black holes, because we would be able to detect the X rays such objects emit. Remember the amount of dark matter is 10 to 100 times the amount of visible matter, so such a large mass of normal matter should be detectable. Many astronomers suspect that the dark matter is not normal matter but is composed of some exotic atomic particle whose properties are not understood.

We will return to the mysterious dark matter in the following chapters when we discuss other galaxies and the universe as a whole. For now it is sufficient to consider the implications of this dark matter for understanding our own Milky Way Galaxy. Simple observations of the motion of stars in our galaxy and a simple theory of orbital motion lead us to the conclusion that the visible part of our galaxy may make up only a small percentage of a much larger object. This is disturbing; it is as if we looked at a ham sandwich and saw only the mayonnaise. Furthermore, we are led to suspect that there are forms of matter unknown to us at present. Our simple quest to find the mass of our galaxy has opened a door to an exciting puzzle—the dark matter.

REVIEW **Critical Inquiry**

If gas and dust block our view, how do we know how big our galaxy is?

The gas and dust in our galaxy block our view only in the plane of the galaxy. When we look away from the plane of the galaxy, we look out of the gas and dust and can see to great distances. The globular clusters are scattered through the halo with a strong concentration in the direction of Sagittarius. When we look above or below the plane of the galaxy, we can see those clusters, and careful observations reveal variable stars in the clusters.

The disk component and spherical component are quite different from each other. In the next section, we will discover that they also differ in chemical composition; and that clue, combined with our knowledge of stellar evolution, suggests a model for the formation of the galaxy.

15-2 The Origin of the Milky Way

Just as paleontologists reconstruct the history of life on Earth from the fossil record, astronomers try to reconstruct our galaxy's past from the fossil it left behind as it formed and evolved. That fossil is the spherical component of the galaxy. The stars we see in the halo formed when the galaxy was young. By studying those stars—their chemical composition and their distribution in the galaxy—we can try to understand how the galaxy formed.

The Age of the Milky Way

Because we know how to find the age of star clusters, we can estimate the age of our galaxy. Nevertheless, uncertainties make our answer hard to interpret.

The oldest open clusters are about 7 billion years old. These ages come from the turnoff point in the H-R diagram (see Chapter 13), but finding the age of an old cluster is difficult because old stars change so slowly. Also, the exact location of the turnoff point depends on chemical composition, which differs among clusters. Finally, open clusters are not strongly bound by their gravity, so even older open clusters may have dissipated as their stars wandered away. Thus, open clusters tell us the galactic disk is at least 7 billion years old.

Globular clusters have faint turnoff points in their H-R diagrams and are clearly old, but finding these ages is difficult. Clusters differ slightly in chemical composition, which must be accounted for in calculating the stellar models from which ages are determined. Also, to find the age of a cluster, astronomers must know the distance to the cluster. Precise parallaxes from the Hipparcos satellite (Chapter 11) have allowed astronomers to more precisely calibrate the Cepheid stars, and careful studies with the newest large telescopes have refined the H-R diagrams. Globular cluster ages seem to range from 11 to 13 billion years. Some of this is the uncertainty in the measurement of ages, but some clusters do seem to be older than others. Nevertheless, these data tell us the halo of our galaxy is at least 11 billion years old.

Both populations and clusters tell us the disk is younger than the halo. We can combine these ages with the origin of the elements to tell the story of our galaxy.

Stellar Populations

In the 1940s, astronomers realized that there were two types of stars in the galaxy. The type they were accustomed to studying was that located in the disk, like those stars near the sun. These they called **population I** stars. The second type, called **population II** stars, are usually found in the halo, in globular clusters, or in the central bulge. In other words, the two stellar populations are associated with the two components of the galaxy.

The stars of the two populations are very similar. They burn nuclear fuels and evolve in nearly identical ways. They differ only in their abundance of atoms heavier than helium, atoms that astronomers refer to collectively as **metals.** (Note that this is not the way the word *metal* is commonly used by nonastronomers.) Population I stars are metal-rich, containing 2 to 3 percent metals, while population II stars are metal-poor, containing only about 0.1 percent metals or less. The metal content of the star defines its population.

Population I stars belong to the disk component of the galaxy and are sometimes called disk population stars. They have circular orbits in the plane of the galaxy and are relatively young stars that formed within the last few billion years. The sun is a population I star.

Population II stars belong to the spherical component of the galaxy and are sometimes called the halo population stars. These stars have randomly tipped, elliptical orbits and are old stars. The metal-poor globular clusters are part of the halo population.

To see how important this difference is, refer back to Figure 15-2c. We can now understand the two kinds of Cepheid variables shown in the period–luminosity diagram. Type I Cepheids are population I stars with metal abundance similar to that of the sun. The Type II Cepheids are population II stars. The difference in metal abundance affects the ease with which radiation

One of the most common organizing themes in science is the concept of a process—a sequence of events that leads to some result or condition. Scientists try to discover the natural processes that shape the universe. Telling the stories of these processes explains why the universe is the way it is. At the same time, recognizing these processes and learning to tell their stories makes science easier to understand.

For example, astronomers try to assemble the sequence of events that led to the formation of the chemical elements. We know that the first stars formed from nearly pure hydrogen and helium and that they made certain heavier elements

in their cores. Supernova explosions made other elements and mixed the elements back into the interstellar medium. New stars formed from this gas, and generation after generation of stars added to the heavier elements that we are made of. Telling that story explains much of the science of the stars.

As you study any science, be alert for processes as organizing themes. Science, at first glance, seems to be nothing but facts, but one of the ways to organize the facts is to fit them into the story of a process. If you study biology and learn the process by which plant cells know whether to become roots, stems, or leaves, you will have organized a great deal of

factual information into a story that not only is easy to remember but also makes sense.

When you see a process in science, ask yourself a few basic questions. What conditions prevailed at the beginning of the process? What sequence of steps occurred? Can some steps occur simultaneously, or must one step occur before another can occur? What is the final state that this process produces? Most important, what evidence do we have that nature really works this way?

Not everything in science is a process, but identifying a process will help you remember and understand a great many details and principles.

flows through the gas, and that changes the balance between gravity and energy flowing outward through the star. A difference in metal abundance may seem like a small detail, but it makes a big difference to a star.

Since the discovery of stellar populations, astronomers have realized that there is a gradation between populations (Table 15-1). Extreme population I stars, like the stars in Orion, are found only in the spiral arms. Slightly less metal-rich population I stars, called intermediate population I stars, are located throughout the disk. The sun is such a star. Stars even less metal-rich, such as stars in the nuclear bulge, belong to the intermediate population II. The most metal-poor stars are those in the halo and in globular clusters. These are extreme population II stars.

Why do the disk and halo stars have different metal abundances? Those chemical differences tell astrono-

mers that the stars have different ages and thus must have formed at different stages in the life of the galaxy. To use this clue to tell the story of our galaxy, we must discuss the cycle of element building.

The Element-Building Cycle

The atoms of which we are made were created in a process that spanned a number of generations of stars. Natural processes are all around us, and we must learn to recognize and understand them if we are to understand nature (Window on Science 15-2). The process that built the chemical elements may be one of the most important processes in the history of our galaxy.

We saw in Chapter 13 how elements heavier than helium but lighter than iron are built up by nuclear reactions inside evolving stars. Atoms more massive than

Table 15-1 Stellar Populations

| | Population I | | Population II | |
	Extreme	Intermediate	Intermediate	Extreme
Location	Spiral arms	Disk	Nuclear bulge	Halo
Metals (%)	3	1.6	0.8	Less than 0.8
Shape of orbit	Circular	Slightly elliptical	Moderately elliptical	Highly elliptical
Average age (yr)	100 million and younger	0.2–10 billion	2–10 billion	10–14 billion

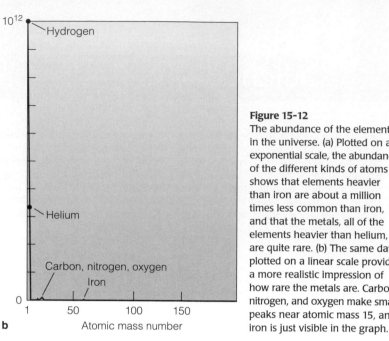

Figure 15-12
The abundance of the elements in the universe. (a) Plotted on an exponential scale, the abundance of the different kinds of atoms shows that elements heavier than iron are about a million times less common than iron, and that the metals, all of the elements heavier than helium, are quite rare. (b) The same data plotted on a linear scale provide a more realistic impression of how rare the metals are. Carbon, nitrogen, and oxygen make small peaks near atomic mass 15, and iron is just visible in the graph.

iron are made by short-lived nuclear reactions that occur during a supernova explosion. This explains why lower-mass atoms like carbon, nitrogen, and oxygen are so common and why atoms more massive than iron—such as gold, silver, platinum, and uranium—are so rare. Figure 15-12a shows the abundance of the chemical elements, but the exponential scale makes the heavy atoms seem unrealistically abundant. Figure 15-12b has a linear scale, which shows how rare the heavy atoms really are.

Most of the matter in the stars is hydrogen and helium, and the metals (including carbon, nitrogen, oxygen, and so on) were cooked up inside stars. When the galaxy first formed, there should have been no metals because stars had not yet manufactured any. The gas from which the galaxy condensed must have been almost pure hydrogen (80 percent) and helium (20 percent). (Where the hydrogen and helium came from is a mystery we will save until Chapter 18.)

The first stars to form from this gas were metal-poor, and now, 10 to 15 billion years later, their spectra show few metal lines. Of course, they may have manu-

factured some atoms heavier than helium, but because the stars' interiors are not mixed, those heavy atoms stay trapped at the centers of the stars where they were produced and do not affect the spectrum (Figure 15-13). The population II stars that we see today in the halo are the survivors of an earlier generation of stars to form in the galaxy.

Most of the first stars evolved and died, and the most massive became supernovae and enriched the interstellar gas with metals. Succeeding generations of stars formed from gas clouds that were more enriched, and each generation added to the enrichment through the supernovae of massive stars. By the time the sun formed, 5 billion years ago, the element-building process had added about 1.6 percent metals. Since then, the metal abundance has increased further, and stars forming now, in the Orion Nebula for example, incorporate 2 to 3 percent metals and represent the metal-rich population I stars. Thus metal abundance varies between populations because of the production of heavy atoms in successive generations of stars.

Figure 15-13
Spectra of population II and population I stars of similar spectral type show hydrogen lines of equal strength, but the lines of heavier atoms are conspicuously weak in the population II star's spectrum. Labeled lines are calcium (Ca) and iron (Fe). *(UCO/Lick Observatory photograph)*

The lack of metals in the spherical component tells us it is very old, a fossil left behind by the galaxy when it was young and drastically different from its present disk shape. Thus, the study of element building and stellar populations leads us to the fundamental question, "How did our galaxy form?"

The Formation of the Galaxy

By 1962, astronomers knew enough about stellar populations, stellar evolution, and the origin of the elements to propose a theory of the formation of the Milky Way Galaxy. At first, that theory seemed adequate, but starting in the 1980s, new observations produced new questions, and astronomers are now reexamining the theory.

The traditional theory, what we might call the top-down theory, proposes that our galaxy formed from a single cloud of gas contracting under its own gravitation. Because the cloud had never formed stars before, it contained hydrogen and helium and very few metals. Thus, the first stars to form were metal-poor. Because the cloud was spherical and turbulent, these metal-poor stars and star clusters took up elliptical, randomly tipped orbits. Today, we see this spherical cloud of stars and the remains of the largest star clusters as the halo stars and the globular clusters. Thus, the halo is a fossil of the first stage in the formation of the Milky Way Galaxy.

According to this top-down theory, the spherical cloud contracted, randomly moving eddies of gas collided, turbulent motions canceled out, and the cloud was left gently rotating. A low-density cloud of gas rotating uniformly around an axis cannot maintain a spherical shape because, unlike a star, it has a very low gas pressure. Gravity pulls the material inward, and the rotation of the cloud can resist the contraction only in the direction perpendicular to the axis of rotation. Like a blob of pizza dough spun into the air, the cloud must flatten into a rotating disk (Figure 15-14).

As the disk of the galaxy formed, star formation continued, and the metal abundance in the gas increased as the most massive stars evolved and died in supernova explosions. In addition, the gas in the disk contained little turbulence, so the stars forming in the disk tended to have circular orbits in the plane of the disk. Thus the disk of the galaxy contains metal-rich stars in circular orbits and nearly all of the gas and dust.

This top-down theory accounts for many of the Milky Way's properties, but advances in technology have improved astronomical observation and computer modeling; and, beginning in the 1980s, contradictions arose. For example, many halo stars have metal abundances like globular clusters and may have escaped from such clusters, but some halo stars are extremely metal-poor. One star contains 30,000 times less metal

Figure 15-14
According to the traditional theory, the galaxy began as a spherical cloud of gas (shaded) in which stars and star clusters (dots) formed. As the rotating gas cloud collapsed into a disk, the halo stars were left behind as a fossil of the early galaxy.

than the sun. These stars may be older than the globular clusters. Also, some of the oldest stars in our galaxy have been found in the central bulge, not in the halo. In other studies, astronomers have carefully measured the ages of many globular clusters, and they find that the oldest are about 4 billion years older than the youngest, a difference that seems real. Furthermore, the younger globular clusters seem to be the ones in the outer halo. According to the top-down theory, the halo should be the oldest part of the galaxy, and all halo stars and clusters should be about the same age.

An alternative theory, which we might call the bottom-up theory, proposes that the bulge of our galaxy and an early form of the disk took shape first from a smaller cloud of gas. The halo formed later as many small clouds of gas fell in over billions of years. This would explain the span of ages in halo stars and globular clusters, and the collapse of this gas into the disk would generate star formation and make the disk look as young as it does today.

In fact, the ages of the halo and disk lie at the heart of the problem. The ages of the globular clusters are very uncertain, as we saw earlier in this chapter, but the youngest globular clusters seem to be billions of years older than the disk. The oldest open clusters are about 7 billion years old. Also, white dwarfs tend to accumulate over the eons, and there aren't as many white dwarfs in the disk as there should be if the disk were as old as the younger globular clusters. Why this gap in ages? A successful theory should account for these ages.

Another problem is the metals in the oldest stars. These stars are metal-poor, but they are not metal-free. There must have been at least a few massive stars living before the formation of the oldest stars we see to create some metals. Otherwise the oldest stars would contain no metals at all.

One solution is a hybrid theory, which combines the top-down and bottom-up theories in a new way. This theory suggests that the galaxy began with the formation of a central bulge and the later accumulation of the halo from gas that had been slightly enriched in metals by a very early generation of a few massive stars. But it adds the possibility that entire galaxies were captured by the growing Milky Way Galaxy. We will see in the next chapter that such mergers do occur. If the Milky Way Galaxy absorbed a few small but partially evolved galaxies, then some of the globular clusters we see in the halo may be hitchhikers picked up during these mergers. The ages of the hitchhikers would not be characteristic of the Milky Way Galaxy as a whole.

The problem of the formation of our galaxy is frustrating because the theories are incomplete. We prefer certainty, but astronomers are still gathering observations and testing hypotheses. We see an older theory that has proven to be inadequate to explain the observations, and we see astronomers attempting to refine the observations and devise new theories. The metal abundances and ages of the stars in our galaxy seem to be important clues, but metal abundance and age do not tell the whole

story. Astronomer Bernard Pagel was thinking of this when he said, "Cats and dogs may have the same age and metallicity, but they are still cats and dogs."

REVIEW **Critical Inquiry**

Why do metal-poor stars have the most elliptical orbits?

Of course, the metal abundance of a star cannot affect its orbit, so our analysis must not confuse cause and effect with the relationship between these two factors. Both chemical composition and orbital shape depend on a third factor—age. The oldest stars are metal-poor because they formed before there had been many supernova explosions to create and scatter metals into the interstellar medium. Those stars formed long ago when the galaxy was young and motions were not organized into a disk. In the top-down hypothesis, the first stars formed before the collapsing galaxy could form a disk; in the bottom-up hypothesis, the oldest stars date from the assembly of the halo from smaller clouds of gas and stars. In either case, those first stars tended to take up randomly shaped orbits, and many of those orbits are quite elliptical. Thus, today we see the most metal-poor stars following the most elliptical orbits.

Nevertheless, even the oldest stars we can find in our galaxy contain some metals. They are metal-poor, not metal-free. How does that affect our theories?

The origin of our galaxy is an astronomical work in progress, but that is not the only mystery hidden in the Milky Way. The next two sections discuss two special problems that astronomers are trying to understand—the nature of spiral arms and the secret of the galactic nucleus.

15-3 Spiral Arms and Star Formation

The sky is filled with distant galaxies, and many have beautiful spiral arms filled with bright stars and glowing nebulae. In the next chapter we will classify the spiral galaxies and see how our galaxy fits in with those that have a bar-shaped nuclear bulge and loosely wound spiral arms. Here we are interested in the spiral arms themselves. We face two problems. First, how do we know our galaxy has spiral arms when gas and dust in the disk block our view? Second, how can spiral arms survive when our galaxy rotates differentially? The solution to both problems involves star formation.

Tracing the Spiral Arms

Studies of other galaxies show us that spiral arms contain hot, blue stars (Figure 15-15a). Thus, one way to study the spiral arms of our own galaxy is to locate these stars. Fortunately, this is not difficult, since O and B stars are often found in associations and, being very bright, are easy to detect across great distances. Unfortunately, at these great distances their parallax is too small to measure, so their distances must be found by other means, usually by spectroscopic parallax.

O and B associations in the sky are not located randomly but reveal parts of three spiral arms near the sun, which have been named for the prominent constellations through which they pass (Figure 15-15b). If we could penetrate the dust clouds, we could locate other O and B associations and trace the spiral arms farther, but like a traveler in a fog, we see only the region near us.

Objects used to map spiral arms are called **spiral tracers.** O and B associations are good spiral tracers because they are bright and easy to see at great distances. Other tracers include young open clusters, clouds of hydrogen ionized by hot stars (emission nebulae), and certain kinds of variable stars.

Notice that all spiral tracers are young objects. O stars, for example, live for only a few million years. If their orbital velocity is about 250 km/s, they cannot have moved more than about 500 pc since they formed. This is less than the width of a spiral arm. Because they don't live long enough to move away from the spiral arms, they must have formed there.

The youth of spiral tracers gives us an important clue to the nature of the arms. Somehow the arms are associated with star formation. Before we can follow this clue, however, we must extend our map of spiral arms to show the entire galaxy.

Radio Maps of Spiral Arms

The dust clouds that block our view at visual wavelengths are transparent at radio wavelengths because radio waves are much longer than the diameter of the dust particles. When we point a radio telescope at a section of

a

b

Figure 15-15

(a) Images of other galaxies, such as NGC1232 shown here, show the spiral arms illuminated by clouds of young stars. The hot O and B stars in these clouds give the arms a blue tint. *(European Southern Observatory)* (b) Segments of spiral arms near the sun appear when we plot the location of O and B associations and the ionized nebulae that often surround such associations. These groups of massive stars have very short lives, so we find them only in regions of recent star formation. Dust clouds block our view, and we can see only a few kiloparsecs in the plane of the Milky Way, but in the region we can see near the sun, O and B associations trace out segments of three spiral arms as diagrammed here.

the Milky Way, we receive 21-cm radio signals coming from cool hydrogen in a number of spiral arms at various distances across the galaxy. Fortunately, the signals can be unscrambled by measuring the Doppler shifts of the 21-cm radiation, except in the direction toward the nucleus. There, the orbital motions of gas clouds are perpendicular to the line of sight, and all of the radial velocities are zero. Thus, the radio map shown in Figure 15-16 reveals spiral arms throughout the disk of the galaxy, except in the wedge-shaped region toward the center.

Radio astronomers can also use the strong radio emission from carbon monoxide (CO) to map the location of giant molecular clouds in the plane of the galaxy. Recall from Chapter 12 that these clouds are sites of active star formation. Maps constructed from such observations reveal that the giant molecular clouds are located in large structures that resemble segments of spiral arms (Figure 15-17).

Radio maps combined with optical data reveal the spiral pattern of our galaxy. The segments we see near the sun are part of a spiral pattern that continues throughout the disk. But the spiral arms are rather irregular and are interrupted by bends, spurs, and gaps. The stars we see in Orion, for example, appear to be a detached segment of a spiral arm. There are significant sources of error in the radio mapping method, but many of the irregularities along the arms seem real, and photographs of nearby spiral galaxies show similar features. One study compared all of the available data on our galaxy's spiral pattern with the patterns seen in other galaxies and constructed a model of what our galaxy probably looks like from a distance (Figure 15-18).

Figure 15-16
A map of our galaxy based on 21-cm observations. The sun is located near the top at center, and the center of the galaxy is located in the center of the map. The conical region around and beyond the center of the galaxy is the region where 21-cm radiation from gas clouds at different distances cannot be unscrambled. Such maps show clear evidence that spiral arms exist throughout our galaxy, but the pattern is neither simple nor clear. *(Adapted from radio map by Gart Westerhout)*

Figure 15-17
Mapping spiral arms using giant molecular clouds. (a) At the wavelength emitted by the CO molecule we find many overlapping giant molecular clouds along the Milky Way. By finding the distance to each cloud from its radial velocity and the rotation of the galaxy, radio astronomers can use the clouds to map spiral arms. (b) This diagram shows the location of giant molecular clouds in the plane of our galaxy as seen from a location 2 kpc directly above the sun. Now we can see that the molecular clouds are located along spiral arms. Angles in this diagram are galactic longitudes measured clockwise from the sun. *(Adapted from a diagram by T. M. Dame, B. G. Elmegreen, R. S. Cohen, and P. Thaddeus)*

a

b

Figure 15-18
(a) An observer a few million light-years away might have this view of our galaxy. Note that the nucleus is elongated and that the sun, marked by a cross, is located on the inside edge of a separate branch of a spiral arm. *(Painting by M. A. Seeds based on a study by G. De Vaucouleurs and W. D. Pence)* (b) Spurs and branches in spiral arms are common. The galaxy M83 appears to have a strong two-armed spiral pattern, but also has spurs and branches. *(Anglo-Australian Telescope Board)*

The most important feature in the radio maps is easy to overlook—spiral arms are regions of higher gas density. Spiral tracers told us that the arms contain young objects, and we suspected active star formation. Now radio maps confirm our suspicion by telling us that the material needed to make stars is abundant in spiral arms.

The Density Wave Theory

Having mapped the spiral pattern, we can ask, What are spiral arms? We can be sure that they are not physically connected structures such as bands of a magnetic field holding the gas in place—a theory that was considered about four decades ago. Like a kite string caught on a spinning hubcap, such arms would be wound up and pulled apart by differential rotation within a few tens of millions of years. Yet spiral arms are common in disk-shaped galaxies and must be reasonably permanent features.

The most popular theory since the 1950s is called the **density wave theory.** It proposes that spiral arms are waves of compression rather like sound waves, which move around the galaxy, triggering star formation. Because these waves move slowly, orbiting gas clouds overtake the spiral arms from behind and create a moving traffic jam within the arms.

We can imagine the density wave as a traffic jam behind a truck moving slowly along a highway. Seen from an airplane overhead, the jam seems a permanent, though slow-moving, feature. But individual cars overtake the truck, move up through the jam, await their opportunity, and pass the truck. So too do clouds of gas overtake the spiral density wave, become compressed in the "traffic jam," and eventually move out in front of the arm, leaving the slower-moving density wave behind.

Mathematical models of this process have been very successful at generating spiral patterns that look like our own and other spiral galaxies. In each case, the density wave takes on a regular two-armed spiral pattern winding outward from the nuclear bulge to the edge of the disk. In addition, the theory predicts that the spiral arms will be stable over a period of a billion years. The differential rotation does not wind them up because they are not physically connected structures.

Of course, star formation will occur where the gas clouds are compressed. Stars pass through the spiral arms unaffected, like bullets passing through a wisp of fog, but large clouds of gas slam into the spiral density wave from behind and are suddenly compressed (Figure 15-19). We saw in Chapter 12 that sudden compression could trigger the formation of stars in a gas cloud. Thus, new star clusters should form along the spiral arms.

The brightest stars, the O and B stars, live such short lives that they never travel far from their birthplace and are found only along the arms. Their presence is what makes the spiral arms glow so brightly. Lower-mass stars, like the sun, live longer and have time to move out of the arms and continue their journey around the galaxy. The sun may have formed in a star cluster about 5 billion years ago when a gas cloud smashed into a spiral arm. Since that time, the sun has escaped from its cluster and made about 20 trips around the galaxy, passing through spiral arms many times.

Figure 15-19
According to the density wave theory, gas clouds overtake the spiral arm from behind and smash into the density wave. The compression triggers the formation of stars. The massive stars (open circles) are so short-lived that they die before they can leave the spiral arm. The less massive stars (dots) emerge from the front of the arm with the remains of the gas cloud.

The density wave theory is very successful in explaining the properties of spiral galaxies, but it has two problems. First, what stimulates the formation of the spiral pattern? Of course, some process must generate the spiral arms in the first place, but the pattern must be restimulated or it would die away over a period of a billion years or so. One possibility is that the galaxy is naturally unstable to certain disturbances, just as a guitar string is unstable to certain vibrations. Any sudden disturbance—the rumble of a passing truck, for example—can set the string vibrating at its natural frequencies. Similarly, minor fluctuations in the galaxy's disk or gravitational interactions with passing galaxies might generate a density wave.

The second problem with the density wave theory involves the spurs and branches in the arms of our own and other galaxies. Computer models of density waves produce regular, two-armed spiral patterns. Some galaxies, called grand-design galaxies, do indeed have symmetric two-armed patterns, but others do not. Other galaxies have a great many short spiral segments, giving them a fluffy appearance. These galaxies have been termed **flocculent,** meaning "woolly." Our galaxy is probably an intermediate galaxy. How can we explain these variations if the density wave theory always produces two-armed, grand-design, spiral arms? Perhaps the solution lies in a process that involves the compressional triggering of star formation.

Star Formation in Spiral Arms

Star formation is a critical process in the creation of the spiral patterns that we see. It makes spiral arms visible and may shape the spiral pattern itself.

The spiral density wave creates spiral arms by the gravitational attraction of the stars and gas flowing through the arms. Even if there were no star formation at all, rotating disk galaxies could form spiral arms. But without star formation to make young, hot, luminous stars, the spiral arms would be difficult to see. It is the star formation that lights up the spiral arms and makes them so prominent. Thus, star formation helps determine what we see when we look at a spiral galaxy.

But star formation can also control the shape of spiral patterns if the birth of stars in a cloud of gas can renew itself and continue making more new stars. Consider a newly formed star cluster with a single massive star. The intense radiation from that hot star can compress nearby parts of the gas cloud and trigger further star formation (Figure 15-20). Also, massive stars evolve so quickly that their lifetimes are only an instant in the history of a galaxy. When they explode as supernovae, the expanding gasses can compress neighboring clouds of gas and trigger more star formation. Examples of such a **self-sustaining star formation** have been found. The

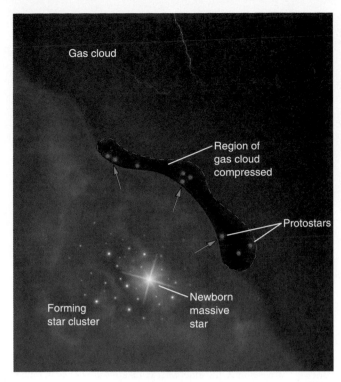

Figure 15-20
The energetic birth of a massive star compesses nearby gas clouds, triggering the gravitational contraction of the gas to form new stars.

Orion complex, consisting of the Great Nebula in Orion and the star formation buried deep in the dark interstellar clouds behind the nebula, is such a region.

Self-sustaining star formation can produce growing clumps of new stars, and the differential rotation of the galaxy can drag the inner edge ahead and let the outer edge lag behind to produce a cloud of star formation shaped like a segment of a spiral arm, a spur. Mathematical models of galaxies filled with such segments of spiral arms do have a spiral appearance, but they lack the bold, two-armed spiral that astronomers refer to as the grand-design spiral pattern (Figure 15-21). Rather, astronomers suspect that self-sustaining star formation can produce the branches and spurs so prominent in flocculent galaxies, but only the spiral density wave can generate the beautiful two-armed spiral patterns.

While the spiral density wave theory has been very successful in explaining spiral arms in galaxies, astronomers still do not understand all of the details. For example, Figure 15-22 shows the spiral pattern in a galaxy extending inward very close to the center of the galaxy. Such examples warn us that we do not yet understand all there is to know about spiral arms.

Our discussion of star formation in spiral structure illustrates the importance of natural processes. The spiral density wave creates graceful arms, but it is the star formation in the arms that makes them stand out so

a

b

Figure 15-21
(a) Some galaxies are dominated by two spiral arms, but even in these galaxies minor spurs and branches are common. The spiral density wave can generate the two-armed grand design pattern, but self-sustained star formation may be responsible for the irregularities.
(b) Many spiral galaxies do not appear to have two dominant spiral arms. Spurs and branches suggest that star formation is proceeding rapidly in such galaxies. *(Anglo-Australian Telescope Board)*

Figure 15-22

This near-infrared image of the center of the Whirlpool Galaxy has been computer-enhanced and given false colors to reveal the spiral pattern. Hidden by dust at visible wavelengths, the innermost spiral arms are visible at longer wavelengths. The image shows the arms extending much closer to the center than current theory predicts. (See Figure 16-16.) *(Courtesy Dennis F. Zaritsky and Hans-Walter Rix)*

prominently. Self-sustaining star formation can act in some galaxies to modify the spiral arms and produce branches and spurs. In some galaxies, it can make the spiral pattern flocculent. By searching out and understanding the details of such natural processes, we can begin to understand the overall structure and evolution of the universe around us.

REVIEW **Critical Inquiry**

Why can't we use solar-type stars as spiral tracers?

Sometimes the timing of events is the critical factor in any analysis. In this case, we must think about the evolution of stars and their orbital periods around the galaxy. Stars like the sun live about 10 billion years, but the sun's orbital period around the galaxy is 240 million years. The sun almost certainly formed when a gas cloud passed through a spiral arm, but since then the sun has circled the galaxy many times and has passed through spiral arms often. Thus, the sun's present location has nothing to do with the location of any spiral arms. An O star, however, lives only a few million years. It is born in a spiral arm and lives out its entire lifetime before it can leave the spiral arm. We find short-lived stars such as O stars only in spiral arms, but G stars are found all around the galaxy.

The spiral arms of our galaxy must make it beautiful in photographs taken from a distance, but we are trapped inside it. How do we know that the spiral arms we can trace out near us actually extend across the disk of our galaxy?

Spiral arms are a puzzle of great beauty, but the center of our galaxy hides a puzzle of great power.

15-4 The Nucleus

The most mysterious region of our galaxy is its very center, the nucleus. At visual wavelengths, this region is totally hidden from us by dust clouds that dim the light by 30 magnitudes (Figure 15-23). If a trillion (10^{12}) photons of light left the center of the galaxy on a journey to Earth, only one would make it through the dust clouds. Thus, visual-wavelength photos tell us nothing about the nucleus. Observations at radio and infrared wavelengths paint a picture of tremendously crowded stars orbiting the center at high velocity. To understand what is happening at the center of our galaxy, we must carefully compare observations and theories.

a

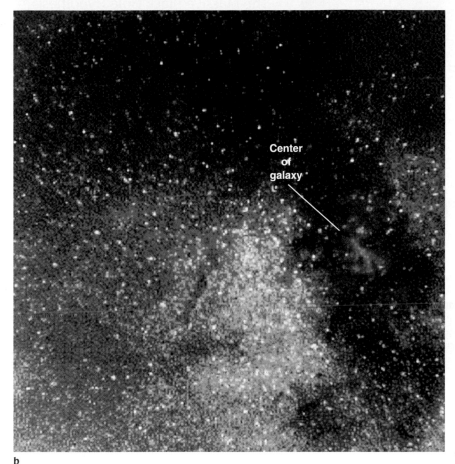

Center
of
galaxy

b

Figure 15-23
(a) The summer Milky Way stretches from horizon to horizon, from Cassiopeia at left to Sagittarius at right. In this visual-wavelength photograph, dark clouds of dust hide the center of the galaxy at extreme right (arrow). *(Steward Observatory)* (b) An infrared photo penetrates the dust clouds and reveals the crowded stars near the galactic center. This image was made by combining images made at the wavelengths 1350 nm, 1680 nm, and 2260 nm and coloring them blue, green, and red. Only at the longest of the three wavelengths can we see the star clouds near the center, and thus they look red in this composite image. *(Courtesy Bradford W. Greeley, Rhodes College)*

Observations

If we observe the Milky Way on a dark night, we might notice a slight thickening in the direction of the constellation Sagittarius, but nothing specifically identifies this as the direction of the heart of the galaxy. Even Shapley's study of globular clusters identified the center only approximately. When radio astronomers turned their telescopes toward Sagittarius, they found a collection of radio sources with one, **Sagittarius A*** (abbreviated Sgr A* and usually pronounced "sadge A-star"), lying at the expected location of the galactic core (Figure 15-24). This single object is smaller than a few astronomical units in diameter but emits five times more radio energy than all the energy radiated by the

sun. Sgr A* is generally thought to be the geometrical center of the galaxy.

Infrared observations show that the stars are tremendously crowded in the central region of the galaxy. Infrared radiation at wavelengths shorter than 2 microns (2000 nm) comes almost entirely from cool stars. Observations at these wavelengths show that the stars are only about 1000 AU apart. Near the sun, stars are about 330,000 AU (1.6 pc) apart. Infrared photons with wavelengths longer than 4 microns (4000 nm) come almost entirely from interstellar dust warmed by stars, and the intensity of this radiation once again confirms that there are large numbers of stars in the central regions.

Radio observations show that the region around Sgr A* is disturbed. A ring of gas clouds dense enough to contain molecules surrounds the center, extending

Figure 15-24

A radio map of the Sagittarius region of the Milky Way (see box in Figure 15-5) reveals an intense radio source, Sagittarius A*, at the expected location of the center. Crosses mark far-infrared sources associated with star formation. Box marks boundaries of Figure 15-25. Angular size of the moon shown for comparison. *(Adapted from observations by W. L. Altenhoff, D. Downes, T. Pauls, and J. Schrami; and by S. F. Odenwalk and G. G. Fazio)*

from a distance of about 25 pc inward to about 6 pc. The inner edge of this ring contains gas ionized by bright ultraviolet radiation from the stars crowded in the central area. Inside this ring of gas lies a cavity in which the gas is 10 to 100 times less dense. Because differences in gas density should smooth out within about 100,000 years, astronomers suspect that a central eruption drove most of the gas outward, forming the cavity, within the last 100,000 years.

The orbital motions of the gas and stars near the center provide further evidence. One infrared study found that only 860 AU from the center, stars orbit with such high velocity that they would complete an orbit in a bit less than 16 years. A simple calculation of the mass inside this orbit using Kepler's third law, a^3 divided by P^2 (By the Numbers 11-4), reveals that the orbit must enclose about two and a half million solar masses. Another infrared study confirms this result. Stars only 275 AU from the center orbit with a period of 2.8 years. Again, Kepler's third law tell us that this small orbit must enclose about two and a half million solar masses.

The center of our galaxy is occupied by an object that is very small but is a powerful source of energy. At the same time, it must have a strong gravitational field to hold rapidly moving stars and gas in a small region. All of these characteristics suggest a supermassive black hole.

The Massive Black Hole Hypothesis

To explain the nature of the center of our galaxy, astronomers must explain how a large mass can occupy a small space, and a massive black hole fills the bill quite well. A black hole of 10 million solar masses, for example, would be only about 20 times larger than the sun—quite small in terms of the galaxy. Also, a massive black hole could produce energy if matter dribbled into it through an accretion disk, and it could produce occasional eruptions if an entire star fell in. Before we can decide, we must examine the hypothesis more carefully, compare it with observations, and consider alternatives.

A massive black hole could supply almost unlimited energy. There is no known limit to the mass of a black hole, and the more massive it is, the more energy it could generate. Also, other energy sources are limited. A supernova, for example, converts very little of its mass into energy before blowing itself apart, and there is a limit to the number of luminous, massive stars capable of producing supernovae that can occupy a small region. For steady energy production from a very small region, a massive black hole works very well.

Such a massive black hole would not be the remains of a single collapsed star but would have developed gradually as gas clouds and stars lost orbital velocity and drifted toward the center of the galaxy. We will see in later chapters that other galaxies may contain massive black holes.

Certain characteristics of Sgr A* are compatible with its being a massive black hole within an accretion disk. Studies of Sgr A* show that it does not seem to be moving at all. While other stars in the central regions rush about their orbits, Sgr A* remains fixed, suggesting that it must have a mass of at least a few hundred solar masses. It can't be a typical compact object of a few solar masses. Radio astronomers using the highest resolution find that Sgr A* is less than 2 AU in diameter and elongated, as we might expect of a massive black hole surrounded by an inclined accretion disk.

Radio maps reveal further structures characteristic of accretion into a black hole. A radio map made at a wavelength of 190 cm has revealed what seems to be a

a

b

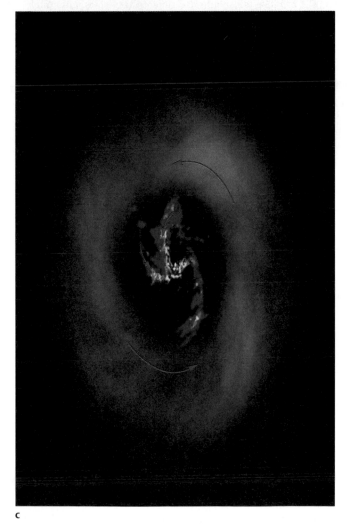

c

Figure 15-25

Sagittarius A*. (a) This VLA map shows filaments of excited gas 50 pc long arching out of the center of our galaxy as if constrained by a magnetic field. Compare with boxed region in Figure 15-24. *(NRAO)* (b) A detailed map of the center, shown roughly by the box in (a), reveals the excited inner edge of a disk of gas and dust with gas flowing into a central object. *(N. Killeen and Kwok-Yung Lo)* (c) The central spiral appears to lie in the open center of the larger disk of neutral gas. It is believed to show matter flowing into a central object.

low-energy jet extending about 30 pc away from the center of the galaxy along the southern axis of rotation. When we discussed SS433, we saw that jets of matter can be ejected from black holes swallowing matter. No such jet is seen extending to the north, but extremely high-resolution radio maps show continuous arcs of hot gas filaments following what appears to be a magnetic field (Figure 15-25a). This suggests to theorists that the core of the galaxy contains a rapidly rotating body of hot gas capable of generating such a magnetic field, perhaps an accretion disk around a black hole.

At the highest resolution, the VLA radio telescope can detect a spiral swirl of matter in the central 3 pc of the galaxy, the hole in the disk of gas and dust that whirls around the center. This spiral is not related to the spiral arms but appears to be matter spiraling inward toward the central object (Figures 15-25b and 15-25c). This inflow appears to add a few tenths of a solar mass per year to the central mass, and accretion at that rate onto a massive black hole could easily provide the observed energy.

A supermassive black hole at the center of our galaxy is an exciting idea, but we must notice the difference between adequacy and necessity. A supermassive black hole is adequate to explain the observed phenomena, but it is not necessary. A few astronomers have thought of other explanations. For example, one idea is that gas flowing into the core forms very massive stars that are then ejected from the core by encounters between stars. Thus, the observational evidence we have does not require a black hole at the center of our galaxy. However, astronomers are finding more and more evidence that other galaxies contain supermassive black holes in their nuclei, so although we can't be sure, most astronomers take the hypothesis very seriously.

REVIEW Critical Inquiry

Why do we think the center of our galaxy contains a large mass?

The best way to measure the mass of an astronomical object is to watch something orbit around it. We can then use Kepler's third law to find the mass inside the orbit. In the case of the nucleus of our galaxy, we can't see individual stars orbiting the center at visual wavelengths, but infrared telescopes can detect these stars and study their orbital motion. For example, matter only 275 AU from the center orbits with such a high velocity that it must complete an entire orbit in only 2.8 years. We can use Kepler's third law to estimate the mass inside this small orbit. Dividing a^3 by P^2, we discover that the mass inside this small orbit must be roughly two and a half million solar masses.

In addition to these orbital motions, the large amount of infrared radiation at wavelengths longer than 4 microns (4000 nm) also implies that vast numbers of stars are crowded into the center. Why?

Despite all the observations and theories, we cannot yet be sure what lies at the center of the Milky Way Galaxy. One way to extend our research is to compare our galaxy with others. We begin that strategy in the next chapter.

distance to the globular clusters and thus discover that the Milky Way Galaxy was much larger than anyone had previously supposed.

The Milky Way Galaxy contains two components, the disk component and the spherical component. Disk stars are metal-rich population I stars moving in circular orbits that lie in the plane of the disk. Because stars that are farther from the center move more slowly in their orbits, the disk rotates differentially.

The spherical component consists of a nuclear bulge at the center and a halo of thinly scattered stars and globular clusters that completely envelops the disk. Halo stars are metal-poor population II stars moving in random, elliptical orbits.

Studies of the motions of stars and star clusters in the outer parts of the galaxy show that it is rotating more rapidly than was expected. This seems to mean that the galaxy is surrounded by a galactic corona extending roughly seven times farther than the traditional edge of the galaxy. Because this mass is nonluminous, astronomers have dubbed it dark matter.

The distribution of populations through the galaxy suggests a way the galaxy could have formed from a spherical cloud of gas that gradually flattened into a disk. The younger the stars, the more metal-rich they are and the more circular and flat their orbits are.

The very youngest objects lie along spiral arms within the disk. They live such short lives they don't have time to move from their place of birth in the spiral arms. Maps of these spiral tracers and cool hydrogen clouds reveal the spiral pattern of our galaxy.

The density wave theory suggests that the spiral arms are regions of compression that move through the disk. When an orbiting gas cloud smashes into the compression wave, the gas cloud forms stars. Another process, self-sustaining star formation, may act to modify the arms as the birth of massive stars triggers the formation of more stars by compressing neighboring clouds.

The nucleus of the galaxy is invisible at visual wavelengths, but radio, infrared, X-ray, and gamma-ray radiation can penetrate the dust clouds. These wavelengths reveal crowded central stars and heated clouds of dust.

The very center of the Milky Way is marked by a radio source, Sagittarius A*, that is also a source of infrared radiation, X rays, and gamma rays. The core must be less than 2 AU in diameter and must contain 2 or 3 million solar masses. Many astronomers believe that this central object is a black hole.

Summary

The Milky Way Galaxy is a star system roughly 75,000 ly in diameter, containing about 200 billion stars. Its true dimensions were not known until early in this century when Harlow Shapley began studying globular clusters. In the clusters, he found a type of variable star that had been studied some years before by Henrietta Leavitt. She had found that the period of variation of these Cepheid variables was related to their luminosity. Shapley used these variable stars to find the

New Terms

variable stars	globular star cluster
instability strip	proper motion
RR Lyrae variable star	calibration
Cepheid	Shapley-Curtis debate
period–luminosity relation	disk component
open star cluster	kiloparsec (kpc)

spiral arm

spherical component

halo

nuclear bulge

differential rotation

rotation curve

dark matter

dark halo

galactic corona

population I

population II

metals

spiral tracer

density wave theory

flocculent galaxy

self-sustaining star
   formation

Sagittarius A*

Review Questions

1. Why isn't it possible to tell from the appearance of the Milky Way that the center of our galaxy is in Sagittarius?

2. Why is there a period–luminosity relation?

3. How can astronomers use variable stars to find distance?

4. Why is it difficult to specify the thickness or diameter of the disk of our galaxy?

5. Why didn't astronomers before Shapley realize how large the galaxy is?

6. How do we know how old our galaxy is?

7. Why do we conclude that metal-poor stars are older than metal-rich stars?

8. How can astronomers find the mass of the galaxy?

9. What evidence do we have that our galaxy has an extended corona of dark matter?

10. How do the orbits of stars around the Milky Way Galaxy help us understand its origin?

11. What evidence contradicts the top-down theory for the origin of our galaxy?

12. Why do spiral tracers have to be short-lived?

13. What evidence do we have that the density wave theory is not fully adequate to explain spiral arms in our galaxy?

14. What evidence do we have that the center of our galaxy is a powerful source of energy?

15. Why is the lack of motion of Sgr A* important evidence in our study of the center of our galaxy?

Discussion Questions

1. How would this chapter be different if interstellar matter didn't absorb starlight?

2. Are there any observations you could make with the Hubble Space Telescope that would allow you to better understand the nature of Sgr A*?

Problems

1. Make a scale sketch of our galaxy in cross section. Include the disk, sun, nucleus, halo, and some globular clusters. Try to draw the globular clusters to scale.

2. Because of dust, we can see only about 5 kpc into the disk of the galaxy. What percentage of the galactic disk can we see? (*Hint:* Consider the area of the entire disk and the area we can see.)

3. If the fastest passenger aircraft can fly 0.45 km/s (1000 mph), how long would it take to reach the sun? the galactic center? (*Hint:* 1 pc = 3×10^{13} km.)

4. If a typical halo star has an orbital velocity of 250 km/s, how long does it take to pass through the disk of the galaxy? Assume that the disk is 1000 pc thick.

5. If the RR Lyrae stars in a globular cluster have apparent magnitudes of 14, how far away is the cluster? (*Hint:* See Figure 15-2c.)

6. If interstellar dust makes an RR Lyrae variable star look 1 magnitude fainter than it should, by how much will we overestimate its distance? (*Hint:* Use the magnitude–distance formula or Table 11-1.)

7. If a globular cluster is 10 minutes of arc in diameter and 8.5 kpc away, what is its diameter? (*Hint:* Use the small-angle formula.)

8. If we assume that a globular cluster 4 minutes of arc in diameter is actually 25 pc in diameter, how far away is it? (*Hint:* Use the small-angle formula.)

9. If the sun is 5 billion years old, how many times has it orbited the galaxy?

10. If the true distance to the center of the galaxy is found to be 7 kpc and the orbital velocity of the sun is 220 km/s, what is the minimum mass of the galaxy? (*Hint:* Use Kepler's third law.)

11. What temperature would interstellar dust have to have to radiate most strongly at 100 μm? (*Hints:* 1 μm = 1000 nm. Use Wien's law, Chapter 4.)

12. Infrared radiation from the center of our galaxy with a wavelength of about 2 μm (2×10^{-6} m) comes mainly from cool stars. Use this wavelength as λ_{max}, and find the temperature of the stars.

13. If an object at the center of our galaxy has a linear diameter of 10 AU, what will its angular diameter be as seen from Earth? (*Hint:* Use the small-angle formula, Chapter 1.)

Critical Inquiries for the Web

1. How is our view of other galaxies affected by our position in the Milky Way? Search the Internet or other sources for sky maps that trace the path of the Milky Way through the constellations. Which constellations does the Milky

Way run through? Choose one of these constellations and another constellation that lies far from the band of the Milky Way. Consult the Web for lists of galaxies visible in these two constellations. In which one have more galaxies been found? Why is this so?

2. Massive stars like Eta Carinae (Figure 13-12) can play a significant role in the formation of stars along a spiral arm. Look for information on the Web related to Eta Carinae, which is one of the most carefully studied stars in the heavens. Briefly summarize the observed behavior of this star, and discuss why it is a good candidate for contributing to self-sustaining star formation in our galaxy.

Exploring *The Sky*

1. View the Milky Way Galaxy in a 180° view. Use Time Skip to simulate Earth's rotation. Note the rift (dust) in region of Cygnus and Aquilla. At what time in September is part of the Milky Way overhead? You might want to switch off the stars by clicking the **Stars** button on the **Object Toolbar.**

2. Most, but not all, of the visible matter in our Milky Way Galaxy lies in the plane of the galaxy. Compare the distribution of globular clusters relative to the Milky Way and the distribution of open clusters relative to the Milky Way. Based on your observations, what can you conclude about how globular clusters and open clusters are distributed throughout our galaxy?

How to proceed: Using the zoom feature, set up *The Sky* to view the whole visible sky centered on your screen. Adjust the time and/or date so that part of the Milky Way is overhead. To view the distribution of only globular clusters, click on **View,** then **Filters . . .,** and check only **Milky Way** and **Globular Clusters.** Note that the distribution of globular clusters is easiest to see if the stars are switched off. To view the distribution of open clusters, repeat the previous procedure, but check only **Milky Way** and **Open Clusters.**

Go to the Brooks/Cole Astronomy Resource Center (www. brookscole.com/astronomy) for critical thinking exercises, articles, and additional readings from InfoTrac College Edition, Brooks/Cole's online student library.

CHAPTER SIXTEEN

Galaxies

The ability to theorize is highly personal; it involves art, imagination, logic, and something more.

Edwin Hubble
The Realm of the Nebulae

Guidepost

We can no more understand galaxies by understanding a single example, the Milky Way, than we could understand humanity by understanding a single person. Whereas the last chapter examined a single galaxy, this chapter generalizes to examine the entire range of objects called galaxies.

We take two lessons from this chapter for use in future chapters. First, galaxies interact with each other in complicated ways, and that will become critical in the next chapter when we study peculiar galaxies. The second lesson is that 90 to 99 percent of the matter in the universe is invisible. That astonishing conclusion comes from the study of galaxies, but we will understand its impact better when we get to Chapter 18, where we will see that it determines the structure and fate of the entire universe.

Figure 16-1
Modern photographs of spiral galaxies, such as this image of M100, show bright stars and star clusters clearly. Before World War I, astronomers using the largest telescopes and the best photographic plates could see only faint nebulosity and vague spiral patterns. Not until the 1920s did we have telescopes large enough and photographic plates sensitive enough to reveal that many of the faint nebulae in the sky are other galaxies. *(Anglo-Australian Telescope Board)*

Science fiction heroes flit effortlessly between the stars, but almost none voyage between the galaxies. As we leave our home galaxy, the Milky Way, behind, we voyage out into the vast depths of the universe, space so deep it is unexplored even in fiction, out among the galaxies.

Less than a century ago, astronomers did not understand that there were galaxies. Nineteenth-century telescopes revealed faint nebulae scattered among the stars, and some were spiral. Astronomers argued about the nature of these nebulae, but it was not until the 1920s that astronomers understood that some were other galaxies much like our own, and it was not until recent decades that astronomical telescopes could reveal the tremendous beauty and intricacy of the galaxies (Figure 16-1).

In this chapter, we will try to understand how galaxies form and evolve, and we will discover that the amount of gas and dust in a galaxy is a critical clue. We will also discover that interactions between galaxies can influence their structure and evolution.

Before we can build theories, however, we must gather some basic data concerning galaxies. We must classify the different kinds of galaxies and discover their basic properties—diameter, luminosity, and mass. Once we know what galaxies are like, once we can characterize them by describing their typical properties, we will be ready to theorize about their origin and evolution.

16-1 Three Types of Galaxies

Often a new branch of science begins with a simple classification system. The shapes of animals can tell a biologist how a species evolved, and the shapes of galaxies can help modern astronomers understand how galaxies evolve. Our discussion of galaxy shapes begins in the 1920s with Mount Wilson astronomer Edwin Hubble (namesake of the Hubble Space Telescope), who obtained some of the first good photographs of other galaxies. He classified the galaxies according to their shapes (Window on Science 16-1) and summarized the classification in a diagram known today as the **tuning fork diagram** (Figure 16-2). We will consider each main shape category in detail.

Elliptical Galaxies

Galaxies that appear to be round or elliptical in shape; have almost no visible gas or dust; lack hot, bright stars; and have no spiral pattern are called **elliptical galaxies**

Figure 16-2
Modern astronomers use the tuning fork diagram as an organizing framework for galaxy classification. Elliptical (E), spiral (S), barred spiral (SB), and irregular (Irr) galaxies are placed from left to right in approximate order of increasing gas, dust, and star formation.

(Figure 16-3). The stars are most crowded toward the center. The largest, the giant ellipticals, are a number of times larger than our galaxy, and the smallest, the dwarf ellipticals, are hardly more luminous than a globular cluster.

Elliptical galaxies are classified according to shape. If we measure on a photograph distances a and b, the largest and smallest diameters through the center of an elliptical galaxy, then the equation

$$10(a - b)/a$$

gives an index of its shape. This index, rounded to the nearest integer, ranges from 0, for circular outlines, to 7, for the most elliptical outlines observed. Thus, elliptical

a

b

Figure 16-3
Elliptical galaxies contain little gas and dust and few bright stars. Thus, they are nearly featureless clouds of stars. (a) M87 is a giant elliptical galaxy surrounded by a swarm of over 500 globular clusters. Objects to the lower right are smaller galaxies. *(NOAO)* (b) The Leo I dwarf elliptical is a small, nearby galaxy and is resolved into individual stars. Note the lack of dust, gas, and hot, bright stars. *(Anglo-Australian Telescope Board)*

galaxies are classified into subgroups ranging from E0 to E7.

The true shape of an elliptical galaxy is not evident from its outline. An elliptical galaxy shaped like a rounded football might appear elliptical if we saw it from the side or round if we saw it from one end. From statistical studies of the number of different types of elliptical galaxies, it seems that such galaxies have a range of shapes. Some are truly spherical, but others are more ellipsoidal. None are known more elliptical than E7.

The spectra of elliptical galaxies reveal that they are composed mainly of population II stars. In fact, the best images of giant ellipticals reveal that they are surrounded by a swarm of globular clusters. A few studies have detected small dust and gas clouds where a few new stars are forming, but for the most part, these galaxies contain a single, old generation of stars. Lack of gas and dust means there is little star formation, so elliptical galaxies cannot make new stars.

Spiral Galaxies

Galaxies that contain an obvious disk component are called **spiral galaxies.** To varying degrees, these galaxies contain gas, dust, and hot, bright, population I stars. As in our own galaxy, the gas, dust, and youngest stars, which include the bright O and B stars, are confined to a thin disk. This is especially obvious in those galaxies we see edge-on (Figure 16-4). We also see nuclear bulges in these galaxies and, in the nearest spiral galaxies, halos filled with population II stars and glob-

ular clusters. Thus, spiral galaxies contain both old and young stars.

Among spiral galaxies we identify three distinct types: S0 galaxies, normal spirals, and barred spirals. Unlike our galaxy, the S0 galaxies show no obvious spiral arms, have very little gas and dust, and contain very few hot, bright stars (Figure 16-5). However, they have an obvious disk component with a large nuclear bulge at the center. They appear to be intermediate between elliptical and spiral galaxies.

Normal spiral galaxies can be further subclassified into three groups according to the size of their nuclear bulge and the degree to which their arms are wound up (Figure 16-6). Spirals that have little gas and dust, larger nuclear bulges, and tightly wound arms are classified Sa. Sc galaxies have large clouds of gas and dust, small nuclear bulges, and very loosely wound arms. The Sb galaxies are intermediate between Sa and Sc. Because there is more gas and dust in the Sb and Sc galaxies, we find more young, hot, bright stars along their arms. The Andromeda Galaxy (see Figure 15-1a) and our own Milky Way Galaxy are Sb galaxies.

A minority of spirals have an elongated nuclear bulge with spiral arms springing from the ends of the bar. These **barred spiral galaxies** are classified SBa, SBb, or SBc, according to the same criteria listed for normal spirals (Figure 16-7). The elongated shape of the nuclear bulge is not well understood, but some astronomers working with computer models have succeeded in imi-

Figure 16-4
Spiral galaxies have an obvious disk with a nuclear bulge and, although it is not visible in most images, a halo. Dust, gas, and hot, bright population I stars are confined to a thin disk, which is most prominent in those galaxies we see edge-on. The dust lanes in this galaxy are quite prominent. *(Anglo-Australian Telescope Board)*

Figure 16-5
Although S0 galaxies have a disk, they are not making many new stars. They have no bright O and B stars to illuminate spiral arms. The S0 galaxies are thought of as intermediate between elliptical and spiral galaxies. Even when we see an S0 galaxy edge-on, as in this image, no dust is visible. Compare this galaxy with that in Figure 16-4. *(Rudolph E. Schild)*

a

b

c

Figure 16-6

(a) The galaxy catalogued as NGC3623 is an Sa galaxy with a large nuclear bulge and tightly wound spiral arms. (b) NGC3627 is an Sb galaxy with a smaller nuclear bulge and more open arms. (c) NGC2997 is an Sc galaxy with a small nuclear bulge and loosely wound spiral arms. *(Anglo-Australian Telescope Board)*

tating the rotating bar structure. It appears to occur when an instability develops in the stellar distribution within the rotating galaxy, altering the orbits of the inner stars and generating a stable, elongated nuclear bulge. Thus, except for the peculiar rotation in the nuclear bulge, barred spiral galaxies are quite similar to normal spirals. In fact, some studies suggest that most spiral galaxies, including our own Milky Way Galaxy, are barred.

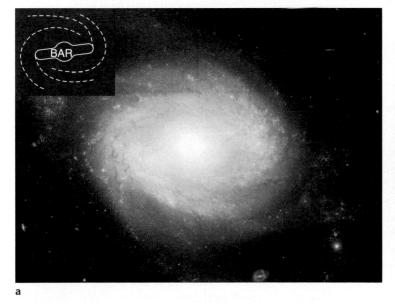

a

b

Figure 16-7

Barred spiral galaxies. (a) The SBb galaxy NGC4639 has a bar-shaped nuclear bulge with spiral arms springing from the ends of the bar (see inset). Notice the bright blue stars and dust lanes in the spiral arms. Notice also the fainter galaxies visible, including a distant barred spiral near the bottom edge of the frame. *(Sandage, Saha, Tammann, Labhardt, Macchetto, Panagia, and NASA)* (b) NGC1365 is an SBc galaxy with a pronounced bar and very open spiral arms (see inset). *(Anglo-Australian Telescope Board)*

a

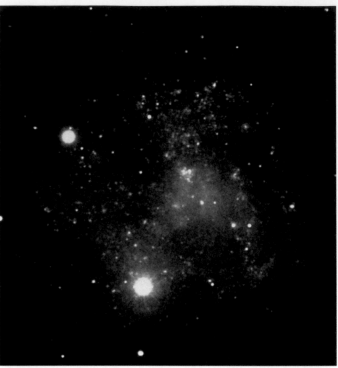

b

Figure 16-8

(a) The Large Magellanic Cloud is an irregular galaxy that is only about 50 kpc away. It and the Small Magellanic Cloud are apparently in orbit around our Milky Way Galaxy. The bright pink emission is from ionized hydrogen gas. Note the Tarantula Nebula at upper left. (See Figure 13-14b) *(Copyright R. J. Dufour, Rice University)* (b) Irregular galaxy IC4182 is typical of the irregular galaxies in that it is small and faint, and it contains both red and blue stars but no obvious spiral arms. *(George Jacoby and Mike Pierce, NOAO, © AURA, Inc.)*

Irregular Galaxies

Certain galaxies that have a chaotic appearance, with large clouds of gas and dust mixed with both young and old stars, are known as **irregular galaxies.** They have no obvious nuclear bulge or spiral arms. The prototypical irregular galaxies are the **Large** and the **Small Magellanic Clouds,** which were first noted by Magellan's men as they sailed through the southern seas in the early 1500s. For observers in the Southern Hemisphere, these nearby galaxies are visible to the naked eye as hazy patches in the sky (Figure 16-8a).

It is difficult to decide how common irregular galaxies are because they are faint (Figure 16-8b). In catalogues of galaxies, about 70 percent are spiral, but that is the result of a selection effect (Window on Science 16-2). Spiral galaxies contain hot, bright stars and are thus very luminous and easy to see. Many ellipticals are faint and harder to notice. Small galaxies, such as irregular galaxies and dwarf ellipticals, are very hard to spot if they are not nearby, so compiling a census of galaxies is difficult. It seems that ellipticals are more common than spirals and that irregulars make up only about 25 percent of all galaxies.

> **REVIEW** Critical Inquiry
>
> **What color are galaxies?**
>
> What we know about stars and populations allows us to analyze this question. Different kinds of galaxies have different colors, depending mostly on how much gas and dust they contain. If a galaxy contains large amounts of gas and dust, it probably contains lots of young stars, and a few of those young stars will be massive, hot, luminous O and B stars. They will give the galaxy a distinct blue tint. In contrast, a galaxy that contains little gas and dust will probably contain few hot stars. Its most luminous stars will be red giants, and they will give the galaxy a red tint. Because the light from a galaxy is a blend of the light from billions of stars, the colors are only tints. Nevertheless, the most luminous stars in a galaxy determine its color. From this we can con-

Many different kinds of science depend on selecting objects to study. Scientists studying insects in the rain forest, for example, must choose which ones to catch. They can't catch every insect they see, so they might decide to catch and study any insect that is red. If they are not careful, a selection effect could bias their data and lead them to incorrect conclusions without their ever knowing it. The reason selection effects are dangerous in science is that they can have powerful influences without being evident. Only by carefully designing a research project can scientists avoid selection effects.

For example, suppose we decide to measure the speed of cars on a highway. There are too many cars to measure every one, so we reduce the workload and measure only the speed of red cars. It is quite possible that this selection criterion will mislead us because people who buy red cars may be more likely to be younger and drive faster. Should we measure only brown cars? No, because we might suspect that only older, more sedate people would buy a brown car. Should we measure the speed of any car in front of a truck? Perhaps we should pick any car following a truck? Again, we may be selecting cars that are traveling a bit faster or a bit slower than normal. Only by very carefully designing our experiment can we be certain that the cars we measure are traveling at representative speeds.

Astronomers face the danger of selection effects quite often. Iron meteorites are easier to find than stone meteorites.

Very luminous stars are easier to see at great distances than faint stars. Spiral galaxies are brighter, bluer, and more noticeable than elliptical galaxies. What we see through a telescope depends on what we notice, and that is powerfully influenced by selection effects.

Scientists engaged in observation must spend considerable time designing their experiments. They must be careful to observe an unbiased sample if they expect to make logical deductions from their results. The scientists in the rain forest, for example, should not catch and study only the red insects. Often, the most brightly colored insects are poisonous. If they are good scientists, they will plan their work with great care and avoid any possible selection effects.

---

clude that elliptical galaxies tend to be red and the disks of spiral galaxies tend to be blue.

Of course, the halo of a spiral galaxy would be both dimmer and redder than the disk. Why?

We end our discussion of the classification of galaxies with the irregulars, and that is convenient because it leads us to two important questions. First, how do we know the properties of galaxies? In the next section, we will try to find the basic properties of galaxies. Second, we must wonder why the irregular galaxies look as if they have been stirred with a big spoon. In the last section of this chapter, we will try to answer that question and see what is happening to the unfortunate Magellanic clouds.

16-2 Measuring the Properties of Galaxies

Unfortunately, the diameter, luminosity, and mass of a galaxy are not obvious in photographs, but to understand galaxies, astronomers must measure these properties. Just as in our study of stellar characteristics (Chapter 11), the first step in our study of galaxies is to find out how far away they are. Once we know a galaxy's distance from us, its size and luminosity are relatively easy to find. Later in this section, we will see that finding mass is more difficult.

Distance

The distances to galaxies are so large that it is not convenient to measure them in light-years, parsecs, or even kiloparsecs. Instead, we will use the unit **megaparsec (Mpc)**, or 1 million pc. One Mpc equals 3.26 million ly, or approximately 2×10^{19} miles.

To find the distance to a galaxy, we must search among its stars for a familiar object whose luminosity or diameter we know. Such objects are called **distance indicators.**

Because their period is related to their luminosity (see Figure 15-2c), Cepheid variable stars are reliable distance indicators. If we know the period of the star's variation, we can use the period–luminosity diagram to learn its absolute magnitude. By comparing its absolute and apparent magnitudes, we can find its distance (see By the Numbers 11-2).

Even with the Hubble Space Telescope, Cepheids are not visible in galaxies much beyond 50 million ly (15 Mpc) (Figure 16-9), so astronomers must search for less common but brighter distance indicators and calibrate them using nearby galaxies containing visible

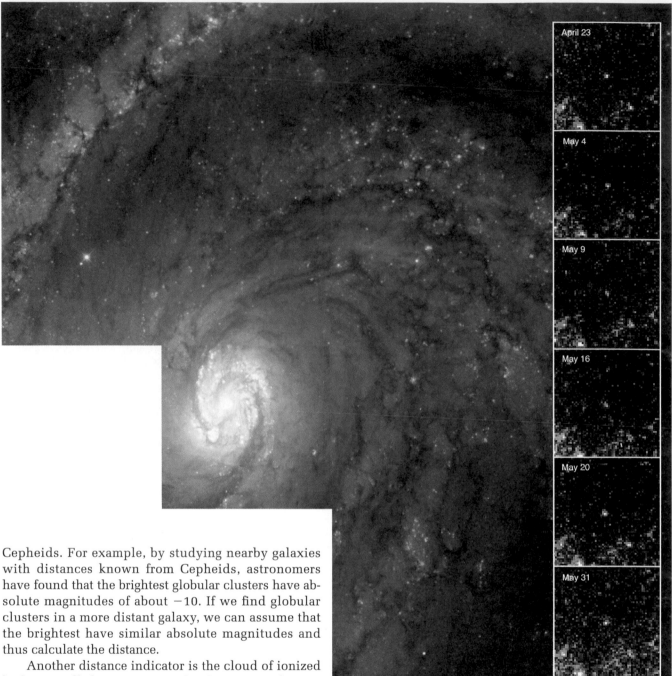

Figure 16-9
The vast majority of spiral galaxies are too distant for Earth-based telescopes to detect Cepheid variable stars. The Hubble Space Telescope, however, can locate Cepheids in some of these galaxies, as it has in the bright spiral galaxy M100. From a series of images taken on different dates, astronomers can locate Cepheids (inset), determine the period of pulsation, and measure the average apparent brightness. They can then deduce the distance to the galaxy—51 million ly for M100. *(J. Trauger, JPL; Wendy Freedman, Observatories of the Carnegie Institution of Washington; and NASA)*

Cepheids. For example, by studying nearby galaxies with distances known from Cepheids, astronomers have found that the brightest globular clusters have absolute magnitudes of about −10. If we find globular clusters in a more distant galaxy, we can assume that the brightest have similar absolute magnitudes and thus calculate the distance.

Another distance indicator is the cloud of ionized hydrogen called an **H II region** that forms around a very hot star. Astronomers have attempted to calibrate these, but they have not been as reliable as Cepheids because some are larger and brighter than others. Also, the size of H II regions seems to depend on the kind of galaxy they occupy, and so astronomers have not been able to calibrate them as accurately as other distance indicators.

Planetary nebulae have proven to be very useful distance indicators, which is surprising considering how faint their central star is. However, the central stars of planetary nebulae seem faint because they are very hot and radiate most of their energy in the ultraviolet. The planetary nebula absorbs this ultraviolet radiation and reradiates it as visible emission lines, and astronomers have been able to calibrate the brightest planetary nebulae by studying nearby galaxies, such as the Andromeda Galaxy. Like brightly colored paper lanterns illuminated from within, planetary nebulae in other galaxies can be identified by their emission at the wavelength of 500.7 nm, the wavelength of an emission line of oxygen; and that identification allows the calculation of the distance to the galaxy.

When a supernova explodes in a distant galaxy, astronomers rush to observe it. Studies show that Type Ia supernovae, those caused by the collapse of a white dwarf, all reach about the same absolute magnitude at maximum. Because these supernovae are much brighter than Cepheids, they can be seen at distances over 1000 Mpc, and thus astronomers can calculate the distances to these galaxies. The drawback to this distance indicator is that supernovae are rare, and none may occur during our lifetime in a galaxy we might be studying.

At the greatest distances, astronomers must calibrate the total luminosity of the galaxies themselves. For example, studies of nearby galaxies show that an average galaxy like our Milky Way Galaxy has a luminosity about 16 billion times the sun's. If we see a similar galaxy far away, we can measure its apparent magnitude and calculate its distance. Of course, it is important to recognize the different types of galaxies, and that is difficult to do at great distances (Figure 16-10). Averaging the brightest galaxies in a cluster can reduce the uncertainty in this method.

Notice how astronomers use calibration (Window on Science 16-1) to build a distance scale reaching from the nearest galaxies to the most distant visible galaxies. Often astronomers refer to this as the distance pyramid or the distance ladder, because each step depends on the steps below it. Of course, the foundation of the distance scale rests on the Cepheid variable stars and our understanding of the luminosities of the stars in the H-R diagram, which ultimately rest on measurements of the parallax of stars within a thousand light-years of Earth.

The most distant visible galaxies are roughly 10 billion ly (3000 Mpc) away, and at such distances we see an effect akin to time travel. When we look at a galaxy millions of light-years away, we do not see it as it is now, but as it was millions of years ago when its light began the journey toward Earth. Thus, when we look at a distant galaxy, we look back into the past by

Figure 16-10
Only the general characteristics of the galaxies are recognizable in this cluster of galaxies 4 to 6 billion light-years away. The larger galaxies in this cluster are about the size of our Milky Way Galaxy. The diagonal line at lower right was produced by an artificial Earth satellite. *(K. Ratnatunga, R. Griffiths, Carnegie Mellon University, and NASA)*

an amount called the **look-back time,** a time in years equal to the distance to the galaxy in light-years.

The look-back time to nearby objects is usually not significant. The look-back time to the moon is 1.3 seconds, to the sun only 8 minutes, and to the nearest star about 4 years. The Andromeda Galaxy has a look-back time of about 2 million years, a mere eye blink in the lifetime of a galaxy. But when we look at more distant galaxies, the look-back time becomes an appreciable part of the age of the universe. We will see evidence in Chapter 18 that the universe began no more than 15 billion years ago. Thus, when we look at the most distant visible galaxies, we are looking back 10 billion years to a time when the universe may have been significantly different. This effect will be important in our discussions in this and the next chapter.

The Hubble Law

Although astronomers find it difficult to measure the distance to a galaxy, they often estimate such distances using a simple relationship.

Early in this century, astronomers noticed that the spectral lines in galaxy spectra were shifted slightly toward longer wavelengths—what astronomers call red shifts. Interpreted as Doppler shifts (Chapter 4), these red shifts implied that the galaxies had large radial velocities and were receding from us.

In 1929, the American astronomer Edwin Hubble published a graph that plotted the velocity of recession

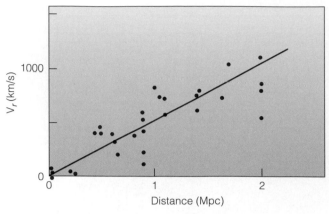

Figure 16-11
Edwin Hubble's first diagram of the velocities and distances of galaxies did not probe very deeply into space. It did show, however, that the galaxies are receding from one another. Because of errors in his distances, his first estimate of the Hubble constant was too large.

versus distance for a number of galaxies. The points in the graph fell along a straight line (Figure 16-11). The straight line relation can be written as the simple equation

$$V_r = Hd$$

Velocity of recession V_r equals distance d in millions of parsecs times a constant H. This relation between red shift and distance is known as the **Hubble law,** and the constant H is known as the **Hubble constant.**

Measurements of H have been uncertain because of the difficulty in measuring the distances to galaxies. Estimates have ranged from 50 to 100 km/s/Mpc.* A long-term study using the Hubble Space Telescope yields a value of about 70. As astronomers gather more observations, this value may be revised slightly, but it seems unlikely that it will be changed by more than 5 percent.

The Hubble law is important because it is commonly interpreted to show that the universe is expanding. In Chapter 18, we will discuss this expansion, but here we will use the Hubble law as a practical way to estimate the distance to a galaxy. Simply stated, a galaxy's distance equals its velocity of recession divided by H. For example, if a galaxy has a radial velocity of 700 km/s, and H is 70 km/s/Mpc, then the distance to the galaxy is 700 divided by 70, or about 10 Mpc. This makes it relatively easy to estimate the distances to galaxies, because a large telescope can photograph the spectrum of a galaxy and determine its velocity of recession even when it is too distant to have visible distance indicators. We will refer to such distances as estimates, because H is not known accurately.

*H has the units of a velocity divided by a distance. These are usually written as km/s/Mpc, meaning km/s per Mpc.

Diameter and Luminosity

The distance to a galaxy is the key to finding its diameter and its luminosity. We can easily photograph a galaxy and measure its angular diameter in seconds of arc. If we know the distance, we can use the small-angle formula (see By the Numbers 1-2) to find its linear diameter. Also, if we measure the apparent magnitude of a galaxy, we can use the distance to find its absolute magnitude and thus its luminosity (see By the Numbers 11-2).

Galaxies differ dramatically in size, mass, and luminosity. Irregular galaxies tend to be small, 1 to 25 percent the size of our galaxy, and of low luminosity. Although they are common, they are easy to overlook. Our Milky Way Galaxy is large and luminous compared with most spiral galaxies, though we know of a few spiral galaxies that are even larger and more luminous—at least 50 percent larger in diameter and about 10 times more luminous. Elliptical galaxies cover a wide range of diameters and luminosities. The largest, called giant ellipticals, are five times the size of our Milky Way. But many elliptical galaxies are very small, dwarf ellipticals, only 1 percent the diameter of our galaxy.

To put galaxies in perspective, we can use an analogy. If our galaxy were an 18-wheeler, the smallest dwarf galaxies would be the size of pocket-size toy cars, and the largest giant ellipticals would be the size of jumbo jets. Even among spiral galaxies that looked similar, the sizes could range from that of a go-cart to that of a small yacht.

Clearly, the diameter and luminosity of a galaxy do not determine its type. Some small galaxies are irregular, and some are elliptical. Some large galaxies are spiral, and some are elliptical. We need more data before we can build a theory for the origin and evolution of galaxies.

Of the three basic parameters that describe a galaxy, we have found two—diameter and luminosity. The third, as was the case for stars, is more difficult to discover.

Mass

Although the mass of a galaxy is difficult to determine, it is an important quantity. It tells us how much matter the galaxy contains, which gives us clues to the galaxy's origin and evolution. In this section, we will examine two fundamental ways to find the masses of galaxies.

One way is called the **rotation curve method.** We begin by photographing the galaxy's spectrum at different points along its diameter and plotting the Doppler shift velocities in a rotation curve like that in Figure 16-12a. This tells us how fast the galaxy is rotating. The sizes of the orbits the stars follow around the galaxy's center are related to the size of the galaxy, easily found from its angular diameter and its distance. If we divide the circumference of the orbit by the orbital velocity of

Figure 16-12

(a) In this artist's conception, the astronomer has placed the galaxy over the slit of the spectrograph (left). Light entering the slit produces a spectrum of which only a short segment is shown. Emission lines in the spectrum are red shifted on one side of the galaxy and blue shifted on the other. Converting these shifts to radial velocities, the astronomer can plot the galaxy's rotation curve (right). (b) Actual data show galaxy NGC 2998 as the astronomer sees it on the spectrograph slit. A short segment of the spectrum includes three emission lines (dark as seen on the photographic plate), red shifted on one side and blue shifted on the other. The rotation curve is plotted for only half of the galaxy. *(Courtesy Vera Rubin)*

the stars there, we find their orbital period. We can then find the mass of the system by using the radius of the orbit and the orbital period, just as we did for binary stars (see By the Numbers 11-4).

The rotation curve is the best method for finding the masses of galaxies, but it works only for the nearer galaxies, whose rotation curves can be observed. More distant galaxies appear so small we cannot measure the radial velocity at different points along the galaxy. Also, recent studies of our own galaxy and others show that the outer parts of the rotation curve do not decline to lower velocities (Figure 16-12b). This shows that the outermost visible parts of some galaxies do not travel more slowly and tells us that the galaxies contain large amounts of mass outside this radius, perhaps in extended galactic coronas like the one that seems to surround our own galaxy (see Chapter 15). Because the ro-

tation curve method can be applied only to nearby galaxies, and because it cannot determine the masses of galactic coronas, we must look at the second way to find the masses of galaxies.

The **cluster method** of finding galactic mass depends on the motions of galaxies within a cluster. If we measure the radial velocities of many galaxies in a cluster, we find that some velocities are larger than others because of the orbital motions of the individual galaxies in the cluster. Given the range of velocities and the size of the cluster, we can ask how massive a cluster of this size must be to hold itself together with this range of velocities. Dividing the total mass of the cluster by the number of galaxies in the cluster yields the average mass of the galaxies. This method contains the built-in assumption that the cluster is not flying apart. If it is, our result is too large. Because it seems likely that most clusters are held together by their own gravity, the method is probably valid.

A related way of measuring a galaxy's mass is called the **velocity dispersion method.** It is really a version of the cluster method. Instead of observing the motions of galaxies in a cluster, we observe the motions of matter within a galaxy. In the spectra of some galaxies, broad lines indicate that stars and gas are moving at high velocities. If we assume the galaxy is bound by its own gravity, we can ask how massive it must be to hold this moving matter within the galaxy. This method, like the one before, assumes that the system is not coming apart.

The structure and evolution of a star are determined by its mass, but this is clearly not so for galaxies. There is a wide range of masses among the elliptical, spiral, and irregular galaxies. The smallest contain about 10^{-6} as much mass as the Milky Way Galaxy, and the largest contain as much as 50 times more.

Dark Matter in Galaxies

In the 1930s, astronomers began to suspect there was more to a galaxy than meets the eye, and the evidence has grown more conclusive with each passing year.

Given the size and luminosity of a galaxy, astronomers can make a rough guess as to the mass it should contain. Studies of the stars in our own galaxy show that much of the light is produced by a few hot, luminous stars, but that most of the mass lies in cool, faint stars. Thus, from the amount of light a galaxy emits, astronomers can estimate the total mass of stars. Again, studies of our own galaxy reveal that some matter is located, not in stars, but in gas and dust between the stars. Taking such studies into account, astronomers can estimate the total amount of matter in stars, gas, and dust in a given galaxy.

Surprisingly, the measured mass of galaxies is much larger than the estimated mass. We discovered this effect in Chapter 15 when we studied the rotation curve of our own galaxy and concluded that it must contain large amounts of dark matter. The measured masses of galaxies amount to 10 to 100 times the mass estimated from the amount of light the galaxy emits. Evidently most galaxies contain large amounts of dark matter.

The rotation curves of galaxies are strong evidence for dark matter, but astronomers have found even more evidence that most of the mass in the universe is invisible. For example, in many clusters of galaxies, the velocities of the galaxies are so high the clusters would fly apart if they contained only the mass we see. These clusters must contain large amounts of dark matter.

X-ray observations reveal more evidence of dark matter. X-ray images of galaxy clusters show that many of them are filled with very hot, low-density gas that would easily leak away if it were not held by a strong gravitational field. To provide that gravity, the cluster must contain much more matter than we can see. The galaxies near the center of the Coma cluster, for instance, amount to only 11 to 35 percent of the total mass (Figure 16-13).

Dark matter is not an insignificant issue. Observations of galaxies and clusters of galaxies tell us that 90 to 99 percent of the matter in the universe is dark matter. The universe we see—the luminous matter—has been compared to the foam on an invisible ocean. Dark matter affects the formation and evolution of galaxies and can affect the evolution of the universe as a whole (Chapter 18).

Dark matter is difficult to detect, and it is even harder to explain. Some astronomers have suggested that dark matter consists of low-luminosity stars and planets scattered through an extended halo around each galaxy. Searches for such objects have not been successful. For example, if the dark matter in our galaxy is hidden in cool, low-mass stars in the halo, then careful searches using the Hubble Space Telescope should reveal many of these stars. Actual photographs looking out into the halo reveal almost none of these low-mass objects, so it seems unlikely that cool, low-mass stars can explain the dark matter in galaxies.

The dark matter can't be hidden in vast numbers of black holes and neutron stars because we don't see the X rays these objects would emit. Remember, we need 10 to 100 times more dark matter than visible matter, and that many black holes should produce X rays that are easy to detect.

Because observations seem to tell us the dark matter can't be hidden in normal objects, some theorists have suggested that the dark matter is made up of unexpected forms of matter, such as neutrinos with mass or exotic subatomic particles, such as WIMPs (Weakly Interacting Massive Particles). So far, no laboratory experiments have been able to measure a mass for the neutrino, and no WIMPs have ever been detected. Thus, dark matter remains one of the fundamental unresolved problems of modern astronomy. We will return to this problem in Chapter 18 when we try to understand how dark matter affects the nature of the universe, its past, and its future.

REVIEW Critical Inquiry

Why does the measured mass of a galaxy depend on its distance?

Often the critical analysis of a problem is complicated by interrelationships between factors, but a bit of care is usually enough to reveal these connections. For example, to find the mass of a galaxy, we compare the mo-

a   b

Figure 16-13
(a) The Coma cluster of galaxies contains at least 1000 galaxies and is especially rich in E and S0 galaxies. Two giant galaxies lie near its center. If the cluster were visible, it would span eight times the diameter of the full moon in the sky. *(L. Thompson © NOAO)* (b) In false colors, this X-ray image of the Coma cluster shows it filled with hot gas. The clumpy distribution suggests that the cluster formed by the merger of smaller clusters, and the temperature of the gas reveals the presence of up to 90 percent dark matter in the cluster. *(Simon D. M. White, Ulrich G. Briel, and J. Patrick Henry)*

tions of stars in the outer parts of a galaxy with velocities predicted by Kepler's third law. Measuring the velocity of stars in a galaxy is simply a matter of observing the Doppler shift in the spectrum, but we also need to know the size of a star's orbit around the galaxy. We can observe the radius of the orbit, the distance from the stars to the center of the galaxy we are observing, in seconds of arc, but Kepler's third law requires we know that distance in some linear unit of distance such as AU. That is where the distance enters. If we know the distance to the galaxy, we can use the small-angle formula to convert the orbital radius from seconds of arc to AU. Then we can use Kepler's third law to find the mass.

Many different measurements in astronomy depend on the calibration of the distance scale. For example, what would happen to our estimate of the diameter and luminosity of galaxies if astronomers discovered that the Cepheid variable stars were slightly more luminous than had been believed?

We have gathered the basic data to help us characterize galaxies. That is, we can describe their general characteristics. Now we turn to the more difficult and more interesting question. How did galaxies get to be the way they are?

16-3 The Evolution of Galaxies

Our goal in this chapter has been to build a theory to explain the evolution of galaxies. In Chapter 15, we developed a theory that described the origin of our own Milky Way Galaxy; and, presumably, other galaxies formed similarly from the contraction of large clouds of gas. But why did some galaxies become spiral, some elliptical, and some irregular? Clues to that mystery lie in the clustering of galaxies.

Clusters of Galaxies

The distribution of galaxies is not entirely random. Galaxies tend to occur in clusters ranging from a few to thousands. Furthermore, modern astronomers have catalogued over 2700 clusters within 4 billion ly, and deep photographs with the largest telescopes reveal even more clusters of galaxies at the limit of our vision. Not only are these vast clusters fascinating, but they can also help us understand the lives of the galaxies.

For our discussion, we can sort clusters into two groups: rich and poor. **Rich galaxy clusters** contain over a thousand galaxies, mostly elliptical, scattered through a spherical volume about 3 Mpc (10^7 ly) in diameter. Such a cluster is very crowded, with the galaxies more concentrated toward the center. Rich clusters often contain one or more giant elliptical galaxies at their centers.

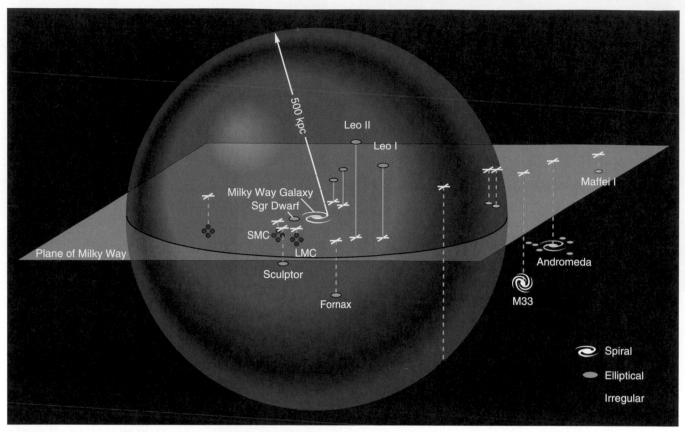

Figure 16-14

The Local Group. Our galaxy is a member of a poor galaxy cluster called the Local Group. In this three-dimensional map, the galaxies are shown with respect to the plane of our galaxy. Most of the galaxies are small ellipticals. Other member galaxies may not have been seen because they lie near the plane of the Milky Way and thus are obscured by dust clouds within our galaxy.

The Coma cluster (located in the constellation Coma Berenices) is an example of a rich cluster (Figure 16-13). It lies over 100 Mpc from us and contains at least 1000 galaxies, mostly E and S0 galaxies. Its galaxies are highly crowded around a central giant elliptical and a large S0.

One of the nearest clusters to us, the Virgo cluster, contains over 2500 galaxies and is, by our definition, a rich cluster. It does contain a giant elliptical galaxy, M87, near its center, but it is not very crowded and contains mostly spiral galaxies. Although the Virgo cluster is exceptional as a rich cluster, it is critically important because it is nearby. From a distance of about 59 million ly (18 Mpc), no Cepheids are visible in Earth-based telescopes. The Hubble Space Telescope, however, has detected Cepheids in the Virgo cluster, which is a critical stepping-stone in calibrating other distance indicators. Thus, improving the distance to the Virgo cluster improves the calibration of the entire distance scale.

Poor galaxy clusters contain fewer than 1000 galaxies and are irregularly shaped. Instead of being crowded toward the center, poor clusters tend to have subcondensations, small groupings within the clusters.

Our Local Group, which contains our Milky Way Galaxy, is a good example of a poor cluster (Figure 16-14). It contains a few dozen members scattered irregularly through a volume slightly over 1 Mpc in diameter. Of the brighter galaxies, fourteen are elliptical, three are spiral, and four are irregular.

The number of galaxies in the Local Group is uncertain for two reasons. Many galaxies are small, faint, and difficult to detect. Also, some galaxies lie in the plane of the Milky Way and are hidden by the obscuring dust. Two galaxies called Maffei 1 and Maffei 2 are a giant elliptical and a spiral. Maffei 1 would be visible to the naked eye were it not obscured by dust. The Dwingeloo 1 spiral galaxy was discovered by radio astronomers, and a galaxy known as the Sagittarius Dwarf was found from studies of the motions of stars. None of these galaxies are easily visible because dust in our galaxy dims them. Still more members of the Local Group may lie hidden behind the dusty plane of our galaxy.

The clumpiness of the Local Group illustrates the subclustering in poor galaxy clusters. The two largest galaxies, the Milky Way Galaxy and the Andromeda Galaxy, are the centers of two subclusters. The Milky Way is accompanied by the Magellanic clouds and three other dwarf galaxies. The Andromeda Galaxy is attended by seven dwarf elliptical galaxies and a smaller spiral known as M33. Thus, the galaxies in a poor cluster are not as concentrated as those in a rich cluster.

We have classified galaxy clusters into rich and poor clusters to reveal a fascinating and suggestive clue to the evolution of galaxies. In general, rich clusters tend to have a larger proportion of E and S0 galaxies than do poor clusters. This suggests that a galaxy's environment is important in determining its structure and has led astronomers to suspect that the secrets to galaxy evolution lie in collisions between galaxies.

Colliding Galaxies

Astronomers are finding more and more evidence to show that galaxies collide, interact, and merge. In fact, collisions among galaxies may dominate their evolution.

We should not be surprised that galaxies collide with each other. The average separation between galaxies is only about 20 times their diameter. Like two elephants blundering about under a circus tent, galaxies should bump into each other once in a while. Stars, on the other hand, almost never collide. In the region of our galaxy near the sun, the average separation between stars is about 10^7 times their diameter. Thus a collision between two stars is about as likely as a collision between two gnats flitting about in a baseball stadium.

Even if two galaxies pass through each other, the stars never collide. The interstellar gas and dust in the galaxies do interact as gas clouds collide, and the sudden compression can trigger rapid star formation. It is also possible for one galaxy to strip gas and dust away from the other galaxy. Thus, bursts of star formation and galaxies stripped of gas and dust are clear consequences of galaxy collisions.

Because collisions between galaxies can last hundreds of millions of years, we cannot see the galaxies change, but computer models of collisions can speed up the process and show how colliding galaxies are twisted and deformed by their gravitational fields to produce peculiar shapes, tails, and bridges. It is not even necessary for the galaxies to penetrate each other—a close encounter is enough to allow tidal forces to distort the galaxies. One of the most famous pairs of colliding galaxies is called the Mice because of the tail-like deformities (Figure 16-15). These tails are flung out as the galaxies whip around their common center of mass.

Collisions between galaxies may trigger rapid changes in a galaxy. Some galaxies contain large amounts of gas and dust but are making few new stars. Such galaxies tend to be isolated, with no nearby galaxies that might disturb them and trigger star formation. Collisions may stimulate the formation of spiral density waves, and observations confirm that flocculent galaxies (Chapter 15), which lack grand-design spiral patterns, tend to be isolated from neighboring galaxies. Galaxies with strong two-armed spiral patterns tend to be located in clusters with lots of nearby neighbors (Figure 16-16).

As two galaxies approach each other, they each carry large amounts of orbital momentum. Computer simulations show that colliding galaxies experience tidal forces that convert some of their orbital motion into random motion among their stars and gas clouds and thus rob them of orbital momentum. It is possible for galaxies to collide, interact, and then go their separate ways, but it is also possible for colliding galaxies to lose enough momentum that they fall together and merge.

a

b

Figure 16-15
(a) The Mice are a pair of interacting galaxies with peculiar tails (shown here in a false-color image). *(NOAO)* (b) A computer simulation of a close encounter between normal galaxies produces similar tails. *(Illustration by Allen Beechel.)*

A small galaxy merging with a larger galaxy will be gradually pulled apart and its stars will spread through the larger galaxy in a process called **galactic cannibalism.** Computer simulations of such mergers show that the galaxies throw off shells of stars as the nuclei of the two galaxies spiral into each other, and such shells of stars have been observed around real galaxies (Figure 16-17). Our own Milky Way Galaxy is interacting with the Large and Small Magellanic clouds and will eventually consume them both.

Figure 16-16
A computer simulation of a collision between galaxies produces a pattern very similar to M51, the Whirlpool Galaxy (bottom) and its small companion. *(Illustration by Allen Beechel; NOAO)*

Figure 16-17
Computer simulations of merging galaxies suggest that they can produce shells of stars as the two galaxies whirl around their common center of mass. Only one galaxy is shown in this sequence. Such shells are seen around galaxies such as NGC3923 (bottom), shown here as a negative image specially enhanced to reveal low-contrast features. *(Model courtesy François Schweizer and Alan Toomre; photo courtesy David Malin, Anglo-Australian Telescope Board)*

Mergers between galaxies are dramatic not only because a merger distorts the galaxies but also because it triggers violent star formation. A pair of galaxies called the Antennae, because they have long curving tails that look like a pair of insect antennae, are passing through each other, and the collision of their gas clouds is triggering rapid star formation, evident from the many hot bright stars, the clouds of dust, and the large numbers of young, massive star clusters (Figure 16-18).

Evidence of mergers is common. For example, the distorted galaxy NGC7252 is rapidly making new stars. The Hubble Space Telescope found that the center of the galaxy contains a small spiral of gas rotating opposite to the rest of the galaxy (Figure 16-19a). Evidently, the distorted galaxy is the result of a merger between two oppositely rotating galaxies that happened about a billion years ago. The elliptical galaxy NGC1316 contains clouds of gas

Figure 16-18
The colliding galaxies NGC4038 and NGC4039 are known as the Antennae because their long curving tails resemble the antennae of an insect. Earth-based photos (inset) show little detail, but this Hubble Space Telescope image reveals the collision of two galaxies that produces thick clouds of dust and raging star formation creating roughly a thousand massive star clusters such as the one at the top (arrow). Such collisions between galaxies are common. *(Brad Whitmore, STScI, and NASA)*

a

b

c

Figure 16-19
Evidence of mergers. (a) The center of this distorted galaxy rotates backward. *(François Schweizer, Carnegie Institution of Washington, and Brad Whitmore, STScI)* (b) Elliptical galaxies should not contain clouds of dust, but this one appears to have gained dust clouds when it devoured a smaller galaxy. *(Carl Grillmair, CalTech, and NASA)* (c) Multiple nuclei in giant elliptical galaxies hint at recent meals. *(Michael J. West)*

Figure 16-20
The Cartwheel Galaxy was once a normal spiral galaxy, but a few hundred thousand years ago one or the other of the two smaller galaxies at the right passed through the center of the larger galaxy. Tidal forces generated an expanding wave of star formation and a central core of newborn stars heavily reddened by dust. Such ring galaxies appear to be the fleeting consequences of bull's-eye collisions. Soon after the collision, the violent star formation exhausts itself, and normal spiral arms may re-form. This is suggested in the Cartwheel Galaxy by the faint spiral structure inside the outer ring. *(Kirk Borne, STScI and NASA)*

that are believed to be the result of a merger about 100 million years ago (Figure 16-19b). Furthermore, some giant elliptical galaxies have multiple nuclei thought to be the undigested bits of recently cannibalized galaxies (Figure 16-19c).

In contrast to mergers, high-speed head-on crashes between galaxies can produce dramatic results. **Ring galaxies** consist of a bright nucleus surrounded by a ring (Figure 16-20). Computer models show that they could be produced by a galaxy passing roughly perpendicularly through the disk of a larger galaxy. Observations show that many ring galaxies have nearby companions.

We have seen that collisions and mergers between galaxies are common and can produce dramatic changes in the structure of the galaxies. We can now propose a hypothesis to help us understand how galaxies evolve.

The Origin and Evolution of Galaxies

The test of any scientific understanding is whether or not we can put all the evidence and theory together to tell the history of the objects we study. Can we describe the origin and evolution of the galaxies? Just a few decades ago, it would have been impossible, but new evidence is helping astronomers understand the history of the galaxies.

We can eliminate a few older ideas immediately. The tuning fork diagram tempts us to think that galaxies evolve from left to right, from elliptical to spiral to irregular. The evidence, however, shows that elliptical galaxies contain no gas and dust to make new stars, and they contain lots of old stars. Ellipticals can't be young. Furthermore, galaxies can't evolve from right to left in the tuning fork diagram, from irregular to spiral to elliptical; irregular galaxies contain both young and old stars and thus can't be young. The tuning fork diagram is not an evolutionary diagram. It tells us that the galaxies have had different histories of star formation, but it does not tell us how galaxies evolve.

Another old idea held that galaxies that form from rapidly rotating clouds of gas would have lots of angular momentum and would contract slowly to form disk-shaped spiral galaxies. Clouds of gas that rotated less rapidly would contract more rapidly, form stars quickly, use up all the gas and dust, and become elliptical galaxies. The evidence clearly shows that galaxies didn't form and evolve in isolation. Collisions and mergers dominate the history of the galaxies.

The ellipticals appear to be the product of galaxy mergers, which triggered star formation that used up the gas and dust. In fact, we see many **starburst galaxies** that are very luminous in the infrared because a collision has triggered a burst of star formation that is heating the dust. The warm dust reradiates the energy in the infrared. The Antennae (Figure 16-18) contain over 15 billion solar masses of hydrogen gas and will become a starburst galaxy as the merger triggers rapid star formation. Supernovae in a starburst galaxy may eventually blow away any remaining gas and dust that doesn't get used up making stars. A few collisions and mergers could leave a galaxy with no gas and dust from which to make new stars. Astronomers now suspect that most ellipticals are formed by the merger of at least two or three galaxies.

In contrast, spirals seem never to have suffered major collisions. Their thin disks are delicate and would be destroyed by tidal forces in a collision with a massive galaxy. Also, they retain plenty of gas and dust and continue making stars.

Other processes can alter galaxies. The S0 galaxies may have lost much of their gas and dust in a burst of star formation, but they still managed to remain disk shaped. Also, galaxies moving through the gas trapped in dense clusters of galaxies may have their own gas and dust blown away. For example, X-ray observations show that the Virgo cluster contains thin, hot gas between the galaxies. A galaxy orbiting through that gas would encounter a tremendous wind blowing its gas and dust away. This could explain the dwarf ellipticals, which are too small to be made of merged spirals. In contrast, the irregular galaxies may be small fragments of galaxies ripped apart by collisions.

a

b

c

Figure 16-21
Some of the most distant objects visible from Earth are the faint galaxies visible in the Hubble Deep Fields. The 10-day-long time exposures of typical spots in the northern sky (a) and the southern sky (b) look strikingly similar. *(Robert Williams and the Hubble Deep Field Team, STScI and NASA)* (c) The most distant objects in these photos are irregular blue clouds of gas and stars smaller than most galaxies. These may be the precursors of galaxies. Each of these three small frames is about one-third the diameter of our Milky Way Galaxy. *(Rogier Windhorst and Sam Pascarelle, Arizona State University, and NASA)*

The evidence for galaxy evolution by merger is quite strong. Observations with the largest telescopes take us to great distances and great look-back times. We see the galaxies as they were long ago, and we discover that there were more spirals then and fewer ellipticals. Also, the ellipticals were smaller than they are now. We can even see that galaxies were closer together long ago; about a third of all distant galaxies are in close pairs, but only 7 percent of nearby galaxies are in pairs. The observational evidence clearly supports the hypothesis that galaxies have evolved by merger.

The Hubble Space Telescope has recorded very long exposure images called deep fields at two seemingly empty spots in the northern and southern sky. Each of these frames shows over a thousand galaxies at such large look-back times that we see them as they were when the universe was very young (Figure 16-21). The entire sky is filled with many billions of faint galaxies.

Among the smallest objects in such deep photos are small, blue clouds of stars and gas containing about 10 times the mass of large globular clusters (Figure 16-21c). These distorted clouds may be the objects that first fell together to form galaxies when the universe was young.

REVIEW **Critical Inquiry**

How did elliptical galaxies get that way?

A growing body of evidence suggests that elliptical galaxies have been subject to collisions in their past and that spiral galaxies have not. During collisions, a galaxy can be driven to use up its gas and dust in a burst of star formation, or a galaxy may be stripped of gas and dust by an encounter. Thus, elliptical galaxies now contain little star-making material. The beautiful disk typical of spiral galaxies is very orderly, with all the stars following similar orbits. When galaxies collide, the stellar orbits get scrambled, and an orderly disk galaxy could be converted into a chaotic swarm of stars typical of elliptical galaxies. Thus, it seems likely that elliptical galaxies have had much more complex histories than spiral galaxies have had.

What evidence do we have to support the story in the preceding paragraph?

Before we can consider the universe as a whole, we must examine the galaxies from a different perspective. Some galaxies are suffering tremendous eruptions in their centers. In the next chapter, we will discover that these peculiar galaxies are closely related to collisions and mergers between galaxies.

Summary

We can divide galaxies into three classes—elliptical, spiral, and irregular—with subclasses specifying the galaxy's shape. The elliptical galaxies contain little gas and dust and few bright, young stars. Spiral and irregular galaxies have large amounts of gas and dust and are actively making new stars.

To measure the properties of galaxies, we must first find out how far away they are. For the nearer galaxies, we can judge distances using distance indicators, objects whose luminosity or diameter is known. The most accurate distance indicators are Cepheid variable stars. Other distance indicators include globular clusters, H II regions, planetary nebulae, and supernovae. In addition, astronomers can estimate the distance to the farthest galaxies by using the average luminosity of the brightest galaxies in a cluster.

The Hubble law shows that the radial velocity of a galaxy is proportional to its distance. Thus, we can use the Hubble law to estimate distances. The galaxy's radial velocity divided by the Hubble constant equals its distance in megaparsecs.

The masses of galaxies can be measured in two basic ways—the rotation curve method and the cluster method. The rotation curve method is more accurate but can be applied only to nearby galaxies. Both methods suggest that galaxies contain 10 to 100 times more dark matter than visible matter.

Galaxies occur in clusters. Our own galaxy is a member of the Local Group, a small cluster. A galaxy in a rich cluster may collide with other galaxies more often than a galaxy in a poor cluster, and such collisions can force a galaxy to form new stars and use up its gas and dust. Collisions can also strip gas out of a galaxy. This may explain why elliptical and S0 galaxies are more common in rich clusters than in poor clusters. Spiral galaxies may be star systems that have not experienced many collisions.

New Terms

tuning fork diagram	megaparsec (Mpc)
elliptical galaxy	distance indicator
spiral galaxy	H II region
barred spiral galaxy	look-back time
irregular galaxy	Hubble law
Large Magellanic Cloud	Hubble constant (H)
Small Magellanic Cloud	rotation curve method

cluster method	galactic cannibalism
velocity dispersion method	ring galaxy
rich galaxy cluster	starburst galaxy
poor galaxy cluster	

Review Questions

1. Why didn't astronomers before the 1920s recognize galaxies for what they are?
2. How can a classification system aid a scientist?
3. Why can't galaxies evolve from E to S to Irr? Why can't they evolve from Irr to S to E?
4. How do selection effects make it difficult to decide how common E, S, and Irr galaxies are?
5. Why are Cepheid variable stars good distance indicators? What about planetary nebulae?
6. Why is it difficult to measure the Hubble constant?
7. How is the rotation curve method related to binary stars and Kepler's third law?
8. What evidence do we have that galaxies contain dark matter?
9. What evidence do we have that galaxies collide and merge?
10. Why are the shells visible around some elliptical galaxies significant?
11. Propose an explanation for the lack of gas, dust, and young stars in elliptical galaxies.

Discussion Questions

1. Why do we believe that galaxy collisions are likely, but star collisions are not?
2. Should an orbiting infrared telescope find irregular galaxies bright or faint in the far infrared? Why? What about elliptical galaxies?

Problems

1. If a galaxy contains a Type I (classical) Cepheid with a period of 30 days and an apparent magnitude of 20, what is the distance to the galaxy?
2. If you find a galaxy that contains globular clusters with apparent magnitudes no brighter than 18th magnitude, how far away is the cluster? (*Hint:* What is the absolute magnitude of a globular cluster?)
3. If a galaxy contains a supernova that at brightest has an apparent magnitude of 17, how far away is the galaxy?

(*Hint:* Assume that the absolute magnitude of the supernova at brightest is -19.)

4. If a galaxy has a radial velocity of 2000 km/s and H is 70 km/s/Mpc, how far away is the galaxy?

5. If you find a galaxy that is 20 minutes of arc in diameter and you measure its distance to be 1 Mpc, what is its diameter in parsecs? (*Hint:* See By the Numbers 1-2.)

6. We have found a galaxy in which the outer stars have orbital velocities of 150 km/s. If the radius of the galaxy is 4 kpc, what is the orbital period of the outer stars? (*Hints:* 1 pc = 3.08×10^{13} km, and 1 yr = 3.15×10^7 s.)

7. Calculate the minimum mass of the galaxy shown in Figure 16-12b. (*Hint:* See By the Numbers 11-4.)

8. A galaxy has been found that is 5 kpc in radius and whose outer stars orbit the center with a period of 200 million years. What is the mass of the galaxy? On what assumptions does this result depend? (*Hint:* See By the Numbers 11-4.)

Critical Inquiries for the Web

1. How far out into the universe can we see Cepheid variables? Research sources on the Internet to find other galaxies whose distances have been found through observation of Cepheids. List the galaxies in which Cepheids have been identified and the distances determined from these data.

2. How does the Milky Way stack up against the other galaxies in the Local Group? Look for information on the other galaxies in our cluster and rank the top six members in order of total mass.

3. Locate a Web page dedicated to the Messier Catalogue—a list of galaxies, clusters, and nebulae that is often used as a list of targets for small telescopes. Be sure that your destination includes images of the objects. For each of the galaxies in the Messier list, determine its Hubble classifi-

cations (see Figure 16-2). You may be given this information at the site, but examine the images to see if the features of these galaxies conform to a particular Hubble type.

Exploring *The Sky*

1. Many galaxies organize themselves into clusters. One of the largest clusters of galaxies is the Virgo cluster. Center this cluster on your screen.
   a. List the Messier objects likely to belong to this cluster. How to proceed: Click on the galaxy button on the **Objects Toolbar.**
   b. Estimate the minimum number of galaxies belonging to this cluster. How to proceed: Center the cluster on your screen, and zoom in to show the maximum number of galaxies.

2. The Andromeda Galaxy, or M31, is the closest spiral neighbor to our Milky Way Galaxy. Find and center M31, and zoom in to view its apparent elliptical shape. Notice two very close and at least partially overlapping galaxies.
   a. Find the Messier numbers of these neighbors to M31.
   b. Find the magnitudes of all three galaxies. Are any of them visible with the unaided eye?
   c. Find the angular distance between M31 and two neighboring galaxies. These neighboring galaxies are actually satellite galaxies physically interacting with M31 (and to a far lesser extent, with each other). From your angular distance determination and the fact that M31 is approximately 2 million light years away, determine the minimum distance between M31 and the two satellite galaxies.

Go to the Brooks/Cole Astronomy Resource Center (www. brookscole.com/astronomy) for critical thinking exercises, articles, and additional readings from InfoTrac College Edition, Brooks/Cole's online student library.

CHAPTER SEVENTEEN

Galaxies with Active Nuclei

Virtually all scientists have the bad habit of displaying feats of virtuosity in problems in which they can make some progress and leave until the end the really difficult central problems.

M. S. Longhair
High Energy Astrophysics

Guidepost

This chapter is important for two reasons. First, it draws together ideas from many previous chapters to show how nature uses the same basic rules on widely different scales. Matter flowing into a protostar, into a white dwarf, into a neutron star, or into the heart of a galaxy must obey the same laws of physics, so we see the same geometry and the same phenomena. The only difference is the level of violence.

Second, this chapter is important because the most distant objects we can see in the universe are the most luminous galaxies, and many of those are erupting in outbursts and are thus peculiar. By studying these galaxies, our attention is drawn out in space to the edges of the visible universe and back in time to the earliest stages of galaxy formation. In other words, we are led to think of the origin and evolution of the universe, the subject of the next chapter.

Explosion? Eruption? Detonation? Colossal? Titanic? Words fail us when we try to describe the fantastic violence that astronomers find blasting out of the center of some galaxies. If a supernova is said to explode, what word shall we use for the eruption of a galaxy? Astonishing it may be, but violence in the centers of galaxies is easily detected across billions of light-years, and our double goal in this chapter is to explain the origin of these eruptions and relate them to the formation and evolution of galaxies.

The adventure of studying active galaxies began when radio astronomers first noticed such objects in the 1950s. In the decades that followed, astronomers used all parts of the electromagnetic spectrum to study these galaxies and piece together the puzzle. Today, astronomers generally agree on how a galaxy can develop an active core. Thus, our study of active galaxies not only reveals how galaxies can erupt but also illustrates how scientists can use evidence, theory, and their own ingenuity to learn about the natural world.

A number of astronomers object to the quotation that opens this chapter, perhaps because it seems to belittle scientists. But it is a realistic description of how scientists attack a problem. Big problems are made of little problems, and scientists begin by solving the little problems, learning from those solutions, and moving ever closer to the center of the big problem. Perhaps no better example can be found than modern astronomy's attack on "the really difficult central problem" in the nuclei of active galaxies.

17-1 Active Galactic Nuclei

In the 1950s, radio astronomers began to notice that some galaxies were powerful sources of radio energy. Of course, all galaxies, including our own Milky Way, emit some radio energy, but some emit over 10 million times more radio energy than does a normal galaxy. Such objects were first known as **radio galaxies.** But newer observations reveal that many of these galaxies also radiate at infrared, ultraviolet, and X-ray wavelengths, so these objects have become known as **active galaxies.** In many cases, the source of the activity lies in the nucleus of the galaxy, and such objects are known as **active galactic nuclei (AGN).** In some cases, however, the radio energy is detected coming from two areas (lobes) on opposite sides of the visible galaxy. We will eventually develop a model that relates these seemingly different kinds of objects.

Double-Lobed Radio Galaxies

Radio astronomers have found many examples of radio sources in which the radio energy comes from two regions, called lobes, located near each other in the sky.

Figure 17-1
Radio telescopes can map a galaxy and produce a false-color image as if we could see radio waves. In this radio image of 3C388, a typical double-lobed radio galaxy, the galaxy lies at the center flanked by bright radio lobes above and below. Here, red represents the strongest radio energy and violet the weakest. *(NRAO; observers J. O. Burns and W. A. Christiansen)*

Photographs of the region between these lobes reveal a galaxy, usually a giant elliptical galaxy. Many of these **double-lobed radio galaxies** are peculiar in some way, with distorted shapes, close companion galaxies, dust clouds, or jets; but some of these galaxies are not themselves unusually bright at radio wavelengths.

Radio lobes are generally much larger than the galaxy they accompany (Figure 17-1). Many are as large as 60 kpc (200,000 ly) in diameter, twice the size of the Milky Way Galaxy. From tip to tip, the radio lobes span hundreds of kiloparsecs. One of the largest-known radio galaxies is 3C236 (the 236th object in the *Third Cambridge Catalogue of Radio Sources*). In fact, it is one of the largest objects in the universe, with radio lobes that reach across 5.8 Mpc (19×10^6 ly).

Radio lobes have two properties that hint at their origin: They radiate synchrotron radiation, and they are often most intense on the side away from the central galaxy. Recall that synchrotron radiation is produced when high-speed electrons spiral through a magnetic field (see Chapter 13). Though the field is at least 1000

Figure 17-2

Cygnus A is a powerful, double-lobed radio galaxy. This radio map produced by the VLA radio telescope shows hot spots as brightenings on the outer edges of the lobes. Note the long, thin jet of matter flowing from the galaxy to the right-hand lobe. The jet is at least 50 kpc long—about 50 percent longer than the diameter of our galaxy. A fainter counterjet is also visible leading to the left lobe. *(NRA Observatory)* In a visible-light photograph, the deformed central galaxy (inset) is roughly the size of the region between the radio lobes. *(Caltech)*

times weaker than Earth's, it fills a tremendous volume and represents vast stored energy. In addition, the high-speed electrons also contain energy. The total energy in a radio lobe equals about 10^{53} J, approximately the energy equivalent of converting 10^6 solar masses directly into energy. Clearly, the process that creates radio lobes is powerful.

The second hint to the nature of the radio lobes is that many radio lobes emit more intensely from the side away from the galaxy (Figure 17-2). Not all radio lobes have these **hot spots,** but many do.

Synchrotron radiation and hot spots suggest an explanation for double-lobed radio galaxies now known as the **double-exhaust model.** According to this model, if an eruption in the core of the galaxy drove jets of high-energy gas outward in opposite directions, they could blow up a pair of radio lobes like bubbles in the thin gas of the intergalactic medium. The ejected gas would be ionized and would drag along part of the galaxy's magnetic field, and

the high-speed electrons in the gas would produce synchrotron radiation. The beams could produce hot spots where they pushed against the far side of each radio lobe (Figure 17-3).

a

b

Figure 17-3

(a) In the double-exhaust model, the hot spots are produced by beams, or jets, of high-speed particles pushing into the intergalactic gas. (b) In this false-color radio map of the galaxy 3C219, an active nucleus (red) ejects a narrow jet pointing toward one of the two radio lobes containing hot spots (yellow). *(NRAO; observers A. H. Bridle and R. A. Perley)*

Figure 17-4

(a) The peculiar galaxy NGC5128 appears to be an elliptical galaxy colliding with a spiral. *(NOAO)* (b) A photo by the Hubble Space Telescope reveals young, bright, blue stars, evidence of star formation triggered by a collision, scattered through the dust disk. *(E. Schreier, STScI, and NASA)* (c) High-resolution radio maps of the center of the galaxy reveal an energetic central object (arrow) and a jet pointing toward one of the radio lobes that lie on either side of the galaxy. *(NRAO)*

The radio map of Cygnus A shown in Figure 17-2 contains a striking detail: A long, narrow jet leads out into the radio lobe at the right. Another jet, slightly harder to see in this map, leads into the left radio lobe. Notice also that the radio lobes contain hot spots at their outer edges. These jets give us direct evidence that the vast structure of the radio lobes is produced by high-energy jets ejected from the nucleus of the galaxy.

More evidence is found in the galaxy NGC5128. It looks like a giant elliptical galaxy, but it is surrounded by a belt of dust. Astronomers believe we are seeing a collision between a giant elliptical and a small spiral (Figure 17-4a). The evidence for this interpretation is strong. Doppler shifts show that the bright, spherical part of the galaxy is rotating about an axis that lies in the plane of the dust ring, but that the dust ring is rotating about an axis perpendicular to the plane of the ring. Also, the best images reveal bright blue stars in the dust ring, stars that must have formed recently (Figure 17-4b). Collisions are known to trigger star formation, so the presence of young stars supports the collision hypothesis.

Radio astronomers know this object by the name Centaurus A. The radio object is two giant radio lobes that span 10° in the sky, bigger than the bowl of the big dipper. Radio inteferometers reveal a small jet leading from the center of the galaxy toward the northern lobe (Figure 17-4c). At the highest resolution, radio maps reveal a central object only 10 light-days in diameter and a jet 1 pc long pointing toward the northern lobe. Clearly the center of this galaxy is ejecting jets of high-velocity gas to inflate the radio lobes. Because this is a collision between two galaxies, we must wonder if the collision could have triggered the eruption in the core.

We can't see the core of NGC5128 at visible wavelengths because of dust, but infrared studies made by the Hubble Space Telescope can penetrate the dust.

a

b

Figure 17-5

(a) Active galaxy 3C449 has spiral patterns in its radio lobes, thought to be produced by the orbital motion of the source of the jets. At visual wavelengths, the galaxy is revealed to be a giant elliptical about 10 times the size of the central spot in this false-color image *(Richard A. Perley)* (b) Head–tail radio galaxy 3C129 has its jets swept back by the motion of the galaxy through the intergalactic medium. Twists in the jets, shown in the mathematical model at right, seem to be produced by the precession of the jets. *(Laurence Rudnick, University of Minnesota, and Jack Burns, University of New Mexico; model courtesy Vincent Icke, University of Minnesota and The Astrophysical Journal, published by the University of Chicago Press, © 1981 American Astronomical Society)*

They reveal a brilliant central object encircled by a disk of gas 130 ly in diameter. Evidently, this small, brilliant object is ejecting the jets and inflating the radio lobes.

Radio lobes are very tenuous. They do not contain strong enough gravity or magnetic fields to hold themselves together. Apparently, the lobes are held together by their collisions with the intergalactic gas.

What other evidence can we find to show that our double-exhaust model is a valid way to think about these galaxies? Radio astronomers have found examples of double-lobed radio galaxies in which the jets of gas flowing out of the galaxy are distorted. Some appear twisted into corkscrew shapes, perhaps because the source of the jets is orbiting around the center of the galaxy or because the source is precessing (Figure 17-5a). Other jets contain irregular blobs suggesting that the jets do not remain "on" all the time. Some of these distorted radio galaxies have been called **head–tail radio galaxies,** because their jets bend together to form a tail that leads away from the galaxy. Actually, the head–tail galaxies appear to be moving relative to the intergalactic gas, and the hot gas spewing outward in their jets is swept away like smoke from twin smokestacks to form the tail. 3C129 in Figure 17-5b contains all these features. Its jets appear to be spiral and irregular and are swept back to form a tail. The mathematical model shown in Figure 15-5b reproduces the structure in the galaxy and gives astronomers added confidence in the model.

All of the evidence suggests that radio lobes are produced by two oppositely directed jets of high-energy gas flowing out of the central galaxy. The double-exhaust model helps us understand the overall geometry, but it does not help us understand the central problem: How does the center of a galaxy produce such eruptions? To pursue that part of the problem, we must consider active galaxies that lack radio lobes.

Galaxies with Erupting Nuclei

Some galaxies have active nuclei but lack radio lobes. **Seyfert galaxies,** for example, are spiral galaxies with unusually bright, tiny cores that fluctuate in brightness (Figure 17-6). Most Seyfert galaxies are powerful sources of infrared radiation, and some are radio sources. Of the roughly 120 known, 25 emit X rays and one is known to emit gamma rays.

Detailed studies reveal two different kinds of Seyfert galaxies. Both kinds have emission lines in their spectra, but Type 1 Seyferts have broad spectral lines typical of gas swirling around at velocities over 1000 km/s. These Type 1 Seyferts are also very bright at ultraviolet and X-ray wavelengths. Type 2 Seyferts have narrow emission lines in their spectra, suggesting the gas emitting the light is not moving so violently, and these galaxies are bright at infrared wavelengths. The differences among Seyfert galaxies will become critical evidence when we try to devise a model to explain what is happening in the hearts of these active galaxies.

Notice that some scientific evidence is statistical. For example, we might argue that Seyfert galaxies are three times more likely among interacting galaxies. This evidence is statistical because we can't be certain that any particular interacting galaxy will be a Seyfert. Yet the probability is higher than if it were an isolated galaxy, and that leads us to suspect that interactions between galaxies may trigger the Seyfert phenomenon. Such statistical evidence can tell us something in general about the cause of Seyfert eruptions.

Statistical evidence is common in science, but it is inadmissible in most courts of law. The American legal system is based on the principle of reasonable doubt, so most judges would not allow statistical evidence. For example, plaintiffs have had great trouble suing tobacco companies for causing their cases of lung cancer even though there is a clear statistical link between smoking and lung cancer. The statistics tell us something in general about smoking but cannot be used to prove that any specific case of lung cancer was caused only by smoking. There are other causes of lung cancer, so there is always a reasonable doubt as to the cause of any specific case.

We can use statistical evidence in science because we do not demand that the statistics demonstrate anything conclusively about any single example. To continue our example of Seyfert galaxies, we don't demand that the statistics predict that any specific interacting galaxy be a Seyfert. Rather, we use the statistical evidence to gain a general insight into the cause of active galactic nuclei. That is, we are trying to understand galaxies as a class, not convict any single galaxy.

Of course, if we surveyed only a few galaxies, our statistics might not be very good, and a critic might be justified in making the common complaint, "Oh, that's only statistics." For example, if we surveyed only four galaxies and found that three had companions, our statistics would not be very reliable. But if we surveyed 1000 galaxies, our statistics could be very good indeed, and our conclusions could be highly significant.

Thus, scientists can use statistical evidence if it passes two tests. It cannot be used to draw conclusions about specific cases, and it must be based on samples large enough that the statistics are significant. With these restrictions, statistical evidence can be a powerful scientific tool.

Figure 17-6
The spiral galaxy NGC7742 has a small, brilliant central core identifying it as a Seyfert galaxy. The thick, lumpy ring surrounding the yellow glow of the core is a region of active star formation about 3000 ly from the center of the galaxy. *(Hubble Heritage Team, AURA, STScI, and NASA)*

Statistics can give us further insights (Window on Science 17-1). Since 2 percent of all spiral galaxies are Seyferts, we can conclude two things: Either active cores occur in 2 percent of all spiral galaxies, or all spiral galaxies erupt as Seyferts 2 percent of the time. The peculiar heart of our own galaxy suggests that all spirals erupt now and then; we may be living in a galaxy that was a Seyfert only a few hundred million years ago. A second statistic tells us that Seyfert galaxies are three times more common among galaxies interacting with other galaxies (Figure 17-7). This statistic suggests that galaxy collisions, interactions, and mergers may trigger activity in the cores of galaxies.

As we gather more and more evidence that the cores of galaxies can erupt violently, we turn from the spiral Seyferts to **BL Lac objects** (also known as *blazars*), which appear to be associated with elliptical galaxies. These objects are named after the first one discovered—object BL in the constellation Lacerta (the Lizard). They have no absorption or emission lines in their spectra, but when the light of the bright core is blocked, we see the faint spectrum of an elliptical galaxy. Evidently, BL Lac objects are the active cores of elliptical galaxies.

Some galaxies with active nuclei and no radio lobes emit jets. M87, for example, is a giant elliptical galaxy in the nearby Virgo cluster of galaxies. Short-exposure photographs reveal a small, brilliant nucleus and a jet

9100 KM/S

NGC 7674
8620 KM/S

8700 KM/S

NGC 7675
8200 KM/S

60"

a

b

Figure 17-7

(a) Seyfert galaxy NGC7674 is distorted with tails to upper right and upper left in this false-color image. Note the companion galaxies. *(John W. Mackenty, Institute for Astronomy, University of Hawaii)* (b) Seyfert galaxy NGC4151 has a brilliant nucleus in visible light images (top). The Space Telescope Imaging Spectrograph reveals Doppler shifts showing that part of the expelled gas has a blue shift (to the right) and is approaching us, and part has a red shift (to the left) and is receding. *(John Hutchings, Dominion Astrophysical Observatory; Bruce Woodgate, GSFC/NASA; Mary Beth Kaiser, Johns Hopkins University; and the STIS Team)*

of ionized matter 1800 pc long squirting out of the nucleus (Figure 17-8a). Radio observations of the jet over a few years show that matter in the jet is traveling at about 48 percent the speed of light.

What could produce such violence in the cores of galaxies? Astronomers have found strong evidence that at least some active galaxies contain supermassive black holes at their centers, and the giant elliptical galaxy M87 (see Figure 16-3a) is one of the best examples. Hubble Space Telescope images of the nucleus reveal that the jet springs from a tiny spinning disk of gas (Figure 17-8b), and spectra show that the disk is rotating very fast. To be so small and spin so fast, it must contain a vast amount of matter at its center to hold the disk together. Only 25 pc from the center, the orbital velocity is 740 km/s, and the orbital period is only about 200,000 years. Kepler's third law tells us the mass inside the orbit is roughly 3 billion solar masses, presumably a supermassive black hole. The gas motions in other galaxies reveal that most contain supermassive black holes, even the nearby Andromeda Galaxy and its companion galaxy M32.

These supermassive black holes are not the remains of a single supernova explosion, as described in Chapters 13 and 14. The giant black holes at the centers of galaxies may have accumulated over billions of years as matter drifted inward, or they may have formed as the galaxies were born billions of years ago. Some astronomers think the giant black holes formed first, and the galaxies formed around them. This is one of the puzzles that supermassive black holes pose for today's astronomers.

We can combine the double-exhaust model with supermassive black holes to build a description of how such an object can eject jets and inflate radio lobes. Whether our description is a model, a hypothesis, or a theory is an interesting question, but astronomers are using the word "model" for this exciting stuggle to understand active galaxies.

The Search for a Unified Model

When a field of research is young, scientists often find many seemingly different phenomena, such as double-lobed radio galaxies, Seyfert galaxies, BL Lac objects,

Figure 17-8

(a) Giant elliptical galaxy M87 (Figure 16-3a) contains mostly cool stars and appears red in this image recorded with the Very Large Telescope (Figure 3-22b). The core of the galaxy is ejecting an 1800-pc-long jet of high-speed gas that is bright at ultraviolet wavelengths (blue in this image). *(European Southern Observatory)* (b) Hubble Space Telescope images show that the core of M87 is located at the center of a small disk of gas spinning at high speed. The rotation of the disk suggests the core contains a supermassive black hole. *(NASA/STScI)*

a

b

be the source of the jets often seen coming out of active galaxy cores, but the process by which jets are generated is not understood. The outer part of the disk, according to calculations, is a fat, dense doughnut (a torus) of dusty gas (Figure 17-9).

What we see when we view the core of an active galaxy must depend on how this disk is tipped with respect to our line of sight. We should note, at this point, that the accretion disk may be tipped at a steep angle to the plane of a galaxy, so just because we see a galaxy face-on doesn't mean we are looking at the accretion disk face-on. Thus the important factor is not the inclination of the galaxy, but the inclination of the accretion disk.

BL Lac objects, for example, are believed to be active cores emitting jets that are pointed directly at Earth. We are looking directly into the brilliant jet, and that drowns out the fainter parts of the galaxy. All we see is a brilliant point of light.

If the accretion disk is tipped slightly, we may not be looking directly into the central cavity, but we may be able to see some of the intensely hot gas there. This region emits broad spectral lines because the gas is orbiting at high velocities and the high Doppler shifts smear out the lines. Seyfert 1 galaxies may be explained by this phenomenon.

If we view the accretion disk edge-on, we cannot see the central area at all because the thick, dusty torus blocks our view. Instead, we see radiation emitted by gas lying above and below the central disk. Because this gas is farther from the center, it orbits more slowly and has smaller Doppler shifts. Thus we see narrow spectral lines. This might account for the Seyfert 2 galaxies.

This unified model is far from complete. The actual structure of accretion disks is poorly understood, and factors other than inclination must also be important. For example, what we see must also depend on

and so on. As the research matures, the scientists begin seeing similarities and eventually are able to unify the diverse phenomena as different aspects of a single process. This is the real goal of science, to organize evidence and theory into logical arguments that explain how nature works. (Window on Science 17-2). Astronomers studying active galaxies are now struggling to find a unifying model of active galaxy cores.

It seems clear that the cores of active galaxies contain supermassive black holes surrounded by accretion disks that are extremely hot near the black hole but cooler further out. Theory predicts that the central cavity in the disk where the black hole lurks is very narrow but that the disk there is "puffed up" and thick. This means the black hole may be hidden deep inside this narrow central well. The hot inner disk seems to

Window on Science 17-2

An argument can be a shouting match, but another definition of the word is "a discourse intended to persuade." Scientists construct arguments as part of the business of science not because they want to persuade others they are right, but because they want to test their ideas. For example, we can construct a scientific argument to show that the radio lobes that flank some active galaxies have their origin in jets coming from the cores of the galaxies. If we do our best and our argument is not persuasive, we begin to suspect we are on the wrong track. But if our argument seems convincing, we gain confidence that we are beginning to understand radio lobes.

A scientific argument is a logical presentation of evidence and theory with interpretations and explanations that help us understand some aspect of nature. Geologists might construct an argument to explain volcanism on mid-ocean islands, and the argument could include almost anything—maps, for example, or mineral samples compared with seismic evidence and mathematical models of subsurface magma motions. Such an argument might involve the physics of radioactive decay or observations of the shapes of bird beaks on the islands. The scientists would be free to include any evidence or theory that helps persuade, but they must observe one fundamental rule of scientific argument: They must be totally honest. The purpose of a scientific argument is to test our understanding, not to win votes or sell soap. Dishonesty in a scientific argument is self-deluding, and scientists consider such dishonesty the worst possible behavior.

On the other hand, scientists are human, and they often behave no better than other humans. They sometimes defend their own theories and attack opposing theories more vigorously than the evidence warrants. They may unintentionally overlook contradictory evidence or unconsciously misinterpret data. Of course, if a scientist did these things intentionally, we would be shocked because he or she would be violating the most serious rules of honesty in science. In the struggle to understand how nature works, however, it is not surprising to see that human passions rise and that scientific debate becomes heated. That is why science is so exciting—it is a struggle. Nonetheless, no matter how scientists behave in the excitement of a controversy, they all aspire to that ideal of the scientific argument—the cool, logical presentation of evidence that leads to a new understanding of nature.

Much of this book or any other scientific book consists of scientific arguments—logical presentations of evidence. As you read about any science, look for the arguments and practice organizing and presenting them to test your own understanding. The "Critical Inquiry" that ends each chapter section in this book is intended to illustrate how arguments are constructed. Use the same plan when you develop your own arguments.

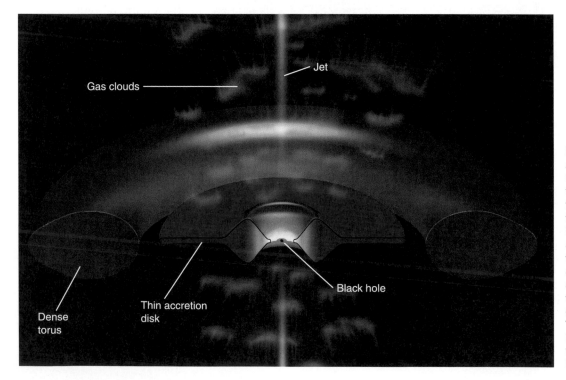

Figure 17-9
An accretion disk around a massive black hole. Matter flowing inward passes first through a large, opaque torus; then into a thinner, hotter disk; and finally into a hot, narrow cavity around the black hole. Gas clouds above and below the disk are excited by radiation from the hot disk. This diagram is not to scale. The central cavity may be only 0.01 pc in radius, while the outer torus may be 1000 pc in radius.

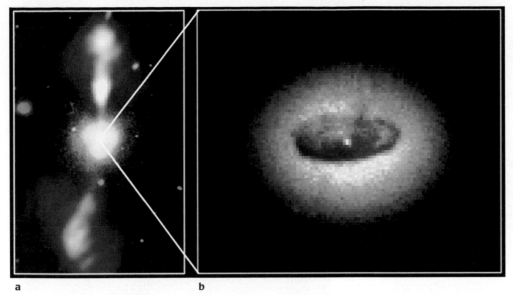

a   b

Figure 17-10
(a) Galaxy NGC4261 looks like a distorted fuzzy blob in an Earth-based photograph at visible wavelengths (white), but radio telescopes reveal jets leading into radio lobes (orange). (b) A Hubble Space Telescope image of the core of the galaxy reveals a bright central spot surrounded by a disk with its axis pointing at the radio lobes. (*L. Ferrarese, Johns Hopkins University, and NASA*)

the mass of the central black hole, the rate at which matter flows in, and the strength of magnetic fields in the disk. Astronomers are now working to build a unified model that incorporates all of the important factors in a single logical argument to explain the active cores of galaxies.

Galaxy Collisions and Active Cores

We must now consider the central knotty problem of active galaxies. What makes a galaxy erupt? The unified model suggests the core contains a supermassive black hole, and evidence is mounting that most galaxies contain such objects. These are not the black holes produced by dying stars but are large masses that probably began accumulating as the galaxy first took shape.

If all galaxies contain supermassive black holes, what triggers an eruption? Observation and theory tell us that mass must flow into the black hole through an accretion disk, and models of colliding galaxies show that during a collision significant amounts of mass are thrown into the center of the galaxies. Thus a collision between galaxies could trigger an eruption by feeding an otherwise dormant black hole. Most galaxies apparently contain supermassive black holes that are inactive because little matter is flowing into them.

There is strong evidence that collisions trigger core eruptions. Seyfert galaxies often have nearby companions, and active galaxies are often distorted as they would be by an encounter with another galaxy. NGC5128 is thought to be two galaxies in the act of colliding. In fact, high-quality images of this galaxy show that it is surrounded by shells of stars, a common by-product of galaxy mergers (Figure 16-17).

Evidence continues to mount. NGC4261 is a distorted galaxy flanked by radio lobes. Images of its core

reveal a disk of cool gas only 100 pc across spinning on an axis that points toward the radio lobes (Figure 17-10). Doppler shifts show that the disk is rotating so fast it must enclose 1.2 billion solar masses—evidently a supermassive black hole. Presumably, matter flows through the cool disk into a smaller, hot accretion disk around the black hole at the center and thus produces the jets we see. The distortion of this galaxy suggests it has suffered a recent collision that would explain the violence erupting in its core.

REVIEW   Critical Inquiry

Construct an argument to show that double-lobed radio galaxies and Seyfert galaxies have similar energy sources.

Observations show that double-lobed radio galaxies are produced by two jets flowing outward from a galaxy's core to inflate the radio lobes with hot gas. In many cases, radio maps actually show these jets. Seyfert galaxies do not have radio lobes, but the unified model of active galactic nuclei suggests that Seyferts are produced by matter flowing into an accretion disk around a supermassive black hole. We have seen that such an accretion disk can produce oppositely directed jets as in the case of the X-ray binary star SS433, so a more massive version in the core of a galaxy could presumably produce the jets we see inflating radio lobes. The same geometry, an accretion disk around a supermassive black hole, can also explain the difference between the two types of Seyfert galaxies, so it seems reasonable to suppose that double-lobed radio galaxies and Seyfert galaxies have similar energy sources at their centers.

Observational evidence and basic theory have led us to the unified model of active galactic nuclei. Although the model seems successful, there are still difficulties, and astronomers continue to struggle to understand the details. Nevertheless, the unified model may help us understand the mysterious quasars.

17-2 Quasars

The largest telescopes detect multitudes of faint, starlike points of light with peculiar emission spectra, objects called **quasars** (also known as *quasi-stellar objects,* or *QSOs*). Although astronomers now recognize quasars as some of the most distant visible objects in the universe, they were a mystery when they were first identified. The discovery of these objects in the 1960s and the struggle to understand them comprise a classic example of how scientists explore hypotheses, gather evidence, and build confidence in new ways to understand nature. We begin this section by telling that story, a story that will lead us to think about the depths of the universe in both space and time.

The Discovery of Quasars

In the early 1960s, radio interferometers (see Chapter 3) showed that a number of radio sources are much smaller than normal radio galaxies. Photographs of the location of these radio sources did not reveal a galaxy, not even a faint wisp, but rather a single starlike point of light. The first of these objects so identified was 3C48, and later the source 3C273 was added. Though these objects emitted radio signals like those from radio galaxies, they were obviously not normal radio galaxies. Even the most distant photographable galaxies look fuzzy, but these objects looked like stars. Their spectra, however, were totally unlike stellar spectra, so the objects were called quasi-stellar objects (Figure 17-11).

For a few years, the spectra of quasars were a mystery. A few unidentifiable emission lines were superimposed on a continuous spectrum. But in 1963, Maarten Schmidt at Hale Observatories tried red-shifting the hydrogen Balmer lines to see if they could be made to agree with the lines in 3C273's spectrum. At a red shift of 15.8 percent, three lines clicked into place (Figure

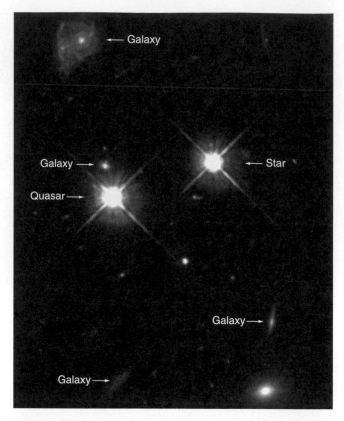

Figure 17-11
Quasars have starlike images clearly different from the images of distant galaxies. The spectra of quasars are unlike the spectra of stars or galaxies. (C. Steidel, Caltech, NASA)

17-12). Other quasar spectra quickly yielded to this approach, revealing even larger red shifts.

The red shift z is the change in wavelength $\Delta\lambda$ divided by the unshifted wavelength λ_0:

$$\text{red shift} = z = \frac{\Delta\lambda}{\lambda_0}$$

Astronomers commonly use the Doppler formula to convert these red shifts into radial velocities. Thus, the large red shifts of the quasars imply large radial velocities.

To understand the significance of these large radial velocities, we must draw on the Hubble law. Recall that the Hubble law states that galaxies have radial velocities proportional to their distances, and thus the distance to a quasar is equal to its radial velocity divided by the Hubble constant. The large red shifts of the quasars imply that they must be at great distances.

The red shift of 3C273 is 0.158, and the red shift of 3C48 is 0.37. These are large red shifts, but not as large as the largest then known for galaxies, about 1.0. Soon, however, quasars were found with red shifts much larger than that of any known galaxy. Some quasars are evidently so far away that galaxies at those distances are very difficult to detect. Yet the quasars are easily photographed. This leads us to the conclusion that qua-

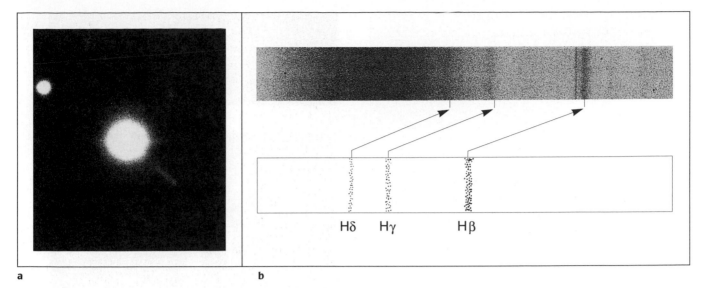

Figure 17-12
(a) This negative image of 3C273 shows the bright quasar at the center surrounded by faint fuzz. Note the jet protruding to lower right. (b) The spectrum of 3C273 (top) contains three hydrogen Balmer lines red-shifted by 15.8 percent. The drawing shows the unshifted positions of the lines. *(Courtesy Maarten Schmidt)*

sars must have 10 to 1000 times the luminosity of a large galaxy. Quasars must be superluminous.

Soon after quasars were discovered, astronomers detected fluctuations in brightness over times as short as a few days. Recall from our discussion of pulsars (Chapter 14) that an object cannot change its brightness appreciably in less time than it takes light to cross its diameter. The rapid fluctuations in quasars showed that they are small objects, not more than a few light-days or light-weeks in diameter.

Thus, by the late 1960s astronomers faced a problem: How could quasars be superluminous but also be very small? Since then, evidence has accumulated that quasars are the active cores of very distant galaxies. The distance to quasars is so great that we need to pay special attention to how astronomers estimate the distance to these objects.

Quasar Distances

Astronomers estimate the distances to quasars by using the Hubble law and dividing their radial velocity by the Hubble constant (Chapter 16). Thus, knowing the distance to a quasar depends critically on knowing its red shift and thus its radial velocity. Many quasars have red shifts greater than 1, and if we put such large red shifts into the Doppler formula (By the Numbers 4-2), we get radial velocities greater than the velocity of light. This seems unreasonable at first, but if we consider the nature of quasar red shifts, we will see that red shifts greater than 1 do make sense.

First we must note that the Doppler formula given in By the Numbers 4-2 is only an approximation good for low velocities. It is correctly called the classical Doppler formula, and it works fine if the velocity is significantly less than the velocity of light. For very high velocities, however, we need to use the **relativistic red shift** formula derived from Einstein's theory of relativity (By the Numbers 17-1). That formula tells us that very rapidly moving objects will appear to have red

By the Numbers 17-1

The Relativistic Red Shift
The classical Doppler formula given in By the Numbers 4-2 is correct only for velocities significantly less than that of light. The theory of relativity gives the following formula for the Doppler shift of an object traveling at any velocity up to the velocity of light. Here V_r is the radial velocity, c is the velocity of light, and z is the red shift:

$$\frac{V_r}{c} = \frac{(z+1)^2 - 1}{(z+1)^2 + 1}, \text{ where } z = \frac{\Delta\lambda}{\lambda_0}$$

Example: If an object has a red shift of 2, what is its velocity? *Solution:* First, $z + 1$ equals 3, and thus $(z + 1)^2$ equals 9. Then the velocity in terms of the velocity of light is

$$\frac{V_r}{c} = \frac{9-1}{9+1} = \frac{8}{10} = 0.8$$

That is, the object has a radial velocity equal to 80 percent the velocity of light.

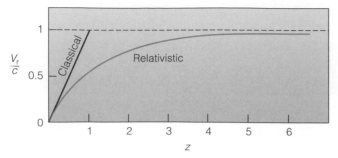

Figure 17-13
At high velocities, the relativistic red shift must be used in place of the classical approximation. Note that no matter how large z gets, the velocity can never equal that of light.

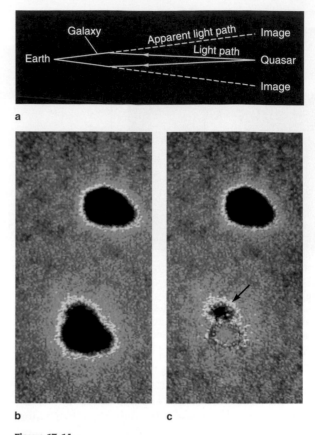

Figure 17-14
The gravitational lens effect. (a) A distant quasar can appear to us as multiple images if its light is deflected and focused by the mass of an intervening galaxy. The galaxy, being much fainter than the quasar, is not easily visible from Earth. (b) The two images of quasar 0957+561 appear as black blobs in this computer-enhanced false-color image. (c) When the upper image is subtracted from the lower image, we are able to detect the faint image of the giant elliptical galaxy (arrow) whose gravity produces the gravitational lens effect. *(Courtesy Alan Stockton, Institute for Astronomy, University of Hawaii)*

shifts greater than 1 even though they are not traveling faster than the velocity of light. Figure 17-13 plots the velocity of an object versus its red shift z, and we see that the true z we observe can become quite large without the object's actually exceeding the velocity of light. Thus, the quasars with red shifts greater than 1 should not give us special concern. They do not violate any rules of physics.

Now that we feel some comfort with quasar red shifts greater than 1, we should note that the red shifts associated with the expansion of the universe are not described precisely by the Doppler effect. In the next chapter, we will discuss the expansion of the universe in detail, and we will conclude that galaxy red shifts are not really Doppler shifts. The exact relation between a quasar's red shift and its expansion velocity is similar to the relativistic red shift formula in that red shifts greater than 1 imply velocities near but not exceeding the velocity of light. But the exact relation depends critically on such parameters as the average density of the universe, parameters that are not well known. Thus, we can't at present convert quasar red shifts into precise distances.

What evidence do we have that quasars really are located at these great distances? A discovery made in 1979 provides a dramatic illustration of the tremendous distances to quasars. The object catalogued as 0957+561 lies just a few degrees west of the bowl of the Big Dipper and consists of two quasars separated by only 6 seconds of arc. The spectra of these quasars are identical and even have the same red shift of 1.4136. Quasar spectra are as different as fingerprints, so when two quasars so close together were discovered to have the same spectra and the same red shift, astronomers concluded that they were separate images of the same quasar.

The two images are formed by a **gravitational lens,** an effect first predicted by Einstein in 1936 but never seen before. The gravitational field of a galaxy between us and the distant quasar acts as a lens, bending the

light from the quasar and focusing it into multiple images (Figure 17-14a). The lensing galaxy itself is much too far away to be easily visible against the glare of the quasar. However, when astronomers used computer techniques to subtract the upper image of the quasar from the bottom image, the faint image of a giant elliptical galaxy appeared (Figure 17-14b and c). This galaxy is probably the brightest member of a cluster of galaxies. The red shift of the galaxy is only 0.36, much less than that of the quasar, 1.4136. As a result, we conclude that the quasar must lie behind the galaxy and that the galaxy is so far away it is hardly visible even though it is a giant elliptical. Other examples of gravitational lensing of quasars have been found, and they provide very strong evidence that the quasars really are distant objects, as implied by their large red shifts.

The quasars are very distant and must be superluminous and quite small. To understand quasars, we must examine their observed characteristics and compare that evidence with the unified model of active galaxies.

Evidence of Quasars in Distant Galaxies

How do astronomers know that quasars are located in distant galaxies? A few thousand quasars have been found, but most are so distant any host galaxy would be invisible. The largest known red shifts are about 5, meaning that the objects are exceedingly far away. The look-back time to these objects is so large we see them as they were when the universe was only 10 percent its present age.

Before the launch of the Hubble Space Telescope and the construction of new, large telescopes on Earth, it was difficult to detect anything near quasars. Lower-red-shift quasars are not quite so far away, however, and fuzzy wisps are visible around such objects. The spectra of this quasar fuzz are the same as spectra of normal spiral and elliptical galaxies, so astronomers began to suspect that quasars were imbedded in galaxies.

Further evidence appeared when a new point of light appeared near quasar QSO 1059+730. The point of light was a supernova exploding in the galaxy that hosts the quasar (Figure 17-15). Although the galaxy is too faint to see, the appearance of the supernova confirms that the quasar lies in a galaxy.

A few quasars have characteristics related to active galaxies. For example, quasar 3C273 seems to be ejecting a jet much like the jet in M87 (see Figure 17-12a). Also, some quasars have radio lobes. This evidence seems to link quasars with active galactic nuclei.

Photographs made with the largest telescopes on Earth and with the Hubble Space Telescope give us dramatic evidence. The galaxies that host quasars are in many cases clearly visible. Furthermore, the galaxies are distorted or have nearby companions (Figure 17-16). The host galaxies appear to be involved in collisions and interactions with other galaxies, or they have the distorted shapes that we recognize as the result of such collisions. Because we suspect that galaxy collisions can cause the core of a galaxy to erupt, we can also suspect that quasars may have been triggered into existence by interactions between galaxies.

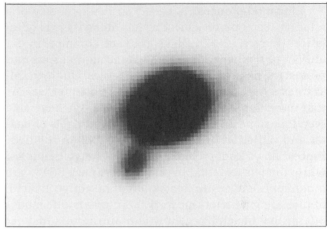

Figure 17-15
Two negative images of the quasar 1059+730. The lower image shows an extra point of light—apparently a supernova that exploded in the galaxy associated with the quasar. *(Bruce Campbell)*

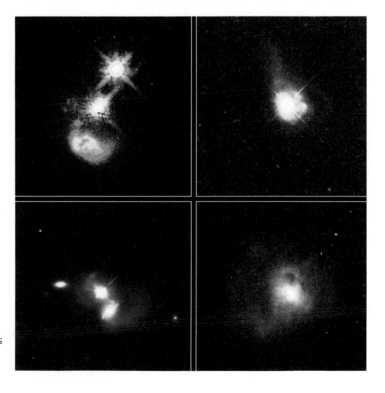

Figure 17-16
The bright object at the center of each of these images is a quasar. Fainter objects near the quasar are galaxies distorted by collisions. Compare the upper left ring-shaped galaxy with Figure 16-20, and compare the tail in the lower left image with the Antennae galaxies in Figure 16-18. *(J. Bahcall, Institute for Advanced Study, Mike Disney, University of Wales, NASA)*

The evidence seems conclusive, and astronomers are now able to combine evidence and theory in a logical argument that describes how a quasar works.

A Model Quasar

The most satisfying part of science is combining evidence with theory to understand some aspect of nature. We have learned to interpret quasar red shifts, and we have reviewed the basic characteristics of quasars. Can we now assemble the evidence and theories in a logical argument that describes quasars? That is, can we create a model of a quasar?

Our model quasar is based on the geometry of a rotating accretion disk ejecting jets along its axis of rotation (Figure 17-17). Disks are very common in the universe. We have discovered rotating disks around protostars, neutron stars, and black holes. Our own galaxy has a disk shape, and in an earlier chapter we saw that our solar system has a disk shape. We have also seen that some disks can produce powerful jets of matter and radiation directed along the axis of rotation. Bipolar flows from protostars are low-energy examples, while the jets expelled by the X-ray binary SS433 are much more powerful. Jets expelled by active galactic nuclei are even more powerful. To create our model quasar, we imagine a galaxy containing this disk-jet structure at its core. Our model is, in fact, just an extension of the unified model of active galactic nuclei discussed earlier in this chapter.

To make the energy typical of a quasar we need a black hole of 10 million to a billion solar masses surrounded by a large, opaque disk. The innermost part of the accretion disk, much less than a light-year in diameter, would be intensely hot. If matter amounting to about 1 solar mass per year flows through the accretion disk into the black hole, the energy released could power a quasar. Exactly how some of this matter and energy are focused into two oppositely directed high-speed jets isn't well understood, but magnetic fields are probably involved. In any case, as matter flows into the black hole, some is ejected in jets and a great deal of energy is radiated away. One astronomer described the situation by commenting, "Black holes are messy eaters."

This model could explain the different kinds of radiation we receive from quasars (see Figure 17-9). A small percentage of quasars are strong radio sources, and the radio radiation may come from synchrotron radiation produced in the high-energy gas and magnetic fields in the jets. Much of the light we see from a quasar is spread in a continuous spectrum, and that is the light that fluctuates quickly. Thus, it must come from a small region. Our model could produce this light in the innermost cavity surrounding the black hole. Because this is a very small region, roughly the size of our solar

Figure 17-17
A model to account for the quasar phenomenon is based on disk-jet geometry, as shown in this artist's conception. In the heart of a galaxy, matter flows into a hot accretion disk around a black hole containing at least a million solar masses. Radiation from the accretion disk and jets ejected along the axis of rotation heat the surrounding gas and turn the galaxy into a quasar. *(Dana Berry, Space Telescope Science Institute)*

system, rapid fluctuations can occur, probably because of random fluctuations in the flow of matter into the accretion disk. The emission lines in quasar spectra don't fluctuate rapidly, and that suggests they are produced in a larger region. In our model, emission lines are emitted by clouds of gas surrounding the core in a region many light-years in diameter and excited by the intense synchrotron radiation streaming out of the central cavity.

Just as in the unified model of active galactic nuclei, the orientation of the accretion disk is important (see Figure 17-9). If the disk faces us so that one of its jets points directly toward us, we may see one kind of quasar. But if the disk is tipped slightly so the jet is not pointing straight at us, we may see a slightly different kind of quasar. Astronomers are now using this unified model to sort out the different kinds of quasars and active galaxies so we can understand how they are related. For example, using infrared radiation to penetrate dust, astronomers observed the core of the double-lobed radio galaxy Cygnus A (see Figure 17-2) and found an object much like a quasar. Astronomers have begun to refer to such objects as buried quasars.

As in the case of active galaxies, many quasars are found in distorted galaxies that are interacting with nearby companions. Such a tidal interaction could throw matter into a central black hole and trigger a quasar outburst. Figure 17-18 shows three quasars that appear to illustrate different stages in such an encounter.

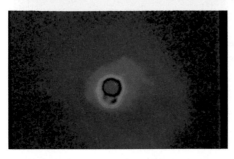

Figure 17-18

These three quasars appear to illustrate three stages in the triggering of a quasar by the tidal interaction between galaxies. The quasar IRAS 00275-2859 (top) is located in a galaxy interacting with a relatively distant companion (not visible in this image) and thus represents an early stage. In the galaxy containing quasar PG1613 (middle), the galaxies are closer together. The nucleus of the companion is visible at the 2 o'clock position, and the outline of the interacting galaxies is distorted. Markarian 231 (bottom) is located in a severely distorted galaxy containing radial streaks and a tail to the lower right. Such features are common when galaxies merge, suggesting that Markarian 231 represents a late stage in the merger of two galaxies. *(NOAO)*

It is important to remember that the look-back time to quasars is very large. We see them as they were when the universe was young. In the next chapter, we will see that the universe appears to have begun about 15 billion years ago, and we can suppose that gas collected to form the first galaxies by the time the universe was a few billion years old. If at least some of those galaxies had supermassive black holes at their centers, then collisions between galaxies could have triggered eruptions in their cores that we would see as quasars. Because the universe was so young then, it had not expanded much, and the galaxies were closer together and would have collided more often. Also, theorists suppose that young galaxies would contain more gas because they would not have used the gas up to make stars, and such gas-rich galaxies might erupt more violently during a collision. Nevertheless, the quasar phenomenon must have been rare; at any one time, few galaxies had quasars in their cores.

Quasars are most common with red shifts of about 2, but they become much less common at red shifts above 2.7. The largest-known red shifts are about 5, and such high red-shift quasars are quite rare. This suggests that when we look at quasars with red shifts of 2 or so, we are looking back to an age when galaxies were actively colliding and merging. In that age in the history of the universe, quasars were about 1000 times more common than they are now. When we look at even more distant quasars, we are looking back to an earlier time before most galaxies formed. We see no quasars with really large red shifts because we would be looking back to an age before there were galaxies.

REVIEW Critical Inquiry

Why are most quasars so far away?

To analyze this question we must combine two factors. First, quasars are the active cores of galaxies, and only a small percentage of galaxies contain active cores. In order to sample a large number of galaxies in our search, we must extend our search to great distances. Most of the galaxies lie far away from us because the amount of space we search increases rapidly with distance. Just as most seats in a baseball stadium are far from home plate, most galaxies are far from our Milky Way Galaxy. Thus, most of the galaxies that might contain quasars lie at great distances.

But a second factor is much more important. The farther we look into space, the farther back in time we look. It seems that there was a time in the distant past when quasars were more common, and thus we see most of those quasars at large look-back times, meaning at large distances. For these two reasons, most quasars lie at great distances.

If our model of quasars is correct, then quasars must be triggered into eruption. What observational evidence supports that argument?

Our study of active galaxies has led us far out in space and back in time to quasars. The light now arriving from the most distant quasars left them when the universe was only a fraction of its present age. The quasars stand at the very threshold of the study of the history of the universe itself, the subject of the next chapter.

Summary

Some galaxies appear to be suffering eruptions in their cores. Such active-core galaxies are powerful sources of energy and appear to be related to the in-fall of matter into central, massive black holes.

Double-lobed radio galaxies emit radio energy from lobes on either side of the galaxy. In some cases, we can see jets leading from the core of the galaxy out into the lobes. The double-exhaust model proposes that the core of the galaxy ejects relativistic particles in two jets that punch into the intergalactic medium and blow up the radio lobes like balloons. The impact of the beams on the intergalactic medium produces the hot spots detected on the outer edges of many radio lobes. We see many active galaxies producing jets, and we can see instances where the jets are blown back by the intergalactic medium to produce head–tail galaxies.

Some galaxies, such as Seyfert galaxies and BL Lac objects, have small, brilliant cores. Giant elliptical galaxies such as M87 have bright cores with jets.

The energy source for active galactic nuclei has been described as the unified model—in which mass flows into an accretion disk around a supermassive black hole at the center of the galaxy. Active galaxies are often distorted by interaction with other galaxies, and such tidal interactions could throw matter into the central region, feed the black hole, and trigger an eruption. What we would see would depend on the inclination of the disk to our line of sight. If the disk were face-on, we would look directly into the hot cavity and see a brilliant BL Lac object. At higher inclinations, we would see a Seyfert 1 galaxy; and if the disk were edge-on, we would see a Seyfert 2 galaxy.

The quasars appear to be related objects. Their spectra show emission lines with very large red shifts. The large red shifts are interpreted to mean the quasars are at great distance and thus must be superluminous. Yet, because they fluctuate in brightness over intervals as short as a week or less, quasars must also be very small. The detection of quasars in clusters of galaxies, gravitational lenses, and the galaxy-like spectra of quasar fuzz show that quasars are indeed located inside galaxies at great distances.

Growing evidence suggests that quasars are the active cores of very distant galaxies. As in the case of active galaxies, quasars are often found in galaxies that are distorted or interacting with other galaxies. The rapid flow of matter into a central, massive black hole could produce so much energy that the quasars would be visible at great distances.

Because the quasars lie at great distances, we see them as they were long ago when the universe was much younger than it is now. Quasars are most common at a red shift of about 2, which suggests that there was an age when galaxies had formed, were still rich in gas and dust, but were close enough together to collide and merge often and thus produce quasars. Few quasars are found with larger red shifts, and no quasars have been found beyond a red shift of about 5 (to date). This suggests that at the largest red shifts we are looking back to a time when the universe had not yet formed galaxies.

New Terms

radio galaxies	head–tail radio galaxies
active galaxies	Seyfert galaxies
active galactic nuclei	BL Lac objects
double-lobed radio galaxies	quasars
hot spots	relativistic red shifts
double-exhaust model	gravitational lens

Review Questions

1. What is the difference between the terms *radio galaxy* and *active galaxy*?

2. What evidence do we have that the energy source in double-lobed radio galaxies lies at the center of the galaxy?

3. How does the peculiar rotation of NGC5128 help us understand the origin of this active galaxy?

4. What statistical evidence suggests that Seyfert galaxies have suffered recent interactions with other galaxies?

5. How does the unified model explain the two kinds of Seyfert galaxies?

6. What observations are necessary to identify the presence of a supermassive black hole at the center of a galaxy?

7. How does the unified model implicate collisions and mergers in triggering active galaxies?

8. Why were quasars first noticed as being peculiar?

9. How do the large red shifts of quasars lead us to conclude they must be very distant?

10. Why do we conclude that quasars are superluminous but must be very small?

11. How do gravitational lenses provide evidence that quasars are distant?

12. What evidence do we have that quasars occur in distant galaxies?

13. How can our model quasar explain the different radiation we receive from quasars?

14. What evidence do we have that quasars must be triggered by collisions and mergers?

15. Why are there few quasars at low red shifts and at high red shifts but many at red shifts of about 2?

Discussion Questions

1. Why do quasars, active galaxies, SS433, and protostars have similar geometry?

2. By custom, astronomers refer to the unified *model* of AGN and not to the unified *hypothesis* or unified *theory*. In your opinion, which of the words seems best?

Problems

1. The total energy stored in a radio lobe is about 10^{53} J. How many solar masses would have to be converted into energy to produce this energy? (*Hints:* Use $E = m_0 c^2$. One solar mass equals 2×10^{30} kg.)

2. The jet in NGC5128 is traveling at 5000 km/s and is 40 kpc long. How long will it take for gas to travel from the core of the galaxy to the end of the jet? (*Hint:* 1 pc equals 3×10^{13} km.)

3. Cygnus A is roughly 225 Mpc away, and its jet is about 50 seconds of arc long. What is the length of the jet in parsecs? (*Hint:* Use the small-angle formula.)

4. If the average giant elliptical galaxy is 50 kpc in diameter and the visible galaxy associated with 3C449 (see Figure 17-5a) is 1 minute of arc in diameter, calculate an estimate of the distance to the galaxy. (*Hint:* Use the small-angle formula.)

5. Use the small-angle formula to find the linear diameter of a radio source with an angular diameter of 0.0015 second of arc and a distance of 3.25 Mpc.

6. If the active core of a galaxy contains a black hole of 10^6 solar masses, what will the orbital period be for matter orbiting the black hole at a distance of 0.33 AU? (*Hint:* See By the Numbers 11-4.)

7. If a quasar is 1000 times more luminous than an entire galaxy, what is the absolute magnitude of such a quasar? (*Hint:* The absolute magnitude of a bright galaxy is about −21.)

8. If a quasar in Problem 7 were located at the center of our galaxy, what would its apparent magnitude be? (*Hint:* Use the magnitude–distance formula.)

9. What is the radial velocity of 3C48 if its red shift is 0.37?

10. The hydrogen Balmer line H-beta has a wavelength of 486.1 nm. It is shifted to 563.9 nm in the spectrum of 3C273. What is the red shift of this quasar? (*Hint:* What is $\Delta\lambda$?)

Critical Inquiries for the Web

1. What object currently holds the distinction as the farthest known galaxy? Search the Web for information on this distant object, and find out its red shift and distance. What is the look-back time for this object?

2. Gravitational lenses were first predicted by Einstein in 1936 but were not observed until recently. Search the Web for instances of gravitational lensing of galaxies and quasars. For a particular case, discuss how the lens effect allows astronomers to determine information about the lensing and/or lensed objects that might not have been available without the alignment.

Go to the Brooks/Cole Astronomy Resource Center (www. brookscole.com/astronomy) for critical thinking exercises, articles, and additional readings from InfoTrac College Edition, Brooks/Cole's online student library.

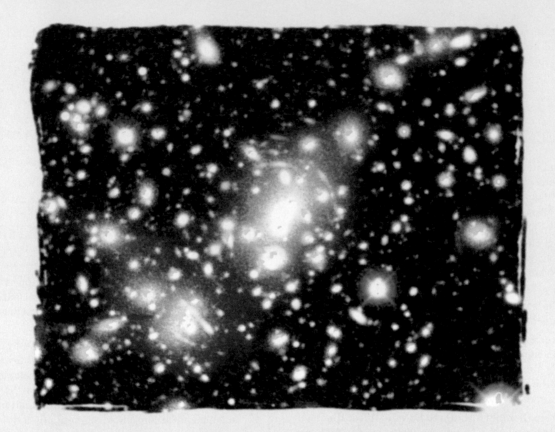

CHAPTER EIGHTEEN

Cosmology

The Universe, as has been observed

before, is an unsettlingly big place, a

fact which for the sake of a quiet life

most people tend to ignore.

Douglas Adams
*The Restaurant at the End
of the Universe*

Guidepost

This chapter is a culmination of our study of astronomy. Since Chapter 1, our discussion has focused on understanding the universe. Our outward voyage has carried us from our planet past other worlds, stars, and galaxies. Now we reach the limit of our voyage in space and time, the origin and evolution of the universe as a whole.

The ideas in this chapter are the biggest and the most difficult in all of science. Our imaginations can hardly grasp such ideas as the edge of the universe and the first instant of time. It is not surprising that the biggest questions are the most challenging.

But this chapter is not an end to our voyage. Once we discuss the nature of the universe as a whole, we are ready to focus on our place as living beings in that universe and the possibility that other beings admire the universe as we do.

What is the biggest number? A billion? How about a billion billion? That is only 10^{18}. How about 10^{100}? That very large number is called a googol, a number that is at least a billion billion times larger than the total number of atoms in all of the galaxies we can observe. Can you name a number bigger than a googol? Try a billion googols, or better yet, a googol of googols. You can go even bigger than that, 10 raised to a googol. (That's called a googolplex.) No matter how large a number you name, we can name a bigger number. That is what infinity means—big without limit.

If the universe is infinite in size, then you can name any distance you want, and the universe is bigger than that. Try a googol to the googol light-years. If the universe is infinite, there is more universe beyond that distance.

Is the universe infinite or finite? In this chapter, we will try to answer that question and others, not by playing games with big numbers, but through **cosmology,** the study of the universe as a whole.

If the universe is not infinite, then we face the edge–center problem. Suppose that the universe is not infinite, and you journey out to the edge of the universe. What do you see: a wall of cardboard? a great empty space? nothing, not even space? What happens if you try to stick your head out beyond the edge? These are almost nonsensical questions, but they illustrate the problem we have when we try to think about an edge to the universe. Modern cosmologists believe the universe cannot have an edge. In this chapter, we will see how the universe might be finite but have no edge.

If the universe has no edge, it can have no center. We find the centers of things—galaxies, globular clusters, oceans, pizzas—by referring to their edges. With no edge, there can be no center.

Finally, we must answer a basic question. How old is the universe? If it is not eternal (infinite in age), it must have had a beginning, and we must think about how the universe began and how it will end. Are a beginning and an ending like edges in time?

These are arguably the most challenging questions in human knowledge, and we start our search for answers by trying to understand the structure of the universe. That leads us, as is always the case in astronomy, to compare observations with theories.

18-1 The Structure of the Universe

We begin our study of cosmology by considering the most basic property of the universe—its geometry in space and time. To help us, we will discuss the two fundamental observations of cosmology. The first of these observations is easy to make: It gets dark at night. That observation can tell us that the universe had to have a beginning.

Olbers' Paradox and the Necessity of a Beginning

We have all noticed that the night sky is dark. However, reasonable assumptions about the geometry of the universe can lead us to the conclusion that the night sky should glow as brightly as a star's surface. This conflict between observation and theory is called **Olbers' paradox** after Heinrich Olbers, a Viennese physician and astronomer who discussed the problem in 1826.

However, Olbers' paradox is not Olbers'. The problem of the dark night sky was first discussed by Thomas Digges in 1576 and was further analyzed by such astronomers as Johannes Kepler in 1610 and Edmond Halley in 1721. Olbers gets the credit through an accident of scholarship on the part of modern cosmologists who did not know of previous discussions. What's more, Olbers' paradox is not a paradox. We will be able to understand why the night sky is dark by revising our assumptions about the nature of the universe.

To begin, let's state the so-called paradox. Suppose the universe is static, infinite, eternal, and uniformly filled with stars. (The aggregation of stars into galaxies makes no difference to our argument.) If we look in any direction, our line of sight must eventually reach the surface of a star (Figure 18-1). Consequently, every point on the surface of the sky should be as bright as the surface of a star, and it should not get dark at night.

Of course, the most distant stars would be much fainter than the nearer stars, but there would be a greater number of distant stars than nearer stars. The intensity of the light from a star decreases according to the inverse square relation, so distant stars would not contribute much light. However, the farther we look in space, the larger the volume we survey. Thus, the number of stars we see at any given distance increases as the square of the distance. The two effects cancel out, and the stars at any given distance contribute as much total light as the stars at any other distance. Then, given our assumptions, every spot on the sky must be occupied by the surface of a star, and it should not get dark at night.

Imagine the entire sky glowing with the brightness of the surface of the sun. The glare would be overpowering. In fact, the radiation would rapidly heat Earth and all other celestial objects to the average temperature of the surface of the stars, roughly a few thousand degrees. Thus, we can pose Olbers' paradox in another way: Why is the universe so cold?

Olbers assumed that the sky was dark because clouds of matter in space absorb the radiation from distant stars. But this interstellar medium would gradually heat up to the average surface temperature of the stars, and the gas and dust clouds would glow as brightly as the stars.

a

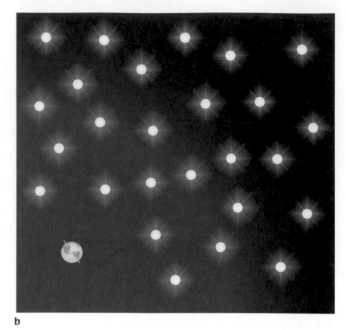

b

Figure 18-1
(a) Every direction we look in a forest eventually reaches a tree trunk, and we cannot see out of the forest. *(Photo courtesy Janet Seeds)* (b) If the universe is infinite and uniformly filled with stars, then any line from Earth should eventually reach the surface of a star. This assumption predicts that the night sky should glow as brightly as the surface of the average star, a puzzle commonly referred to as Olbers' paradox.

Today, cosmologists believe they understand why the sky is dark. Olbers' paradox makes the incorrect prediction that the night sky should be bright because it is based on two incorrect assumptions: The universe is neither static nor infinitely old.

In Chapter 16, we saw that the galaxies are receding from us. The distant stars in these galaxies are receding from Earth at high velocity, and their light is Doppler-shifted to long wavelengths. We can't see the light from these stars because their light is red-shifted, and the energy of the photons is reduced to levels we cannot detect. Expressed in another way, the universe is cold because the photons from very distant stars arrive with such low energy they cannot warm objects by any significant amount. Although this explains part of the problem, the red shifts of the distant galaxies are not enough to make the night sky as dark as it appears. The red shifts explain no more than 10 percent of the problem.

The second part of the explanation was first stated by Edgar Allan Poe in 1848. He proposed that the night sky was dark because the universe was not infinitely old but had been created at some time in the past. The more distant stars are so far away that light from them has not reached us yet. That is, if we look far enough, the look-back time is greater than the age of the universe, and we look back to a time before stars began to

shine. Thus, the night sky is dark because the universe is not infinitely old.

This is a powerful idea because it clearly illustrates the difference between the universe and the observable universe. The universe is everything that exists, and it could be infinite. But the **observable universe** is the part that we can see. We will learn later that the universe is about 15 billion years old. In that case, the observable universe has a radius of 15 billion light-years. Do not confuse the observable universe, which is finite, with the universe as a whole, which could be infinite.

The assumptions that we made when we described Olbers' paradox were at least partially in error. This illustrates the importance of assumptions in cosmology and serves as a warning that our commonsense expectations are not dependable. All of astronomy is reasonably unreasonable—that is, reasonable assumptions often lead to unreasonable results. That is especially true in cosmology, so we must always examine our assumptions with particular care.

Basic Assumptions in Cosmology

Although we could make many assumptions, three are basic—homogeneity, isotropy, and universality.

Homogeneity is the assumption that matter is uniformly spread throughout space. Obviously, this is not true on the small scale, because we can see matter concentrated in planets, stars, and galaxies. Homogeneity refers to the large-scale distribution. If the universe is homogeneous, we should be able to ignore individual galaxies and think of matter as material evenly spread through space. Observation of the distribution of galaxies reveals that the universe is homogeneous only on the largest scales, a subject we will discuss later in this chapter.

Isotropy is the assumption that the universe looks the same in every direction, that it is isotropic. On the small scale this is not true, but if we ignore local variations like galaxies and clusters of galaxies, the universe should look the same in any direction. For example, we should see roughly the same number of galaxies in every direction. Again, observations suggest that the universe is isotropic only on the largest scale.

The most easily overlooked assumption is **universality,** which holds that the physical laws we know on Earth, such as the law of gravity, apply everywhere in the universe. Although this may seem obvious at first, some astronomers challenge universality by pointing out that when we look out in space we look back in time. If the laws of physics change with time, we may see peculiar effects when we look at distant galaxies. For now we will assume that the physical laws observed on Earth apply everywhere in the universe.

The assumptions of homogeneity and isotropy lead to an assumption so fundamental it is called the **cosmological principle.** According to this principle, any observer in any galaxy sees the same general features of the universe. For example, all observers should see the same kinds of galaxies. As in previous assumptions, we ignore local variations and consider only the overall appearance of the universe, so the fact that some observers live in galaxies in clusters and some live in isolated galaxies is only a minor irregularity.

Evolutionary changes are not included in the cosmological principle. If the universe is expanding and the galaxies are evolving, observers living at different times may see galaxies at different stages. The cosmological principle says that once observers correct for evolutionary changes, they should see the same general features.

The cosmological principle is actually an extension of the **Copernican principle.** Copernicus said that Earth is not in a special place; it is just one of a number of planets orbiting the sun. The cosmological principle says that there are no special places in the universe. Local irregularities aside, one place is just like another. Our location in the universe is typical of all other locations.

If we accept the cosmological principle, then we may not imagine that the universe has an edge or a center. Such locations would not be the same as all other locations.

The first fundamental observation of cosmology, that the night sky is dark, has led us to deep insights about the nature of the universe. The second fundamental observation will be equally revealing.

The Expansion of the Universe

The second fundamental observation of cosmology is that the spectra of galaxies contain red shifts that are proportional to their distances (Figure 18-2). These red shifts are commonly interpreted as Doppler shifts due to the recession of the galaxies, which is why we say the universe is expanding.

Edwin P. Hubble discovered the velocity–distance relationship in 1929 using the spectra of only 46 galaxies (see Figure 16-11). Since then, the Hubble law, as it's known, has been confirmed for large numbers of galaxies out to great distances. This law is clear evidence that the universe is expanding uniformly and has no center of expansion.

To see how the Hubble law implies uniform, centerless expansion, we can think about an analogy (Window on Science 18-1). Imagine that we make a loaf of

Figure 18-2
Galaxies at different distances are shown at left, with their spectra shown at the right. The two vertical blue lines mark the unshifted location of the H and K lines of once-ionized calcium. The expansion of the universe causes a red shift that moves the spectral lines to longer wavelengths, as shown by the red arrows. More distant galaxies have larger red shifts, expressed as radial velocities in kilometers per second. *(Caltech)*

"The economy is overheating and it may seize up," an economist might say. Economists like to talk in analogies because economics is often abstract, and one of the best ways to think about abstract problems is to find a more approachable analogy. Rather than discuss the details of the national economy, we might make conclusions about how the economy works by thinking about how a gasoline engine works. Of course, if our analogy is not a good one, it can mislead us.

Much of astronomy is abstract, and cosmology is the most abstract subject in astronomy. Furthermore, cosmology is highly mathematical, and unless we are prepared to learn some of the most difficult mathematics known, we must use analogies, such as raisin bread or an ant on an orange, to discuss the nature of the universe and the curvature of space–time.

Reasoning by analogy is a powerful technique. An analogy can reveal unexpected insights and lead us to further discoveries. But carrying an analogy too far can be misleading. We might compare the human brain to a computer, and that would help us understand how data flow in and are processed and how new data flow out. But our analogy is flawed. Data in computers are stored in specific locations, but memories are stored in the brain in a distributed form. No single brain cell holds a specific memory. If we carry the analogy too far, it can mislead us. Whenever we reason by analogy, we should be alert for potential problems.

As you study any science, be alert for analogies. They are tremendously helpful, but we must always take care not to carry them too far.

a

b

Figure 18-3
The uniform expansion of the universe can be represented by raisins in a loaf of raisin bread (a). As the dough rises (b), raisins originally near each other move apart more slowly, and raisins originally farther apart move away from each other more rapidly. A colony of bacteria living on any raisin will find that the velocities of recession of the other raisins are proportional to their distances.

raisin bread (Figure 18-3). As the dough rises, the expansion pushes the raisins away from one another. Two raisins that were originally separated by only 1 cm move apart rather slowly, but two raisins that were originally separated by 4 cm of dough are pushed apart more rapidly. The uniform expansion of the dough causes the raisins to move away from each other at velocities proportional to their distances. According to the Hubble law, the larger the distance between two galaxies, the faster they recede from each other. This is exactly the result we expect from uniform expansion.

The raisin bread also shows that the expansion has no identifiable center. Bacteria living on one of the raisins would see themselves surrounded by a universe of receding raisins. (Here we must assume that they cannot see to the edge of the loaf—that is, they must not be able to see to the edge of their universe.) It does not matter on which raisin the bacteria live. The bacterial astronomers would measure the distances and velocities of recession and derive a bacterial Hubble law showing that the velocities of recession are proportional to the distances. So long as they cannot see the edge of the loaf, they cannot identify any raisin as the center of the expansion.

Similarly, we see galaxies receding from us, but no galaxy or point in space is the center of the expansion. Any observer in any galaxy would see the same expansion that we see.

Although astronomers and cosmologists commonly refer to these red shifts as Doppler shifts and often speak of the recession of the galaxies, relativity provides a more elegant explanation. Einstein's theory of general relativity explains the expansion of the universe as an expansion of space–time itself. A photon traveling through this space–time is stretched as space–

time expands, and the photon arrives with a longer wavelength than it had when it left (Figure 18-4). Photons from distant galaxies travel for a longer time and are stretched more than photons from nearby galaxies. Thus, the expansion of the universe is an expansion of the geometry and not just a simple recession of the galaxies.

This explains, by the way, why the equation for the relativistic red shift given in By the Numbers 17-1 is not quite correct for galaxies. That equation comes from the special theory of relativity and expresses the Doppler shift caused by the motion of an object through space. But the galaxies, in receding from one another, are not moving through space any more than the raisins in our raisin bread are swimming through the bread dough. The galaxies are carried away from each other by the uniform expansion of space itself.

We need not worry about the actual form of the red-shift equation; it is different for different assumptions about the nature of space and time. Rather, we must focus on the central concept of modern cosmology, the curvature of space–time.

The Geometry of Space–Time

How can the universe expand if it does not have any extra space to expand into? How can it be finite if it doesn't have an edge? The properties we have ascribed to the universe seem to violate common sense, so we must proceed with caution. Our everyday intuition can be misleading. For example, everything we touch and see every day has an edge, so thinking about an edgeless universe can be a challenge. One powerful tool is reasoning by analogy, but we must also use words carefully so they do not mislead us (Window on Science 18-2). With such cautions in mind, we can use general relativity to understand the geometry of space–time.

Einstein's theory of general relativity proposes that the presence of mass can curve space–time and that we sense this curvature as a gravitational field. This theory explains Earth's gravity and that of a black hole, but it also predicts that the entire universe can have a general curvature that determines the geometry and expansion of the universe. Thus, modern cosmological theories are based on general relativity and curved space–time.

To discuss curvature, we can use a two-dimensional analogy for our three-dimensional universe. Suppose an ant lived on an orange and was a perfectly two-dimensional creature (Figure 18-5). That is, the ant could move in two dimensions over the surface of the orange but could not move in the third dimension perpendicular to the surface of the orange. Furthermore, suppose that light could not travel perpendicular to the surface of the orange. Then the ant would never know there was a third dimension, and it could wander over the

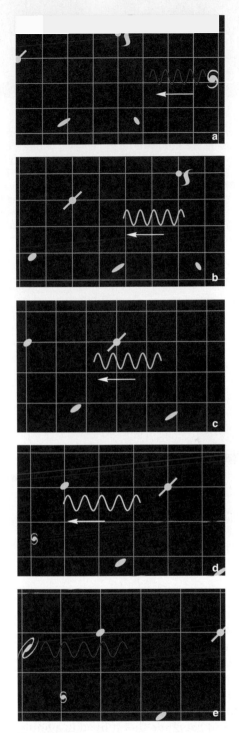

Figure 18-4

General relativity explains that the red shift of light from distant galaxies is produced by the stretching of space–time (shown here as an expanding grid) between the time the photon is emitted (a) and the time it arrives at Earth (e). (Notice that the wavelength in frame e is longer than that in frame a.) Photons from distant galaxies spend more time traveling through space and are thus stretched more than photons from nearby galaxies. Thus, distant galaxies have a larger red shift.

There are certain words we should never say, not even as a joke. We should never call our friend "fool," for example, because we might begin to think of our friend as a fool. Words lead thoughts. It works in advertising and politics, and it can work in science, too. Using words carelessly can lead us to think carelessly.

In science, there are certain ways to say things, not because scientists are sticklers for good grammar, but because if we say things wrong we begin to think things

wrong. For example, a biologist would never let us say a beehive knows it must store food. "No, don't say it that way," the biologist would object. "The hive is just a collection of individual creatures and it can't 'know' anything. Even the individual bees don't really 'know' things. The instinctive behavior of individual bees causes food to accumulate in the hive. That's the way to say it."

All scientists are careful of language because careless words can mislead us.

We would never refer to the center of the universe, for example, but we must also be careful not to say things like "galaxies flying away from the big bang." Those words imply a center to the expansion of the universe, and we know the expansion must be centerless. Rather, we should say "galaxies flying away from each other."

Whatever science you study, notice the customary ways of using words. It is not just a matter of convention. It is a matter of careful thought.

Figure 18-5
An ant confined to the two-dimensional surface of an orange could explore the entire surface without coming to an edge. Were it to leave dirty footprints, it might realize that its two-dimensional universe was finite but unbounded.

surface of the orange thinking it was moving on a horizontal surface that had no edge. If the ant left dirty footprints, however, it might eventually say, "I have visited every square centimeter in my two-dimensional universe, so it must be finite, but I can't find any edge or any center. My universe is finite but it is unbounded." Of course, if the ant lived on a sheet of paper it would discover the edges and conclude that it lived in a finite, bounded universe. Our two-dimensional analogy shows that curvature can allow a universe to be finite but have no edge. If we expand our analogy to three dimensions, we see that our universe could be finite in volume if it is curved such that it is unbounded.

To continue our discussion of curved space–time, we can extend our analogy of the ant. There are three possible ways that a universe could be curved. It could be flat (zero curvature), it could be spherical (positive curvature), or it could be saddle-shaped (negative cur-

vature). If an ant lived on such a two-dimensional universe, it might be unaware of any curvature; but it could detect the curvature of its universe if it could draw circles and measure their areas (Figure 18-6). On the flat surface, a circle would always have an area of exactly πr^2, but on the positively curved surface its area would be less than πr^2, and on the negatively curved surface it would be more. Drawing small circles would not suffice, because the area of small circles would not differ noticeably from πr^2, but if the ant could draw big enough circles, it could actually measure the curvature of its two-dimensional universe.

We are three-dimensional creatures, but our universe might still be curved, and we could measure that curvature by measuring the volume of spheres. If space–time is flat, then no matter how big a sphere we measure, its volume will always be $\frac{4}{3}\pi r^3$. But in positively curved space–time, spheres would have less volume than this; in negatively curved space–time, they would have more. The spheres must be many megaparsecs in radius for the difference to be detectable.

We could measure the volume of such spheres by counting the number of galaxies within a certain distance r from Earth. If galaxies are homogeneously scattered through space, the number within distance r should be proportional to the volume of the sphere. If, as we count to greater and greater distances, we find the number of galaxies increasing proportional to $\frac{4}{3}\pi r^3$, space–time is flat. However, if we find an excess of distant galaxies, space–time is negatively curved, and if we find a deficiency, space–time is positively curved.

By thinking of the two-dimensional analogy shown in Figure 18-6, astronomers usually refer to a universe with positive curvature as a **closed universe.** A zero-curvature universe is said to be a **flat universe,** and a

Positive curvature | Zero curvature | Negative curvature
Closed | Flat | Open

Area of circle less than πr^2 | Area of circle πr^2 | Area of circle more than πr^2

Distant galaxies too few | Uniform galaxy distribution | Distant galaxies too many

Figure 18-6
In two-dimensional space, curvature distorts the area of a circle; in three-dimensional space, it distorts the volume of a sphere. We could measure distortion and detect curvature by counting galaxies.

universe with negative curvature is an **open universe.** Notice that flat and open universes are infinite, but closed universes are finite.

A closed universe has a finite volume but no edge. Like the ant in Figure 18-5, we might explore our universe and visit every location but never find an edge. Of course, an infinite universe also has no edge. Thus, we can answer our question about the edge of the universe. Whether it is open or closed, the universe can have no boundary.

We must not try to visualize the expansion of the universe as an outer edge moving into previously unoccupied space. Open, flat, or closed, the universe has no edge, so it does not need additional room to expand. The universe contains all of the volume that exists, and the expansion is a change in the nature of space–time that causes that volume to increase.

REVIEW Critical Inquiry

Why do we say the universe can't have an edge or center?

The quick answer is that an edge or a center would violate the cosmological principle, which says that every place in the universe is similar in its general properties to every other place. Then a place at the edge or center would be different, so there can be no edge and no center. Of course, this isn't critical thinking; it is only an appeal to the arbitrary authority of cosmological principle. What is the real reason we conclude that there can be no edge and no center?

Imagine that the universe had an edge and you went there. What would you see? Of course, you might imagine an edge to the distribution of matter with empty space beyond, but that would not be a real edge. Imagine an edge beyond which there is no space. Could you stick your head beyond the edge and look around? Thinking about an edge to the universe leads us to logical problems we cannot resolve. From this, we conclude that an edge is impossible. Of course, if there is no edge, then there is no center because we define the centers of things by reference to their edges.

We believe the universe has no edge and no center for good logical reasons, not because of some arbitrary rule. But then why do we believe the cosmological principle is correct? What evidence or assumptions support it?

According to the evidence and consistent with general relativity, we live in an expanding universe. Our problem is to review the evidence and understand how an expanding universe could have begun.

18-2 The Big Bang

If the universe were a videotape, we could run it backward and see the universe contracting as the galaxies approached each other and eventually merged into a hot, high-density gas. As we watched, we would see the volume of space decrease and trap the matter and energy in a high-temperature, high-density state, the state from which the universe began expanding—the **big bang.**

The name, big bang, was coined by cosmologist Fred Hoyle as a sardonic comment on the believability of the theory, and it is often the butt of jokes and cartoons (Figure 18-7). At first glance, it is difficult to imagine how the big bang could have occurred, but it is the dominant theory for the origin of the universe. Our challenge is to stretch our imaginations to encompass the beginning of everything.

The Beginning

The big bang occurred long ago, and our instinct is to think of it as a historical event, like the Gettysburg Address—something that ended long ago and happened in a particular place. As we will discover, the big bang isn't over, and it is happening everywhere.

Because light travels at a finite speed, we see distant galaxies, not as they are now, but as they were when the light left them to begin its journey to Earth. The Andromeda Galaxy is about 2 million light-years away, so the look-back time to the Andromeda Galaxy

Figure 18-7
The population of Cosmos, Minnesota, is apparently in flux; someone has painted out the numbers on the city limits sign. This sign reminds us of the big bang theory. Although the theory seems at first sight difficult to believe and is sometimes the object of jokes and unwarranted criticism, it is currently the only widely accepted theory for the origin of the universe. The evidence that there was a big bang is overwhelming. *(Richard Fluck)*

Figure 18-8
One of the most distant galaxies ever photographed (arrow) has a red shift of 4.25. The look-back time to this galaxy is so large we see it as it was when the universe was only 7 to 10 percent of its present age. The galaxy was first mapped by radio astronomers; this image was made using the 10-m Keck Telescope atop Mauna Kea in Hawaii. The two brightest objects in the field of view are less distant galaxies. *(H. Spinrad, A. Dey, and J. Graham, University of California, Keck Observatory)*

is 2 million years. The look-back time to more distant galaxies is larger. The largest telescopes can detect galaxies with look-back times that are a large fraction of the age of the universe (Figure 18-8). Thus, the look-back time allows us to see what the universe was like when it was young.

Suppose that we look even farther, past the most distant objects, backward in time to the fiery clouds of matter that filled the universe during the big bang. From these great distances we receive light that was emitted by the hot gas soon after the universe began. That means the big bang is all around us.

Although our imaginations try to visualize the big bang as a localized event, we must keep firmly in mind that the big bang did not occur at a single place but filled the entire volume of the universe. The matter of which we are made was part of that big bang, so we are inside the remains of that event, and the universe continues to expand around us. We cannot point to any particular place and say "The big bang occurred over there." The big bang occurred everywhere, and in whatever direction we look, at great distance we look back to the age when the universe was filled with hot gas (Figure 18-9). Why can't we see the glow of that hot gas?

Radiation from the hot gas of the big bang comes to us from tremendous distance, so it has a large red shift. The most distant known galaxies have red shifts well over 6 and perhaps as high as 10. The radiation from the hot gas of the big bang has a red shift of about 1000. Thus the energy that the gas emitted as visible light arrives at Earth as infrared radiation and short radio waves. Unlike the Gettysburg Address, the big bang isn't over. We are part of it, and we should be able to observe it if we can detect the radiation it emitted.

The Cosmic Background Radiation

If radiation is now arriving from the big bang, then it should be detectable. The story of that discovery begins in the mid-1960s when two Bell Laboratories physicists, Arno Penzias and Robert Wilson, were using a horn antenna to measure the radio brightness of the sky (Figure 18-10a). Their measurements showed a peculiar noise in the system, which they at first attributed to pigeons living inside the antenna. After relocating the birds and cleaning out their droppings, the scientists still detected the noise. Perhaps they would have enjoyed cleaning the antenna more if they had known they would win the 1978 Nobel Prize for physics for the discovery they were about to make.

The explanation for the noise goes back to 1948, when George Gamow predicted that the early big bang would be very hot and would emit copious black body radiation. A year later, physicists Ralph Alpher and Robert Herman pointed out that the large red shift of the big bang gas clouds would lengthen the wavelengths of the radiation and make the hot gas clouds seem very cold. There was no way to detect this radiation until the mid-1960s, when Robert Dicke at Princeton con-

a A region of the universe during the big bang

b A region of the universe now

Milky Way Galaxy

c The present universe as it appears from our galaxy

Figure 18-9
Three views of a small region of the universe centered on our galaxy. (a) During the big bang explosion, the region is filled with hot gas and radiation. (b) Later, the gas forms galaxies, but we can't see the universe this way because the look-back time distorts what we see. (c) Near us we see galaxies, but farther away we see young galaxies (dots), and at a great distance we see radiation (arrows) coming from the hot clouds of the big bang explosion.

a

$T = 2.735 \pm 0.06$ K

Intensity

0.05 0.1 0.5 1

Wavelength (cm)

b

Figure 18-10
(a) Robert Wilson (left) and Arno Penzias pose before the horn antenna with which they discovered the cosmic background radiation. *(AT&T Archives)* (b) The cosmic background radiation from the big bang is observed in the infrared and radio part of the spectrum. The critical measurements near the peak of the curve had to be made from balloons and rockets until the COBE satellite observed from orbit. These COBE data show that the radiation fits a black body curve with a temperature of 2.735 K.

cluded the radiation should be just strong enough to detect with newly developed techniques. Dicke and his team began building a receiver.

When Penzias and Wilson heard of Dicke's work, they recognized the noise they had detected as radiation from the big bang, the **cosmic background radiation.**

The background radiation was measured at many wavelengths as astronomers tried to confirm that it did indeed follow a black body curve. Some measurements could be made from the ground, but balloon and rocket observations were critical and difficult. A few disturbing observations suggested that the radiation departed inexplicably from a black body curve and thus might not come from the big bang.

In November 1989, the orbiting Cosmic Background Explorer (COBE) began taking the most accurate measurements of the background radiation ever made. Within 2 months, the verdict was in (Figure 18-10b). To an accuracy of better than 1 percent, the radiation follows a black body curve with a temperature of 2.735 ± 0.06 K—in good agreement with accepted theory.

It may seem strange that the hot gas of the big bang seems to have a temperature of only 2.7K, but recall the tremendous red shift. We see light that has a red shift of about 1000—that is, the wavelengths of the photons are about 1000 times longer than when they were emitted. The gas clouds that emitted the photons had a temperature of about 3000 K, and they emitted black body radiation with a λ_{max} of about 1000 nm (see By the Numbers 4-1). Although this is in the near infrared, the gas would also have emitted enough visible light to glow orange-red. But the red shift has made the wavelengths about 1000 times longer, so λ_{max} is about 1 million nm (equivalent to 1 mm). Thus, the hot gas of the big bang seems to be 1000 times cooler, about 2.7 K.

Although the radiation is almost perfectly isotropic, precise observations reveal small departures from complete isotropy. The background radiation has slightly shorter wavelengths—it seems hotter—in one direction and has slightly longer wavelengths—it seems cooler—in the opposite direction. This difference is caused by the motion of our Milky Way Galaxy along with the Local Group at about 540 km/s toward the massive Virgo cluster. This motion causes a slight blue shift in the direction of motion and that makes the background radiation appear slightly hotter. In the opposite direction there is a red shift and the radiation looks cooler. (Smaller irregularities in the background radiation will be discussed later in this chapter.)

The greatest importance of the cosmic background radiation lies in its widely accepted interpretation as radiation from the big bang. If that interpretation is correct, the radiation is evidence that a big bang really did occur. In fact, the evidence is so strong that an alternative theory of the universe was abandoned. The **steady state theory,** popular with some astronomers during the 1950s and 1960s, held that the universe was eternal and unchanging. If the universe had no beginning, then it could never have had a big bang. According to the theory, new matter continuously appeared from nothing to maintain the density of the expanding, steady state universe. The discovery of the cosmic background radiation in 1965 led astronomers to abandon the steady state theory within a few years. We know there was a big bang because, using radio and infrared telescopes, we can see it happening.

A History of the Big Bang

Modern cosmologists have been able to reconstruct the history of the early universe to reveal how energy and matter interacted as the universe began. As we review that history, remember that the big bang did not occur in a specific place. The big bang filled the entire volume of the universe from the first moment.

We cannot begin our history at time zero, because we do not understand the physics of matter and energy under such extreme conditions, but we can come close. If we could visit the universe when it was very young, only 10 millionths of a second old, for instance, we would find it filled with high-energy photons having a temperature well over 1 trillion (10^{12}) K and a density greater than 5×10^{13} g/cm^3, nearly the density of an atomic nucleus. When we say the photons had a given temperature, we mean that the photons were the same as black body radiation emitted by an object of that temperature. Thus, the photons in the early universe were gamma rays of very short wavelength and therefore very high energy. When we say that the radiation had a certain density, we refer to Einstein's equation $E = m_0c^2$. We can express a given amount of energy in the form of radiation as if it were matter of a given density.

If a photon has enough energy, it can decay and convert its energy into a pair of particles—a particle of normal matter and a particle of **antimatter.** When an antimatter particle meets its matching particle of normal matter—when an antiproton meets a normal proton, for example—the two particles annihilate each other and convert their mass into energy in the form of gamma rays. In the early universe, the photons had enough energy to produce proton–antiproton pairs; but when the particles collided, they converted their mass back into energy. Thus, the early universe was a dynamic soup of energy flickering from photons into particles and back again.

While all of this went on, the universe was expanding, and the wavelengths of the photons were lengthened by the expansion. This lowered the energy of the gamma rays, and the universe cooled. By the time the universe was 0.0001 second old, its temperature had fallen to 10^{12} K. By this time, the average energy of the gamma rays had fallen below the energy equivalent to the mass of a proton or a neutron, so the gamma rays could no longer produce such heavy particles. The particles combined with their antiparticles and quickly converted most of the mass into photons.

It would seem from this that all the protons and neutrons should have been annihilated with their antiparticles, but for quantum-mechanical reasons a small excess of normal particles existed. For every billion protons annihilated by antiprotons, one survived with no antiparticles to destroy it. Thus, we live in a world of normal matter, and antimatter is very rare.

Although the gamma rays did not have enough energy to produce protons and neutrons, they could produce electron–positron pairs, which are about 1800 times less massive than protons and neutrons. This continued until the universe was about 4 seconds old, at which time the expansion had cooled the gamma rays to the point where they could no longer create electron–positron pairs. Thus, the protons, neutrons, and electrons of which our universe is made were produced during the first 4 seconds of its history.

This soup of hot gas and radiation continued to cool and eventually began to form atomic nuclei. High-energy gamma rays can break up a nucleus, so the formation of such nuclei could not occur until the universe had cooled somewhat. By the time the universe was about 2 minutes old, protons and neutrons could link to form deuterium, the nucleus of a heavy hydrogen atom; and, by the end of the next minute, further reactions began converting deuterium into helium. But no heavier atoms could be built because no stable nuclei exist with atomic weights of 5 or 8 (in units of the hydrogen atom). Cosmic element building during the big bang had to proceed rapidly, step by step, like someone hopping up a flight of stairs (Figure 18-11). The lack of stable nuclei at atomic weights of 5 and 8 meant there were missing steps in the stairway, and the step-by-step reactions could not jump over these gaps.

By the time the universe was 30 minutes old, it had cooled sufficiently that nuclear reactions had stopped. About 25 percent of the mass was helium nuclei, and the rest was in the form of protons—hydrogen nuclei. This is the cosmic abundance we see today in the oldest stars. (The heavier elements, remember, were built by nucleosynthesis inside many generations of massive stars.) The cosmic abundance of helium was fixed during the first minutes of the universe.

For the first million years, the universe was dominated by radiation (Figure 18-12). The gamma rays in-

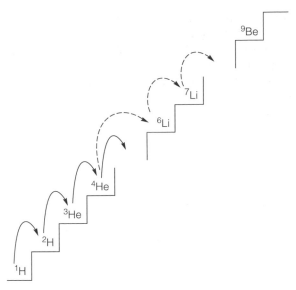

Figure 18-11
Cosmic element building. During the first few minutes of the big bang, temperatures and densities were high, and nuclear reactions built heavier elements. Because there are no stable nuclei with atomic weights of 5 or 8, the process built very few atoms heavier than helium.

teracted continuously with the matter, and they cooled together as the universe expanded. The gas was ionized because it was too hot for the nuclei to capture electrons to form neutral atoms, and the free electrons made the gas very opaque. A photon could not travel very far before it collided with an electron and was deflected. Thus, radiation and matter were locked together.

When the universe reached an age of about 1 million years, it was cool enough for nuclei and electrons to form neutral atoms. The free electrons were captured by atomic nuclei, the gas became transparent, and the radiation was free to travel through the universe. The temperature at the time of this **recombination** was about

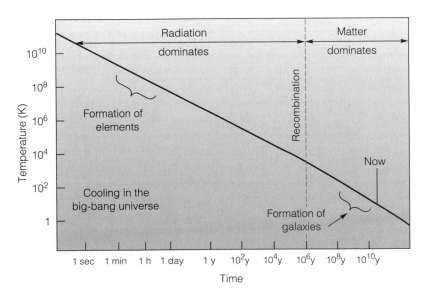

Figure 18-12
During the first few minutes of the big bang, some hydrogen became deuterium and helium and a few heavier atoms. Later, after recombination, when radiation was no longer dominant, galaxies formed, and nuclear reactions inside stars made the rest of the chemical elements.

3000 K. We see these photons arriving now as cosmic background radiation.

After recombination, the universe was no longer dominated by radiation. Instead, matter was free to move under the influence of gravity, so we say the universe after recombination is dominated by matter. How the matter cooled and collected into clouds and eventually gave birth to galaxies in clusters and superclusters is not well understood. We will return to this problem at the end of this chapter.

For now, we must turn our attention away from the first moments of the big bang and instead consider its final moments—the ultimate fate of the universe.

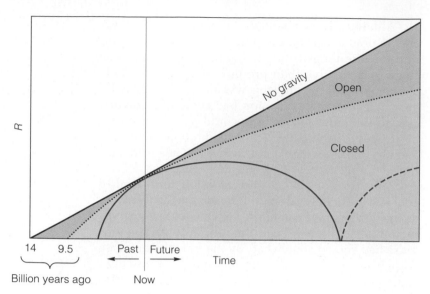

Figure 18-13

The expansion of the universe as a function of time. Models of the universe can be represented as curves in a graph of R, a measure of the size of the universe, and time. Open-universe models expand without end, and the corresponding curves fall in the region shaded orange. Closed models expand and then contract back to a high-density state (solid curve) from which they might expand again. Curves representing closed models fall in the region shaded blue. The dotted line represents a flat universe, the dividing line between open and closed models. Note that the estimated age of the universe depends on the rate at which the expansion is slowing down. (This figure assumes $H = 70$ km/s/Mpc.)

The End of the Universe: A Question of Density

Will the universe ever end? Will it go on expanding forever with stars burning out, galaxies exhausting their star-forming gas and becoming cold, and dark systems expanding forever through an endless, dark universe? How else could the universe end?

Earlier, we decided that the universe could be either open, flat, or closed. The geometry of the universe determines how it will end, so we must consider these three possible universes.

The general curvature of the entire universe is determined by its density. If the average density of the universe is equal to the **critical density,** or 4×10^{-30} g/cm^3, space–time will be flat. If the average density of the universe is less than the critical density, the universe is negatively curved and open. If the average density is greater than the critical density, the universe is positively curved and closed.

We decided earlier that an open universe and a flat universe are both infinite. Both of these universes will expand forever. The gravitational field of the material in the universe, present as a curvature of space–time, will cause the expansion to slow; but, if the universe is open, it will never come to a stop (Figure 18-13). If the universe is flat, it will barely slow to a stop after an infinite time. Thus if the universe is open or flat, it will expand forever, and the galaxies will eventually become black, cold, solitary islands in a universe of darkness.

If the universe is closed, however, its fate will be quite different. In a closed universe, the gravitational field, present as curved space–time, is sufficient to slow the expansion to a stop and make the universe contract. Eventually, the contraction will compress all matter and energy back into the high-energy, high-density state from which the universe began. This end to the universe, a big bang in reverse, has been called the big

crunch. Nothing in the universe could avoid being destroyed by this "crunch."

Some theorists have suggested that the big crunch will spring back to produce a new big bang and a new expanding universe. This theory, called the **oscillating universe theory,** predicts that the universe undergoes alternate stages of expansion and contraction. Theoretical work, however, suggests that successive bounces of an oscillating universe would be smaller until the oscillation ran down, and the theory is no longer taken seriously.

The fate of the universe depends on its density, so astronomers have attempted to measure the density of the universe. This turns out to be quite difficult because, as we saw in Chapters 15 and 16, there is clear evidence that the universe contains dark matter. If we include all of the matter we can detect—all of the stars and galaxies and gas and dust—we get only a few percent of the density needed to produce a closed universe.

Does dark matter make up all the rest of the density of the universe and thus produce a closed universe? Many astronomers will admit that they prefer to believe the universe is flat. But without very large amounts of dark matter, the universe must be open. To go further in our study of cosmology, we must try to understand the nature of the dark matter.

Dark Matter in Cosmology

There is more to the universe than meets the eye. In Chapter 16, we discovered that much of the mass of the universe was invisible, the dark matter. Galaxies have invisible, massive coronas, and clusters of galaxies contain more mass than the sum of the visible galaxies. This is not a small correction. Apparently 90 to 99 percent of the mass of the universe is dark matter. This is a big effect. To correct for the dark matter, we would not just add a small percentage to the visible matter. We have to multiply by a *factor* between 10 and 100.

If we have any doubt how insubstantial the visible universe is, we have only to look at gravitational lensing of distant galaxies. In Chapter 17 we saw how light from distant quasars could be bent by the gravitational field of an intervening galaxy to create false images of the quasar. Using the largest telescopes in the world, astronomers are finding that light from the most distant galaxies can be bent by the mass of a nearby cluster of galaxies. The distant galaxies are extremely faint and blue, perhaps because of rapid star formation. As the light from these faint galaxies passes through a nearby cluster of galaxies, the images of the distant galaxies can be distorted into short arcs like blue rainbows (Figure 18-14). From such arcs, astronomers can calculate the mass of the cluster of galaxies. For instance, the arcs in Figure 18-14 show that over 90 percent of the mass in the cluster is dark matter.

Looking at the universe of visible galaxies is like looking at a ham sandwich and seeing only the mayonnaise. Most of the universe is invisible, and most astronomers now believe that invisible matter is not the normal matter of which you and the stars are made. During the first few minutes of the big bang, nuclear reactions converted some protons into helium and a small amount of other heavier elements. How much of these elements was created depends critically on the density of the material, so astronomers have tried to measure the abundance of these elements. Deuterium, for example, is an isotope of hydrogen in which the nucleus contains a proton and a neutron. This element is so easily converted into helium that none can be produced in stars. In fact, stars destroy what deuterium they have by converting it into helium. Nevertheless, deuterium lines do appear in spectra of gas clouds in space, and astronomers using the giant Keck telescopes have observed gas clouds near quasars at large look-back times. At such large look-back times, we would see how much deuterium existed when the universe was much younger—before stars could destroy significant amounts of deuterium. The gas clouds contain roughly 25 deuterium

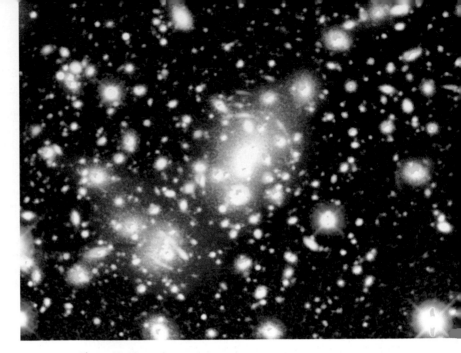

Figure 18-14
Galaxy cluster Abell 2218 (orange images) distorts the images of background galaxies to produce short blue arcs. This gravitational lens effect outlines regions of high mass concentration in the galaxy cluster. *(Courtesy Tony Tyson, AT&T Bell Labs)*

atoms for every million hydrogen atoms, and that tells us something about the big bang.

The amount of deuterium in the universe depends critically on the density of normal matter during the big bang. If there were a lot of normal particles like protons and neutrons, then particles would have collided often, and most of the deuterium particles would have been converted into helium nuclei. If the density of normal particles was lower, there would have been fewer collisions, and more of the deuterium nuclei would have survived. The observed amount of deuterium seems to tell us that the normal matter in the big bang wasn't very dense—no more than about 5 percent of the mass needed to make a flat universe. Yet many observations show that dark matter is abundant, so modern astronomers believe that the dark matter can't be normal matter. That is, it can't be made of protons and neutrons and the other atomic particles that make up our bodies and the stars.

Theorists have proposed that the dark matter might be made up of exotic particles that don't often interact with normal matter. If they don't interact very often, then we would not notice them. Axions and photinos, for example, are predicted by some nuclear particle theories but have never been detected in the laboratory. If they exist, their mass could exert gravitational forces, and they could be part of the dark matter. Data from a large particle detector hint that related particles called WIMPs, Weakly Interacting Massive Particles, may indeed exist, but the result is controversial and is still

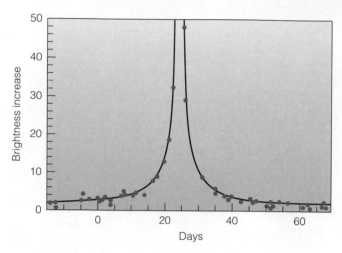

Figure 18-15
Measurements of the brightness of a distant star (dots) increased and then decreased exactly as predicted by general relativity (curve) for a star's passage behind a gravitational lens caused by a MACHO. In this example, the star brightened by a factor of over 40.

being tested. How much WIMPs might contribute to the dark matter is unknown.

More promising are neutrinos. As described in Chapter 10, observations made in underground detectors suggest that neutrinos oscillate, and to do that, they must have some mass. Even if that mass is very small, there are over 10^8 neutrinos in the universe for every particle of normal matter, so if neutrinos prove to have even a small mass, they could be an important part of the dark matter. However, theories of galaxy formation suggest that the dark matter can't be mostly neutrinos. If the universe is flat, something else must contribute to the dark matter.

A contrasting theory holds that each galaxy is orbited by large numbers of very-low-mass stars or planets too faint to see. These **MACHOs,** for Massive Compact Halo Object, can be detected if they pass between our telescopes on Earth and a distant star. The gravitational field of the MACHO acts as a gravitational lens, focusing the light of the star and making it grow brighter for a period of a few weeks as the MACHO passes between the star and the Earth. Extensive searches have detected such events (Figure 18-15) but not in the large numbers expected. It does not appear that MACHOs can make up enough dark matter to produce a flat universe.

Although there is good evidence that dark matter exists, it has proven difficult to identify. No form of dark matter has been found that is abundant enough to provide the density needed if the universe is flat. Thus astronomers continue to struggle to understand the dark matter. It is one of modern astronomy's most important problems.

How do we know there was a big bang?

The cosmic background radiation consists of photons emitted by the hot gas of the big bang, so when we detect those photons, we are "seeing" the big bang. Of course, all scientific evidence must be interpreted, so we must understand how the big bang could produce isotropic radiation before we can accept the background radiation as evidence. First, we must recognize that the big bang event filled all of the universe with hot, dense gas. The big bang didn't happen in a single place; it happened everywhere. At recombination, the expansion of the universe reached the point where the matter became transparent and the radiation was free to travel through space. Today we see that radiation from the age of recombination arriving from all over the sky. It is isotropic because we are part of the big bang event, and as we look out into space to great distance, we look back in time and see the hot gas in any direction we look. We can't see the radiation as light because of the large red shift that has lengthened the wavelengths by a factor of 1000 or so, but we can detect the radiation as infrared and radio photons.

With this interpretation, the cosmic background radiation is powerful evidence that there was a big bang. That tells us how the universe began. How can the density of the universe tell us how it will end?

Whether the universe is open or closed, the big bang theory has been phenomenally successful in explaining the origin and evolution of the universe. At present, no other theory is a serious contender. Yet there are important questions to be resolved about the big bang theory.

18-3 Refining the Big Bang: Theory and Observation

Many problems remain to puzzle cosmologists, but the four we will examine here will illustrate some important trends in modern cosmology. They range from the nature of matter to the nature of space, from the origin of the smallest subatomic particles to the origin of the largest structures in the universe.

The Quantum Universe

A theory developed in 1980 that combined general relativity and quantum mechanics may be able to tell us how the big bang began and why the universe is in the particular state that we observe. To introduce the new theory, we can consider two unsolved problems in the current big bang theory.

One of the problems is called the **flatness problem.** The universe seems to be balanced near the boundary between an open and a closed universe. That is, it seems nearly flat. Given the vast range of possibilities, from zero to infinite, it seems peculiar that the density of the universe is within a factor of 10 of the critical density that would make it flat. If dark matter is as common as it seems, the density may be even closer than a factor of 10.

Even a small departure from critical density when the universe was young would be magnified by subsequent expansion. To be so near critical density now, the density of the universe during its first moments must have been within 1 part in 10^{49} of the critical density. So the flatness problem is, Why is the universe so nearly flat?

Another problem with the big bang theory is the isotropy of the cosmic background radiation. When we correct for the motion of our galaxy, we see the same background radiation in all directions to at least 1 part in 1000. Yet when we look at background radiation coming from two points in the sky separated by more than a degree, we look at two parts of the big bang that were not causally connected when the radiation was emitted. That is, when recombination occurred and the gas of the big bang became transparent to the radiation, the universe was not old enough for any signal to have traveled from one of these regions to the other. Thus, the two spots we look at did not have time to exchange heat and even out their temperatures. Then how did every part of the entire big bang universe get to be so nearly the same temperature by the time of recombination? This is called the **horizon problem** because the two spots are said to lie beyond their respective light–travel horizons.

The key to these two problems and to others involving subatomic physics may lie with the newer theory. It has been called the **inflationary universe** because it predicts a sudden expansion when the universe was very young, an expansion even more extreme than that predicted by the big bang theory.

To understand the inflationary universe, we must consider recent attempts to unify the four forces of nature. Recall from Chapter 10 that physicists know of only four forces—gravity, the electromagnetic force, the strong nuclear force, and the weak nuclear force.

For many years, theorists have tried to unify these forces; that is, they have tried to describe the forces

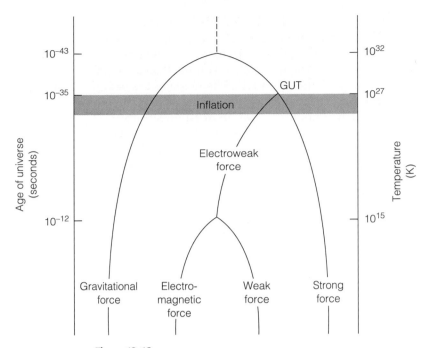

Figure 18-16
When the universe was very young and hot (top), the four forces of nature were indistinguishable. As the universe began to expand and cool, the forces separated and triggered a sudden inflation in the size of the universe.

with a single mathematical law (Figure 18-16). A century ago, James Clerk Maxwell showed that the electric force and the magnetic force were really the same effect, and we now count them as a single electromagnetic force. In the 1960s, theorists succeeded in unifying the electromagnetic force and the weak force in what they called the electroweak force, effective only for processes at very high energy. At lower energies, the electromagnetic force and the weak force behave differently. Now theorists have found ways of unifying the electroweak force and the strong force at even higher energies. These new theories are called **grand unified theories,** or **GUTs.**

Studies of GUTs suggest that the universe expanded and cooled until about 10^{-35} second after the big bang, when it became so cool that the forces of nature began to separate. This released tremendous energy, which suddenly inflated the universe by a factor between 10^{20} and 10^{30}. At that time, the part of the universe that we can see now, the entire observable universe, was no larger than the volume of an atom, but it suddenly inflated to the volume of a cherry pit and then continued its slower expansion to its present extent.

That sudden inflation can solve the flatness problem and the horizon problem. The sudden inflation of the universe would have forced whatever curvature it had toward zero, just as inflating a balloon makes a small spot on its surface flatter. Thus, we now see the universe nearly flat because of that sudden inflation long ago. In addition, because the part of the universe we can see was once no larger in volume than an atom, it had plenty of time to equalize its temperature before the inflation occurred. Now we see the same temperature for the background radiation in all directions.

The inflationary theory predicts that the universe is almost perfectly flat. That is, the true density must equal the critical density. A theory can never be used as evidence, but the beauty of the inflationary theory has given many cosmologists confidence that the universe must be flat and that the dark matter must make up the difference between the matter we can see and the critical density.

The newest observations, however, seem to tell us that the universe is not dense enough to be flat, so we must now turn from theory to evidence. We begin by trying to find the age of the universe.

The Age of the Universe

In 1929, Edwin Hubble announced that the universe was expanding and reported the rate of expansion, the number now known as the Hubble constant (H). Within a few years, astronomers had proposed the idea that the universe had a violent beginning and they used H to extrapolate backward to find the age of the universe. But the value of H that Hubble reported, 530 km/s/Mpc, led them to conclude that the universe was only half as old as Earth. Something was wrong.

In the decades that followed, astronomers figured out that Hubble's distances to galaxies were systematically too small and thus his value of H was too large (Figure 16-11). Modern measurements of H yield an age of the universe that is at least twice the age of Earth, but controversy still rages.

The calculation of the age of the universe from its expansion is really very simple. We make such calculations whenever we take a drive and divide distance by speed. We say, "It is 100 miles to the city, and I can average 50 miles an hour, so I can get there in 2 hours." We know the distance to a galaxy and its radial velocity, so we can divide distance by velocity to find out how long the expansion of the universe has taken to separate the galaxies to their present distance. That is the age of the universe.

The Hubble constant makes this calculation easy. We need to divide distance by velocity, and the Hubble constant is just velocity divided by distance. The reciprocal of the Hubble constant ($\frac{1}{H}$) must be proportional to the age of the universe. The only problem is the units. Astronomers always express H as a velocity in kilometers per second divided by a distance in megaparsecs. To make the division give us an age, we must convert megaparsecs to kilometers, and then the distances will cancel out and we will have an age in seconds. To get years, we must divide by the number of seconds in a year. If we make these simple changes in units, the age of the universe in years is approximately 10^{12} divided by H in its normal units, km/s/Mpc.

$$T \approx \frac{1}{H} \times 10^{12} \text{ years}$$

This is known as the **Hubble time.** For example, if H is 70 km/s/Mpc, then the age of the universe can be no older than $10^{12}/70$, which equals 14 billion years.

When we calculate the Hubble time, we are assuming the galaxies have receded at a constant velocity since the big bang, but gravity would slow the expansion of the universe. Then the galaxies must have traveled faster in the past, and the Hubble time is just an upper limit. The universe is no older than one Hubble time, but it could be younger. To know how much younger, we must know the density of the universe. If the density exactly equals the critical density (if the universe is flat), then the true age equals two-thirds of a Hubble time ($\frac{2}{3}\frac{1}{H}$). If the density of the universe is less than the critical density, the universe is older; and if the density is higher, the universe is younger.

Astronomers have measured the distances to galaxies in different ways and obtained different values for H ranging from 50 km/s/Mpc to 100 km/s/Mpc. This generated heated controversy because a value of H higher than about 80 predicts that the universe is younger than the globular clusters. It is obviously impossible for objects to be older than the universe.

Two spacecraft have helped solve the problem. High-precision parallaxes from the Hipparcos satellite have recalibrated globular cluster H-R diagrams, and astronomers now estimate that the clusters are about 11 billion years old, give or take a billion years. Hipparcos data have also allowed the recalibration of the period–luminosity diagram for Cepheid variable stars, slightly increasing the distance to galaxies and decreasing the observed value of H.

The Hubble Space Telescope has yielded more accurate distances to galaxies by detecting Cepheids in galaxies as far away as the Virgo cluster of galaxies. Also, astronomers have been able to calibrate the brightness of supernovae caused by the collapse of a white dwarf. These Type Ia supernovae all have about the same maximum luminosity, so they can be used to find the distances to galaxies and thus the value of H.

The most recent values of H are about 70 km/s/Mpc. This implies a Hubble time of about 14 billion years, just large enough to permit the formation of the galaxies and the globular clusters.

The controversy over the Hubble constant is easing as better observations are made, but a new controversy is brewing about the actual expansion of the universe.

The Cosmological Constant

When Einstein published his theory of general relativity in 1916, he recognized that his equations describing space and time implied that space had to expand or contract. Thinking such motion unreasonable, he introduced a constant called the **cosmological constant** represented by an upper-case lambda (Λ). The constant plays the mathematical role of a force of repulsion that balances the gravitation in the universe so it does not have to expand or collapse. Thirteen years later, in 1929, Edwin Hubble announced that the universe was expanding, and Einstein said introducing the cosmological constant was his biggest blunder. Modern astronomers aren't so sure.

The ages of the oldest star clusters and the age of the universe derived from H are uncomfortably close. Since 1929, astronomers have generally believed that the cosmological constant was zero and so has no effect, but the age conflict could be resolved if the cosmological constant was not zero. Then the repulsion represented by the cosmological constant would allow the universe to expand more slowly at first and gradually accelerate to its present rate. If this is true, then the universe could be older than implied by H.

Two international teams of astronomers working independently announced in 1998 that their data show that the expansion of the universe is accelerating, a result that requires that the cosmological constant be nonzero. Both teams calibrated Type Ia supernovae as distance indicators. They then used such supernovae seen in very distant galaxies to measure the distance to the galaxies. The astronomers expected to measure the Hubble constant and detect the rate at which the expansion is slowed by gravity. Instead, both teams got the same amazing result. The expansion is accelerating (Figure 18-17).

If the expansion of the universe is being accelerated, then the cosmological constant is not zero, and that tells us that empty space contains tremendous energy. Because energy and matter are equivalent, the energy of empty space acts as a mass, and that mass interests astronomers. Although the inflationary theory predicts that the universe must be exactly flat, astronomers find that normal matter makes up only a few percent of the matter needed to make the universe flat. Dark matter may make up roughly a quarter of the necessary matter. The mass equivalent of the cosmological constant may make up the rest. Thus the universe may indeed be flat although the expansion is accelerating.

For 80 years, Einstein's cosmological constant was out of fashion, but it seems to have returned dramatically. More observations and more theoretical work are needed to help us understand this new cosmology. Nevertheless, this is not the only excitement in cosmology. While some astronomers struggle to understand the behavior of space and time, others are trying to understand how the hot gas of the big bang made galaxies.

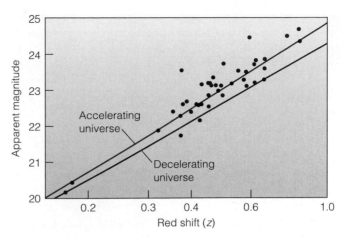

Figure 18-17

This plot of calibrated magnitude versus red shift for Type Ia supernovae in distant galaxies shows evidence that the expansion of the universe is accelerating. Distant supernovae are fainter (higher in the diagram) than they should be in a decelerating universe. The black line shows the theoretical relationship for a decelerating, flat universe dominated by dark matter. The red line shows the relationship for an accelerating, flat universe dominated by a cosmological constant.

The Origin of Structure

When astronomers study the distribution of galaxies, they find them in clusters and in clusters of clusters called **superclusters.** The Local Supercluster, in which we live, is a roughly disk-shaped swarm of galaxy clusters 50 to 75 Mpc in diameter. Furthermore, the superclusters are distributed in long, narrow filaments and thin walls that outline great voids that are nearly empty of galaxies (Figure 18-18). Astronomers refer to all of this as structure, and they wonder how the galaxies, clusters, superclusters, filaments, walls, and voids formed from the hot gas of the big bang.

This structure is a problem for cosmologists because the primordial background radiation is extremely uniform, and thus the gas of the big bang must have been extremely uniform at the time of recombination. If the universe was so uniform when it was young, how did it get to be so lumpy now?

The problem of the origin of structure has developed slowly since the 1960s, when the background radiation was first discovered. Then, the most distant galaxies visible had red shifts no greater than 1. These galaxies had look-back times no greater than 4 billion years, so it was not difficult to imagine how the uniform gas of the big bang could eventually form galaxies. That is, there seemed to be plenty of time between recombination and the origin of galaxies.

Through the 1970s and 1980s, however, that age of mystery narrowed. Astronomers recognized quasars as

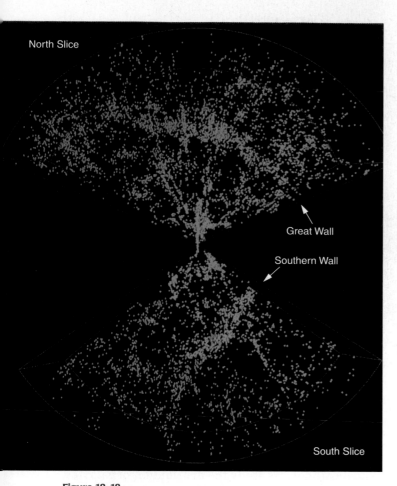

North Slice

Great Wall

Southern Wall

South Slice

Figure 18-18
By measuring the red shifts and positions of thousands of galaxies, astronomers have been able to create this map of the location of galaxies in two slices of the universe looking north and south out of our Milky Way Galaxy. Filaments, walls, and voids are clearly evident, with the largest, the Great Wall, stretching from left to right across the northern slice, while the Southern Wall reaches diagonally across the southern slice. *(Margaret J. Geller, Luis A. N. da Costa, John P. Huchra, and Emilio E. Falco, Smithsonian Astrophysical Observatory © 1994)*

the active nuclei of galaxies much farther away than 4 billion light-years. Galaxies at such great distances are almost beyond detection, but the brilliant quasars assure us that such galaxies really exist. These galaxies must have formed early in the history of the universe. At present, the most distant quasars have look-back times that are 93 percent of the age of the universe. Thus, the age of mystery when the galaxies formed must be less than 7 percent of the age of the universe.

How did all of this structure originate? The universe had to be very uniform during recombination and then clump quickly to form the structure we see. Cosmologists now work on a two-pronged solution: Dark matter provides the gravitation, and galaxy seeds provide nuclei around which structure grows.

We have seen clear evidence that dark matter must exist, and cosmologists are now trying to build models that show how the gravitation of dark matter could cause the matter of the early universe to clump quickly into large structures. One kind of dark matter is called hot dark matter. If it exists, it would consist of particles that travel at nearly the speed of light, such as neutrinos with nonzero masses. Cold dark matter, on the other hand, would consist of slow-moving particles such as WIMPs, axions, and similar as-yet-undetected particles. Cold dark matter, because its particles travel more slowly, seems better able to clump together quickly (Figure 18-19).

Something must trigger the clumping of matter to make a galaxy, and these clumping centers have been called **galaxy seeds.** Theorists say these irregularities had their start before inflation as unavoidable quantum fluctuations in space. Inflation could have magnified these from smaller than an atom to large enough to stimulate the formation of galaxies, clusters, filaments, walls—all of the structure we see today in the distribution of galaxies.

Figure 18-19
A computer simulation of the evolution of a small region of a universe containing dark matter shows small irregularities growing into filaments, walls, and voids. The area visible in each frame remains the same as the universe expands. Galaxy formation presumably occurs at about the third panel. Although this model spans the history of the universe from soon after the big bang to the present, the time steps between frames are not uniform. The first two frames show the young universe, and the last three frames show more recent times. *(Courtesy James Gelb and Edmund Bertschinger, MIT)*

Figure 18-20

Over 70 million measurements by the COBE satellite went into this all-sky map of the cosmic background radiation. Patchy areas are irregularities in the background radiation that suggest the gas of the big bang was not perfectly uniform. The gravitational influence of galaxy seeds and dark matter could have forced these denser regions to grow into the clouds of galaxies we see today. *(NASA/Goddard Space Flight Center)*

Testing such theories is difficult, but one way to do so is to study the irregularities in the primordial background radiation. In 1992, the Cosmic Background Explorer (COBE) satellite mapped the background radiation over the entire sky and detected small fluctuations (Figure 18-20). These are evidently the kind of fluctuations that existed in the gas of the big bang at the time of recombination, and thus they tell us about the geometry of the expansion of the universe. More detailed studies of size of these fluctuations fit theoretical calculations for a universe that is flat.

Both theory and observation seem to point to a flat universe. The inflation predicts a flat universe, and the structure in the primordial background radiation supports a flat universe. Also, the acceleration in the expansion of the universe tells us that empty space contains energy. Although normal matter makes up only a few percent of the critical density and dark matter is not abundant, astronomers now believe the energy of empty space makes up the rest. That is, the cosmological constant supplies enough density to make the universe flat.

While some astronomers struggle to understand the overall structure of the universe, others are studying the structure of very distant galaxies. The most distant galaxies seem small, distorted, and filled with new stars (Figure 18-21). Perhaps these are the first star clouds, wracked by collisions and bursts of star formation, that fell together to form the galaxies we see today. How the galaxies were born and evolved is one of the frontiers of modern astronomy.

The four subjects we have reviewed here—inflation, the age of the universe, the cosmological constant, and the origin of structure—in no way invalidate the big bang theory. They are refinements that extend our understanding of how our universe has evolved since that first moment we call the big bang.

REVIEW Critical Inquiry

Why do we have to understand dark matter to understand the origin of galaxies?

The big bang created a universe filled with gas too hot to collect and form galaxies, so astronomers believe the galaxies began to form some billions of years after the beginning when the expansion of the universe had cooled the gas enough. Somehow the gas must have collected to form galaxies, and the inflationary theory provides the possibility that small irregularities could have acted

Age of the Universe

Today: 14 Billion Years 9 Billion Years 5 Billion Years 2 Billion Years

Elliptical

Spiral

Figure 18-21

Because modern telescopes such as the Hubble Space Telescope can photograph very distant galaxies, we can look back in time and see that the galaxies formed quickly after the big bang and have evolved since then. Images at the left represent an elliptical and a spiral galaxy as they are today. Images to the right show typical galaxies when the universe was younger than it is today. Images at the far right show galaxies only 2 billion years after the big bang. Elliptical galaxies appear to have formed quickly and to have taken on a modern appearance, but early spiral galaxies are often distorted, further evidence that galaxies have evolved as the universe has aged. *(A. Dressler, Carnegie Institutions of Washington; M. Dickinson, STScI; D. Macchetto, ESA/STScI; M. Giavalisco, STScI; and NASA)*

as seeds around which gas collected to form galaxies in clusters, superclusters, filaments, and walls. Just the visible matter in the universe, however, does not appear to have enough gravity to have pulled itself together quickly, so the added gravity of the dark matter may have been important in aiding the first contractions. Once galaxies and galaxy clusters formed, the presence of dark matter stabilized them as the galaxies continued to collide, merge, and evolve. One of the unknown factors in this story is the age of the universe. The current best estimates of the Hubble constant seem to make the universe just old enough for the matter to have contracted to form the structure we see.

Cosmology is definitely a work in progress; but astronomers have made dramatic progress in understanding the history of the universe, and they have done it by comparing observation with theory. As an example, how would you compare observation with the inflationary theory to explain the horizon problem?

Although we have traced the origin of the universe, the origin of the elements, and the birth and death of stars, we have left out something very important—life. Is there life on many worlds or only on Earth? How do we thinking beings fit into the astronomical scheme of origins? That is the subject of the next chapter.

Summary

The fact that the night sky is dark shows that the universe is not infinitely old. If it were infinite in extent and age, then every spot on the sky would glow as brightly as the surface of a star. This problem, known incorrectly as Olbers' paradox, illustrates how important assumptions are in cosmology.

The basic assumptions of cosmology are homogeneity, isotropy, and universality. Homogeneity says that matter is spread uniformly through the universe. Isotropy says the universe looks the same in any direction. Both deal only with general features. Universality assumes that the laws of physics known on Earth apply everywhere. In addition, the cosmological principle asserts that the universe looks the same from any location.

The Hubble law implies that the universe is expanding. Tracing this expansion backward in time, we come to an initial high-density state commonly called the big bang. From the Hubble constant (about 70 km/s/Mpc) we can conclude that the expansion began no more than 14 billion years ago.

The universe can be infinite or finite, but it cannot have an edge or a center. Such regions would violate the cosmological principle. Instead, we assume that the universe occupies curved space–time and thus could be finite but unbounded. Depending on the average density of the universe, it could be open, flat, or closed. Measurements of the density of the universe are uncertain because of the presence of dark matter, which we cannot detect easily.

During the first few minutes of the big bang, about 25 percent of the matter became helium, and the rest remained hydrogen. Very few heavy atoms were made. As the matter expanded, instabilities caused the formation of clusters of galaxies, which are still receding from each other.

Whether the universe expands forever or slows to a stop and begins to contract depends on the amount of matter in the universe. If the average density is greater than the critical density of 4×10^{-30} g/cm^3, it will provide enough gravity to slow the expansion and force the universe to collapse. The collapse will smash all matter back to a high density from which a new big bang may emerge. Such a universe is termed "closed," because the curvature of space is positive. If the average density is less than 4×10^{-30} g/cm^3, gravity will be unable to stop the expansion, and it will continue forever. Such a universe is termed "open," because the space curvature is negative.

The inflationary universe theory combines quantum mechanics, general relativity, and cosmology. It predicts that the universe underwent a sudden inflation when it was only 10^{-35} seconds old. This seems to explain why the universe is so flat and why the cosmic background radiation is so isotropic.

Studies of Type Ia supernovae suggest that the expansion of the universe is accelerating, which means that Einstein's cosmological constant must not be zero. Empty space contains energy that forces the acceleration. The mass equivalent of that energy, combined with the small amount of normal matter in the universe and the dark matter, may provide enough density to make the universe flat. The inflationary theory predicts that the universe must be flat.

Although the cosmic background radiation is highly isotropic, it is not exactly uniform. Small irregularities in the background radiation appear to be the remains of quantum fluctuations magnified by inflation. These irregularities may be the galaxy seeds around which galaxies, clusters, filaments, and walls took shape. The size of these fluctuations fits cosmological theories that are flat, so astronomers take this as further evidence that the universe is flat.

The irregular shapes of the most distant galaxies suggest that they formed by collision and interaction as they merged to build bigger galaxies.

New Terms

cosmology	antimatter
Olbers' paradox	recombination
observable universe	critical density
homogeneity	oscillating universe theory
isotropy	MACHOs
universality	flatness problem
cosmological principle	horizon problem
Copernican principle	inflationary universe
closed universe	grand unified theories (GUTs)
flat universe	
open universe	Hubble time
big bang	cosmological constant (Λ)
cosmic background radiation	superclusters
	galaxy seeds
steady state theory	

Review Questions

1. Would the night sky be dark if the universe were only 1 billion years old and was contracting instead of expanding? Explain your answer.

2. How is the cosmological principle related to the Copernican principle?

3. How can we be located at the center of the observable universe if we say the universe has no center?

4. Why can't an open universe have a center? Why can't a closed universe have a center?

5. What evidence do we have that there was a big bang?

6. Why do astronomers say the hot gases of the big bang have a temperature of 2.7 K?

7. Why couldn't atomic nuclei exist when the universe was younger than 3 minutes?

8. Why do we conclude that elements heavier than helium were made in generations of stars and not during the big bang?

9. Why is it difficult to determine the present density of the universe?

10. Why does the age of the universe depend on the density?

11. How does the inflationary universe theory resolve the flatness problem? the horizon problem?

12. Why do we conclude that the universe must have been very uniform during its first million years?

13. How is the presence of structure in the early universe a problem for cosmologists?

14. What is the difference between hot dark matter and cold dark matter? What difference does it make to cosmology?

Discussion Questions

1. Do you think Copernicus would have accepted the cosmological principle? Why or why not?

2. Suppose a civilization lives in a galaxy that we see as a very distant quasar. What would the universe look like to them?

Problems

1. Use the data in Figure 18-2 to plot a velocity–distance diagram, find H, and determine the approximate age of the universe.

2. If a galaxy is 8 Mpc away from us and recedes at 456 km/s, how old is the universe, assuming that gravity is not slowing the expansion? How old is the universe if it is flat?

3. If the temperature of the big bang had been 10^6 K at the time of recombination, what wavelength of maximum would the cosmic background radiation have as seen from Earth?

4. If the average distance between galaxies is 2 Mpc and the average mass of a galaxy is 10^{11} solar masses, what is the average density of the universe? (Hints: The volume of a sphere is $\frac{4}{3}\pi r^3$. The mass of the sun is 2×10^{33} g.)

5. Figure 18-13 is based on an assumed Hubble constant of 70 km/s/Mpc. How would you change the diagram to fit a Hubble constant of 100 km/s/Mpc?

6. Hubble's first estimate of the Hubble constant was 530 km/s/Mpc. If his distances were too small by a factor of 7, what answer should he have obtained?

7. What is the maximum age of the universe predicted by Hubble's first estimate of the Hubble constant?

8. High-resolution radio observations show that a galaxy contains a supernova remnant that is expanding 0.00004 second of arc per year, and spectra show that the remnant is expanding at 10,000 km/s. How far away is the galaxy?

Critical Inquiries for the Web

1. Will the universe go on expanding forever? Search the Web for information on recent investigations that shed light on the question of the density of matter in the universe. What predictions do these studies make about the fate of the universe? What kinds of observations were necessary to make these predictions?

2. The steady state theory was once a rival cosmology of the big bang. Search for Web sites that provide information on steady state cosmology. (Be careful to locate legitimate sites that discuss the theory, rather than sites where individuals use steady state ideas as part of nonscientific arguments on cosmology.) What were the key predictions of steady state cosmology? How has recent evidence led to its decline?

 Go to the Brooks/Cole Astronomy Resource Center (www.brookscole.com/astronomy) for critical thinking exercises, articles, and additional readings from InfoTrac College Edition, Brooks/Cole's online student library.

Life on Other Worlds

Did I solicit thee from darkness

to promote me?

John Milton
Paradise Lost

Guidepost

This chapter is either unnecessary or critical depending on our point of view. If we believe that astronomy is the study of the physical universe above the clouds, then this chapter does not belong here. But if we believe that astronomy is the study of our position in the universe, not only our physical position but also our role as living beings in the origin and evolution of the universe, then everything else in this book is just preparation for this chapter.

Astronomy is the only science that truly acts as a mirror. In studying the universe up there, we learn what we are down here. Astronomy is not really about planets, stars, and galaxies; it is about us.

Science is a way of understanding the world around us, and the heart of that understanding is the explanations that science gives us for natural phenomena. Whether we call these explanations stories, histories, theories, or hypotheses, they are attempts to describe how nature works based on fundamental rules of evidence and intellectual honesty. While we may take these explanations as factual truth, we should understand that they are not the only explanations that satisfy the rules of logic.

A separate class of explanations involves religion, and those explanations can be quite logical. The Old Testament description of the creation of the world, for instance, does not fit scientific observations; but if we accept the existence of an omnipotent being, then the biblical explanation is internally logical and acceptable. Of course, it is not a scientific explanation; but religion is a matter of faith and not subject to the rules of evidence. Religious explanations follow their own logic, and we would be wrong to demand that they follow the rules of evidence that govern scientific explanations, just as we would be wrong to demand that scientific explanations accept certain religious beliefs on faith. Both kinds of explanations are logical, but the rules are different.

If scientific explanations are not the only logical explanations, then why do we give them such weight? First, we must notice the tremendous success of scientific explanations in producing technological advances in our daily lives. Smallpox is a disease of the past thanks to the application of scientific explanations to modern medicine. The power of science to shape our world can lead us to think its explanations are unique. But second, the process we call science depends on the use of evidence to test and perfect our explanations, and the logical rigor of this process gives us great confidence in our conclusions.

Scientific explanations have given us tremendous insight into the workings of nature, and consequently both scientists and nonscientists tend to forget there can be other kinds of explanations. The so-called conflict between science and religion has been symbolized for centuries by the trial of Galileo. That conflict is easier to understand when we consider the nature of scientific explanations and the role of evidence in testing scientific understanding.

As living things, we have been promoted from darkness. We are made of heavy atoms that could not have formed at the beginning of the universe. Successive generations of stars fusing light elements into heavier elements have built the atoms so important to our existence. When a dark cloud of interstellar gas enriched in these heavy atoms fell together to form our sun, a small part of the cloud gave birth to the planet we inhabit.

Are there intelligent beings living on other planets? That is the last and perhaps the most challenging question in our study of astronomy. We will try to answer it in three steps, each dealing with a different aspect of life.

First, we must decide what we mean by *life*. Life is not so much a form of matter as it is a behavior. Living matter extracts energy from its environment to modify itself and its surroundings so as to preserve itself and to create offspring. Thus, life is based on information, the recipe for the process of survival and reproduction.

Second, we must study the origin of life on Earth. If we can understand how life began on our planet, then we can try to estimate the likelihood that it has originated on other worlds as well.

Third, we must try to understand how the first primitive living organism on Earth could give rise to the tremendously diverse life forms that now inhabit our world. By understanding how living things evolve to fit their environment, we will see evidence that intelligence is a natural development in the evolution of life.

If life can originate on other worlds, and if intelligence is a natural result of the evolution of life forms, then we might expect that other intelligent races inhabit other worlds. Visits between worlds seem impossible, but perhaps we can detect their existence by radio.

Alien life forms could be quite different from us, but if they are alive, then they must share with us certain characteristics. Our goal in this chapter is to use our knowledge of science to explain the greatest of mysteries—the origin and evolution of life on Earth and on other worlds (Window on Science 19-1).

19-1 The Nature of Life

What is life? Philosophers have struggled with that question for thousands of years, so it is unlikely that we will answer it here. But we must agree on a working model of life before we can speculate on its occurrence on other worlds. To that end, we will identify in living things two important aspects: a physical basis and a unit of controlling information.

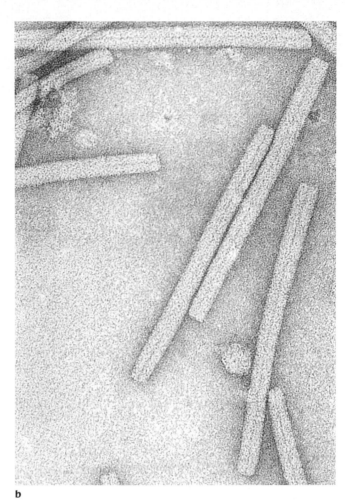

The Physical Basis of Life

On Earth, the physical basis of life is the carbon atom (Figure 19-1). Because of the way this atom bonds to other atoms, it can form long, complex, stable chains that are capable of extracting, storing, and utilizing energy. Other chemical bases of life may exist. Science fiction stories and movies abound with silicon creatures, living things whose body chemistry is based on silicon rather than carbon. However, silicon forms weaker bonds than carbon does, and it cannot form double bonds as easily. Consequently, it cannot form the long, complex, stable chains that carbon can. Silicon is 135 times more common on Earth than carbon is, yet there are no silicon creatures among us. All Earth life is carbon-based. Thus, the likelihood that distant planets are inhabited by silicon people seems small, but we cannot rule out life based on noncarbon chemistry.

In fact, nonchemical life might be possible. All that nature requires is some mechanism capable of supporting the extraction and utilization of energy that we have identified as life. One could at least imagine life based on electromagnetic fields and ionized gas. No one has ever met such a creature, but science fiction writers conjure up all sorts.

Clearly, we could range far in space and time, theorizing about different bases for alien life, but to make progress we must discuss what we know best—carbon-based life on Earth. How can a lump of carbon-rich mat-

Figure 19-1
All living things on Earth are based on carbon chemistry. Even the long molecules that carry genetic information, DNA and RNA, have a framework defined by chains of carbon atoms. (a) Katie, a complex mammal, contains about 30 AU of DNA. *(Michael Seeds)* (b) Each rod of the tobacco mosaic virus contains a single spiral strand of RNA about 0.01 mm long. All life on Earth stores its genetic information in such carbon-chain molecules. *(Courtesy Lee D. Simon)*

ter live? The answer lies in the information that guides its life processes.

Information Storage and Duplication

The key to understanding life is information—the information the organism uses to control its utilization of energy. We must discover how life stores and uses that information and how the information changes and thus preserves the species.

The unit of life on Earth is the cell (Figure 19-2), the self-contained factory capable of absorbing nourishment from its surroundings, maintaining its own existence, and performing its task within the larger organism. The foundation of the cell's activity is a set of patterns that describe how it is to function. This information must

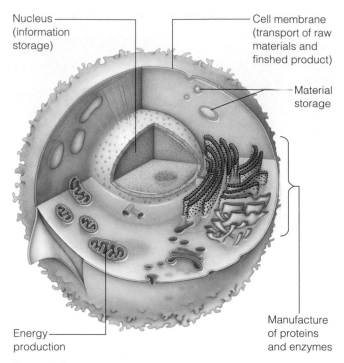

Figure 19-2
A living cell is a self-contained factory that absorbs raw materials from its surroundings and uses them to maintain itself and manufacture finished products for the use of the organism as a whole.

Nucleus (information storage)

Cell membrane (transport of raw materials and finished product)

Material storage

Manufacture of proteins and enzymes

Energy production

Figure 19-3
The DNA molecule consists of two rails of sugars and phosphates (dark) and rungs made of bases: adenine (A), cytosine (C), guanine (G), thymine (T).

Figure 19-4
The DNA molecule can duplicate itself by splitting in half (top), assembling matching bases, sugars, and phosphates (center), and thus producing two DNA molecules (bottom). The actual duplication process is significantly more complex than in this schematic diagram.

be stored in the cell in some safe location, yet it must be passed on easily to new cells and be used readily to guide the cell's activity. To understand how matter can be alive, we must have an understanding of how the cell stores, reproduces, and uses this information.

The information is stored in long carbon-chain molecules called **DNA (deoxyribonucleic acid),** most of which reside in the cell nucleus. The structure of DNA resembles a long, twisted ladder. The rails of the ladder are made of alternating phosphates and sugars; the rungs are made of pairs of molecules called bases (Figure 19-3). Only four kinds of bases are present in DNA, and the order in which they appear on the DNA ladder represents the information the cell needs to function. One human cell stores about 1.5 m of DNA, containing about 4.5 billion pairs of bases. Thus 4.5 billion pieces of information are available to run a human cell. That is enough to record all of the works of Shakespeare over 200 times. Because the human body contains about 60×10^{12} cells, the total DNA in a single human adult would stretch 9×10^{13} m, about 600 AU.

Storing all this data in each cell does the organism no good unless the data can be reproduced and passed on to new cells. The DNA molecule is specially adapted for duplicating itself by splitting its ladder down the center of the rungs, producing two rails with protruding bases (Figure 19-4). These quickly bond with the proper bases, phosphates, and sugars to reconstruct the missing part of the molecule, and presto—the cell has two complete copies of the critical information. One set goes to each of the newly forming cells. Thus, the DNA is the genetic information passed from parent to offspring.

Segments of the DNA molecules are patterns for the production of **proteins.** Many proteins are structural molecules—the cell might make protein to repair its cell wall, for example. **Enzymes** are special proteins that control other processes—growth, for example. Thus, the DNA molecule contains the recipes to make all of the different molecules required in an organism.

Actually, the cell does not risk its precious DNA patterns by involving them directly in the manufacture of protein. The DNA stays safely in the cell nucleus, where it produces a copy of the patterns by assembling a long carbon-chain molecule called **RNA (ribonucleic acid).** This RNA carries the information out of the nucleus and then assembles the proteins from simple molecules called **amino acids,** the basic building blocks of protein. Thus the RNA acts as a messenger, carrying copies of the necessary plans from the central office to the construction site.

Though the information coded on the DNA must be preserved for the survival of the organism, it must be changeable or the species will become extinct. To see why, we must study evolution, the process that rewrites the data in the DNA.

Modifying the Information

If living things are to survive for many generations, then the information stored in their DNA must change as the environment gradually changes. A slight warming of the climate, for example, might kill a species of plant, in turn starving the rabbits, deer, and other plant eaters, and leaving the hawks, wolves, and mountain lions with no prey. If the information stored in DNA could never change, then these life forms would become extinct. If a species is to survive in a changing world, then the information in its DNA must change as well. The species must evolve.

Species evolve by **natural selection.** Each time an organism reproduces, its offspring receive the data stored in the DNA, but some variation is possible. For example, most of the rabbits in a litter may be normal, but it is possible for one to get a DNA recipe that gives it stronger teeth. If it has stronger teeth, it may be able to eat something other than the plant the others depend on, and if that plant is becoming scarce, the rabbit with stronger teeth has a survival advantage. It can eat other plants and so will be healthier than its littermates and have more offspring. Some of these offspring may also have stronger teeth, as the altered DNA data are handed down to the new generation. Thus nature selects and preserves those attributes that contribute to the survival of the species. Those that are unfit die. Natural selection is merciless to the individual, but it gives the species the best possible chance to survive.

The only way nature can obtain new DNA patterns from which to select the best is to alter actual DNA molecules. This can happen through chance mismatching of base pairs—errors—in the reproduction of the DNA molecule. Another way this can occur is through damage to reproductive cells from exposure to radioactivity. Cosmic rays or natural radioactivity in the soil might perform this function. In any case, an offspring born with altered DNA is called a **mutant.** Most mutations are fatal, and the individual dies long before it can have offspring of its own. But in rare cases, a mutation may give a species a new survival advantage. Then natural selection makes it likely that the new DNA message will survive and be handed down, making the species more capable of surviving.

REVIEW Critical Inquiry

Why can't the information in DNA be permanent?

The information stored in a creature's DNA provides the recipes that make the creature what it is. For example, the DNA in a starfish must contain all the recipes for making the various kinds of proteins needed to consume and digest food. That information must be passed on to offspring starfish, or they will be unable to survive. But the information must be changeable, because our environment is changeable. Ice ages come and go, mountains rise, lakes dry up, and ocean currents shift. If the environment changes in some way, one or more of the recipes may no longer work. In our example, a change in the temperature of the ocean water may kill off the shellfish the starfish eat. If they can't digest other shellfish, the entire species will become extinct. Natural variation in DNA means that among all the infant starfish in any generation, some of the recipes are different; if the environment changes, all of the old-style starfish may die, but a few—those with the different DNA—can carry on.

The survival of life depends on this delicate balance between reliable reproduction and the introduction of small variations in DNA information. What are some of the ways these small changes in DNA can arise?

Life is based not only on information, but also on the duplication of information. Today, that process seems so complex that it is hard to imagine how it could have begun.

Figure 19-6
Among the oldest fossils known, this microscopic filament resembles modern bacterial forms (artist's reconstruction at right). This fossil was found in the 3.5-billion-year-old chert of the Pilbara Block in north-western Australia. *(Courtesy J. William Schopf)*

Figure 19-5
Trilobites made their first appearance in the Cambrian oceans about 600 million years ago. This example, about the size of a human hand, lived 400 million years ago in an ocean floor that is now a limestone deposit in Pennsylvania. *(Grundy Observatory photograph)*

19-2 The Origin of Life

If life on Earth is based on the storage of information in these long, complex, carbon-chain molecules, how could it have ever gotten started? Obviously, 4.5 billion chemical bases didn't just happen to drift together to form the DNA formula for a human being. The key is evolution. Once a life form begins to reproduce itself, natural selection preserves the most advantageous traits. Over long periods of time spanning thousands, perhaps millions of generations, the life form becomes more fit to survive. This nearly always means the life form becomes more complex. Thus life could have begun as a very simple process that gradually became more sophisticated as it was modified by evolution.

We begin our search on Earth, where fossils and an intimate familiarity with carbon-based life give us a glimpse of the first living matter. Once we discover how Earthly life could have begun, we can look for signs that life began on other planets in our solar system. Finally, we can speculate on the chances that other planets, orbiting other stars, have conditions that give rise to life.

The Origin of Life on Earth

The oldest fossils hint that life began in the oceans. The oldest easily identified fossils appear in sedimentary rocks that formed between 0.6 and 0.5 billion years ago—the **Cambrian period.** Such Cambrian fossils were simple ocean creatures, the most complex of which were trilobites (Figure 19-5), but there are no Cambrian fossils of land plants or animals. Evidently, land surfaces were totally devoid of life until only 400 million years ago.

Precambrian deposits contain no obvious fossils, but microscopes reveal microfossils that were the ancestors of the Cambrian creatures. Fig tree chert* in South Africa is 3.0 to 3.3 billion years old, and chert from the Pilbara Block in northwestern Australia is 3.5 billion years old. Both contain structures that appear to be microfossils of bacteria or simple algae such as those that live in water (Figure 19-6). Apparently, life was already active in Earth's oceans a billion years after the planet formed.

The key to the origin of this life may lie in an experiment performed by Stanley Miller and Harold Urey in 1952. This **Miller experiment** sought to reproduce the conditions on Earth under which life began (Figure 19-7). In a closed glass container, the experimenters placed water (to represent the oceans), the gases hydro-

*Chert is a rock form that resembles flint.

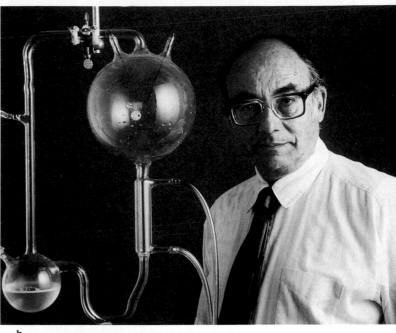

a

b

Figure 19-7

The Miller experiment (a) circulated gases through water in the presence of an electric arc. This simulation of primitive conditions on Earth produced amino acids, the building blocks of proteins. (b) Stanley Miller with a Miller apparatus. *(Courtesy Stanley Miller)*

gen, ammonia, and methane (to represent the primitive atmosphere), and an electric arc (to represent lightning bolts). The apparatus was sterilized, sealed, and set in operation.

After a week, Miller and Urey stopped the experiment and analyzed the material in the flask. Among the many compounds the experiment produced, they found four amino acids that are common building blocks in protein, various fatty acids, and urea, a molecule common to many life processes. Evidently, the energy from the electric arc had molded the atmospheric gases into some of the basic components of living matter. Other energy sources such as hot silica (to simulate hot lava spilling into the sea) and ultraviolet radiation (to simulate sunlight) give similar results.

Recent studies of the composition of meteorites and models of planet formation suggest that Earth's first atmosphere did not resemble the gases used in the Miller experiment. Earth's first atmosphere was probably composed of carbon dioxide, nitrogen, and water vapor. This finding, however, does not invalidate the Miller experiment. When such gases are processed in a Miller apparatus, a tarry gunk rich in organic molecules soon coats the inside of the chamber.

The Miller experiment did not create life, nor did it necessarily imitate the exact conditions on the young Earth. Rather, it is important because it shows that com-

plex organic molecules form naturally in a wide variety of circumstances. The chemical deck is stacked to deal nature a hand of complex molecules. If we could travel back in time, we would probably find Earth's oceans filled with a rich mixture of organic compounds in what some have called the **primordial soup.**

The next step on the journey toward life is for the compounds dissolved in the oceans to link up and form larger molecules. Amino acids, for example, can link together to form proteins. This linkage occurs when amino acids join together end to end and release a water molecule (Figure 19-8). For many years, experts have assumed that this process must have happened in sun-warmed tidal pools where evaporation concentrated the broth. But recent studies suggest that the young Earth was subject to extensive volcanism and large meteorite impacts that periodically modified the climate enough to destroy any life forms exposed on the surface. Thus, the early growth of complex molecules likely took place among the hot springs along the midocean ridges. The heat from such springs could have powered the growth of long protein chains. Deep in the oceans, they would have been safe from climate changes.

Although these proteins might have contained hundreds of amino acids, they would not have been alive. Not yet. Such molecules would not have reproduced, but would have merely linked together and broken apart at random. Because some molecules are more stable

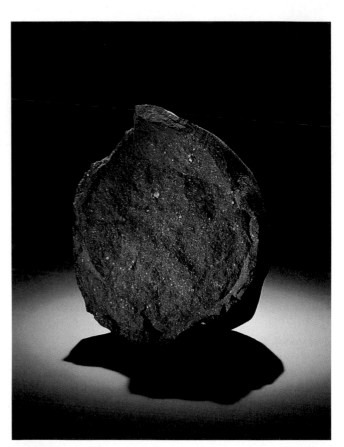

Figure 19-8
Amino acids can link together through the release of a water molecule to form long carbon-chain molecules. The amino acid in this hypothetical example is alanine, one of the simplest.

than others, however, and because some molecules bond together more readily than others, this blind **chemical evolution** would have led to the concentration of the varied smaller molecules into the most stable larger forms. Eventually, somewhere in the oceans, a molecule took shape that could reproduce itself. At that point the chemical evolution of molecules became the biological evolution of living things.

An alternative theory proposes that primitive living things such as reproducing molecules did not originate on Earth but came here in meteorites or comets. Radio astronomers have found a wide variety of organic molecules in the interstellar medium, and some studies have found similar compounds inside meteorites (Figure 19-9). Such molecules form so readily that we would be surprised if they were not present in space. A few investigators, however, have speculated that living, reproducing molecules originated in space and came to Earth as a cosmic contamination. If this is true, every planet in the universe is contaminated with the seeds of life. However entertaining this theory may be, it is presently untestable, and an untestable theory is of little use in science. Experts studying the origin of life proceed on the assumption that life began as reproducing molecules in Earth's oceans.

Which came first, reproducing molecules or the cell? Because we think of the cell as the basic unit of life, this question seems to make no sense, but in fact the cell may have originated during chemical evolution. If a dry mixture of amino acids is heated, the acids form long, proteinlike molecules that, when poured into water, collect to form microscopic spheres that function in ways similar to cells (Figure 19-10). They have a thin membrane surface, they can absorb material from their surroundings, they grow in size, and they can divide and bud just as cells do. They contain no large molecule that copies itself, however. Thus the structure of the cell may have originated first and the reproducing molecules later.

An alternative theory proposes that the replicating molecule developed first. Such a molecule would have been exposed to damage if it had been bare, so the first to manufacture or attract a protective coating of protein would have had a significant survival advantage. If this

was the case, the protective cell membrane was a later development of biological evolution.

The first living things must have been single-celled organisms much like modern bacteria and simple algae. Some of the oldest fossils known are **stromatolites,** structures produced by communities of blue-green algae or bacteria that grew in mats and, year by year, deposited

Figure 19-9
A sample of the Murchison meteorite, a carbonaceous chondrite that fell in 1969 near Murchison, Australia. Analysis of the interior of the meteorite revealed evidence of amino acids. Whether the first building blocks of life originated in space is unknown, but the amino acids found in meteorites illustrate how commonly amino acids and other complex molecules occur through nonorganic means. *(Courtesy Chip Clark, National Museum of Natural History)*

Figure 19-10

Single amino acids can be assembled into long proteinlike molecules. When such material cools in water, it can form microspheres, microscopic spheres with double-layered boundaries similar to cell membranes. Microspheres may have been an intermediate stage in the evolution of life between complex molecules and cells holding molecules reproducing genetic information. *(Courtesy Sidney Fox and Randall Grubbs)*

layers of minerals that were later fossilized. One of the oldest such fossils known is 3.5 billion years old (Figure 19-11). If such algae were common when Earth was young, it may have been able to produce a small amount of oxygen in the early atmosphere. Recent studies suggest that only 0.1 percent would have been sufficient to provide an ozone screen that would protect organisms from the sun's ultraviolet radiation.

How evolution shaped these creatures to live in the ancient oceans, molded them into multicellular organisms, and caused them to develop sexual reproduction, photosynthesis, and respiration is a fascinating story, but we cannot explore it in detail here. We can see that life could have begun through simple chemical reactions building complex molecules, and that once some DNA-like molecule formed, it protected its own survival with selfish determination. Over billions of years, the genetic information stored in living things kept those qualities that favored survival and discarded the rest. As Samuel Butler said, "The chicken is the egg's way of making another egg." In that sense, all living matter on Earth is merely the physical expression of DNA's mindless determination to continue its existence.

Perhaps this seems harsh. Human experience goes far beyond mere reproduction. *Homo sapiens* has art, poetry, music, philosophy, religion, science. Perhaps

Figure 19-11

A 3.5-billion-year-old fossil stromatolite from western Australia is one of the oldest known fossils (below). Stromatolites were formed, layer by layer, by mats of blue-green algae or bacteria living in shallow water. Such algae may have been common in shallow seas when Earth was young (right). Stromatolites are still being formed today in similar environments. *(Mural by Peter Sawyer; photo courtesy Chip Clark, National Museum of Natural History)*

all of the great accomplishments of our intelligence represent more than mere reproduction of DNA. Nevertheless, intelligence, the ability to analyze complex situations and respond with appropriate action, must have begun as a survival mechanism. For example, a fixed escape strategy stored in the DNA is a disadvantage for a creature that frequently moves from one environment to another. A rodent that always escapes from predators by automatically climbing the nearest tree would be in serious jeopardy if it met a hungry fox in a treeless clearing. Even a faint glimmer of intelligence might allow the rodent to analyze the situation and, finding no trees, to choose running over climbing. Thus, intelligence, of which *Homo sapiens* is so proud, may have developed in ancient creatures as a way of making them more versatile.

Any discussion of the evolution of life seems to involve highly improbable coincidences until we consider how many years have passed in the history of the Earth. We read the words—4.6 billion years—so easily, but it is in truth hard to grasp the meaning of such a long period of time.

Geologic Time

Humanity is a very new experiment on planet Earth. We can fit all of the evolution that leads from the primitive life forms in the oceans of the Cambrian period 600 million years ago to fishes, amphibians, reptiles, and mammals into a single chart such as Figure 19-12. We can take comfort in thinking that human-like creatures have walked the earth for roughly 3 million years, but when we add our history to the chart, we discover that the entire history of the human race makes up no more than a thin line at the top. In fact, if we tried to represent the entire 4.6-billion-year history of the earth on the chart, the portion describing the rise of life on the land would be an unreadably small segment.

One way to represent the evolution of life is to compress the 4.6-billion-year history of Earth into a 1-year-long videotape. In such a program, Earth forms as the video begins on January 1, and through all of January and February it cools and is cratered and the first oceans form. Search as we might, we would find no trace of life in these oceans until sometime in March or early April, when the first living things develop. The slow development of these simplest of living forms grinds on slowly through the spring and summer of our videotape. The entire 4-billion-year history of Precambrian evolution lasts until the video reaches mid-November, when the primitive ocean life begins to evolve into more complex organisms such as trilobites.

While our year-long videotape plays on and on, we might amuse ourselves by looking at the land instead of the oceans, but we would be disappointed. The land is a lifeless waste with no plants or animals of any kind.

Not until November 28 in our video does life appear on the land; but once it does, it evolves rapidly into a wide range of plants and animals. Dinosaurs, for example, appear about December 12 and vanish by Christmas evening as mammals and birds flourish.

Throughout the 1-year run of our video there have been no humans; and even during the last days of the year, as the mammals rise to dominate the landscape, there are no people. In the early evening of December 31, vaguely human forms move through the grasslands, and by late evening they begin making stone tools. The Stone Age lasts till about 11:45 PM, and the first signs of civilization, towns and cities, do not appear until 11:54 PM. The Christian era begins only 14 seconds before the New Year, and the Declaration of Independence is signed with but 1 second to spare.

By converting the history of the Earth into a year-long videotape, we have placed the rise of life in perspective. Tremendous amounts of time were needed to create the first simple living things in the oceans, and even more time was needed to develop complex creatures that could colonize the land. As life became more complex, it evolved and diversified faster and faster, as if evolution were drawing on a growing library of solutions that had been previously invented with great effort to solve earlier problems. The burst of diversity on land led slowly to the rise of intelligent creatures like us, a process that has taken 4.6 billion years.

The vast scope of geologic time can be illustrated in another way. The atoms in your body have a wonderful history, and we can now tell their story. The matter was born in the big bang, but it was almost entirely hydrogen and helium. The heavier atoms in your body, such as carbon, nitrogen, and oxygen, were created by nuclear fusion inside stars. At least two or three generations of stars must have lived and died to create the heavier atoms that were present in the nebula that gave birth to the sun and Earth. The atoms in your body were present in the first oceans and were part of the first organisms. When life emerged onto the land, some of your atoms were present. Most of the carbon atoms in your body have been inside at least one dinosaur. We only borrow our atoms from Earth. They will have further adventures in the future.

Astronomers take a cosmic view of life. The long sweep of geologic time is only part of a longer story of stars and atoms.

If life could originate on Earth and develop into intelligent creatures, perhaps the same thing could have happened on other planets. This raises three questions. First, could life originate on another world if conditions were suitable? Our study of the molecular origin of life seems to tell us that life is inevitable. As Carl Sagan said, "Life is in the cards." The second question is, if life begins on a planet, will it evolve toward intel-

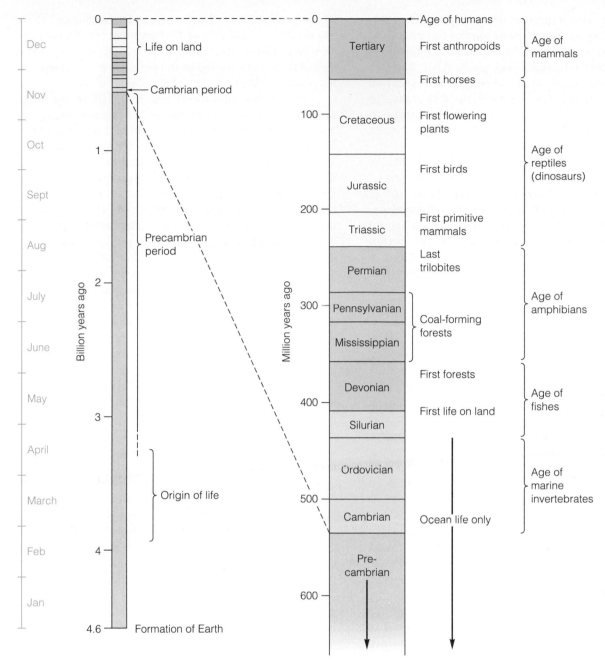

Figure 19-12
Complex life has developed on Earth only recently. If the entire history of Earth were represented in a time line (left), we would have to magnify the end of the line to see details such as life leaving the oceans and dinosaurs appearing. The age of humans would still be only a thin line at the top of our diagram. If the history of Earth were a year-long videotape, humans would not appear until the last hours of December 31.

ligence? We have seen that the process of evolution is directed toward survival, which should lead to versatility and intelligence. Nevertheless, we should recall that intelligence seems to take a long time to develop. Human intelligence is very recent on Earth.

But what of the third question: Are suitable conditions so rare that life almost never gets started? The only way to answer that is to search for life on other worlds. We begin, in the next section, with the other planets and moons in our solar system.

Life in Our Solar System

Although we can imagine life based on something other than carbon chemistry, we know of no examples to tell us how such life might arise and survive. We must limit our discussion to life as we know it and the conditions it requires. The most important requirement is the presence of liquid water, not only as part of the chemical reactions of life, but also as a medium to transport nutrients and wastes within the organism. Also, it seems that life on Earth began in the oceans and developed there for over 4 billion years before it was able to emerge onto the land. Certainly, any world where we hope to find life must have liquid water, and that means it must have moderate temperatures.

The water requirement automatically eliminates many worlds in our solar system. The moon is airless, and although some data suggest ice frozen in the soil at its poles, it has never had liquid water on its surface. In the vacuum of the lunar surface, liquid water would boil away rapidly. Mercury too is airless and cannot have had liquid water on its surface for long periods of time. Venus has some traces of water vapor in its atmosphere, but it is much too hot for liquid water to survive. If there were any lakes or oceans of water on its surface when it was young, they must have evaporated quickly. Even if life began there, no traces would be left now.

The inner solar system seems too hot, and the outer solar system seems too cold. The Jovian planets have deep atmospheres, and, at a certain level, they have moderate temperatures where water might condense into liquid droplets. But it seems unlikely that life could begin there. The Jovian planets have no surfaces where oceans could nurture the beginning of life, and currents in the atmosphere seem destined to circulate gas and water droplets from regions of moderate temperature to other levels that are much to hot or too cold for life to survive.

A few of the satellites of the Jovian planets might have suitable conditions for life. Jupiter's moon Europa seems to have a liquid water ocean below its icy crust, and minerals dissolved in that water would provide a rich broth of possibilities for chemical evolution. Nevertheless, Europa is not a promising site to search for life because conditions may not have remained stable for the billions of years needed for life to evolve beyond the microscopic stage. The subsurface ocean is kept from freezing by tidal heating. If Jupiter's moons interact gravitationally and modify their orbits, Europa may have been frozen solid at some points in its history. Such periods of freezing would probably prevent life from developing. Drilling through the icy crust of Europa to search for life in its ocean will be a wonderful adventure for future generations, but Europa does not seem to be a good bet to harbor life.

Saturn's moon Titan has a complex atmosphere of nitrogen, argon, and methane and may have oceans of liquid methane and ethane on its surface. We saw in Chapter 8 how sunlight can convert the methane in the atmosphere to organic smog particles that settle to the surface. The chemistry of life that might crawl or swim on such a world is unknown, but life there may be un-

likely because of the temperature. The surface of Titan is a deadly −179°C (−290°F). Chemical reactions occur slowly or not at all at such low temperatures, so the chemical evolution needed to begin life may never have occurred on Titan.

Mars is the most likely place for life in our solar system. The evidence, however, is not encouraging. In 1976, two robotic spacecraft, Viking 1 and Viking 2, landed at two different places on the Martian surface. The spacecraft scooped up soil samples, and subjected them to tests for the presence of living organisms. For example, they gave some soil samples a dose of nutrient-rich water and watched for signs the nutrients were taken up by biological processes. The results seem negative. Although some peculiar chemical processes were detected, no evidence was found that clearly indicated the presence of any living things in the soil.

The chemical experiments performed by the Viking landers were sophisticated, but they could only search for traces of currently living things. They could not search for signs of past life, such as fossils. For that, we need to send a geologist to Mars or bring rocks from Mars back to Earth.

Nature has performed the latter task for us. Meteorite ALH84001 (Figure 19-13a) was found on the Antarctic ice in 1984 and only recently recognized as one of a class of meteorites that originated on Mars. How do we know they came from Mars? About a dozen meteorites are classified as SNC meteorites because they have very similar physical and chemical properties. One of those meteorites contains gases trapped in small, glassy inclusions. Analysis of these gases show that they have the same abundance of oxygen isotopes as the gases in the atmosphere of Mars as measured by the two Viking landers. The abundances of the isotopes are like fingerprints, and they match exactly. It seems conclusive that the SNC meteorites are rocks that were blasted off the surface of Mars at some point long ago when a comet or an asteroid struck the planet. From traces of radioactive elements in ALH84001, scientists conclude it left the surface of Mars about 16 million years ago, orbited the sun for a long time, and then fell to Earth in Antarctica about 13,000 years ago.

In 1996, NASA scientists announced that they had found three kinds of evidence that life once existed in ALH84001. Part of the evidence consisted of carbonate deposits that resemble those produced by some Earthly bacteria (Figure 19-13b). A second part of the evidence was complex organic molecules that could be the decayed remains of ancient bacteria. The third piece of evidence was revealed in electron microscope images that show what seem to be small fossils of living things (Figure 19-13c).

This discovery was received with great excitement by the press and the public. It would, indeed, be dramatic evidence if true, because it would show that life could begin on another world.

a

Scientists, being professionally skeptical (see Window on Science 5-2), began testing these results immediately, and in many cases the tests did not confirm the conclusion that life once existed in the meteorite. There is evidence that the water that deposited the carbonates was much too hot for living things to survive. Also, the organic molecules found in the meteorite may have formed from nonbiological processes, or they may be contamination. There is conclusive evidence that the interior of the meteorite has been contaminated by chemical compounds such as amino acids that exist in the Antarctic ice. Furthermore, the microscopic fossils may be natural mineral features of the rock samples and not the remains of living things. Although studies of ALH84001 continue, it does not give us unchallenged evidence that life once existed on Mars.

The issue is not closed. Some scientists continue to defend the evidence of life found in ALH84001, and other meteorites are being studied. Spacecraft now visiting Mars may help us understand the past history of water there and paint a more detailed picture of present conditions. Landers are planned that will eventually land on Mars, collect rocks, and return them to Earth. Nevertheless, conclusive evidence may have to wait until a geologist in a space suit can wander the dry streambeds of Mars cracking open rocks and searching for fossils.

We are left to conclude that, so far as we know, our solar system, except for Earth, is bare of life. Consequently our search for life in the universe takes us to other planetary systems.

Life in Other Planetary Systems

Might life exist in other solar systems? To consider this question, let us try to decide how common planets are and what conditions a planet must fulfill for life to originate and evolve to intelligence. The first question is

b

c

Figure 19-13
(a) ALH84001 is one of a dozen meteorites known to have originated on Mars. (b) Microscopic carbonate globules found inside the meteorite resemble similar deposits made by Earthly bacteria. (c) Concentrations of features that resemble Earthly bacteria have been found inside the meteorite. *(NASA)*

Figure 19-14
The life zone around a star (green) is the region where a planet would have a moderate temperature. M stars are not good candidates because they are cool and their life zone is small. F stars have a large life zone but evolve too quickly to allow the rise of complex life. K and G stars like our sun seem the best candidates in a search for intelligent life.

astronomical; the second is biological. Our ability to discuss the problem of life outside our solar system is severely limited by our lack of experience.

In Chapter 5, we concluded that planets form as a natural result of star formation. In addition, the process that gives rise to planets is probably related to the process that forms binary star systems. Had Jupiter been 100 times more massive, it would have been a star instead of a planet. Because about half of all stars are members of binary or multiple-star systems, it seems that this process is very common, implying that planetary systems are also common.

If a planet is to become a suitable home for life, it must have a stable orbit around its sun. This is simple in a solar system like our own, but in a binary system most planetary orbits are unstable. Most planets in such systems would not last long before they were swallowed up by one of the stars or ejected from the system.

Thus, single stars are the most likely to have planets suitable for life. Because our galaxy contains about 10^{11} stars, half of which are single, there should be roughly 5×10^{10} planetary systems in which we might look for life.

A few million years of suitable conditions does not seem to be enough time to originate life. Our planet required at least 0.5 to 1 billion years to create the first cells and 4.6 billion years to create intelligence. Clearly, conditions on a planet must remain acceptable over a long time. This eliminates giant stars that change their luminosity rapidly as they evolve. It also eliminates massive stars that remain stable on the main sequence for only a few million years. If life requires a few billion years to originate and evolve to intelligence, no star hotter than about F5 will do. This is not really a serious restriction, because upper-main-sequence stars are rare anyway.

In previous sections, we decided that life, at least as we know it, requires liquid water. That requirement

defines a **life zone** (or ecosphere) around each star, a region within which a planet has temperatures that permit the existence of liquid water.

The size of the life zone depends on the temperature of the star (Figure 19-14). Hot stars have larger life zones because the planets must be more distant to remain cool. But the short main-sequence lives of these stars make them unacceptable. M stars have small life zones because they are extremely cool—only planets very near the star receive sufficient warmth. However, planets that are close to a star would probably become tidally coupled, keeping the same side toward the star. This might cause the water and atmosphere to freeze in the perpetual darkness of the planet's night side and end all chance of life. Also, M stars are subject to sudden flares that might destroy life on a planet close to the star. Thus, the life zone restricts our search for life to main-sequence G and K stars. Some of the cooler F stars and warmer M stars might also be good candidates.

Even a star on the main sequence is not perfectly stable. Main-sequence stars gradually grow more luminous as they convert their hydrogen to helium, and thus the life zone around a star gradually moves outward. A planet might form in the life zone and life might begin and evolve for billions of years only to be destroyed as the slowly increasing luminosity of its star moves the life zone outward, evaporates the planet's oceans, and drives off its atmosphere. If a planet is to remain in the life zone for 4 to 5 billion years, it must form on the outer edge of the zone. This may be the most serious restriction we have yet discussed.

If all these requirements are met, will life begin? Early in this chapter we decided that life could begin through simple chemical reactions, so perhaps we should change our question and ask, What could prevent life from beginning? Given what we know about life, it should arise whenever conditions permit, and our galaxy should be filled with planets that are inhabited with living creatures.

What evidence do we have that life is at least possible on other worlds?

Our evidence is limited almost entirely to Earth, but it is promising. The fossils we find show that life originated in Earth's oceans almost 4 billion years ago, and biologists have proposed relatively simple chemical processes that could have created these first reproducing molecules. The fossils show that life developed very slowly at first into more and more complex creatures that filled the oceans. The pace of evolution quickened dramatically about 600 million years ago, the beginning of the Cambrian period, when life began taking on complex forms; and later, when life emerged onto the surface of the land, it evolved rapidly to produce the tremendous diversity we see around us. Human intelligence has been a very recent development; it is only a few million years old.

If this process occurred on Earth, then it seems reasonable that it could have occurred on other worlds as well. Thus, the evidence suggests we can expect that life might begin and evolve to intelligent forms on any world where conditions are right. What are the conditions we should expect of other worlds that host life?

It is both easy and fun to speculate about life on other worlds, but it leads to a simple question: Is there really life beyond Earth? If we can't leave Earth and visit other worlds, then the only life we can detect will be beings intelligent enough to communicate with us. Why haven't we heard from them?

19-3 Communication with Distant Civilizations

If other civilizations exist, perhaps we can communicate with them in some way. Sadly, travel between the stars is more difficult in real life than in science fiction—and may in fact be impossible. If we can't physically visit, perhaps we can communicate by radio. Again, nature places restrictions on such conversations, but the restrictions are not too severe. The real problem lies with the nature of civilizations.

Travel Between the Stars

Practically speaking, roaming among the stars is tremendously difficult because of three limitations: distance, speed, and fuel. The distances between stars are almost beyond comprehension. It does little good to explain that if we use a Ping-Pong ball in New York City to represent the sun, the nearest star would be another Ping-Pong ball in Chicago. It is only slightly better to note that the fastest commercial jet would take about 4 million years to reach the nearest star.

The second limitation is a speed limit—we cannot travel faster than the speed of light. Though science fiction writers invent hyperspace drives so their heroes can zip from star to star, the speed of light is a natural and unavoidable limit that we cannot exceed. This, combined with the large distances between stars, makes interstellar travel very time-consuming.

The third limitation is that we can't even approach the speed of light without using a fantastic amount of fuel. Even if we ignore the problem of escaping from Earth's gravity, we must still use energy stored in fuel to accelerate to high speed and to decelerate to a stop when we reach our destination. To return to Earth, assuming we wish to, we have to repeat the process. These changes in velocity require a tremendous amount of fuel. If we flew a spaceship as big as a large yacht to a star 5 light-years (1.5 pc) away and wanted to get there in only 10 years, we would use 40,000 times as much energy as the United States consumes in a year.

Travel for a few individuals might be possible if we accept very long travel times. That would require some form of suspended animation (currently unknown) or colony ships that carry a complete, though small, society in which people are born, live, and die generation after generation. Whether the occupants of such a ship would retain the social characteristics of humans over a long voyage is questionable.

These three limitations not only make it difficult for us to leave our solar system, they would also make it difficult for aliens to visit Earth. Reputable scientists have studied "unidentified flying objects" (UFOs) and related phenomena and have never found any evidence that Earth is being visited or has ever been visited by aliens from other worlds (Window on Science 19-2). Thus, humans are unlikely ever to meet an alien face to face. The only way we can communicate with other civilizations is via radio.

Radio Communication

Nature places two restrictions on our ability to communicate with distant societies by radio. One has to do with simple physics, is well understood, and merely makes the communication difficult. The second has to do with the fate of technological civilizations, is still unresolved, and may severely limit the number of societies we can detect by radio.

Radio signals are electromagnetic waves that travel at the speed of light. Because even the nearest civilizations must be a few light-years away, this limits our ability to carry on a conversation with distant beings. If

If we discuss life on other worlds, then we might be tempted to use UFO sightings and supposed visits by aliens from outer space as evidence to test our hypotheses. We don't do so for two reasons, both related to the reliability of these observations.

First, the reputation of UFO sightings and alien encounters does not give us confidence that these data are reliable. Most people hear of such events in the grocery store tabloids, daytime talk shows, or sensational "specials" on viewer-hungry cable networks. We must take note of the low reputation of the media that report UFOs and space aliens. Most of these reports are simply made up for the sake of sensation, and we cannot use them as reliable evidence.

Second, the remaining UFO sightings, those not simply made up, do not survive careful examination. Most are mistakes and unconscious misinterpretations of natural events made by honest people. In short, there is no dependable evidence that Earth has been visited by aliens from space.

That's too bad. A confirmed visit by intelligent creatures from beyond our solar system would answer many of our questions. It would be exciting, enlightening, and, like any real adventure, a bit scary. But none of the UFO sightings are dependable, and we are left with no direct evidence of life on other worlds.

---

we ask a question of a creature 4 light-years away, we will have to wait 8 years for a reply. Clearly, the give-and-take of normal conversation will be impossible.

Instead, we could simply broadcast a radio beacon of friendship to announce our presence. Such a beacon would have to consist of a pattern of pulses obviously designed by intelligent beings to distinguish it from natural radio signals emitted by nebulae, pulsars, and so on. For example, pulses counting off the first dozen prime numbers would do. In fact, we are already broadcasting a recognizable beacon. Short-wavelength radio signals such as TV and FM have been leaking into space for the last 50 years or so. Any civilization within 50 light-years might already have detected us.

If we intentionally broadcast a signal, we can anticode it. A coded message is arranged to prevent strangers from reading it, but an anticoded message is arranged to make it as easy as possible to read even by strangers who do not know our language. To create an anticoded message to broadcast into space, we could transmit pulses to represent 1s and gaps to represent 0s. A message counting through the first few prime numbers would distinguish our signal from natural sources of radio noise and attract the attention of intelligent listeners on other worlds. We could even transmit a picture by sending a string of 1s and 0s that can be arranged in only two ways. One way produces nonsense, but the other way produces a meaningful picture (Figure 19-15).

In 1974, at the dedication of the 1000-ft radio telescope at Arecibo, radio astronomers transmitted such a signal toward the globular cluster M13, which is located

Figure 19-15
An anticoded message is designed for easy decoding. Here a string of 35 radio pulses, represented as 1s and 0s, can be arranged in only two ways, as 5 rows of 7 or 7 rows of 5. The second way produces a friendly message. Any number of pulses can be used so long as it is the product of two prime numbers. Then the pulses can be arranged in only two ways.

Start-of-number markers

Binary numbers 1 to 10

Atomic numbers of hydrogen, carbon, nitrogen, oxygen, and phosphorus

Formulas for sugars and bases in DNA

DNA double helix

Number of units in DNA

Start-of-number markers

Human figure

Start-of-number marker

Population of Earth

Height of human in wavelengths

Aricebo radio dish transmitting signal

Sun and planets with Earth offset

Diameter of dish in wavelengths

Start-of-number marker

Figure 19-16
The Arecibo message to M13 begins by counting from 1 to 10 and goes on to describe our solar system and the biochemistry of life on Earth (color added for clarity). Binary numbers give the height of the human figure (1110) and the diameter of the telescope dish (100101111110) in units of the wavelength of the signal, 12.3 cm. *(NASA)*

in the constellation Hercules 26,000 ly from Earth. When the signal finally arrives, any aliens who detect it will be able to arrange its 1679 pulses in only two ways, as 23 rows of 73 or 73 rows of 23. The first arrangement will produce only a jumble, but the second arrangement will form a picture that describes life on Earth (Figure 19-16). Whether humans will still survive on Earth in 52,000 years when any reply to the message arrives is open to debate.

It took only minutes to transmit the Arecibo message. If more time were taken, a more detailed picture could be sent, and if we were sure our radio telescope was pointed at a listening civilization, we could send a long series of pictures. With pictures we could teach aliens our language and tell them all about our life, our difficulties, and our accomplishments.

If we can think of sending such signals, aliens can think of it too. If we point our radio telescopes in the right direction and listen at the right wavelength, we might hear other intelligent races calling out to one another. This raises two questions: Which stars are the best candidates, and what wavelengths are most likely? We have already answered the first question. Main-sequence G and K stars have the most favorable characteristics. But the second question is more complex.

Only certain wavelengths are useful for communication. We cannot use wavelengths longer than about 30 cm, because the signal would be lost in the background radio noise from our galaxy. Nor can we go to wavelengths much shorter than 1 cm, because of ab-

sorption within our atmosphere. Thus only a certain range of wavelengths, a radio window, is open for communication (Figure 19-17).

This communications window is very wide, so a radio telescope would take a long time to tune over all the wavelengths searching for intelligent signals. Nature may have given us a way to narrow the search, however. Within the communications window lie the 21-cm

Figure 19-17
Radio noise from various sources makes it difficult to detect signals from distant civilizations at wavelengths longer than 30 cm or shorter than 1 cm. In this range, radio emission from H atoms and from OH mark a small wavelength range dubbed the water hole, which may be a likely place for communication.

line of neutral hydrogen and the 18-cm line of OH. The interval between these two lines has been dubbed the **water hole,** because the combination of H and OH yields water (H_2O). Water is the fundamental solvent in our life form, so it might seem natural for similar water creatures to call out to each other at wavelengths in the water hole. But even silicon creatures would be familiar with the 21-cm line of hydrogen.

This is not idle speculation. A number of searches for extraterrestrial radio signals have been made, and some major searches are now under way. The field has become known as **SETI,** Search for Extra-Terrestrial Intelligence, and it has generated heated debate. Some scientists and philosophers argue that life on other worlds can't possibly exist, so it is a waste of money to search. Others argue that life on other worlds is common and could be detected with present technology. Congress funded a NASA search for a short time but then ended support because political leaders feared public reaction. In fact, the annual cost of a major search is only about as much as a single Air Force attack helicopter. The controversy may spring in part from the theological and philosophical controversy that would result from the discovery of intelligent life on another world.

In spite of the controversy, some searches have been made, and others are continuing. The first was Project Ozma in 1960. It listened to two stars at a wavelength of 21 cm. Since then, numerous small-scale searches have been conducted, but in order to have some assurance of success a search must scan the entire sky at millions of closely spaced frequency bands. Only recently has technology made that possible, and as technology improves, the searches can scan more sky and more frequencies.

Since 1985, META, Megachannel Extra-Terrestrial Assay, has been searching the entire sky at millions of frequency bands in the water hole. Funded by the Planetary Society, the search uses radio antennas at Harvard and near Buenos Aires, Argentina. The project is now instituting a second, even more efficient search called BETA. Since it began, META has detected dozens of candidate signals. That is, it has found dozens of radio signals that satisfy its most basic criteria. Unfortunately, none of these signals has proved to be continuous, and no signal has been detected when the radio telescopes were redirected at any of the candidate stars. Many of these signals are probably unusual noise from sources on Earth, but the candidate signals are still under study.

One ingenious search is called SERENDIP, Search for Extraterrestrial Radio Emission from Nearby Developed Intelligent Populations. Rather than monopolize an entire radio telescope, SERENDIP rides piggyback on the 305-m Arecibo Telescope. Wherever the radio astronomers point the telescope, the receiver samples the signal, looking for intelligent signals over millions of frequency bands. If candidate signals are found, the receivers note the position of the radio telescope for future investigation.

The NASA search originally called SETI was renamed High Resolution Microwave Survey to reduce criticism, but Congress cut off all of its funds in the fall of 1993. The search continues, however, with private funding, as Project Phoenix. It uses an 8.4-million-channel receiver on major radio telescopes and plans to survey the entire sky in 10 years. In addition, it will listen carefully to 800 nearby solar-type stars. The search officially began on October 12, 1992, 500 years after the landing of Columbus in the New World.

What should we expect if signals are found? By international agreement, those searching for signals have agreed to share candidate signals and obtain reliable confirmation before making public announcements. False alarms would be embarrassing. The chances of success depend on the number of inhabited worlds in our galaxy, and that number is difficult to estimate.

How Many Inhabited Worlds?

The technology exists, and given enough time the searches will find other inhabited worlds, assuming there are at least a few out there. If intelligence is common, then we should find the signals soon—in the next few decades—but if intelligence is rare in the universe, it may be a very long time before we confirm that we are not alone.

Simple arithmetic can give us an estimate of the number of technological civilizations with which we might communicate. The formula for the number of communicative civilizations in a galaxy, N_c, is:

$$N_c = N^* \cdot f_P \cdot n_{LZ} \cdot f_L \cdot f_I \cdot F_S$$

N^* is the number of stars in a galaxy, and f_P represents the probability that a star has planets. If all single stars have planets, f_P is about 0.5. The factor n_{LZ} is the average number of planets in a solar system suitably placed in the life zone, f_L is the probability that life will originate if conditions are suitable, and f_I is the probability that the life form will evolve to intelligence. These factors can be roughly estimated, but the remaining factor is much more uncertain.

F_S is the fraction of a star's life during which the life form is communicative. Here we assume that a star lives about 10 billion years. If a society survives at a technological level for only 100 years, our chances of communicating with it are small. But a society that stabilizes and remains technological for a long time is much more likely to be in the communicative phase at

Table 19-1 The Number of Technological Civilizations per Galaxy

	Variables	Estimates	
		Pessimistic	Optimistic
$N*$	Number of stars per galaxy	2×10^{11}	2×10^{11}
f_P	Fraction of stars with planets	0.01	0.5
n_{LZ}	Number of planets per star that lie in life zone for longer than 4 billion years	0.01	1
f_L	Fraction of suitable planets on which life begins	0.01	1
f_I	Fraction of life forms that evolve to intelligence	0.01	1
F_S	Fraction of star's life during which a technological society survives	10^{-8}	10^{-4}
N_c	Number of communicative civilizations per galaxy	2×10^{-5}	10×10^6

the proper time to signal to us. If we assume that technological societies destroy themselves in about 100 years, F_S is 100 divided by 10 billion, or 10^{-8}. But if societies can remain technological for a million years, then F_S is 10^{-4}. The influence of the factors in the formula is shown in Table 19-1.

If the optimistic estimates are true, there may be a communicative civilization within a few dozen light-years of us, and we could locate it by searching through only a few thousand stars. On the other hand, if the pessimistic estimates are correct, we may be the only planet in our galaxy capable of communication. We may never know until we understand how technological societies function.

REVIEW   Critical Inquiry

Why does the number of inhabited worlds we might hear from depend on how long civilizations survive at a technological level?

When we turn radio telescopes toward the sky and scan millions of frequency bands, we take a snapshot of the universe at a particular time when we are living and able to build radio telescopes. To detect other civilizations, they must be in a similar technological stage so they will be broadcasting either intentionally or accidentally. If we search for decades and detect no other signal, it may mean not that life is rare but rather that civilizations do not survive at a technological level for very long. Only a few decades ago, our civilization was threatened by nuclear war, and now we are threatened by the pollution of our environment. If nearly all civilizations in our galaxy are either on the long road up

from primitive life forms in oceans or on the long road down from nuclear war or environmental collapse, there may be no one transmitting during the short interval when we are capable of building radio telescopes with which to listen.

Radio communication between inhabited worlds is limited because it requires very fast computers to search many frequency intervals. Why must we search so many frequencies when we suspect that the water hole would be a good place to listen?

Are we the only thinking race? If we are, we bear the sole responsibility to understand and admire the universe. Then we are the sole representatives of that state of matter called intelligence. The mere detection of signals from another civilization would demonstrate that we share the universe with others. Although we might never leave our solar system, such communication would end the self-centered isolation of humanity and stimulate a reevaluation of the meaning of our existence. We may never realize our full potential as humans until we communicate with nonhuman intelligent life.

Summary

To discuss life on other worlds, we must first understand something about life in general, life on Earth, and the origin of life. In general, we can identify three properties of living things: a process, a physical basis, and a controlling unit of information. The process must extract energy from the surroundings, maintain the organism, and modify the surroundings to promote the organism's survival. The physical basis is the arrangement of matter and energy that implements the life process. On Earth, all life is based on carbon chemistry. The controlling information is the data necessary to maintain the organism's function. Data for Earth life are stored in long carbon-chain molecules called DNA.

The DNA molecule stores information in the form of chemical bases linked together like the rungs of a ladder. When these patterns are copied by RNA molecules, they can direct the manufacture of proteins and enzymes. Thus the DNA information is the chemical formulas the cell needs to function. When a cell divides, the DNA molecule splits lengthwise and duplicates itself so that each of the new cells has a copy of the information. Errors in the duplication or damage to the DNA molecule can produce mutants, organisms that contain new DNA information and have new properties. Natural selection determines which of these new organisms are best suited to survive, and the species evolves to fit its environment.

The Miller experiment duplicated conditions in Earth's primitive environment and suggests that energy sources such as lightning could have caused amino acids and other complex molecules to form. Chemical evolution would have connected these together in larger and more complex, but not yet living, molecules. When a molecule acquired the ability to produce copies of itself, natural selection perfected the organism through biological evolution. Though this may have happened in the first billion years, life did not become diverse and complex until the Cambrian period, about 0.6 billion years ago. Life emerged from the oceans about 0.4 billion years ago, and humanity developed only a few million years ago.

It seems unlikely that there is now life on other planets in our solar system. Most of the planets are too hot or too cold; Mars may have been suitable in the past, but it is now a cold, dry desert. In 1996, evidence of ancient life was found in a meteorite known to have come from Mars, but that evidence has now been cast into doubt.

To find life, we must look beyond our solar system. Because we suspect that planets form from the leftover debris of star formation, we suspect that most stars have planets. The rise of intelligence may take billions of years, however, so short-lived massive stars and binary stars with unstable planetary orbits must be discarded. The best candidates are G and K main-sequence stars.

The distances between stars are too large to permit travel, but communication by radio could be possible. A certain wavelength range called a radio window is suitable, and a small range between the radio signals of H and OH, the so-called water hole, is especially likely.

We now have the technology to search for other intelligent life in the universe. Although such searches are controversial, a number of radio astronomers are now searching for radio signals from extraterrestrial civilizations. If life is common in the universe, success seems inevitable.

New Terms

DNA (deoxyribonucleic acid)	Cambrian period
protein	Miller experiment
enzyme	primordial soup
RNA (ribonucleic acid)	chemical evolution
amino acid	stromatolite
natural selection	life zone
mutant	water hole
	SETI

Review Questions

1. If life is based on information, what is that information?
2. What would happen to a life form if the information handed down to offspring was always the same? How would that endanger the future of the life form?
3. How does the DNA molecule produce a copy of itself?
4. Give an example of natural selection acting on new DNA patterns to select the most advantageous characteristics.
5. Why do we believe that life on Earth began in the sea?
6. Why do we think that liquid water is necessary for the origin of life?
7. What is the difference between chemical evolution and biological evolution?
8. What was the significance of the Miller experiment?
9. How does intelligence make a creature more likely to survive?
10. How do astronomers know that meteorite ALH84001 really came from Mars?
11. Discuss the evidence pro and con that there is or has been life on Mars.
12. Why are upper-main-sequence stars unlikely sites for intelligent civilizations?
13. Why do we suspect that travel between stars is nearly impossible?
14. How does the stability of technological civilizations affect the probability that we can communicate with them?
15. What is the water hole, and why would it be a good place to look for other civilizations?

Discussion Questions

1. What would you change in the Arecibo message if humanity lived on Mars instead of Earth?

2. What do you think it would mean if decades of careful searches for radio signals for extraterrestrial intelligence turned up nothing?

Problems

1. A single human cell encloses about 1.5 m of DNA containing 4.5 billion base pairs. What is the spacing between these base pairs in nanometers? That is, how far apart are the rungs on the DNA ladder?

2. If we represent the history of Earth by a line 1 m long, how long a segment would represent the 400 million years since life moved onto the land? How long a segment would represent the 3-million-year history of humanity?

3. If a human generation (the time from birth to childbearing) is 20 years, how many generations have passed in the last 1 million years?

4. If a star must remain on the main sequence for at least 5 billion years for life to evolve to intelligence, how massive could a star be and still harbor intelligent life on one of its planets? (*Hint:* See Chapter 12.)

5. If there are about 1.4×10^{-4} stars like the sun per cubic light-year, how many lie within 100 ly of Earth? (*Hint:* The volume of a sphere is $\frac{4}{3}\pi r^3$.)

6. The mathematician Karl Gauss suggested planting forests and fields in a gigantic geometric proof to signal to possible Martians that intelligent life exists on Earth. If Martians had telescopes that like ours could resolve details no smaller than 1 second of arc, how large would the smallest element of Gauss's proof have to be? (*Hint:* Use the small-angle formula.)

7. If we detected radio signals with an average wavelength of 20 cm and suspected that they came from a civilization on a distant planet, roughly how much of a change in wavelength should we expect to see because of the orbital motion of the distant planet? (*Hint:* Use the Doppler formula in By the Numbers 4-2.)

8. Calculate the number of communicative civilizations per galaxy from your own estimates of the factors in Table 19-1.

Critical Inquiries for the Web

1. The popular movie *Contact* focused interest on the SETI program by profiling the work of a radio astronomer dedicated to the search for extraterrestrial intelligence. Visit Web sites that give information about the movie, SETI programs, and radio astronomy, and discuss how realistic the movie was in capturing how such research is done.

2. Where outside the solar system would you look for habitable planets? NASA has increasingly focused its interest on this question. Look for information online about programs dedicated to detecting which planets might support life as we know it. What criteria are used to choose targets for the planned searches? What methods will be used to carry out the searches?

 Go to the Brooks/Cole Astronomy Resource Center (www. brookscole.com/astronomy) for critical thinking exercises, articles, and additional readings from InfoTrac College Edition, Brooks/Cole's online student library.

Afterword

Supernatural is a null word.

Robert A. Heinlein
The Notebooks of Lazarus Long

Our journey is over, but before we part company, there is one last thing to discuss—the place of humanity in the universe. Astronomy gives us some comprehension of the workings of stars, galaxies, and planets, but its greatest value lies in what it teaches us about ourselves. Now that we have surveyed astronomical knowledge, we can better understand our own position in nature.

To some, the word *nature* conjures up visions of furry rabbits hopping about in a forest glade dotted with pastel wildflowers. To others, nature is the blue-green ocean depths filled with creatures swirling in a mad struggle for survival. Still others think of nature as windswept mountaintops of gray stone and glittering ice. As diverse as these images are, they are all Earth-bound. Having studied astronomy, we can view nature as a beautiful mechanism composed of matter and energy interacting according to simple rules to form galaxies, stars, planets, mountaintops, ocean depths, and forest glades.

Perhaps the most important astronomical lesson is that we are a small but important part of the universe. Most of the universe is lifeless. The vast reaches between the galaxies appear to be empty of all but the thinnest gas, and the stars, which contain most of the mass, are much too hot to preserve the chemical bonds that seem necessary to allow life to survive and develop. Only on the surfaces of a few planets, where temperatures are moderate, could atoms link together to form living matter.

If life is special, then intelligence is precious. The universe must contain many planets devoid of life, planets where the wind has blown unfelt for billions of years. There may also be planets where life has developed but has not become complex, planets on which the wind stirs wide plains of grass and rustles dark forests. On some planets, insects, fish, birds, and animals may watch the passing days unaware of their own existence. It is intelligence, human or alien, that gives meaning to the landscape.

Science is the process by which intelligence tries to understand the universe. Science is not the invention of new devices or processes. It does not create home computers, cure the mumps, or manufacture plastic spoons—that is engineering and technology, the adaptation of scientific understanding for practical purposes. Science is understanding nature, and astronomy is understanding nature on the grandest scale. Astronomy is the science by which the universe, through its intelligent lumps of matter, tries to understand its own existence.

As the primary intelligent species on this planet, we are the custodians of a priceless gift—a planet filled with living things. This is especially true if life is rare in the universe. In fact, if Earth is the only inhabited planet, our responsibility is overwhelming. In any case, we are the only creatures who can take action to preserve the existence of life on Earth, and ironically, it is our own actions that are the most serious hazards.

The future of humanity is not secure. We are trapped on a tiny planet with limited resources and a population growing faster than our ability to produce food. In our efforts to survive, we have already driven some creatures to extinction and now threaten others. If our civilization collapses because of starvation, or if our race destroys itself somehow, the only bright spot is that the rest of the creatures on Earth will be better off for our absence.

But even if we control our population and conserve and recycle our resources, life on Earth is doomed. In 5 billion years, the sun will leave the main sequence and swell into a red giant, incinerating Earth. Earth will be lifeless long before that, however. Within the next few billion years, the growing luminosity of the sun will first alter Earth's climate and then boil away its atmosphere and oceans. Our Earth is, like everything else in the universe, only temporary.

To survive, humanity must leave Earth and search for other planets. Colonizing the moon and other planets of our solar system will not save us, since they face the same fate as Earth when the sun dies. But travel to other stars is tremendously difficult and may be impossible with the limited resources we have in our small solar system. We and all the living things that depend on us for survival may be trapped.

This is a depressing prospect, but a few factors are comforting. First, everything in the universe is temporary. Stars die; galaxies die; perhaps the entire universe will fall back in a "big crunch" and die. That our distant future is limited only assures us that we are a part of a much larger whole. Second, we have a few billion years to prepare, and a billion years is a very long time. Only a few million years ago, our ancestors were learning to walk erect and communicate with one another. A billion years ago our ancestors were microscopic organisms living in the primeval oceans. To suppose that a billion years hence we humans will still be human, or that we will still be the dominant species on Earth, or that we will still be the sole intelligence on Earth, is the ultimate conceit.

Our responsibility is not to save our race for all eternity, but to behave as dependable custodians of our planet, preserving it, admiring it, and trying to understand it. That will call for drastic changes in our behavior toward other living things and a revolution in our attitude toward our planet's resources. Whether we can change our ways is debatable—humanity is far from perfect in its understanding, abilities, or intentions. We must not imagine, however, that we and our civilization are less than precious. We have the gift of intelligence, and that is the finest thing this planet has ever produced.

We must not cease from exploration and the end of all our exploring will be to arrive where we began and to know the place for the first time.

—T. S. Eliot

APPENDIX A

Units and Astronomical Data

Introduction

The metric system is used worldwide as the system of units not only in science but also in engineering, business, sports, and daily life. Developed in 18th-century France, the metric system has gained acceptance in almost every country in the world because it simplifies computations.

A system of units is based on the three fundamental units for length, mass, and time. Other quantities, such as density and force, are derived from these fundamental units. In the English (or British) system of units (commonly used in the United States) the fundamental unit of length is the foot, composed of 12 inches. The metric system is based on the decimal system of numbers, and the fundamental unit of length is the meter, composed of 100 centimeters.

To see the advantage of having a decimal-based system, try computing the volume of a bathtub that is 5'9" long, 1'2" deep, and 2'10" wide. In the metric system the length of the tub is 1 m and 75 cm or 1.75 m. The other dimensions are 0.35 m and 0.86 m, and the volume is just $1.75 \times 0.35 \times 0.86$ m^3. To make the computation in English units, we must first convert inches to feet by dividing by 12. We can convert centimeters to meters by the simpler process of moving the decimal point. Thus, the computation is much easier if we measure the tub in meters and centimeters instead of feet and inches.

Because the metric system is a decimal system, it is easy to express quantities in larger or smaller units as is convenient. We can express distances in centimeters, meters, kilometers, and so on. The prefixes specify the relation of the unit to the meter. Just as a cent is $\frac{1}{100}$ of a dollar, a centimeter is $\frac{1}{100}$ of a meter. A kilometer is 1000 m, and a kilogram is 1000 g. The meanings of the commonly used prefixes are given in Table A-1.

The SI Units

Any system of units based on the decimal system would be easy to use, but by international agreement, the preferred set of units, known as the *Système International d'Unités* (SI units) is based on the meter, kilogram, and second. These three fundamental units define the rest of the units as given in Table A-2.

The SI unit of force is the newton (N), named after Isaac Newton. It is the force needed to accelerate a 1 kg mass by 1 m/s^2, or the force roughly equivalent to the weight of an apple at Earth's surface. The SI unit of energy is the joule (J), the energy produced by a force of 1 N acting through a distance of 1 m. A joule is roughly the energy in the impact of an apple falling off a table.

Exceptions Units help us in two ways. They make it possible to make calculations, and they help us to conceive of certain quantities. For calculations the metric system is far superior, and we will use it for our calculations throughout this book.

But Americans commonly use the English system of units, so for conceptual purposes we can express quantities in English units. Instead of saying the average person would weigh 133 N on the moon, we could express the weight as 30 lb. Thus, the text commonly gives quantities in metric form followed by the English form in parentheses: The radius of the moon is 1738 km (1080 miles).

In SI units, density should be expressed as kilograms per cubic meter, but no human can enclose a cubic meter in his or her hand, so this unit does not help us grasp the significance of a given density. This book refers to density in grams per cubic centimeter. A gram is roughly the mass of a paperclip, and a cubic centimeter is the size of a small sugar cube, so we can conceive of a density of 1 g/cm^3, roughly the density of water. This is not a bothersome departure from SI units because we will not make complex calculations using density.

Conversions

To convert from one metric unit to another (from meters to kilometers, for example), we have only to look at the prefix. However, converting from metric to

Table A-1 Metric Prefixes

Prefix	Symbol	Factor
Mega	M	10^6
Kilo	k	10^3
Centi	c	10^{-2}
Milli	m	10^{-3}
Micro	μ	10^{-6}
Nano	n	10^{-9}

Table A-2 SI Metric Units

Quantity	SI Unit	English Unit
Length	Meter (m)	Foot
Mass	Kilogram (kg)	Slug (sl)
Time	Second (s)	Second (s)
Force	Newton (N)	Pound (lb)
Energy	Joule (J)	Foot-pound (fp)

English or English to metric is more complicated. The conversion factors are given in Table A-3.

Example: The radius of the moon is 1738 km. What is this in miles? Table A-3 indicates that 1 mile equals 1.609 km, so:

$$1738 \text{ km} \times \frac{1 \text{ mile}}{1.609 \text{ km}} = 1080 \text{ miles}$$

Temperature Scales

In astronomy, as in most other sciences, temperatures are expressed on the Kelvin scale, although the centigrade (or Celsius) scale is also used. The Fahrenheit scale commonly used in the United States is not used in scientific work.

Temperatures on the Kelvin scale are measured from absolute zero, the temperature of an object that contains no extractable heat. In practice, no object can be as cold as absolute zero, although laboratory apparatuses have reached temperatures less than 10^{-6} K. The scale is named after the Scottish mathematical physicist William Thomson, Lord Kelvin (1824–1907).

The centigrade scale refers temperatures to the freezing point of water (0°C) and to the boiling point of water (100°C). One degree centigrade is 1/100th the temperature difference between the freezing and boiling points of water. Thus the prefix *centi*. The centigrade scale is also called the Celsius scale after its inventor, the Swedish astronomer Anders Celsius (1701–1744).

The Fahrenheit scale fixes the freezing point of water at 32°F and the boiling point at 212°F. Named after the German physicist Gabriel Daniel Fahrenheit (1686–1736), who made the first successful mercury thermometer in 1720, the Fahrenheit scale is used only in the United States.

It is easy to convert temperatures from one scale to another using the information given in Table A-4.

Powers of 10 Notation

Powers of 10 make writing very large numbers much simpler. For example, the nearest star is about 43,000,000,000,000 km from the sun. Writing this number as 4.3×10^{13} km is much easier.

Very small numbers can also be written with powers of 10. For example, the wavelength of visible light is about 0.0000005 m. In powers of 10 this becomes 5×10^{-7} m.

The powers of 10 used in this notation appear in the list below. The exponent tells us how to move the decimal point. If the exponent is positive, we move the decimal point to the right. If the exponent is negative, we move the decimal point to the left. Thus, 2×10^3 equals 2000.0, and 2×10^{-3} equals 0.002.

$$\vdots$$

$$
\begin{aligned}
10^5 &= 100,000 \\
10^4 &= 10,000 \\
10^3 &= 1,000 \\
10^2 &= 100 \\
10^1 &= 10 \\
10^0 &= 1 \\
10^{-1} &= 0.1 \\
10^{-2} &= 0.01 \\
10^{-3} &= 0.001 \\
10^{-4} &= 0.0001
\end{aligned}
$$

$$\vdots$$

If you use scientific notation in calculations, be sure you correctly enter numbers into your calculator. Not all calculators accept scientific notation, but those that can have a key labeled EXP, EEX, or perhaps EE that allows you to enter the exponent of ten. To enter a number such as 3×10^8 we press the keys 3 EXP 8. To enter

Table A-3 Conversion Factors

1 inch = 2.54 centimeters	1 centimeter = 0.394 inch
1 foot = 0.3048 meter	1 meter = 39.36 inches = 3.28 feet
1 mile = 1.6093 kilometers	1 kilometer = 0.6214 mile
1 slug = 14.594 kilograms	1 kilogram = 0.0685 slug
1 pound = 4.4482 newtons	1 newton = 0.2248 pound
1 foot-pound = 1.35582 joules	1 joule = 0.7376 foot-pound
1 horsepower = 745.7 joules/s	1 joule/s = 1 watt

Table A-4 Temperature Scales

	Kelvin (K)	Centigrade (°C)	Fahrenheit (°F)
Absolute zero	0 K	−273°C	−459°F
Freezing point of water	273 K	0°C	32°F
Boiling point of water	373 K	100°C	212°F

Conversions:

$$K = °C + 273$$
$$°C = \tfrac{5}{9}(°F - 32)$$
$$°F = \tfrac{9}{5}°C + 32$$

Table A-5 Units Used in Astronomy

1 angstrom (Å)	$= 10^{-8}$ cm
	$= 10^{-10}$ m
1 astronomical unit (AU)	$= 1.495979 \times 10^{11}$ m
	$= 92.95582 \times 10^{6}$ miles
1 light-year (ly)	$= 6.3240 \times 10^{4}$ AU
	$= 9.46053 \times 10^{15}$ m
	$= 5.9 \times 10^{12}$ miles
1 parsec (pc)	$= 206{,}265$ AU
	$= 3.085678 \times 10^{16}$ m
	$= 3.261633$ ly
1 kiloparsec (kpc)	$= 1000$ pc
1 megaparsec (Mpc)	$= 1{,}000{,}000$ pc

a number with a negative exponent, we must use the change-sign key, usually labeled +/− or CHS. To enter the number 5.2×10^{-3} we press the keys 5.2 EXP +/− 3. Try a few examples.

To read a number in scientific notation from a calculator we must read the exponent separately. The number 3.1×10^{25} may appear in a calculator display as 3.1 25 or on some calculators as 3.1 10^{25}. Examine your calculator to determine how such numbers are displayed.

Astronomy, and science in general, is a way of learning about nature and understanding the universe we live in. To test hypotheses about how nature works, scientists use observations of nature. The tables that follow contain some of the basic observations that support our best understanding of the astronomical universe. Of course, these data are expressed in the form of numbers, not because science reduces all understanding to mere numbers, but because the struggle to understand nature is so demanding we must use every tool available. Quantitative thinking, reasoning mathematically, is one of the most powerful tools ever invented by the human brain. Thus the tables that follow are not nature reduced to mere number, but number supporting humanity's growing understanding of the natural world around us.

Table A-6 Constants

astronomical unit (AU)	$= 1.495979 \times 10^{11}$ m
parsec (pc)	$= 206{,}265$ AU
	$= 3.085678 \times 10^{16}$ m
	$= 3.261633$ ly
light-year (ly)	$= 9.46053 \times 10^{15}$ m
velocity of light (c)	$= 2.997925 \times 10^{8}$ m/s
gravitational constant (G)	$= 6.67 \times 10^{-11}$ N · m²/kg²
mass of Earth ($M_\oplus$)	$= 5.976 \times 10^{24}$ kg
equatorial radius of Earth ($R_\oplus$)	
	$= 6378.164$ km
mass of sun ($M_\odot$)	$= 1.989 \times 10^{30}$ kg
radius of sun ($R_\odot$)	$= 6.9599 \times 10^{8}$ m
solar luminosity ($L_\odot$)	$= 3.826 \times 10^{26}$ J/s
mass of moon	$= 7.350 \times 10^{22}$ kg
radius of moon	$= 1738$ km
mass of H atom	$= 1.67352 \times 10^{-27}$ kg

Table A-7 The Nearest Stars

Name	Absolute Magnitude (M_v)	Distance (ly)	Spectral Type	Apparent Visual Magnitude (m_v)
Sun	4.83		G2	−26.8
Proxima Cen	15.45	4.28	M5	11.05
α Cen A	4.38	4.3	G2	0.1
B	5.76	4.3	K5	1.5
Barnard's Star	13.21	5.9	M5	9.5
Wolf 359	16.80	7.6	M6	13.5
Lalande 21185	10.42	8.1	M2	7.5
Sirius A	1.41	8.6	A1	−1.5
B	11.54	8.6	white dwarf	7.2
Luyten 726-8A	15.27	8.9	M5	12.5
B (UV Cet)	15.8	8.9	M6	13.0
Ross 154	13.3	9.4	M5	10.6
Ross 248	14.8	10.3	M6	12.2
ε Eri	6.13	10.7	K2	3.7
Luyten 789–6	14.6	10.8	M7	12.2
Ross 128	13.5	10.8	M5	11.1
61 CYG A	7.58	11.2	K5	5.2
B	8.39	11.2	K7	6.0
ε Ind	7.0	11.2	K5	4.7
Procyon A	2.64	11.4	F5	0.3
B	13.1	11.4	white dwarf	10.8
Σ 2398 A	11.15	11.5	M4	8.9
B	11.94	11.5	M5	9.7
Groombridge 34 A	10.32	11.6	M1	8.1
B	13.29	11.6	M6	11.0
Lacaille 9352	9.59	11.7	M2	7.4
τ Ceti	5.72	11.9	G8	3.5
BD + 5° 1668	11.98	12.2	M5	9.8
L 725–32	15.27	12.4	M5	11.5
Lacaille 8760	8.75	12.5	M0	6.7
Kapteyn's Star	10.85	12.7	M0	8.8
Kruger 60 A	11.87	12.8	M3	9.7
B	13.3	12.8	M4	11.2

Table A-8 Properties of Main-Sequence Stars

Spectral Type	Absolute Visual Magnitude (M_v)	Luminosity*	Temp. (K)	λ max (nm)	Mass*	Radius*	Average Density (g/cm³)
05	−5.8	501,000	40,000	72.4	40	17.8	0.01
B0	−4.1	20,000	28,000	100	18	7.4	0.1
B5	−1.1	790	15,000	190	6.4	3.8	0.2
A0	+0.7	79	9900	290	3.2	2.5	0.3
A5	+2.0	20	8500	340	2.1	1.7	0.6
F0	+2.6	6.3	7400	390	1.7	1.4	1.0
F5	+3.4	2.5	6600	440	1.3	1.2	1.1
G0	+4.4	1.3	6000	480	1.1	1.0	1.4
G5	+5.1	0.8	5500	520	0.9	0.9	1.6
K0	+5.9	0.4	4900	590	0.8	0.8	1.8
K5	+7.3	0.2	4100	700	0.7	0.7	2.4
M0	+9.0	0.1	3500	830	0.5	0.6	2.5
M5	+11.8	0.01	2800	1000	0.2	0.3	10.0
M8	+16	0.001	2400	1200	0.1	0.1	63

*Luminosity, mass, and radius are given in terms of the sun's luminosity, mass, and radius.

Table A-9 The Brightest Stars

Star	Name	Apparent Visual Magnitude (m_v)	Spectral Type	Absolute Visual Magnitude (M_v)	Distance (ly)
α CMa A	Sirius	−1.47	A1	1.4	8.7
α Car	Canopus	−0.72	F0	−3.1	98
α Cen	Rigil Kentaurus	−0.01	G2	4.4	4.3
α Boo	Arcturus	−0.06	K2	−0.3	36
α Lyr	Vega	0.04	A0	0.5	26.5
α Aur	Capella	0.05	G8	−0.6	45
β Ori A	Rigel	0.14	B8	−7.1	900
α CMi A	Procyon	0.37	F5	2.7	11.3
α Ori	Betelgeuse	0.41	M2	−5.6	520
α Eri	Achernar	0.51	B3	−2.3	118
β Cen AB	Hadar	0.63	B1	−5.2	490
α Aql	Altair	0.77	A7	2.2	16.5
α Tau A	Aldebaran	0.86	K5	−0.7	68
α Cru	Acrux	0.90	B2	−3.5	260
α Vir	Spica	0.91	B1	−3.3	220
α Sco A	Antares	0.92	M1	−5.1	520
α PsA	Fomalhaut	1.15	A3	2.0	22.6
β Gem	Pollux	1.16	K0	1.0	35
α Cyg	Deneb	1.26	A2	−7.1	1600
β Cru	Beta Crucis	1.28	B0.5	−4.6	490

Table A-10 Greatest Elongations of Mercury[†]

Evening Sky	Morning Sky
June 9, 2000	July 27, 2000
Oct. 6, 2000	Nov. 15, 2000
Jan. 28, 2001	March 11, 2001
May 22, 2001	July 9, 2001
Sept. 18, 2001	Oct. 29, 2001*
Jan. 11, 2002	Feb. 21, 2002
May 4, 2002	June 21, 2002
Sept. 1, 2002	Oct. 13, 2002*
Dec. 26, 2002	Feb. 4, 2003
April 16, 2003*	June 3, 2003
Aug. 14, 2003	Sept. 27, 2003*
Dec. 9, 2003	Jan. 17, 2004
March 29, 2004*	May 14, 2004
July 27, 2004	Sept. 9, 2004*
Nov. 21, 2004	Dec. 29, 2004
March 12, 2005*	April 26, 2005
July 9, 2005	Aug. 23, 2005
Nov. 3, 2005	Dec. 12, 2005

Table A-11 Greatest Elongations of Venus[†]

Evening Sky	Morning Sky
Jan. 17, 2001	June 8, 2001
Aug. 22, 2002	Jan. 11, 2003
March 29, 2004	Aug. 17, 2004
Nov. 3, 2005	March 25, 2006
June 9, 2007	Oct. 28, 2007
Jan. 14, 2009	June 5, 2009
Aug. 10, 2010	Jan. 8, 2011

*Most favorable elongations.

[†]Elongation: The angular distance from the sun to a planet.

Table A-12 Meteor Showers

Shower	Dates	Hourly Rate	Radiant R.A.	Radiant Dec.	Associated Comet
Quadrantids	Jan. 2–4	30	15^h24^m	50°	
Lyrids	April 20–22	8	18^h4^m	33°	1861 I
η Aquarids	May 2–7	10	22^h24^m	0°	Halley?
δ Aquarids	July 26–31	15	22^h36^m	−10°	
Perseids	Aug. 10–14	40	3^h4^m	58°	1982 III
Orionids	Oct. 18–23	15	6^h20^m	15°	Halley?
Taurids	Nov. 1–7	8	3^h40^m	17°	Encke
Leonids	Nov. 14–19	6	10^h12^m	22°	1866 I Temp
Geminids	Dec. 10–13	50	7^h28^m	32°	

Table A-13 Properties of the Planets

PHYSICAL PROPERTIES (EARTH = ⊕)

Planet	Equatorial Radius (km)	Equatorial Radius (⊕ = 1)	Mass (⊕ = 1)	Average Density (g/cm³)	Surface Gravity (⊕ = 1)	Escape Velocity (km/s)	Sidereal Period of Rotation	Inclination of Equator to Orbit
Mercury	2439	0.382	0.0558	5.44	0.378	4.3	58.646^d	0°
Venus	6052	0.95	0.815	5.24	0.903	10.3	244.3^d	177°
Earth	6378	1.00	1.00	5.497	1.00	11.2	$23^h56^m04.1^s$	23°27′
Mars	3398	0.53	0.1075	3.94	0.379	5.0	$24^h37^m22.6^s$	25°19′
Jupiter	71,494	11.20	317.83	1.34	2.54	61	$9^h55^m30^s$	3°5′
Saturn	60,330	9.42	95.147	0.69	1.16	35.6	$10^h13^m59^s$	26°24′
Uranus	25,559	4.01	14.54	1.19	0.919	22	17^h14^m	97°55′
Neptune	24,750	3.93	17.23	1.66	1.19	25	16^h3^m	28°48′
Pluto	1185	0.19	0.0022	2.0	0.06	1.2	$6^d9^h21^m$	122°

ORBITAL PROPERTIES

Planet	(AU)	(10⁶ km)	(y)	(days)	Average Orbital Velocity (km/s)	Orbital Eccentricity	Inclination to Ecliptic
Mercury	0.3871	57.9	0.24084	87.969	47.89	0.2056	7°0′16″
Venus	0.7233	108.2	0.61515	224.68	35.03	0.0068	3°23′40″
Earth	1	149.6	1	365.26	29.79	0.0167	0°
Mars	1.5237	227.9	1.8808	686.95	24.13	0.0934	1°51′09″
Jupiter	5.2028	778.3	11.867	4334.3	13.06	0.0484	1°18′29″
Saturn	9.5388	1427.0	29.461	10,760	9.64	0.0560	2°29′17″
Uranus	19.18	2869.0	84.013	30,685	6.81	0.0461	0°46′23″
Neptune	30.0611	4497.1	164.793	60,189	5.43	0.0100	1°46′27″
Pluto	39.44	5900	247.7	90,465	4.74	0.2484	17°9′3″

Table A-14 The Greek Alphabet

A, α	alpha	H, η	eta	N, ν	nu	T, τ	tau
B, β	beta	Θ, θ	theta	Ξ, ξ	xi	Y, υ	upsilon
Γ, γ	gamma	I, ι	iota	O, o	omicron	Φ, ϕ	phi
Δ, δ	delta	K, κ	kappa	Π, π	pi	X, χ	chi
E, ε	epsilon	Λ, λ	lambda	P, ρ	rho	Ψ, ψ	psi
Z, ζ	zeta	M, μ	mu	Σ, σ	sigma	Ω, ω	omega

Table A-15 Principal Satellites of the Solar System

Planet	Satellite	Radius (km)	Distance from Planet (10^3 km)	Orbital Period (days)	Orbital Eccentricity	Orbital Inclination
Earth	Moon	1738	384.4	27.322	0.055	5°8′43″
Mars	Phobos	14 × 12 × 10	9.38	0.3189	0.018	1°.0
	Deimos	8 × 6 × 5	23.5	1.262	0.002	2°.8
Jupiter	Metis	20	126	0.29	0.0	0°.0
	Adrastea	12 × 8 × 10	128	0.294	0.0	0°.0
	Amalthea	135 × 100 × 78	182	0.4982	0.003	0°.45
	Thebe	50	223	0.674	0.0	1°.3
	Io	1820	422	1.769	0.000	0°.3
	Europa	1565	671	3.551	0.000	0°.46
	Ganymede	2640	1071	7.155	0.002	0°.18
	Callisto	2420	1884	16.689	0.008	0°.25
	Leda	~8	11,110	240	0.146	26°.7
	Himalia	~85	11,470	250.6	0.158	27°.6
	Lysithea	~20	11,710	260	0.12	29°
	Elara	~30	11,740	260.1	0.207	24°.8
	Ananke	15	21,200	631	0.169	147°
	Carme	22	22,350	692	0.207	163°
	Pasiphae	35	23,300	735	0.40	147°
	Sinope	20	23,700	758	0.275	156°
Saturn	Pan	10	133.570	0.574	0.000	0°
	Atlas	20 × 15 × 15	137.7	0.601	0.002	0°.3
	Prometheus	70 × 40 × 50	139.4	0.613	0.003	0°.0
	Pandora	55 × 35 × 50	141.7	0.629	0.004	0°.05
	Epimetheus	70 × 50 × 50	151.42	0.694	0.009	0°.34
	Janus	110 × 80 × 100	151.47	0.695	0.007	0°.14
	Mimas	196	185.54	0.942	0.020	1°.5
	Enceladus	250	238.04	1.370	0.004	0°.0
	Tethys	530	294.67	1.888	0.000	1°.1
	Calypso	17 × 11 × 12	294.67	1.888	0.0	~1°.?
	Telesto	12	294.67	1.888	0.0	~1°?
	Dione	560	377	2.737	0.002	0°.0
	Helene	20 × 15 × 15	377	2.74	0.005	0°.15
	Rhea	765	527	4.518	0.001	0°.4
	Titan	2575	1222	15.94	0.029	0°.3
	Hyperion	205 × 130 × 110	1484	21.28	0.104	~0°.5
	Iapetus	720	3562	79.33	0.028	14°.72
	Phoebe	110	12,930	550.4	0.163	150°
Uranus	Cordelia	20	49.8	0.3333	~0	~0°
	Ophelia	15	53.8	0.375	~0	~0°
	Bianca	25	59.1	0.433	~0	~0°
	Cressida	30	61.8	0.462	~0	~0°
	Desdemona	30	62.7	0.475	~0	~0°
	Juliet	40	64.4	0.492	~0	~0°

Planet	Satellite	Radius (km)	Distance from Planet (10^3 km)	Orbital Period (days)	Orbital Eccentricity	Orbital Inclination
Uranus	Portia	55	66.1	0.512	~0	~0°
(continued)	Rosalind	30	69.9	0.558	~0	~0°
	Belinda	30	75.2	0.621	~0	~0°
	S/1986 U10	20	76.2	0.638	~0	~0°
	Puck	85 ± 5	85.9	0.762	~0	~0°
	Miranda	242 ± 5	129.9	1.414	0.017	3°.4
	Ariel	580 ± 5	190.9	2.520	0.003	0°
	Umbriel	595 ± 10	266.0	4.144	0.003	0°
	Titania	805 ± 5	436.3	8.706	0.002	0°
	Oberon	775 ± 10	583.4	13.463	0.001	0°
	Caliban	40	7164	579	0.082	139.2
	Sycorax	80	12,174	1284	0.509	152.7
	S/1999 U2	~20	~10,000?			
	S/1999 U3	~20	~10,000?			
	S/1999 U1	~20	~25,000?			
Neptune	Naiad	30	48.2	0.296	~0	~0°
	Thalassa	40	50.0	0.312	~0	~0°
	Despina	90	52.5	0.333	~0	~0°
	Galatea	75	62.0	0.396	~0	~0°
	Larissa	95	73.6	0.554	~0	~0°
	Proteus	205	117.6	1.121	~0	~0°
	Triton	1352	354.59	5.875	0.00	160°
	Nereid	170	5588.6	360.125	0.76	27.7°
Pluto	Charon	626	19.7	6.38718	~0	122°

Table A-16 Periodic Table of the Elements

Group

Atomic masses are based on carbon-12. Numbers in parentheses are mass numbers of most stable or best-known isotopes of radioactive elements.

Atomic number ⟶ 11
Symbol ⟶ Na
Atomic mass ⟶ 22.99

Noble Gases (18)

	IA(1)																IIIA(13)	IVA(14)	VA(15)	VIA(16)	VIIA(17)	
1	1 H 1.008	IIA(2)																					2 He 4.003
2	3 Li 6.941	4 Be 9.012																5 B 10.81	6 C 12.01	7 N 14.01	8 O 16.00	9 F 19.00	10 Ne 20.18
3	11 Na 22.99	12 Mg 24.31	IIIB(3)	IVB(4)	VB(5)	VIB(6)	VIIB(7)	(8)	VIII (9)	(10)	IB(11)	IIB(12)						13 Al 26.98	14 Si 28.09	15 P 30.97	16 S 32.06	17 Cl 35.45	18 Ar 39.95
4	19 K 39.10	20 Ca 40.08	21 Sc 44.96	22 Ti 47.90	23 V 50.94	24 Cr 52.00	25 Mn 54.94	26 Fe 55.85	27 Co 58.93	28 Ni 58.7	29 Cu 63.55	30 Zn 65.38						31 Ga 69.72	32 Ge 72.59	33 As 74.92	34 Se 78.96	35 Br 79.90	36 Kr 83.80
5	37 Rb 85.47	38 Sr 87.62	39 Y 88.91	40 Zr 91.22	41 Nb 92.91	42 Mo 95.94	43 Tc 98.91	44 Ru 101.1	45 Rh 102.9	46 Pd 106.4	47 Ag 107.9	48 Cd 112.4						49 In 114.8	50 Sn 118.7	51 Sb 121.8	52 Te 127.6	53 I 126.9	54 Xe 131.3
6	55 Cs 132.9	56 Ba 137.3	57* La 138.9	72 Hf 178.5	73 Ta 180.9	74 W 183.9	75 Re 186.2	76 Os 190.2	77 Ir 192.2	78 Pt 195.1	79 Au 197.0	80 Hg 200.6						81 Tl 204.4	82 Pb 207.2	83 Bi 209.0	84 Po (210)	85 At (210)	86 Rn (222)
7	87 Fr (223)	88 Ra 226.0	89** Ac (227)	104 Rf (261)	105 Db (262)	106 Sg (263)	107 Bh (262)	108 Hs (265)	109 Mt (266)	110 Uun (269)	111 Uuu (272)	112 Uub (277)						114 Uuq (285)		116 Uuh (289)			118 Uuo (293)

Period

⟵ Transition Elements ⟶

Inner Transition Elements

Lanthanide Series 6 *	58 Ce 140.1	59 Pr 140.9	60 Nd 144.2	61 Pm (145)	62 Sm 150.4	63 Eu 152.0	64 Gd 157.3	65 Tb 158.9	66 Dy 162.5	67 Ho 164.9	68 Er 167.3	69 Tm 168.9	70 Yb 173.0	71 Lu 175.0
Actinide Series 7 **	90 Th 232.0	91 Pa 231.0	92 U 238.0	93 Np 237.0	94 Pu (244)	95 Am (243)	96 Cm (247)	97 Bk (247)	98 Cf (251)	99 Es (252)	100 Fm (257)	101 Md (258)	102 No (259)	103 Lr (260)

APPENDIX B

Observing The Sky

Observing the sky with the naked eye is of no more importance to modern astronomy than picking up pretty pebbles is to modern geology. But the sky is a natural wonder unimaginably bigger than the Grand Canyon, the Rocky Mountains, or any other natural wonder that tourists visit every year. To neglect the beauty of the sky is equivalent to geologists neglecting the beauty of the minerals they study. This supplement is meant to act as a tourist's guide to the sky. We analyzed the universe in the regular chapters, but here we will admire it.

The brighter stars in the sky are visible even from the centers of cities with their air and light pollution. But in the countryside only a few miles beyond the cities, the night sky is a velvety blackness strewn with thousands of glittering stars. From a wilderness location, far from the city's glare, and especially from high mountains, the night sky is spectacular.

Using Star Charts

The constellations are a fascinating cultural heritage of our planet, but they are sometimes a bit difficult to learn because of Earth's motion. The constellations above the horizon change with the time of night and the seasons.

Because Earth rotates eastward, the sky appears to rotate around us westward. A constellation visible in the southern sky soon after sunset will appear to move westward, and in a few hours it will disappear below the horizon. Other constellations will rise in the east, so the sky changes gradually through the night.

In addition, Earth's orbital motion makes the sun appear to move eastward among the stars. Each day the sun moves about twice its own diameter eastward along the ecliptic, and each night at sunset, the constellations are about 1° farther toward the west.

Orion, for instance, is visible in the southern sky in January, but as the days pass, the sun moves closer to Orion. By March, Orion is difficult to see in the southwest sky soon after sunset. By June, the sun is so

Figure B-1
To use the star charts in this book, select the appropriate chart for the date and time. Hold it overhead, and turn it until the direction at the bottom of the chart is the same as the direction you are facing.

close to Orion it sets with the sun and is invisible. Not until late July is the sun far enough past Orion for the constellation to become visible rising in the eastern sky just before dawn.

Clearly Earth's rotation and orbital motion require that we use more than one star chart to map the sky. Which chart we select depends on the time of night and the time of year. The charts given at the end of this book show the evening sky for each month.

To use the charts, select the chart for the appropriate month and hold it overhead as shown in Figure B-1. If you face south, turn the chart until the words *Southern Horizon* are at the bottom. If you face other directions, turn the chart appropriately.

Northern Hemisphere Sky

JANUARY

Early in Month	9 P.M.
Midmonth	8 P.M.
End of Month	7 P.M.

Months along the ecliptic show the location of the sun during the year.

Numbers along the celestial equator show right ascension.

Northern Hemisphere Sky

FEBRUARY

Early in Month	9 P.M.
Midmonth	8 P.M.
End of Month	7 P.M.

Months along the ecliptic show the location of the sun during the year.

Numbers along the celestial equator show right ascension.

**Northern
Hemisphere Sky**

MARCH

Early in Month	9 P.M.
Midmonth	8 P.M.
End of Month	7 P.M.

Months along the ecliptic show the
location of the sun during the year.

Numbers along the celestial equator
show right ascension.

**Northern
Hemisphere Sky**

APRIL

Early in Month	9 P.M.
Midmonth	8 P.M.
End of Month	7 P.M.

Months along the ecliptic show the
location of the sun during the year.

Numbers along the celestial equator
show right ascension.

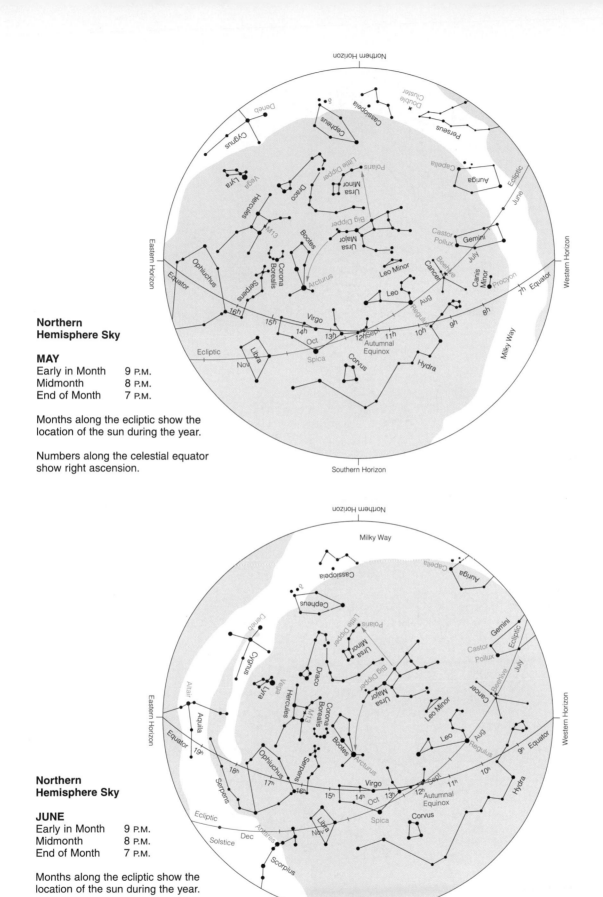

Northern Hemisphere Sky

MAY

Early in Month	9 P.M.
Midmonth	8 P.M.
End of Month	7 P.M.

Months along the ecliptic show the location of the sun during the year.

Numbers along the celestial equator show right ascension.

Northern Hemisphere Sky

JUNE

Early in Month	9 P.M.
Midmonth	8 P.M.
End of Month	7 P.M.

Months along the ecliptic show the location of the sun during the year.

Numbers along the celestial equator show right ascension.

Northern Hemisphere Sky

JULY

Early in Month	9 P.M.
Midmonth	8 P.M.
End of Month	7 P.M.

Months along the ecliptic show the location of the sun during the year.

Numbers along the celestial equator show right ascension.

Northern Hemisphere Sky

AUGUST

Early in Month	9 P.M.
Midmonth	8 P.M.
End of Month	7 P.M.

Months along the ecliptic show the location of the sun during the year.

Numbers along the celestial equator show right ascension.

Northern Hemisphere Sky

SEPTEMBER

Early in Month	9 P.M.
Midmonth	8 P.M.
End of Month	7 P.M.

Months along the ecliptic show the location of the sun during the year.

Numbers along the celestial equator show right ascension.

Northern Hemisphere Sky

OCTOBER

Early in Month	9 P.M.
Midmonth	8 P.M.
End of Month	7 P.M.

Months along the ecliptic show the location of the sun during the year.

Numbers along the celestial equator show right ascension.

**Northern
Hemisphere Sky**

NOVEMBER

Early in Month	9 P.M.
Midmonth	8 P.M.
End of Month	7 P.M.

Months along the ecliptic show the
location of the sun during the year.

Numbers along the celestial equator
show right ascension.

**Northern
Hemisphere Sky**

DECEMBER

Early in Month	9 P.M.
Midmonth	8 P.M.
End of Month	7 P.M.

Months along the ecliptic show the
location of the sun during the year.

Numbers along the celestial equator
show right ascension.

Southern Hemisphere Sky

JANUARY

Early in Month	9 P.M.
Midmonth	8 P.M.
End of Month	7 P.M.

Months along the ecliptic show the location of the sun during the year.

Numbers along the celestial equator show right ascension.

Southern Hemisphere Sky

FEBRUARY

Early in Month	9 P.M.
Midmonth	8 P.M.
End of Month	7 P.M.

Months along the ecliptic show the location of the sun during the year.

Numbers along the celestial equator show right ascension.

Southern Hemisphere Sky

MARCH

Early in Month	9 P.M.
Midmonth	8 P.M.
End of Month	7 P.M.

Months along the ecliptic show the location of the sun during the year.

Numbers along the celestial equator show right ascension.

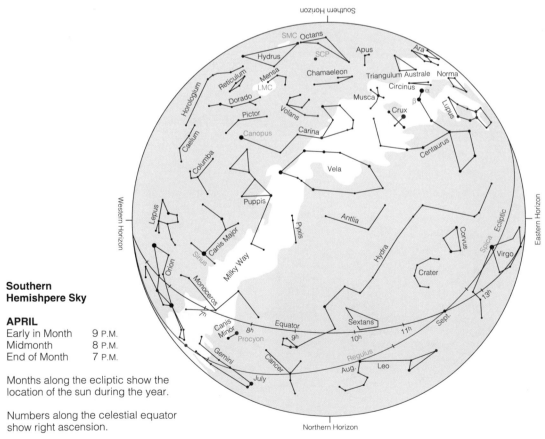

Southern Hemishpere Sky

APRIL

Early in Month	9 P.M.
Midmonth	8 P.M.
End of Month	7 P.M.

Months along the ecliptic show the location of the sun during the year.

Numbers along the celestial equator show right ascension.

Southern Hemisphere Sky

MAY

Early in Month	9 P.M.
Midmonth	8 P.M.
End of Month	7 P.M.

Months along the ecliptic show the location of the sun during the year.

Numbers along the celestial equator show right ascension.

Southern Hemisphere Sky

JUNE

Early in Month	9 P.M.
Midmonth	8 P.M.
End of Month	7 P.M.

Months along the ecliptic show the location of the sun during the year.

Numbers along the celestial equator show right ascension.

**Southern
Hemisphere Sky**

JULY

Early in Month	9 P.M.
Midmonth	8 P.M.
End of Month	7 P.M.

Months along the ecliptic show the location of the sun during the year.

Numbers along the celestial equator show right ascension.

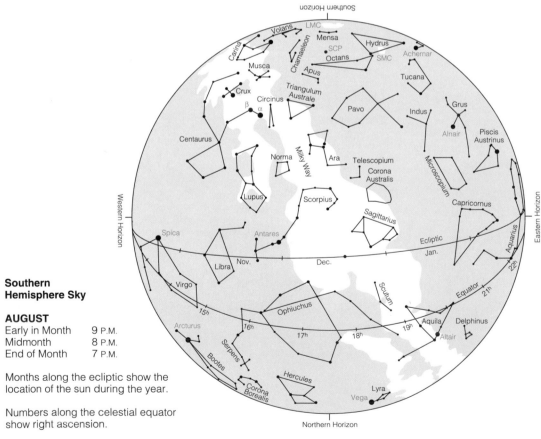

**Southern
Hemisphere Sky**

AUGUST

Early in Month	9 P.M.
Midmonth	8 P.M.
End of Month	7 P.M.

Months along the ecliptic show the location of the sun during the year.

Numbers along the celestial equator show right ascension.

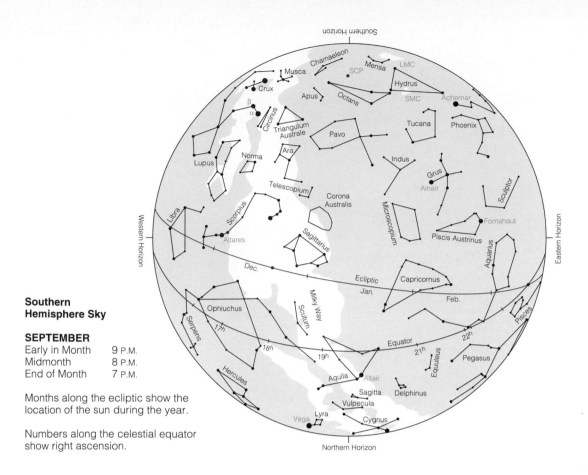

Southern Hemisphere Sky

SEPTEMBER

Early in Month	9 P.M.
Midmonth	8 P.M.
End of Month	7 P.M.

Months along the ecliptic show the location of the sun during the year.

Numbers along the celestial equator show right ascension.

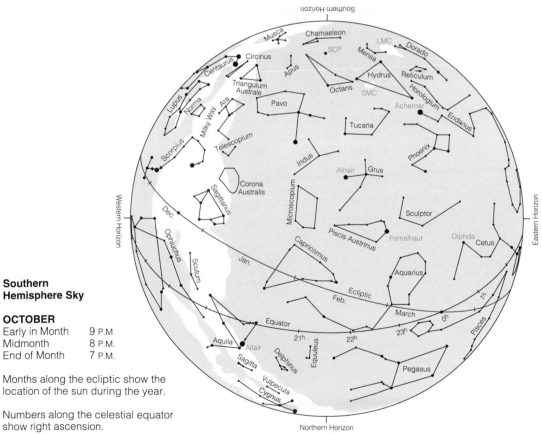

Southern Hemisphere Sky

OCTOBER

Early in Month	9 P.M.
Midmonth	8 P.M.
End of Month	7 P.M.

Months along the ecliptic show the location of the sun during the year.

Numbers along the celestial equator show right ascension.

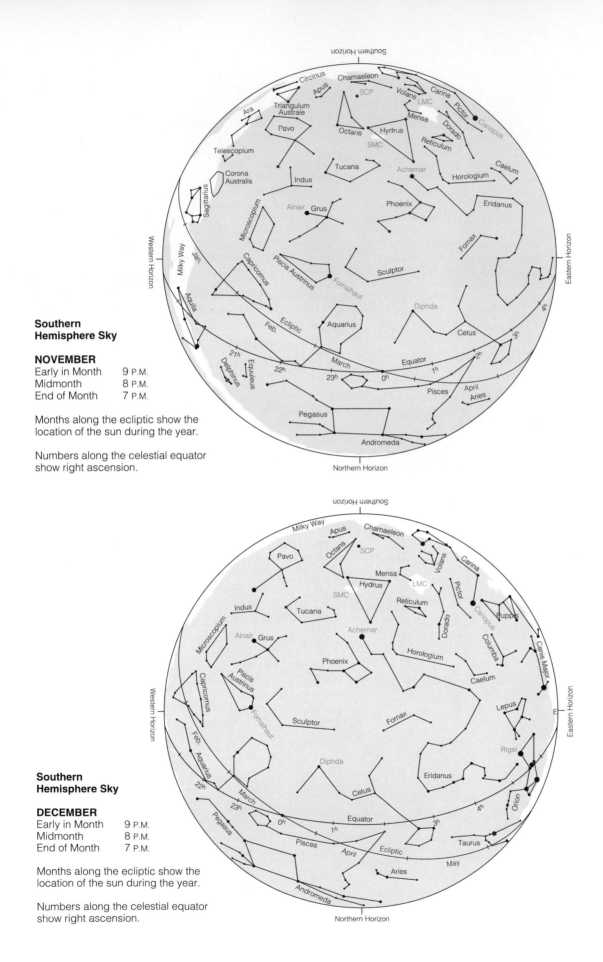

Southern Hemisphere Sky

NOVEMBER

Early in Month	9 P.M.
Midmonth	8 P.M.
End of Month	7 P.M.

Months along the ecliptic show the location of the sun during the year.

Numbers along the celestial equator show right ascension.

Southern Hemisphere Sky

DECEMBER

Early in Month	9 P.M.
Midmonth	8 P.M.
End of Month	7 P.M.

Months along the ecliptic show the location of the sun during the year.

Numbers along the celestial equator show right ascension.

Answers to Even-Numbered Problems

Introduction
2. 1.3 seconds
4. 1.8×10^7 km

Chapter 1
2. about 2800
4. The sun is 400,000 times brighter than the full moon.
6. about 190,000
8. approximately 21 minutes of arc.
10. After one saros cycle of 18 yr $11\frac{1}{3}$ days, the eclipse occurred on 22 July 1990 and was visible from Finland and Siberia. Three saros cycles after July 10, 1972, the eclipse will occur again in Canada on 12 August 2026.

Chapter 2
2. Ratio of greatest distance to closest distance is 6.1 to 1. No; the ratio in the diagram is about 1.5 to 1.
4. 8 years
6. 39.4 AU

Chapter 3
2. short radio waves
4. The 10-m Keck telescope has a light-gathering power that is 1.56 million times greater than the human eye.
6. No; his resolving power should have been about 5.8 seconds of arc at best.
8. 0.013 m = 1.3 cm
10. about 50 cm (From 400 km above, a human is about 0.25 seconds of arc from shoulder to shoulder.)

Chapter 4
2. 150 nm
4. by a factor of $2^4 = 16$
6. 123 km/s; receding

Chapter 5
2. It will look $264,600^2 = 7.0 \times 10^{10}$ times fainter, which is about 27 magnitudes fainter.
4. about 3 times the half-life, or 3.9 billion years
6. large amounts of methane and water ices
8. about 13

Chapter 6
2. 210 million years
4. 3%
6. An object cools by radiating from its surface, and the area of a sphere is proportional to its radius squared (r^2). But the amount of energy an object contains is proportional to its volume, and the volume of a sphere is proportional to its radius cubed (r^3). How fast an object cools is proportional to its area divided by its volume, which equals $1/r$. Thus, the larger an object is, the slower it cools.
8. They would be visible in photos from orbit around the moon.
10. 1.1 seconds of arc
12. No; their angular size was only 0.00054 seconds of arc.

Chapter 7
2. 47.9 km/s
4. 61 seconds of arc
6. 20 km
8. 962 seconds of arc, or about 16 minutes of arc; 0.074 seconds of arc

Chapter 8
2. 12,700 m/s
4. 19°; this is approximately 38 times larger than our moon
6. 14.3 hr
8. 7.3 seconds
10. 4.1 hours

Chapter 9
2. one billion
4. 80 km in radius, 160 km in diameter
6. 0.8 seconds of arc; no
8. 2.5×10^5 seconds, or about 2.9 days
10. 2×10^{26} kg, or about 33 earth masses

Chapter 10
2. 730 km
4. about 3.6 times brighter
6. 400,000 yr
8. 9.0×10^{16} J
10. 0.2 kg

Chapter 11

2. 62.5 pc; 2
4. B; more specifically, B6—but difficult to determine from the graph
6. 6.25 solar luminosity
8. a, c, c, d
10. In Figure 11-14b, Star B is about twice the mass of star A. In Figure 11-15, Sirius A is about 2.3 times the mass of Sirius B.
12. 1.4×10^6 km

Chapter 12

2. 20,000 AU
4. 100 K
6. about 0.07 of the sun's total mass
8. 9.8×10^6 years
10. 75 nm

Chapter 13

2. 0.96 seconds of arc
4. about 17,000 years
6. about 940 years ago (approximately 1060 A.D.)
8. about 300 years ago
10. about 310 million years

Chapter 14

2. 0.00096 AU if the total mass is 2 solar masses
4. 3 m; 1.1×10^{-25} m
6. 8.1 minutes

Chapter 15

2. about 19%
4. 3.8×10^6 years
6. overestimate by a *factor* of 1.58

8. about 21 kpc
10. 7.8×10^{10} solar masses
12. 1500 K

Chapter 16

2. 4 Mpc
4. 28.6 Mpc
6. 165 million years
8. 2.7×10^{10} $M_{\odot}$; this assumes that there is no mass beyond a distance of 5 kpc

Chapter 17

2. 7.8×10^6 years
4. 170 Mpc
6. 1.9×10^{-4} years or 1.7 hours
8. −13.5 (slightly brighter than the full moon, assuming no interstellar extinction of the quasar)
10. 0.16

Chapter 18

2. 17.5 billion years; 11.7 billion years
4. 1.6×10^{-30} gm/cm^3
6. 76 km/s/Mpc
8. 52.6 Mpc

Chapter 19

2. 8.9 cm; 0.67 mm
4. about 1.3 solar masses
6. 380 km
8. pessimistic, 2×10^{-5}; optimistic, 10^7. Answers will vary greatly. Using intermediate values from the table, such as the following numbers: 2×10^{11}, 0.1, 0.1, 0.1, 0.1, 10^{-6}, yields an answer of 20.

Glossary

Numbers refer to the page where the term is first discussed in the text.

absolute age (131)   An age determined in years as from radioactive dating (cf. **relative age**).

absolute visual magnitude (M_v) (243)   Intrinsic brightness of a star. The apparent visual magnitude the star would have if it were 10 pc away.

absolute zero (93)   The theoretical lowest possible temperature at which a material contains no extractable heat energy. Zero on the Kelvin temperature scale.

absorption line (94)   A dark line in a spectrum. Produced by the absence of photons absorbed by atoms or molecules.

absorption spectrum (dark-line spectrum) (94)   A spectrum that contains absorption lines.

accretion (115)   The sticking together of solid particles to produce a larger particle.

accretion disk (312)   The whirling disk of gas that forms around a compact object such as a white dwarf, neutron star, or black hole as matter is drawn in.

achondrite (ā·kŏn′drīt) (200)   Stony meteorite containing no chondrules or volatiles.

achromatic lens (64)   A telescope lens composed of two lenses ground from different kinds of glass and designed to bring two selected colors to the same focus and correct for chromatic aberration.

active galactic nuclei (AGN) (383)   The centers of active galaxies that are emitting large amounts of excess energy. See also **active galaxy**.

active galaxy (383)   A galaxy whose center emits large amounts of excess energy, often in the form of radio emission. Active galaxies are suspected of having massive black holes in their centers into which matter is flowing.

active optics (73)   Thin telescope mirrors that are controlled by computers to maintain proper shape as the telescope moves.

active region (223)   An area on the sun where sunspots, prominences, flares, and the like occur.

AGN   See **active galactic nuclei.**

alt-azimuth mounting (66)   A telescope mounting that allows the telescope to move in altitude (perpendicular to the horizon) and in azimuth (parallel to the horizon). See also **equatorial mounting.**

amino acid (ŭ·mē′nō) (427)   Carbon-chain molecule that is the building block of protein.

Angstrom (Å) (ăng′strŭm) (62)   A unit of distance. 1 Å = 10^{-10} m. Commonly used to measure the wavelength of light.

angular momentum (312)   A measure of the tendency of a rotating body to continue rotating. Mathematically, the product of mass, velocity, and radius.

annular eclipse (30)   A solar eclipse in which the solar photosphere appears around the edge of the moon in a bright ring, or annulus. The corona, chromosphere, and prominences cannot be seen.

anorthosite (ăn·ôr′thŭ·sīt) (135)   Rock of aluminum and calcium silicates found in the lunar highlands.

antimatter (410)   Matter composed of antiparticles, which upon colliding with a matching particle of normal matter annihilate and convert the mass of both particles into energy. The antiproton is the antiparticle of the proton, and the positron is the antiparticle of the electron.

aphelion (ŭ·fē′le·ŭn) (22)   The orbital point of greatest distance from the sun.

apogee (30)   The point farthest from Earth in the orbit of a body circling Earth.

apparent visual magnitude (m_v) (12)   The brightness of a star as seen by human eyes on Earth.

association (272)   Group of widely scattered stars (10 to a few 100) moving together through space. Not gravitationally bound into clusters.

asterism (11)   A named grouping of stars that is not one of the recognized constellations, such as the Big Dipper or the Pleiades.

asteroid (109)   Small, rocky world. Most asteroids lie between Mars and Jupiter in the asteroid belt.

astronomical unit (AU) (4)   Average distance from Earth to the sun; 1.5 × 10^8 km, or 93 × 10^6 mi.

atmospheric window (63)   Wavelength region in which our atmosphere is transparent—at visual, infrared, and radio wavelengths.

aurora (ô·rôr′ŭ) (172)   The glowing light display that results when a planet's magnetic field guides charged particles toward the north and south magnetic poles, where they strike the upper atmosphere and excite atoms to emit photons.

autumnal equinox (19)   The point on the celestial sphere where the sun crosses the celestial equator going southward. Also, the time when the sun reaches this point and autumn begins in the Northern Hemisphere—about September 22.

Babcock model (225) A model of the sun's magnetic cycle in which the differential rotation of the sun winds up and tangles the solar magnetic field in a 22-year cycle. This is thought to be responsible for the 11-year sunspot cycle.

Balmer lines (244) Spectral lines of hydrogen. The three with longest wavelengths are visible to human eyes.

barred spiral galaxy (364) A spiral galaxy with an elongated nucleus resembling a bar from which the arms originate.

basalt (125) Dark igneous rock characteristic of solidified lava.

belt–zone circulation (173) The atmospheric circulation typical of Jovian planets. Dark belts and bright zones encircle the planet parallel to its equator.

big bang (407) The theory that the universe began in a high-density, high-temperature state from which the expanding universe of galaxies formed.

binary stars (253) Pairs of stars that orbit around their common center of mass.

binding energy (89) The energy needed to pull an electron away from its atom.

bipolar flow (275) Jets of gas flowing away from a central object in opposite directions. Usually applied to protostars.

birth line (274) In the H-R diagram, the line above the main sequence where protostars first become visible.

black body radiation (91) Radiation emitted by a hypothetical perfect radiator. The spectrum is continuous, and the wavelength of maximum emission depends on the body's temperature.

black dwarf (301) The end state of a white dwarf that has cooled to low temperature.

black hole (327) A mass that has collapsed to such a small volume that its gravity prevents the escape of all radiation. Also, the volume of space from which radiation may not escape.

BL Lac objects (387) Objects that resemble quasars. Thought to be highly luminous cores of distant active galaxies.

blue shift (95) A decrease in wavelength caused by the Doppler effect in electromagnetic radiation from an object approaching the observer.

Bok globule (277) Small, dark cloud only about 1 ly in diameter that contains 10 to 1000 solar masses of gas and dust. Believed to be related to star formation.

breccia (brěch′ē·ŭ) (135) Rock composed of fragments of earlier rocks bonded together.

bright-line spectrum See **emission spectrum**.

brown dwarf (284) A star whose mass is too low to ignite nuclear fusion. Heated by contraction.

calibration (339) The establishment of the relationship between a parameter that is easily determined and a parameter that is more difficult to determine. For example, the periods of Cepheid variables have been calibrated to reveal absolute magnitudes, which can then be used to find distance. Thus astronomers say Cepheids have been calibrated as distance indicators.

Cambrian period (kăm′brē·ŭn) (428) A geological period 0.6–0.5 billion years ago during which life on Earth became diverse and complex. Cambrian rocks contain the oldest easily identifiable fossils.

capture hypothesis (139) The theory that Earth's moon formed elsewhere in the solar nebula and was later captured by Earth.

carbonaceous chondrite (kär·bŭ·nā′ shŭs kŏn′drīt) (200) Stony meteorite that contains both chondrules and volatiles. These chondrites may be the least-altered remains of the solar nebula still present in the solar system.

carbon detonation (302) The explosive ignition of carbon burning in some giant stars. A possible cause of some supernova explosions.

carbon–nitrogen–oxygen (CNO) cycle (278) A series of nuclear reactions that use carbon as a catalyst to combine four hydrogen atoms to make one helium atom plus energy. Effective in stars more massive than the sun.

Cassegrain telescope (kăs′ŭ·grān) (65) A reflecting telescope in which the secondary mirror reflects light back down the tube through a hole in the center of the objective mirror.

CCD See **charge-coupled device**.

celestial equator (15) The imaginary line around the sky directly above Earth's equator.

celestial pole (north or south) (15) One of the two points on the celestial sphere directly above Earth's poles.

celestial sphere (13) An imaginary sphere of very large radius surrounding Earth and to which the planets, stars, sun, and moon seem to be attached.

center of mass (254) The balance point of a body or system of masses. The point about which a body or system of masses rotates in the absence of external forces.

Cepheid (sě·fē′ĭd) (337) Variable star with a period of 60 days. Period of variation related to luminosity.

Chandrasekhar limit (shăn′drä·sā′kär) (301) The maximum mass of a white dwarf, about 1.4 solar masses. A white dwarf of greater mass cannot support itself and will collapse.

charge-coupled device (CCD) (74) An electronic device consisting of a large array of light-sensitive elements used to record very faint images.

chemical evolution (430) The chemical process that led to the growth of complex molecules on primitive Earth. This did not involve the reproduction of molecules.

chondrite (kŏn′drīt) (200) A stony meteorite that contains chondrules.

chondrule (kŏn′drōōl) (200) Round, glassy body found in some stony meteorites. Believed to have solidified very quickly from molten drops of silicate material.

chromatic aberration (krō·măt′ĭk) (64) A distortion found in refracting telescopes because lenses focus different colors at slightly different distances. Images are consequently surrounded by color fringes.

chromosphere (krō′mŭ·sfîr) (29) Bright gases just above the photosphere of the sun. Responsible for the emission lines in the flash spectrum.

circumpolar constellation (north or south) (16) A constellation so close to one of the celestial poles that it never sets or never rises as seen from a particular latitude.

closed universe (406) A model universe in which the average density is great enough to stop the expansion and make the universe contract.

cluster method (371) The method of determining the masses of galaxies based on the motions of galaxies in a cluster.

CNO cycle See **carbon–nitrogen–oxygen cycle.**

co-accretion hypothesis (139) The theory that Earth and the moon formed together.

coma (207) The glowing head of a comet.

comet (109) One of the small, icy bodies that orbit the sun and produce tails of gas and dust when they approach the sun.

comparative planetology (123) The study of planets in relation to one another.

comparison spectrum (76) A spectrum of known spectral lines used to identify unknown wavelengths in an object's spectrum.

condensation (114) The growth of a particle by addition of material from surrounding gas, atom by atom.

condensation sequence (114) The sequence in which different materials condense from the solar nebula as we move outward from the sun.

conservation of energy (280) One of the basic laws of stellar structure. The amount of energy flowing out of the top of a shell must equal the amount coming in at the bottom plus whatever energy is generated within the shell.

conservation of mass (280) One of the basic laws of stellar structure. The total mass of the star must equal the sum of the masses of the shells, and the mass must be distributed smoothly through the star.

constellation (10) One of the stellar patterns identified by name, usually of mythological gods, people, animals, or objects. Also, the region of the sky containing that star pattern.

continuous spectrum (93) A spectrum in which there are no absorption or emission lines.

convection (218) Circulation in a fluid driven by heat. Hot material rises and cool material sinks.

Copernican principle (403) The belief that Earth is not in a special place in the universe.

corona (29, 152) On the sun, the faint outer atmosphere composed of low-density, high-temperature gas. On Venus, large, round geological faults in the crust caused by the intrusion of magma below the crust.

coronal hole (229) An area of the solar surface that is dark at X-ray wavelengths. Thought to be associated with divergent magnetic fields and the source of the solar wind.

cosmic background radiation (409) Radiation from the hot clouds of the big bang explosion. Because of its large red shift, it appears to come from a body whose temperature is only 2.7 K.

cosmological constant (Λ) (417) Einstein's constant that represents a repulsion in space to oppose gravity.

cosmological principle (403) The assumption that any observer in any galaxy sees the same general features of the universe.

cosmology (401) The study of the nature, origin, and evolution of the universe.

Coulomb barrier (kōō·lŏn) (232) The electrostatic force of repulsion between bodies of like charge. Commonly applied to atomic nuclei.

Coulomb force (89) The electrostatic force of repulsion or attraction between charged bodies.

critical density (412) The average density of the universe needed to make its curvature flat.

dark halo (343) The low-density extension of the halo of our galaxy believed to be composed of dark matter.

dark-line spectrum See **absorption spectrum.**

dark matter (343) The undetected matter that some astronomers believe makes up the missing mass of the universe.

deferent (dĕf'ŭr·ŭnt) (39) In the Ptolemaic theory, the large circle around Earth along which the center of the epicycle was thought to move.

degenerate matter (295) Extremely high-density matter in which pressure no longer depends on temperature due to quantum mechanical effects.

density wave theory (351) Theory proposed to account for spiral arms as compressions of the interstellar medium in the disk of the galaxy.

diamond ring effect (29) During a total solar eclipse, the momentary appearance of a spot of photosphere at the edge of the moon, producing a brilliant glare set in the silvery ring of the corona.

differential rotation (225, 342) The rotation of a body in which different parts of the body have different periods of rotation. This is true of the sun, the Jovian planets, and the disk of the galaxy.

differentiation (116) The separation of planetary material according to density.

diffraction fringe (67) Blurred fringe surrounding any image, caused by the wave properties of light. Because of this, no image detail smaller than the fringe can be seen.

dirty snowball model (110) The theory that the nuclei of comets are kilometer-sized chunks of ice with imbedded silicates in long, elliptical orbits around the sun.

disk component (340) All material confined to the plane of the galaxy.

distance indicator (367) Object whose luminosity or diameter is known. Used to find the distance to a star cluster or galaxy.

distance modulus ($m–M_v$) (243) The difference between the apparent and absolute magnitude of a star. A measure of how far away the star is.

DNA (deoxyribonucleic acid) (426) The long carbon-chain molecule that records information to govern the biological activity of the organism. DNA carries the genetic data passed to offspring.

Doppler effect (95) The change in the wavelength of radiation due to relative radial motion of source and observer.

double-exhaust model (384) The theory that double radio lobes are produced by pairs of jets emitted in opposite directions from the centers of active galaxies.

double-lobed radio galaxy (383) A galaxy that emits radio energy from two regions (lobes) located on opposite sides of the galaxy.

dynamo effect (125)   The process by which a rotating, convecting body of conducting matter, such as Earth's core, can generate a magnetic field.

east point (15)   The point on the eastern horizon exactly halfway between the north point and the south point. Exactly east.

eclipsing binary (257)   A binary star system in which the stars eclipse each other.

ecliptic (19)   The apparent path of the sun around the sky.

ejecta (133)   Pulverized rock scattered by meteorite impacts on a planetary surface.

electromagnetic radiation (61)   Changing electric and magnetic fields that travel through space and transfer energy from one place to another; examples are light or radio waves.

electron (88)   Low-mass atomic particle carrying negative charge.

ellipse (49)   A closed curve around two points called the foci such that the total distance from one focus to the curve and back to the other focus remains constant.

elliptical galaxy (362)   A galaxy that is round or elliptical in outline and contains little gas and dust, no disk or spiral arms and few hot, bright stars.

emission line (94)   A bright line in a spectrum caused by the emission of photons from atoms.

emission nebula (269)   A cloud of glowing gas excited by ultraviolet radiation from hot stars.

emission spectrum (bright-line spectrum) (94)   A spectrum containing emission lines.

energy level (90)   One of a number of states an electron may occupy in an atom, depending on its binding energy.

energy transport (281)   Flow of energy from hot regions to cooler regions by one of three methods: conduction, convection, or radiation.

enzyme (427)   Special protein that controls processes in an organism.

epicycle (ĕp´ŭ·sī·kŭl) (39)   The small circle followed by a planet in the Ptolemaic theory. The center of the epicy-

cle follows a larger circle (the deferent) around Earth.

equant (ē´kwŭnt) (39)   In the Ptolemaic theory, the point off center in the deferent from which the center of the epicycle appears to move uniformly.

equatorial mounting (66)   A telescope mounting that allows motion parallel to and perpendicular to the celestial equator.

escape velocity (157)   The initial velocity an object needs to escape from the surface of a celestial body.

event horizon (327)   The boundary of the region of a black hole from which no radiation may escape. No event that occurs within the event horizon is visible to a distant observer.

evolutionary track (273)   The path a star follows in the H-R diagram as it gradually changes its surface temperature and luminosity.

excited atom (90)   An atom in which an electron has moved from a lower to a higher energy level.

extra-solar planet (102)   A planet orbiting a star other than the sun.

eyepiece (64)   A short-focal-length lens used to enlarge the image in a telescope. The lens nearest the eye.

false-color image (74)   A representation of graphical data with added or enhanced color to reveal detail.

filtergram (219)   A photograph (usually of the sun) taken in the light of a specific region of the spectrum—for example, an H-alpha filtergram.

fission hypothesis (139)   The theory that the moon and Earth formed when a rapidly rotating protoplanet split into two pieces.

flare (228)   A violent eruption on the sun's surface.

flat universe (406)   A model of the universe in which space–time is not curved.

flatness problem (415)   In cosmology, the peculiar circumstance that the early universe must have contained almost exactly the right amount of matter to make space–time flat.

flocculent galaxy (352)   A galaxy whose spiral arms have a woolly or fluffy appearance.

flux (237)   A measure of the flow of energy out of a surface. Usually applied to light.

focal length (63)   The focal length of a lens is the distance from the lens to the point where it focuses parallel rays of light.

folded mountain chain (126)   A long range of mountains formed by the compression of a planet's crust.

forward scattering (175)   The optical property of finely divided particles to preferentially direct light in the original direction of the light's travel.

galactic cannibalism (376)   The theory that large galaxies absorb smaller galaxies.

galactic corona (343)   The extended, spherical distribution of low-luminosity matter believed to surround the Milky Way and other galaxies.

galaxy cluster (373–374)   A group of galaxies numbering from a few to thousands.

galaxy seeds (418)   The first irregularities in the universe, which stimulated the formation of the galaxies.

Galilean satellites (găl·ŭ·lē´ŭn) (109)   The four largest satellites of Jupiter, named after their discoverer Galileo.

geocentric universe (38)   A model universe with Earth at the center, such as the Ptolemaic universe.

giant star (251)   Large, cool, highly luminous star in the upper right of the H-R diagram. Typically 10 to 100 times the diameter of the sun.

globular star cluster (337)   A star cluster containing 100,000 to 1 million stars in a sphere about 75 ly in diameter. Generally old, metal-poor, and found in the spherical component of the galaxy.

grand unified theories (GUTs) (415)   Theories that attempt to unify (describe in a similar way) the electromagnetic, weak, and strong forces of nature.

granulation (218)   The fine structure of bright grains covering the sun's surface.

grating (76)   A piece of material in which numerous microscopic parallel lines are scribed. Light encountering a grating is dispersed to form a spectrum.

gravitational collapse (113)   The process by which a forming body such as a planet gravitationally captures gas from its surroundings.

gravitational lens (394)   The focusing of light from a distant galaxy or quasar by an intervening galaxy to produce multiple images of the distant body.

gravitational red shift (329)   The lengthening of the wavelength of a photon due to its escape from a gravitational field.

greenhouse effect (128)   The process by which a carbon dioxide atmosphere traps heat and raises the temperature of a planetary surface.

grooved terrain (178)   Regions of the surface of Ganymede consisting of parallel grooves. Believed to have formed by repeated fracture and refreezing of the icy crust.

ground state (91)   The lowest permitted electron energy level in an atom.

GUTs   See **grand unified theories.**

H II region (368)   A region of ionized hydrogen around a hot star.

half-life (111)   The time required for half of the atoms in a radioactive sample to decay.

halo (341)   The spherical region of a spiral galaxy, containing a thin scattering of stars, star clusters, and small amounts of gas.

head–tail radio galaxy (386)   A radio galaxy with a contour consisting of a head and a tail. Believed to be caused by the motion of an active galaxy through the intergalactic medium.

heat of formation (116)   In planetology, the heat released by in-falling of matter during the formation of a planetary body.

heavy bombardment (118)   The intense cratering during the first 0.5 billion years in the history of the solar system.

heliocentric universe (42)   A model of the universe with the sun at the center, such as the Copernican universe.

helioseismology (221)   The study of the interior of the sun by the analysis of its modes of vibration.

helium flash (296)   The explosive ignition of helium burning that takes place in some giant stars.

Herbig-Haro objects (275)   Small nebulae that vary irregularly in brightness. Believed to be associated with star formation.

Hertzsprung-Russell (H-R) diagram (hĕrt′sprŭng·rŭs·ŭl) (249)   A plot of the intrinsic brightness versus the surface temperature of stars. It separates the effects of temperature and surface area on stellar luminosity. Commonly plotted as absolute magnitude versus spectral type, but also as luminosity versus surface temperature or color.

homogeneity (402)   The assumption that on the large scale, matter is uniformly spread through the universe.

horizon (13)   The line that marks the apparent intersection of the Earth and the sky.

horizon problem (415)   In cosmology, the circumstance that the primordial background radiation seems much more isotropic than can be explained by the standard big bang theory.

hot spot (384)   In geology, a place on Earth's crust where volcanism is caused by a rising convection cell in the mantle below. In radio astronomy, a bright spot in a radio lobe.

H-R diagram   See **Hertzsprung-Russell diagram.**

Hubble constant (H) (370)   A measure of the rate of expansion of the universe. The average value of velocity of recession divided by distance. Presently believed to be about 70 km/s/Mpc.

Hubble law (370)   The linear relation between the distances to galaxies and their velocity of recession.

Hubble time (416)   The age of the universe, equivalent to 1 divided by the Hubble constant. The Hubble time is the age of the universe if it has expanded since the big bang at a constant rate.

hydrostatic equilibrium (280)   The balance between the weight of the material pressing downward on a layer in a star and the pressure in that layer.

hypothesis (48)   A conjecture, subject to further tests, that accounts for a set of facts.

inflationary universe (415)   A version of the big bang theory that includes a rapid expansion when the universe was very young. Derived from grand unified theories.

infrared cirrus (270)   Wispy network of cold dust clouds discovered by the Infrared Astronomy Satellite.

instability strip (337)   The region of the H-R diagram in which stars are unstable to pulsation. A star passing through this strip becomes a variable star.

intercrater plains (146)   The relatively smooth terrain on Mercury.

interstellar medium (269)   The gas and dust distributed between the stars.

interstellar reddening (270)   The process in which dust scatters blue light out of starlight and makes the stars look redder.

inverse square relation (56)   A rule that the strength of an effect (such as gravity) decreases in proportion as the distance squared increases.

ion (88)   An atom that has lost or gained one or more electrons.

ionization (88)   The process in which atoms lose or gain electrons.

irregular galaxy (366)   A galaxy with a chaotic appearance, large clouds of gas and dust, and both population I and II stars, but without spiral arms.

isotopes (88)   Atoms that have the same number of protons but a different number of neutrons.

isotropy (ī·sŏt′rŭ·pē) (403)   The assumption that in its general properties the universe looks the same in every direction.

joule (J) (jōōl) (92)   A unit of energy equivalent to a force of 1 newton acting over a distance of 1 m. One joule per second equals 1 watt of power.

Jovian planet (106)   Jupiter-like planet with a large diameter and low density.

jumbled terrain (137)   Strangely disturbed regions of the moon opposite the locations of the Imbrium Basin and Mare Oriental.

kiloparsec (kpc) (340)   A unit of distance equal to 1000 pc or 3260 ly.

Kirchhoff's laws (94)   A set of laws that describe the absorption and emission of light by matter.

Kuiper belt (210)   The collection of icy planetesimals believed to orbit in a region from just beyond Neptune out to 100 AU or more.

Lagrangian point (310) Points of stability in the orbital plane of a planet or moon. One is located 60° ahead and one 60° behind the orbiting bodies.

large-impact hypothesis (139) The theory that the moon formed from debris ejected during a collision between Earth and a large planetesimal.

Large Magellanic Cloud (366) The larger of two irregular galaxies that are companions to the Milky Way galaxy.

life zone (436) A region around a star within which a planet can have temperatures that permit the existence of liquid water.

light curve (258) A graph of brightness versus time commonly used in analyzing variable stars and eclipsing binaries.

light-gathering power (67) The ability of a telescope to collect light. Proportional to the area of the telescope's objective lens or mirror.

lighthouse theory (319) The description of a pulsar as a neutron star producing pulses of radiation by sweeping radio beams around the sky as it spins.

light pollution (69) The illumination of the night sky by waste light from cities and outdoor lighting, which prevents the observation of faint objects.

light-year (ly) (5) A unit of distance. The distance light travels in one year.

limb (130) The edge of the apparent disk of a body, as in "the limb of the moon."

liquid metallic hydrogen (108) A form of liquid hydrogen that is a good electrical conductor found in the interiors of Jupiter and Saturn.

lobate scarp (lō′bāt·skärp) (146) A curved cliff such as those found on Mercury.

Local Group (6) The small cluster of a few dozen galaxies that contains our Milky Way Galaxy

look-back time (369) The amount by which we look into the past when we look at a distant galaxy. A time equal to the distance to the galaxy in light-years.

luminosity (L) (244) The total amount of energy a star radiates in 1 second.

luminosity class (252) A category of stars of similar luminosity. Determined by the widths of lines in their spectra.

lunar eclipse (27) The darkening of the moon when it moves through Earth's shadow.

MACHOs (414) Massive Compact Halo Objects. Low-luminosity objects such as planets and brown dwarfs that contribute to the mass of the halo.

Magellanic clouds (măj·ŭ·lăn′ĭk) (366) Small, irregular galaxies that are companions to the Milky Way. Visible in the southern sky.

magnetar (325) A class of neutron stars that have exceedingly strong magnetic fields.

magnetosphere (măg·nē′tō·sfîr) (173) The volume of space around a planet within which the motion of charged particles is dominated by the planetary magnetic field rather than the solar wind.

magnifying power (69) The ability of a telescope to make an image larger.

magnitude scale (11) The astronomical brightness scale. The larger the number, the fainter the star.

main sequence (250) The region of the H-R diagram running from upper left to lower right, which includes roughly 90 percent of all stars.

mantle (125) The layer of dense rock and metal oxides that lies between the molten core and Earth's surface. Also, similar layers in other planets.

mare (sea) (mä′rā) (130) One of the lunar lowlands filled by successive flows of dark lava.

mass (55) A measure of the amount of matter making up an object.

mass–luminosity relation (261) The more massive a star is, the more luminous it is.

Maunder butterfly diagram (môn′dŭr) (224) A graph showing the latitude of sunspots versus time. First plotted by W. W. Maunder in 1904.

Maunder minimum (môn′dŭr) (224) A period of less numerous sunspots and other solar activity between 1645 and 1715.

megaparsec (Mpc) (367) A unit of distance equal to 1,000,000 pc.

metals (344) In astronomical usage, all atoms heavier than helium.

meteor (111) A small bit of matter heated by friction to incandescent vapor as it falls into Earth's atmosphere.

meteorite (111) A meteor that survives its passage through the atmosphere and strikes the ground.

meteoroid (111) A meteor in space before it enters Earth's atmosphere.

meteor shower (200) A multitude of meteors that appear to come from the same region of the sky. Believed to be caused by comet debris.

micrometeorite (135) Meteorite of microscopic size.

midocean rift (125) Chasms that split the midocean rises where crustal plates move apart.

midocean rise (125) One of the undersea mountain ranges that push up from the seafloor in the center of the oceans.

Milankovitch hypothesis (22) Suggestion that Earth's climate is determined by slow periodic changes in the shape of its orbit, the angle of its axis, and precession.

Milky Way Galaxy (6) The spiral galaxy containing our sun. Visible in the night sky as the Milky Way.

Miller experiment (428) An experiment that reproduced the conditions under which life began on Earth and manufactured amino acids and other organic compounds.

millisecond pulsar (322) A pulsar with a pulse period of only a few milliseconds.

minute of arc (15) An angular measure. Each degree is divided into 60 minutes of arc.

model (14) A tentative description of a phenomenon for use as an aid to understanding.

molecular cloud (271) A dense interstellar gas cloud in which atoms are able to link together to form molecules such as H_2 and CO.

molecule (89) Two or more atoms bonded together.

multiringed basins (133) Large impact features (craters) containing two or more concentric rims formed by fracturing of the planetary crust.

mutant (427) Offspring born with altered DNA.

nanometer (62) A unit of distance equaling one billionth of a meter (10^{-9} m).

natural law (48)   A theory that is almost universally accepted as true.

natural selection (427)   The process by which the best traits are passed on, allowing the most able to survive.

neap tide (26)   Ocean tide of low amplitude occurring at first- and third-quarter moon.

nebula (269)   A glowing cloud of gas or a cloud of dust reflecting the light of nearby stars.

neutrino (233)   A neutral, massless atomic particle that travels at the speed of light.

neutron (88)   An atomic particle with no charge and about the same mass as a proton.

neutron star (317)   A small, highly dense star composed almost entirely of tightly packed neutrons. Radius about 10 km.

Newtonian telescope (65)   A reflecting telescope in which a diagonal mirror reflects light out the side of the telescope tube for easier access.

node (32)   The points where an object's orbit passes through the plane of Earth's orbit.

north celestial pole (13)   The point on the celestial sphere directly above Earth's North Pole.

north circumpolar constellations (16) Constellations located so close to the north celestial pole they do not rise or set.

north point (15)   The point on the horizon directly below the north celestial pole. Exactly north.

nova (293)   From the Latin, meaning "new," a sudden brightening of a star making it appear as a new star in the sky. Believed to be associated with eruptions on white dwarfs in binary systems.

nuclear bulge (341)   The spherical cloud of stars that lies at the center of spiral galaxies.

nuclear fission (231)   Reactions that break the nuclei of atoms into fragments.

nuclear fusion (231)   Reactions that join the nuclei of atoms to form more massive nuclei.

nucleus (of an atom) (88)   The central core of an atom containing protons and neutrons. Carries a net positive charge.

objective lens (63)   In a refracting telescope, the long-focal-length lens that forms an image of the object viewed. The lens closest to the object.

objective mirror (63)   In a reflecting telescope, the principal mirror (reflecting surface) that forms an image of the object viewed.

oblateness (181)   The flattening of a spherical body. Usually caused by rotation.

observable universe (402)   The part of the universe that we can see from our location in space and in time.

Olbers' paradox (ôl'bŭrs) (401)   The conflict between observation and theory about why the night sky should or should not be dark.

Oort cloud (ōrt) (210)   The hypothetical source of comets. A swarm of icy bodies believed to lie in a spherical shell extending to 100,000 AU from the sun.

opacity (281)   The resistance of a gas to the passage of radiation.

open star cluster (337)   A cluster of 100 to 1000 stars with an open, transparent appearance. The stars are not tightly grouped. Usually relatively young and located in the disk of the galaxy.

open universe (407)   A model of the universe in which the average density is less than the critical density needed to halt the expansion.

oscillating universe theory (412) The theory that the universe begins with a big bang, expands, is slowed by its own gravity, and then falls back to create another big bang.

outflow channel (162)   Geological features produced by the rapid motion of flood waters. Applied to features on Mars.

outgassing (116)   The release of gases from a planet's interior.

ovoid (189)   The oval features found on Miranda, a satellite of Uranus.

paradigm (44)   A commonly accepted set of scientific ideas and assumptions.

parallax (p) (38)   The apparent change in position of an object due to a change in the location of the observer. Astronomical parallax is measured in seconds of arc.

parsec (pc) (pär'sĕk) (241)   The distance to a hypothetical star whose parallax is 1 second of arc. 1 pc = 206,265 AU = 3.26 ly.

penumbra (pǐ·nŭm'brŭ) (27)   The portion of a shadow that is only partially shaded.

perigee (30)   The point closest to Earth in the orbit of a body circling Earth.

perihelion (pĕr·ŭ·hē'lē·ŭn) (22)   The orbital point of closest approach to the sun.

period–luminosity relation (337) The relation between period of pulsation and intrinsic brightness among Cepheid variable stars.

permitted orbit (89)   One of the energy levels in an atom that an electron may occupy.

photon (61)   A quantum of electromagnetic energy. Carries an amount of energy that depends inversely on its wavelength.

photosphere (29)   The bright visible surface of the sun.

planetary nebula (298)   An expanding shell of gas ejected from a star during the latter stages of its evolution.

planetesimal (plăn·ŭ·tĕs'ŭ·mŭl) (114) One of the small bodies that formed from the solar nebula and eventually grew into protoplanets.

plastic (125)   A material with the properties of a solid but capable of flowing under pressure.

plate tectonics (125)   The constant destruction and renewal of Earth's surface by the motion of sections of crust.

polar axis (66)   In an equatorial telescope mounting, the axis that is parallel to Earth's axis.

poor galaxy cluster (374)   An irregularly shaped cluster that contains fewer than 1000 galaxies, many spiral, and no giant ellipticals.

population I (344)   Stars rich in atoms heavier than helium. Nearly always relatively young stars found in the disk of the galaxy.

population II (344)   Stars poor in atoms heavier than helium. Nearly always relatively old stars found in the halo, globular clusters, or the nuclear bulge.

precession (17)   The slow change in the direction of Earth's axis of rotation. One cycle takes nearly 26,000 years.

primary lens or **mirror** (63) The largest lens or mirror in an astronomical telescope.

prime focus (65) The point at which the objective mirror forms an image in a reflecting telescope.

primeval atmosphere (127) Earth's first air.

primordial soup (429) The rich solution of organic molecules in Earth's first oceans.

prominence (29, 227) Eruption on the solar surface. Visible during total solar eclipses.

proper motion (337) The rate at which a star moves across the sky. Measured in seconds of arc per year.

protein (427) Complex molecule composed of amino acid units.

proton (88) A positively charged atomic particle contained in the nucleus of atoms. The nucleus of a hydrogen atom.

proton–proton chain (232) A series of three nuclear reactions that builds a helium atom by adding together protons. The main energy source in the sun.

protoplanet (115) Massive object resulting from the coalescence of planetesimals in the solar nebula and destined to become a planet.

protostar (273) A collapsing cloud of gas and dust destined to become a star.

pulsar (318) A source of short, precisely timed radio bursts. Believed to be spinning neutron stars.

P **wave** (125) A pressure wave. A type of seismic wave produced in Earth by the compression of the material.

quantum mechanics (89) The study of the behavior of atoms and atomic particles.

quasar (quasi-stellar object, or QSO) (kwā′zär) (392) Small, powerful source of energy believed to be the active core of a very distant galaxy.

radial velocity (V_r) (95) That component of an object's velocity directed away from or toward Earth.

radiation pressure (118) The force exerted on the surface of a body by its absorption of light. Small particles floating in the solar system can be blown outward by the pressure of the sunlight.

radio galaxy (383) A galaxy that is a strong source of radio signals.

radio interferometer (77) Two or more radio telescopes that combine their signals to achieve the resolving power of a larger telescope.

rays (133) Ejecta from meteorite impacts forming white streamers radiating from some lunar craters.

recombination (411) The stage within 1 million years of the big bang, when the gas became transparent to radiation.

red dwarf (263) A faint, cool, low-mass main-sequence star.

red shift (95) An increase in wavelength caused by the Doppler effect in electromagnetic radiation from an object receding from the observer.

reflecting telescope (63) A telescope that uses a concave mirror to focus light into an image.

reflection nebula (269) A nebula produced by starlight reflecting off of dust particles in the interstellar medium.

refracting telescope (63) A telescope that forms images by bending (refracting) light with a lens.

regolith (135) A soil made up of crushed rock fragments.

relative age (131) The age of a geological feature referred to other features. For example, relative ages tell us the lunar maria are younger than the highlands.

relativistic red shift (393) The red shift due to the Doppler effect for objects traveling at speeds near the speed of light.

resolving power (67) The ability of a telescope to reveal fine detail. Depends on the diameter of the telescope objective.

resonance (144) The coincidental agreement between two periodic phenomena; commonly applied to agreements between orbital periods, which can make orbits more or less stable.

retrograde motion (38) The apparent backward (westward) motion of planets as seen against the background of stars.

rich galaxy cluster (373) A cluster containing over 1000 galaxies, mostly elliptical, scattered over a volume about 3 Mpc in diameter.

rift valley (126) A long, straight, deep valley produced by the separation of crustal plates.

ring galaxy (378) A galaxy that resembles a ring around a bright nucleus. Believed to be the result of a head-on collision of two galaxies.

RNA (ribonucleic acid) (427) Long carbon-chain molecules that use the information stored in DNA to manufacture complex molecules necessary to the organism.

Roche limit (rōsh) (176) The minimum distance between a planet and a satellite that holds itself together by its own gravity. If a satellite's orbit brings it within its planet's Roche limit, tidal forces will pull the satellite apart.

Roche lobe (310) The volume of space a star controls gravitationally within a binary system.

rolling plains (150) On Venus, the relatively smooth lowlands that comprise the largest proportion of the planetary surface.

rotation curve (342) A graph of orbital velocity versus radius in the disk of a galaxy.

rotation curve method (370) A method of determining a galaxy's mass by observing the orbital velocity and orbital radius of stars in the galaxy.

RR Lyrae variable star (är·är·lī′rē) (337) Variable star with periods of from 12 to 24 hours. Common in some globular clusters.

runoff channel (162) Geological feature resembling a streambed produced by the regular drainage of water from a region. Applied to features on Mars.

Sagittarius A* (355) The powerful radio source located at the core of the Milky Way Galaxy.

Saros cycle (sě′rōs) (32) An 18-year, 11-day period after which the pattern of lunar and solar eclipses repeats.

Schmidt-Cassegrain telescope (65) Cassegrain telescope that uses a thin corrector plate at the entrance to the tube. A popular design for small telescopes.

Schwarzschild radius (R_s) (schwôrts' shēld) (328) The radius of the event horizon around a black hole.

scientific model (14) A tentative description of a phenomenon for use as an aid to understanding.

scientific notation (4) The system of recording very large or very small numbers by using powers of 10.

secondary atmosphere (127) The gases outgassed from a planet's interior; rich in carbon dioxide.

secondary crater (133) Impact crater formed by debris ejected from a larger impact.

second of arc (15) An angular measure. Each minute of arc is divided into 60 seconds of arc.

seeing (69) Atmospheric conditions on a given night. When the atmosphere is unsteady, producing blurred images, the seeing is said to be poor.

self-sustaining star formation (352) The process by which the birth of stars compresses the surrounding gas clouds and triggers the formation of more stars. Proposed to explain spiral arms.

SETI (440) The Search for Extra-Terrestrial Intelligence.

Seyfert galaxy (sē'fûrt) (386) An otherwise normal spiral galaxy with an unusually bright, small core that fluctuates in brightness. Believed to indicate the core is erupting.

Shapley-Curtis debate (339) A historic debate held in 1920 between Harlow Shapley and Heber Curtis concerning the existence of other galaxies besides our own.

shepherd satellite (183) A satellite that, by its gravitational field, confines particles to a planetary ring.

shield volcano (152) Wide, low-profile volcanic cone produced by highly liquid lava.

shock wave (271) A sudden change in pressure that travels as an intense sound wave.

sidereal drive (sī·dîr'ē·ŭl) (66) The motor and gears on a telescope that turn it westward to keep it pointed at a star.

sidereal period (24) The time a celestial body takes to turn once on its axis or revolve once around its orbit relative to the stars.

singularity (327) The object of zero radius into which the matter in a black hole is believed to fall.

sinuous rille (130) A narrow, winding valley on the moon caused by ancient lava flows along narrow channels.

Small Magellanic Cloud (366) The smaller of two irregular galaxies that are companions to the Milky Way galaxy.

smooth plains (146) Apparently young plains on Mercury formed by lava flows at or soon after the formation of the Caloris Basin.

solar eclipse (28) The event that occurs when the moon passes directly between Earth and the sun, blocking our view of the sun.

solar nebula theory (102) The theory that the planets formed from the same cloud of gas and dust that formed the sun.

solar system (3) The sun and its planets, asteroids, comets, and so on.

solar wind (110) Rapidly moving atoms and ions that escape from the solar corona and blow outward through the solar system.

south celestial pole (15) The point on the celestial sphere directly above Earth's South Pole.

south circumpolar constellations (16) Constellations located so close to the south celestial pole they do not rise or set.

south point (15) The point on the horizon directly above the south celestial pole. Exactly south.

spectral class or type (246) A star's position in the temperature classification system O, B, A, F, G, K, M. Based on the appearance of the star's spectrum.

spectral sequence (246) The arrangement of spectral classes (O, B, A, F, G, K, M) ranging from hot to cool.

spectrograph (74) A device that separates light by wavelengths to produce a spectrum.

spectroscopic binary (257) A star system in which the stars are too close together to be visible separately. We see a single point of light, and only by taking a spectrum can we determine that there are two stars.

spectroscopic parallax (253) The method of determining a star's distance by comparing its apparent magnitude with its absolute magnitude as estimated from its spectrum.

spherical component (341) The part of the galaxy including all matter in a spherical distribution around the center (the halo and nuclear bulge).

spicule (spĭk'yōōl) (220) A small, flamelike projection in the chromosphere of the sun.

spiral arm (6, 340) Long spiral pattern of bright stars, star clusters, gas, and dust. Spiral arms extend from the center to the edge of the disk of spiral galaxies.

spiral galaxy (364) A galaxy with an obvious disk component containing gas; dust; hot, bright stars; and spiral arms.

spiral tracer (349) Object used to map the spiral arms—for example, O and B associations, open clusters, clouds of ionized hydrogen, and some types of variable stars.

spring tide (26) Ocean tide of high amplitude that occurs at full and new moon.

starburst galaxy (378) A galaxy experiencing an episode of extremely rapid star formation.

steady state theory (410) The theory (now generally abandoned) that the universe does not evolve.

stellar model (282) A table of numbers representing the conditions in various layers within a star.

stellar parallax (p) (241) A measure of stellar distance. See also **parallax.**

stromatolite (430) A layered fossil formation caused by ancient mats of algae or bacteria, which build up mineral deposits season after season.

strong force (231) One of the four forces of nature. The strong force binds protons and neutrons together in atomic nuclei.

subduction zone (126) A region of a planetary crust where a tectonic plate slides downward.

subsolar point (148) The point on a planet that is directly below the sun.

summer solstice (20) The point on the celestial sphere where the sun is at its most northerly point. Also, the time when the sun passes this point, about June 22, and summer begins in the Northern Hemisphere.

sunspot (223)   Relatively dark spot on the sun that contains intense magnetic fields.

supercluster (417)   A cluster of galaxy clusters.

supergiant star (251)   Exceptionally luminous star whose diameter is 10 to 1000 times that of the sun.

supergranule (220)   Very large convective features in the sun's surface.

supernova (293)   The violent explosion of a massive star or a collapsing white dwarf.

supernova remnant (306)   The expanding shell of gas marking the site of a supernova explosion.

S **wave** (125)   A shear wave. A type of seismic wave produced in Earth by the lateral motion of the material.

synchrotron radiation (sĭn′krŭ·trŏn) (307)   Radiation emitted when high-speed electrons move through a magnetic field.

synodic period (sĭ·nŏd′ĭk) (25)   The time a solar system body takes to orbit the sun once and return to the same orbital relationship with Earth. That is, orbital period referenced to Earth.

terminator (130)   The dividing line between daylight and darkness on a planet or moon.

terrestrial planet (106)   An Earthlike planet—small, dense, rocky.

theory (48)   A system of assumptions and principles applicable to a wide range of phenomena that have been repeatedly verified.

tidal coupling (129)   The locking of the rotation of a body to its revolution around another body.

tidal heating (178)   The heating of a planet or satellite because of friction caused by tides.

time dilation (328)   The slowing of moving clocks or clocks in strong gravitational fields.

triple-alpha process (278)   The nuclear fusion process that combines three helium nuclei (alpha particles) to make one carbon nucleus.

T Tauri star (tôrē) (274)   Young star surrounded by gas and dust. Believed to be contracting toward the main sequence.

tuning fork diagram (362)   A system of classification for elliptical, spiral, and irregular galaxies.

turnoff point (310)   The point in an H-R diagram at which a cluster's stars turn off of the main sequence and move toward the red-giant region, revealing the approximate age of the cluster.

umbra (ŭm′brŭ) (27)   The region of a shadow that is totally shaded.

uncompressed density (114)   The density a planet would have if its gravity did not compress it.

uniform circular motion (38)   The classical belief that the perfect heavens could only move by the combination of uniform motion along circular orbits.

universality (403)   The assumption that the physical laws observed on Earth apply everywhere in the universe.

variable star (337)   A star whose brightness changes periodically.

velocity dispersion method (372)   A method of finding a galaxy's mass by observing the range of velocities within the galaxy.

vernal equinox (19)   The place on the celestial sphere where the sun crosses the celestial equator moving northward. Also, the time of year when the sun crosses this point, about March 21, and spring begins in the Northern Hemisphere.

vesicular basalt (vŭ·sĭk′yŭ·lŭr) (135)   A porous rock formed by solidified lava with trapped bubbles.

visual binary (255)   A binary star system in which the two stars are separately visible in the telescope.

water hole (440)   The interval of the radio spectrum between the 21-cm hydrogen radiation and the 18-cm OH radiation. Likely wavelengths to use in the search for extraterrestrial life.

wavelength (61)   The distance between successive peaks or troughs of a wave. Usually represented by λ.

wavelength of maximum intensity (λ_{max}) (92)   The wavelength at which a perfect radiator emits the maximum amount of energy. Depends only on the object's temperature.

weak force (231)   One of the four forces of nature. The weak force is responsible for some forms of radioactive decay.

west point (15)   The point on the western horizon exactly halfway between the north point and the south point. Exactly west.

white dwarf star (251)   Dying star that has collapsed to the size of Earth and is slowly cooling off. At the lower left of the H-R diagram.

Widmanstätten pattern (wĭd′mŭn·stä·tŭn) (200)   Bands in iron meteorites due to large crystals of nickel–iron alloys.

winter solstice (20)   The point on the celestial sphere where the sun is farthest south. Also the time of year when the sun passes this point, about December 22, and winter begins in the Northern Hemisphere.

Zeeman effect (223)   The splitting of spectral lines into multiple components when the atoms are in a magnetic field.

zenith (13)   The point on the sky directly above the observer.

zero-age main sequence (ZAMS) (284)   The locus in the H-R diagram where stars first reach stability as hydrogen-burning stars.

zodiac (19)   The band around the sky centered on the ecliptic within which the planets move.

Index